AF778899

Welding and Joining of Aerospace Materials

Woodhead Publishing Series in Welding and other Joining Technologies

Welding and Joining of Aerospace Materials

Second Edition

Edited by

Mahesh Chaturvedi

Distinguished Professor Emeritus, University of Manitoba, Winnipeg, Canada

ELSEVIER

WP WOODHEAD PUBLISHING

An imprint of Elsevier

Woodhead Publishing is an imprint of Elsevier
The Officers' Mess Business Centre, Royston Road, Duxford, CB22 4QH, United Kingdom
50 Hampshire Street, 5th Floor, Cambridge, MA 02139, United States
The Boulevard, Langford Lane, Kidlington, OX5 1GB, United Kingdom

Copyright © 2021 Elsevier Ltd. All rights reserved.

No part of this publication may be reproduced or transmitted in any form or by any means, electronic or mechanical, including photocopying, recording, or any information storage and retrieval system, without permission in writing from the publisher. Details on how to seek permission, further information about the Publisher's permissions policies and our arrangements with organizations such as the Copyright Clearance Center and the Copyright Licensing Agency, can be found at our website: www.elsevier.com/permissions.

This book and the individual contributions contained in it are protected under copyright by the Publisher (other than as may be noted herein).

Notices
Knowledge and best practice in this field are constantly changing. As new research and experience broaden our understanding, changes in research methods, professional practices, or medical treatment may become necessary.

Practitioners and researchers must always rely on their own experience and knowledge in evaluating and using any information, methods, compounds, or experiments described herein. In using such information or methods they should be mindful of their own safety and the safety of others, including parties for whom they have a professional responsibility.

To the fullest extent of the law, neither the Publisher nor the authors, contributors, or editors, assume any liability for any injury and/or damage to persons or property as a matter of products liability, negligence or otherwise, or from any use or operation of any methods, products, instructions, or ideas contained in the material herein.

Library of Congress Cataloging-in-Publication Data
A catalog record for this book is available from the Library of Congress

British Library Cataloguing-in-Publication Data
A catalogue record for this book is available from the British Library

ISBN: 978-0-12-819140-8 (print)
ISBN: 978-0-12-819141-5 (online)

For information on all Woodhead publications visit our website at https://www.elsevier.com/books-and-journals

Publisher: Matthew Deans
Acquisitions Editor: Carrie Bolger
Editorial Project Manager: Mariana L. Kuhl
Production Project Manager: Vijayaraj Purushothaman
Cover Designer: Vicky Pearson Esser

Typeset by SPi Global, India

Working together to grow libraries in developing countries

www.elsevier.com • www.bookaid.org

Contents

Contributors

A.C. Addison TWI Ltd., Cambridge, United Kingdom

M.M. Attallah University of Birmingham, Birmingham, United Kingdom

R. Benedictus Delft University of Technology, Delft, The Netherlands

I. Bhamji TWI Ltd., Cambridge, United Kingdom

S. Błacha Lukasiewicz - Institute of Welding, Gliwice, Poland

J. Blackburn TWI Ltd., Cambridge, United Kingdom

A. Elrefaey Dortmund University of Technology, Dortmund, Germany

Richard Freeman TWI Ltd., Cambridge, United Kingdom

J. Hofstede Delft University of Technology, Delft, The Netherlands

P. Johnson Liverpool John Moores University, Liverpool, United Kingdom

A. Kwakernaak Delft University of Technology, Delft, The Netherlands

Ho-Sung Lee Korea Aerospace Research Institute, Daejeon, Republic of Korea

Gang Li Aerospace Research Centre, National Research Council Canada, Ottawa, ON, Canada

Min Liao Aerospace Research Centre, National Research Council Canada, Ottawa, ON, Canada

O.A. Ojo Department of Mechanical and Manufacturing Engineering, University of Manitoba, Winnipeg, MB, Canada

R.A. Pethrick Deceased

A. Phillips Cambridge Vacuum Engineering, Denny Industrial Centre Pembroke Avenue Waterbeach, Cambridge, United Kingdom

J. Poulis Delft University of Technology, Delft, The Netherlands

M. Preuss University of Manchester, Manchester, United Kingdom; Monash University, Melbourne, VIC, Australia

Guillaume Renaud Aerospace Research Centre, National Research Council Canada, Ottawa, ON, Canada

N.L. Richards Department of Mechanical and Manufacturing Engineering, University of Manitoba, Winnipeg, MB, Canada

P.L. Threadgill TWI Ltd., Cambridge, United Kingdom (Retired)

H.L. Tsai Missouri University of Science and Technology, Rolla, MO, United States

K.R. Vishwakarma Department of Mechanical and Manufacturing Engineering, University of Manitoba, Winnipeg, MB, Canada

P.C. Wang General Motors Corporation, Alpharetta, GA, United States

M.St. Weglowski Lukasiewicz - Institute of Welding, Gliwice, Poland

T.T. Zhang Taiyuan University of Technology, Taiyuan, China

J. Zhou Pennsylvania State University, The Behrend College, Erie, PA, United States

New welding techniques for aerospace materials

1

Richard Freeman
TWI Ltd., Cambridge, United Kingdom

1.1 Introduction

Aircraft have been manufactured for decades using of a wide variety of welding and joining techniques. There have been significant developments in techniques over the last 15–20 years, and this has also led to the adoption of even more appropriate and stringent non-destructive inspection methods.

This chapter will focus on examples of how four different welding and joining technologies were developed by large aerospace companies, and approved by the regulatory authorities. The differing qualification criterion used to develop friction stir welding, laser beam welding, laser blown powder and laser powder bed additive manufacturing will be referenced. This will be followed by examples of welding and joining technologies which are under development for use in the manufacture of future aircraft. This will include the further development of the friction stir welding of aluminum and alloys, linear friction welding and stationary shoulder friction stir welding of titanium alloys, hybrid laser/arc welding of aluminum alloys, reduced pressure electron beam welding and electron beam texturing, reduced spatter MIG welding and further developments in arc welding.

1.2 Airworthiness implications of new welding and joining technologies

To enable a new welding and joining process to be approved for use in the manufacture of parts for a civil aircraft, it is necessary for an Original Equipment Manufacturer (OEM) with a Part 21 approval in Europe "Certification of aircraft and related products, parts and appliances and of design and product organizations" to work with the European Aviation Safety Agency (EASA) to qualify this procedure to the satisfaction of the authority. In the USA the company would work with the Federal Aviation Administration (FAA) in an identical manner, in accordance with the appropriate specification. The major regulatory authorities have agreements with each other, to allow information on the approval of new designs and manufacturing processes to be shared, so that identical qualification approval tests are not carried out in several different countries.

Welding and Joining of Aerospace Materials. https://doi.org/10.1016/B978-0-12-819140-8.00001-8
Copyright © 2021 Elsevier Ltd. All rights reserved.

1.2.1 The use of friction stir welding (FSW) in the eclipse 500 aircraft

The Eclipse Aviation 500 was a small six-seat business jet aircraft manufactured by Eclipse Aviation, based in Albuquerque, New Mexico, USA. The Eclipse 500 became the first of a new class of Very Light Jets (VLJ) when the first jet was delivered in late 2006. Production of the Eclipse 500 was halted in mid-2008 due to lack of funding after the delivery of 260 aircraft, the company entered Chapter 11 bankruptcy protection on 25 November 2008, and then was forced into Chapter 7 liquidation on 24 February 2009. The demise of the company was caused by a number of issues, not least of which was the collapse of DayJet, who had 1400 aircraft on order out of a claimed order book of about 2500, representing 58% of all Eclipse's that had been ordered.

After a lengthy Chapter 7 procedure, Eclipse Aerospace was confirmed as the new owner of the assets from former Eclipse Aviation on 20 August 2009 and opened for business on 1 September 2009. In October 2011 Eclipse Aerospace announced a new version of the aircraft, the Eclipse 550, which replaced the 500 in production, with deliveries starting in 2013. In April 2015 Eclipse Aerospace was merged with Kestrel Aircraft to form One Aviation.

The method of qualifying the FSW process with the FAA for the manufacture of the fuselage and wing sections is a very interesting story. FSW, which was described in some detail in a previous chapter, was initially approved by the FAA for the Eclipse 500 aircraft in March 2002, and it was the first civil aircraft to use this technology.

The Eclipse design was based on the use of FSW to join thin stringers (7055 aluminum alloy) to skin material (2024 aluminum alloy) in a lap configuration, with the main challenges being corrosion protection of the mating surfaces, control of distortion in the thin sheet material and control of interface deformation. Working closely with FAA officers and the South West Research Institute facility, both based in San Antonio, Texas, and the NASA Langley facility in Hampton, Virginia, Eclipse designed a comprehensive test programme to evaluate FSW against riveted aluminum to generate data on static FSW allowables (Type I and II), S/N curves (Type I, II and III), crack growth (da/dn), corrosion and barrel panel testing [1, 2].

The tensile strength results of 7055-T76 friction stir welded to 2024-T3 material of 470–480 MPa proved to be higher than the 2024-T3 riveted equivalent of 440 MPa. The fatigue results were also excellent, with tests running to over 4 million cycles without failure at the aircraft operating load levels. A large number of barrel samples were also tested to 8.33 psi simulating cabin pressure at 41,000 ft altitude. Artificially induced cracks of 50.8 mm (2 in.) length were introduced into certain test panels to look at crack growth behavior. The results showed that the first naturally occurring fatigue cracks were detected at 371,000 cycles or 18.5 lifetimes, and the cracks did not stay in the welds and propagated to machined pockets, which was a desirable outcome. The FSW joint performance exceeded design requirements with considerable margin. In addition, a fluorine based sealant was used between the stringer and skin to protect against crevice corrosion. Trials were carried out to ensure it was possible to friction stir weld through this sealant when making the lap joint welds.

128 m (5,040 in.) of weld (263 welds in total) were made per aircraft in the production of the cabin, aft fuselage and wing sections, replacing 6982 rivets. The FSW tools were routinely replaced after 77 m (3,000 in.) of welding as part of the Total Preventative Maintenance (TPM) system, even though they were capable of more work. 20% of the welds were inspected by an eddy current phased array system, as part of the production process.

1.2.2 The use of laser beam welding for Airbus aircraft

Initial development work in the early 1990s concentrated on the laser welding of 2024 aluminum stringers to the same skin material. However 2000 series aluminum alloys can be crack sensitive when fusion welded, and despite encouraging results, the process was finally developed and qualified on AlMgSiCu aluminum alloys (6013, 6110A, 6056) by Airbus Innovation Works [3]. Panels were constructed using different numbers of stringers welded to skin material, in order to develop the welding procedure for minimal distortion. After significant development of the laser welding procedure including parallel sided welding of the stringers by robot, panels were subjected to an extensive test programme to look at static and fatigue strength including fatigue crack growth rate. The technique is now being used by Airbus in the lower fuselage panels of the A318 single aisle aircraft (Fig. 1.1), the A340–600 dual aisle aircraft, and the A380 very large aircraft to replace riveted structures.

Lower fuselage panels must meet a combination of compression, shear and hoop stress requirements to provide static strength and buckling stability. However taking

Fig. 1.1 Laser beam welding of Airbus 318 fuselage panels.

into account the damage tolerance load scenarios of the upper fuselage applications, the analysis of fatigue crack growth and residual strength tests made at Airbus and Alcoa revealed that the 6000 series aluminum alloy application would not provide sufficient strength advantages over riveted designs. The upper fuselage panels are subject to longitudinal hoop and tension, and the design must provide higher residual strength performance and improved crack propagation performance.

There have been significant developments in aluminum alloys for airframe applications in the last decade or so, from the major aluminum raw material manufacturers. Constellium have pushed their Airware® Al—Li alloys into the market for fuselage skins, stringers, floor structures and seat tracks, window frames, and large internal wing and fuselage components. Arconic has developed their Al—Li Alloy C460 (recently registered as AA2099) for use in aerospace and high performance applications requiring high strength, low density, high stiffness, superior damage tolerance, excellent corrosion resistance and weldability. Aleris are also pioneering the Al-Mg-Sc (AA5028) alloy as a replacement for 2024 alloy for stringer reinforced aircraft fuselage components with improved weldability.

1.2.3 The use of laser blown powder additive manufacturing for the repair of turbine seal segments

Turbine seal segments for aero engines present many design challenges due to their need to maintain performance at high temperatures under abrasive conditions. In addition to this, cost and environmental considerations demand that such components continue to deliver fuel efficient performance during their lifecycle. High pressure (HP) and Intermediate pressure (IP) gas turbines blades under normal operating conditions are in rubbing contact against a turbine seal which is manufactured in circumferential segments [4]. The segments are manufactured as single piece castings in CMSX4 alloy (a single crystal nickel superalloy), chosen to mitigate the oxidizing conditions present at the blade tip. The seal is fabricated using an electrode discharge method (EDM) to machine a metal lattice structure in which the design is chosen to minimize the damage to the turbine blades from rubbing contact while simultaneously allowing sufficient metal in the lattice to resist the high temperatures at the blade tip. The lattice is subsequently filled with an abradable ceramic sintered product to maintain a gas tight seal. During service, the lattice material is slowly worn down and at certain shop visits has to be replaced with new seals, as there was no practical repair available. Rolls-Royce has been investigating a number of different options to manufacture the lattice structures in a different manner. Trials using a brazed-on cast lattice proved unsuccessful in engine tests with break-up of the reformed feature. Replacement costs are significant because the EDM process is slow and environmental considerations lead to minimizing the use of rare elements. With the availability of the repair technique, then up to 95% of the segment by weight could be re-used with cost and environmental benefits.

For these reasons, additive manufacturing methods that were capable of building up the feature with the correct functional and geometric properties were investigated.

Laser blown powder was downselected as the most promising method for further evaluation as the technology was nearing maturity for other build-ups on precision structures for aerospace repair. It utilizes a fine powder (nickel based in this application) that is blown in a cone shape to a fine focus from a nozzle. A high intensity laser beam is passed through the powder focus and the powder is melted and welds to the workpiece to build up a 3D form. The process is repeated to build up the desired shape, with intricate patterns possible with minimal distortion and very small dilution between the feature and the substrate. The laser is moved using a five axis machine system, and the pattern is produced using a toolpath programmed from the CAD model to ensure accuracy.

The development process was managed through the Rolls-Royce Production System Process from the original concept at the University of Birmingham, through process development at TWI and into shop floor use. Lattice structures were manufactured for use in burner rig, abradability and engine testing trials, with the equipment at TWI modified to maximize the powder usage efficiency, minimize the processing time and achieve acceptable levels of porosity and cracking in the final product (Fig. 1.2).

Rolls-Royce is now looking to adopt this process for original equipment manufacture, as bespoke lattice structures can be produced on cheaper substrate materials or the alloy composition can be changed throughout the shape of the component by grading the powder to avoid abrupt changes in material properties.

1.2.4 The use of laser powder bed fusion additive manufacturing for the manufacture of the LEAP engine fuel nozzle

GE Aviation has made several recent investments in AM technology. They purchased over 75% stakes in two large AM companies: Arcam AB and Concept Laser, the latter of which is a leading producer of laser powder bed fusion machines. Not only has GE heavily invested into AM, it has also been using internal funding to implement AM into its own designs and products. GE fairly recently developed a jet engine called LEAP, which is an improved version of its previous CFM56 single-aisle aircraft

Fig. 1.2 Rolls-Royce Trent engine seal segment manufactured by the laser blown powder AM technique.

engine. The LEAP engine now has a 3D printed fuel nozzle using the laser powder bed fusion technique, with GE working hand in hand with the FAA to gain approval. Powder-bed fusion techniques use a build platform containing powder material. In this process a laser (or electron beam) is used to selectively melt and sinter particles together at specific points and once a layer of the object is completed the platform lowers and more powder is rolled on. The process then begins again until the 3D structure is completed. As of October 2018, 30,000 Leap X nozzles have been manufactured using this technique.

1.3 Future developments and trends

This section will cover a number of welding and joining processes that have great potential for use in future aircraft component manufacture.

1.3.1 Friction stir welding of aluminum alloys

Despite the problems associated the Eclipse Aviation business, the FAA [5] and other regulatory bodies are very aware of the process and several companies have been investigating the process for use on civil and military aircraft. Airbus made significant strides toward the adoption of the process for the front fuselage underbelly section of the A340–500 high gross weight aircraft, and were planning to FSW weld the Al—Li A350 fuselage before the aircraft was redesigned to incorporate a composite fuselage in the XWB design. It is pure speculation at present but future high volume single aisle civil aircraft and business jets, manufactured primarily from aluminum, could utilize the technology in the next 20 years.

FSW is used to manufacture the barrier beams for the Boeing 747 and 777 freighter aircraft, and its use has also been investigated for the floor sections of the C17 (Boeing) and the C130 (Lockheed Martin) military transporter aircraft. However for existing military programmes, where future volumes of aircraft build are not easy to predict, it is more difficult to make a case to change the manufacturing process despite the apparent benefits. The fact that it is being proposed for the floor sections of the Airbus A400M military transporter aircraft, is due to the consideration of the process at the aircraft design stage, and the ability to amortize costs across the lifetime of the aircraft programme.

The FSW process is also approved by the following international surveying bodies and classification societies for the production of panels for high speed ferries, hovercraft and cruise ships: American Bureau of Shipping (ABS), Det Norske Veritas (DNV), Germanischer Lloyd (GL), Lloyds Registry of Shipping (LR) and Registro Italiano Navale (RINA) [6].

New developments in the market include the simultaneous double sided friction stir welding of thick section aluminum alloy for the airframe market (Figs. 1.3 and 1.4). This have been developed using TWI's Powerstir FSW machine located at its Yorkshire Technology Centre near Sheffield. This multifunctional gantry-type machine is fitted with opposed, synchronized spindles, and is capable of running tooling in excess of

Fig. 1.3 Thick section friction stir welding of aluminum alloys.

50 mm penetration on each head. A fixture featuring a slot in line with the tools' traversal path both supports the component being welded and enables the two heads to be used simultaneously. TWI performed trials in 50 mm-thick aluminum alloy plates and was able to produce sound welds at a rate that far exceeds those created using the weld-flip-weld technique in which the workpiece is joined in two passes. The majority of the time saving resulted from the huge reduction in processing time, with no need to remove, dress and reposition the component between welds, but this technique also enabled a faster welding speed. This is because two tools operating simultaneously generate more heat in the component than one, creating a softer plasticised region around the tools that reduces resistance and enables them to pass through the material more quickly. This benefit was compounded by the slotted design of the fixture, which helped minimize the rate at which the heat was conducted away from the weld zone.

1.3.2 Friction stir welding of titanium and nickel alloys

As described in the section on Friction Stir Welding, the technique is well developed for aluminum alloys and has been adopted in production in several industry sectors. When attempting to friction stir weld metals with a much higher melting point than aluminum, it is impossible to use the same tool materials. The use of titanium and

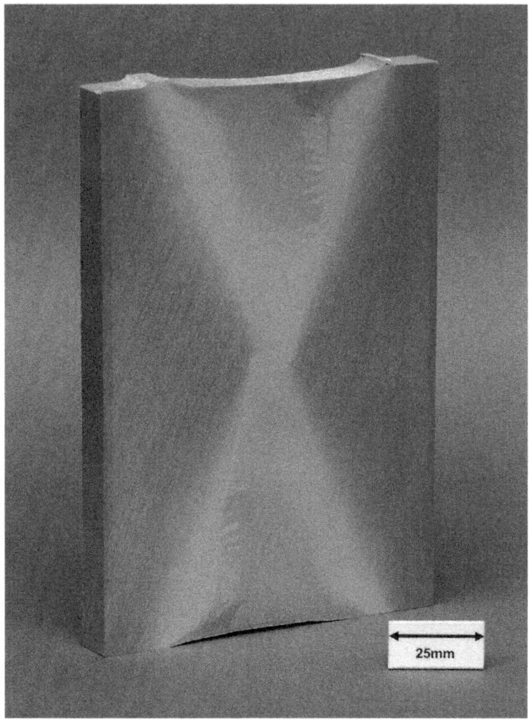

Fig. 1.4 Macro section of thick section friction stir weld of in aluminum alloy.

nickel alloys in aero engine manufacture, has led to developments in the FSW process to allow these materials to be welded, for repair and original equipment applications. However the use of higher strength tool materials, combined with the higher forces required to weld the metallic substrates can lead to tool wear issues. Titanium is also a poor conductor of heat, which means that FSW is even more challenging with this material.

The development of the stationary shoulder friction stir welding technique for titanium means that the tool pin material is making the weld, while the shoulder slides across the top of the material significantly reducing heat input. This has allowed welds of much greater length than can be produced by conventional FSW to be made (Fig. 1.5) although wear of the tungsten based tool material is still too high to consider the process an economic manufacturing route. Work continues at several prominent R&D facilities around the world to address the problem. Similarly, early stage developments in the friction stir welding of nickel alloys, using silicon nitride tooling, are progressing but will need to develop much further before the process can be considered.

In 2014 the European Space Agency (ESA) initiated a programme to investigate the capability of friction stir welding (FSW) as a manufacturing method for making titanium propellant tanks for its future space programmes. TWI, in partnership with Airbus Defence and Space in the UK developed a proposal to conduct a feasibility study of FSW for titanium alloys suitable for launch vehicle propellant tanks. The 2-year project comprised two phases of work, with the main aim to raise the

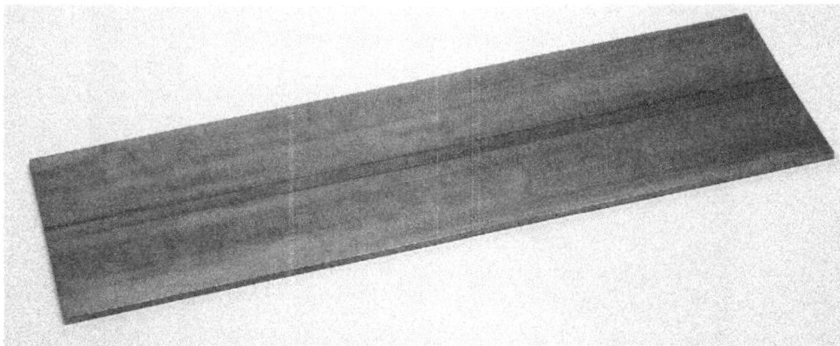

Fig. 1.5 Stationary shoulder FSW of Ti-6-4 alloy.

technology readiness level of FSW of titanium alloys from TRL3 to TRL6 in conjunction with a potential reduction of costs against the current conventional approach for manufacturing propellant tanks. Four demonstration tanks were manufactured using stationary shoulder friction stir welding to join the Ti-6-4 castings (Figs. 1.6 and 1.7). This is the first example of FSW of Ti for the space sector, and offers a cost effective solution to the current method of purchasing expensive Ti forgings up to 30 mm thick, machining them down to 10% of their original thickness and welding them with a resultant high buy-to-fly ratio.

1.3.3 Linear friction welding (LFW)

This process is described in some detail in the Appendix of this publication, but it is important to comment that the process is receiving a lot of attention in the aerospace industry, both in the manufacture of bladed discs (blisks) and in near net shape forming

Fig. 1.6 Friction stir welding of Ti-6-4 propellant tank for European Space Agency.

Fig. 1.7 Completed FSW Ti-6-4 propellant tank.

to assist in the reduction of machining costs. An example of a blisk manufactured by Rolls-Royce using LFW is shown in Fig. 1.8. In 2016–17 TWI worked closely with GKN Aerospace to produce the world's largest LFW component, an aluminum wing rib for commercial aircraft. Comprising 22 rib feet joined to a 2.5 m long baseplate, the wing rib is designed to strengthen an aircraft's wings (Fig. 1.9 shows a linear friction welded and part machined wing rib that was exhibited a the 2016 Farnborough Air Show). The wing rib is manufactured from aluminum AA7050 alloy and is pre-formed to a near net shape, which greatly reduces material wastage and thereby machining time and costs. With 70% less material wastage and a 43% reduction in machining time, the selective material placement greatly reduces buy-to-fly ratios while enhancing performance. Furthermore, the grain orientation of the rib feet can be tailored for directional strengthening, which may allow for further reductions in local thicknesses. Discussions are ongoing with aircraft manufacturers to develop this technology for future aircraft.

There has also been significant R&D work supported by Innovate UK, to develop LFW of Ti for near net shape manufacturing. The £2.6 m project was designed to show how near net shape manufacture could boost productivity by using tailored blanks to produce lightweight titanium alloy parts. Its main aim was to develop an industrial scale demonstrator tooling system and components, and the following parts were produced for the main UK partner CAV Advanced Technologies (Fig. 1.10).

Fig. 1.8 Rolls-Royce bladed disc with blades attached by linear friction welding.

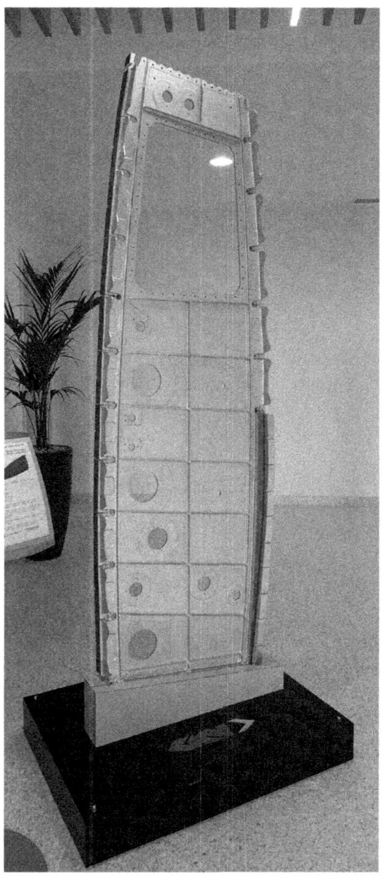

Fig. 1.9 Wing rib with feet attached to rib by linear friction welding.

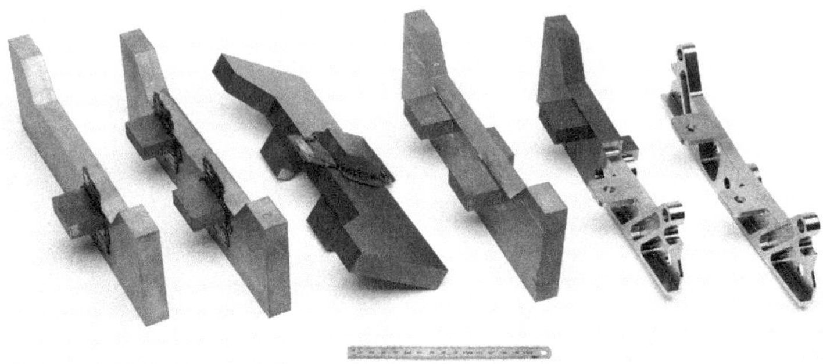

Fig. 1.10 Near net shaped parts produced by linear friction welding.

1.3.4 Hybrid laser arc welding

In the last decade significant developments in the fiber laser industry has led to the production of more powerful lasers with higher efficiencies. This has meant that thicker materials can be welded using the process, and alongside this development hybrid laser arc welding has attracted widespread industrial interest. In this process, an electric arc is introduced into the same molten pool as the laser, resulting in the capability to weld thicker materials or faster welding speeds to be achieved, as well as offering improved joint gap tolerance and weld quality.

Recent work has sown that full penetration can be achieved in a 12.7 mm thickness Al-Zn-Mg-Cu 7000 series aluminum alloy using 7 kW of Yb-fiber power in either an autogenous set-up or in a hybrid combination with a MIG arc [7]. Both the autogenous laser and the hybrid laser-MIG process are capable of producing a level of weld metal porosity in accordance with the most stringent weld quality classes defined in BS EN 13919–2 and AWS D17.1, by considering shielding gas supply and material surface preparation prior to welding, and selecting an appropriate laser spot size and welding speed (Figs. 1.11 and 1.12).

1.3.5 Reduced pressure electron beam welding

Non-vacuum electron beam welding has been researched in many countries over the last 30–40 years for the welding of thick section materials. In more recent years some companies, including TWI, have been involved in the development of reduced pressure electron beam welding (RPEBW). It has been demonstrated that operating the electron beam process in the pressure range 0.1–10 mbar in preference to high vacuum ($\sim 10^{-3}$ mbar) offers the possibility of eliminating the need for a vacuum chamber by permitting the practical use of local sealing and pumping [8]. Using the TWI system, welds have been successfully made in 22 mm thick Hastelloy C-22 nickel alloy and 50 mm thick Ti-6-4 material. TWI also ran a precompetitive research project on the RPEBW of Ti-6-4, which proved that EB Welding of Ti 6Al 4V alloy at reduced pressure was feasible in the thickness range 6.35–50.8 mm. The use of a reduced pressure

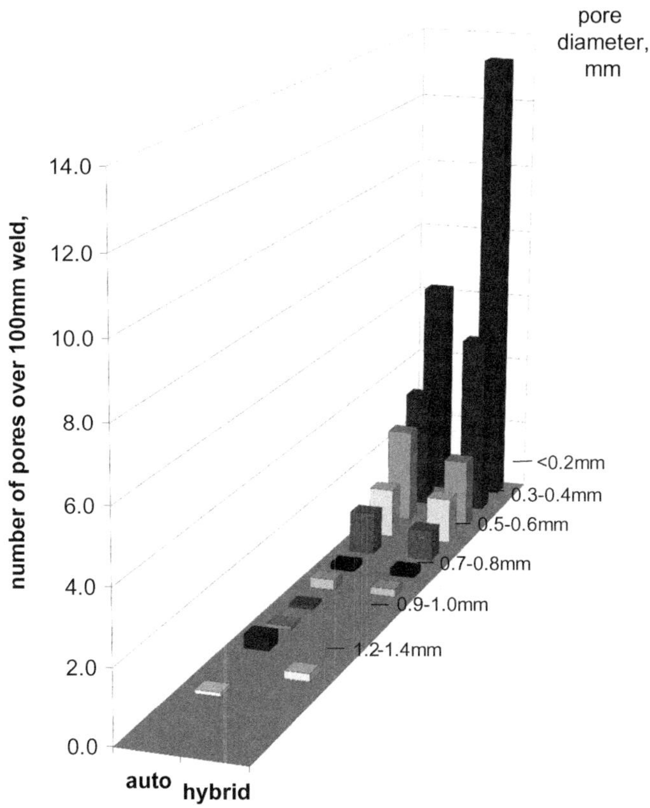

Fig. 1.11 Comparison of autogenous laser welding and hybrid laser arc welding.

atmosphere caused no evidence of bulk weld metal gas contamination but some increase in surface oxidation when compared with EB welds made at high vacuum. The use of helium over pressure gas was also beneficial in this respect. Sufficient mechanical property data would need to be generated in the future before aircraft designers, materials engineers and stress analysts would consider moving from a full vacuum alternative, but significant cost savings could be made if the data was positive enough.

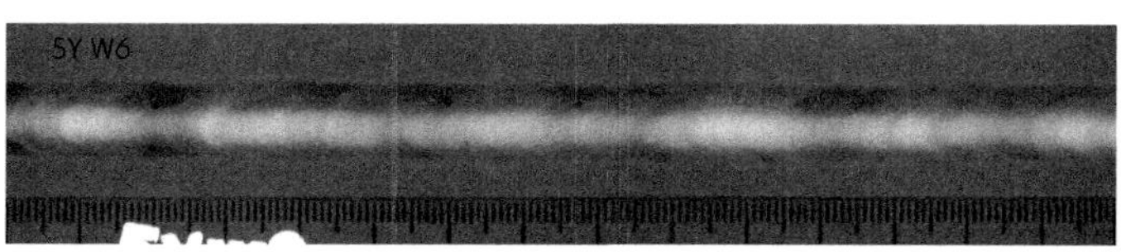

Fig. 1.12 Radiograph of hybrid weld with low porosity levels.

1.3.6 Electron beam texturing (EBT)

An electron beam can be modified to allow welding and hardening operations and even texturing of a surface. The EBT process can operate at extremely high speed, and allows the generation of a range of surface textures. Typically these textures have ~1:1 aspect ratio, and may include overhangs to give re-entrant features suitable for bonding. All this may be implemented with just a single beam, to give (typically) 500–2000 features per second. In some cases even higher speeds are possible. The EBT process is now being industrially applied, with several suitably retro-fitted EB machines as well as dedicated new builds equipped for the task. This transition from laboratory to production necessitated the development of improved equipment, both in beam generation and control. With the first EBT equipment, it was not normally possible to achieve a well-controlled process that incorporated repeated beam visits to a single location on the work. However, the development of more sophisticated controls suited to industrial EBT helped in the development of different processing strategies, eventually leading to the development of the *Surfi-Sculpt*® process by TWI. In the *Surfi-Sculpt* process, typically repeated visits of the beam to overlapping or adjoining locations on the work are used to create a wide range of features [9].

Using the beam probing system, trialing and refinement of new gun designs was readily achieved, and their consistency was tested and proven. With the improved electron gun system, practical implementation of revised gun design prototypes became both easy and precise. Finally, the improved beam deflection systems and associated software allowed complex processes to be programmed and implemented without difficulty. Each of the above improvements, taken alone, would be a significant benefit to process development. Together, they were more than additively beneficial to further progress; each compounded the value of the other. With the development of improved electron gun geometries, it was possible to generate beams with approximately twice the intensity of the original EBT gun. The technology is being evaluated by industrial companies for a range of uses including a precursor for composite to metal bonding, manufacture of aerodynamically enhanced surfaces (Fig. 1.13) and preparation of surfaces prior to coating application for improved coating performance.

1.3.7 Reduced spatter MIG welding of titanium alloys

Historically, due to its poor welding characteristics, MIG welding of titanium has been restricted to low quality applications, and is rarely used in the aerospace industry except for ground support equipment. The surface finish of conventional titanium filler wires can cause arc instability and contact tip wear problems, which can result in excessive spatter and porosity. Recently, a novel titanium wire has been produced by Daido Steel in Japan that was claimed to result in improved arc stability and reduced spatter [10]. This has been achieved by the use of a novel wire production process that modifies the wire surface, together with the use of optimized pulsed MIG parameters. Limited work carried out by TWI as part of the Core Research Programme, using the novel wire has substantiated these claims, when used to make butt

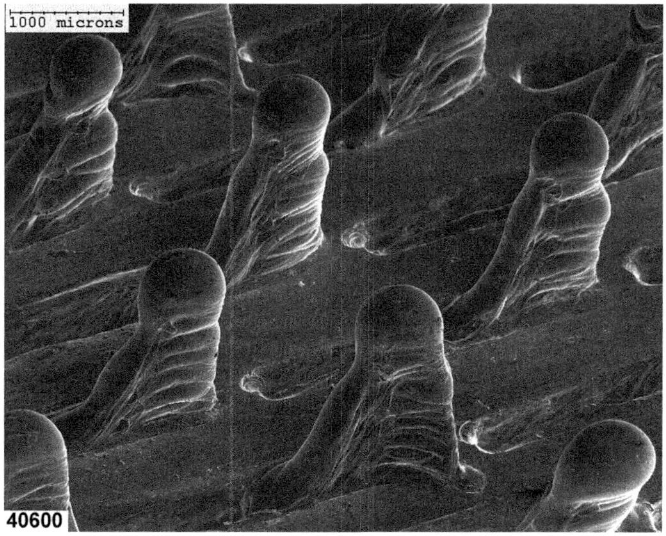

Fig. 1.13 Aerodynamically enhanced features produced by the Surfi-Sculpt® process.

welds in 6 mm thick CP titanium. Particulate fume and ozone rates for the novel wire were also shown to be less than for conventional filler wire [11]. This is being developed further in an industrially funded R&D project, and initial results are again very promising.

Two prominent companies have also altered the arc characteristics to allow MIG welds to be made with much reduced spatter. The Austrian based company Fronius has developed the Cold Metal Transfer (CMT) system, which is carried out exclusively using digital inverter power sources. The welding system basically uses the same latest state of the art hardware as a MIG/MAG system, while at the same time taking certain specific requirements into account. When the power source detects a short circuit, the welding current drops and the filler wire starts to retract. This system is being evaluated by a leading aero engine manufacturer for future manufacture of original equipment and repair work.

OTC Daihen has developed the SynchroFeed low spatter robotic MIG welding systems, and has now added aluminum to the list of materials possible using the process, although whether it will see adoption in the aerospace MRO sector remains to be seen at this time. Lincoln Electric in the USA has also pioneered the Surface Tension Transfer® (STT®) technique, although this aimed more at the pipe fabrication market to the production of single sided low hydrogen root welds.

1.3.8 Additive manufacturing (AM)

Two additive manufacturing case studies were covered in Section 1.2, and there are many developments ongoing in the additive manufacturing arena, with the large aerospace airframe, engine and Tier 1 manufacturers all extremely active in this space.

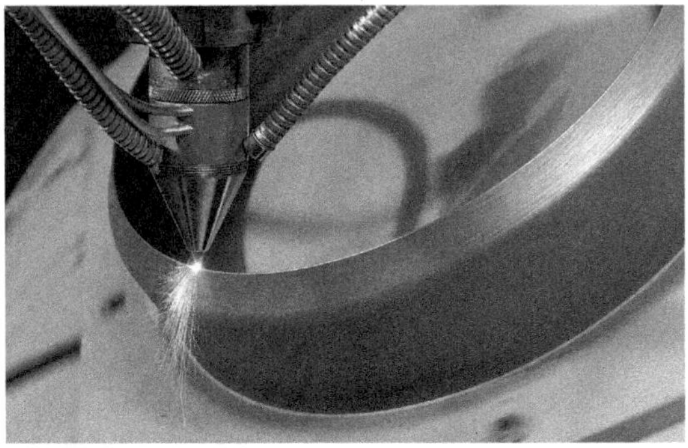

Fig. 1.14 Manufacture of 3 mm thick IN718 helicopter casing by the laser blown powder process.

Powder bed machines form the bulk of AM equipment sold around the world, with several laser powder bed manufacturers and one major EB powder bed manufacturer at present. While you can manufacture many modestly sized parts in one powder bed chamber its size does limit larger component manufacture, and the "open architecture" AM processes such as laser blown powder, EB wire feed, wire and arc fusion processes allow the manufacture of much larger component structures in an open, but controlled environment with inert gas coverage.

Fig. 1.14 shows the process to manufacture a 3 mm thick IN 718 helicopter casing, with the finished part show in Fig. 1.15. This was part of €7 m European collaborative project called MERLIN, to reduce the environmental impact of air transport using Additive Manufacturing (AM) techniques in the manufacture of civil aero engines, involving the likes of Rolls-Royce, MTU, Pratt & Whitney Poland, Airbus, Safran and GKN Aerospace along with several RTO's (including TWI) and technical universities in Europe. The development of more complex features on laser blown parts is the next stage and Fig. 1.16 shows a part manufactured by TWI for the €2.2 m E-Music

Fig. 1.15 Completed IN718 helicopter casing.

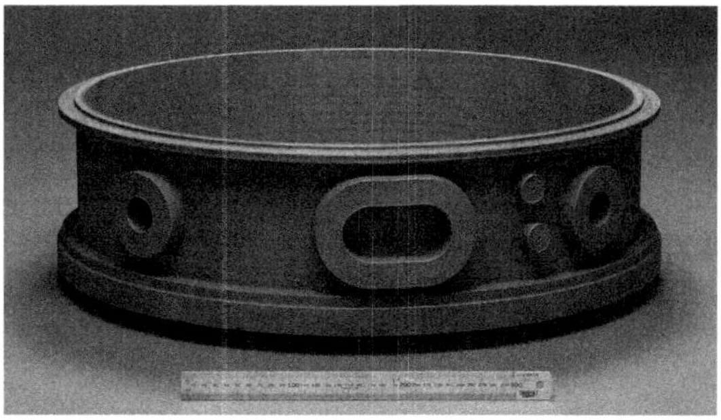

Fig. 1.16 Laser blown powder engine casing.

project (Efficient Manufacturing for Aerospace Components USing Additive Manufacturing, Net Shape HIP and Investment Casting) involving Rolls-Royce, ITP, Collins Aerospace and Airbus and several other R&D entities.

The development of the wire and arc AM (WAAM) technique at Cranfield University, and the creation of the spin-off company WAAM3D, has led to the manufacture of many large items. Fig. 1.17 shows a structure in Ti-6-4 using this process. There have also been many developments in the Electron Beam Additive Manufacturing (EBAM) and Electron Beam Free Form Fabrication (EBFFF) processes, with Lockheed Martin and NASA very visibly pioneering this technology for future space hardware with reduced buy to fly ratios.

Norsk Titanium have also developed their Rapid Plasma Deposition (RPD) additive manufacturing technology for Boeing structural titanium parts, some of which are destined for the 787 aircraft.

Fig. 1.17 Wire and arc AM (WAAM) technique for producing large Ti-6-4 parts.

1.4 Review of welding processes

The aerospace industry is known to be risk averse, due to strict safety requirements, but many welding developments are ongoing, particularly with relation to the use of reduced spatter MIG welding processes in the maintenance, repair and overhaul (MRO) areas.

The push to accept fiction based processes (particularly friction stir welding and linear friction welding) continues with many companies developing cost benefit analysis techniques to look at opportunities to introduce the technology on a life cycle cost basis. This will not be a quick process, given the significant investments already made in another competing technologies and the time taken to qualify a process but there is much interest in reducing "buy to fly" ratios in the push toward more environmentally friendly manufacturing techniques in the future.

Major developments in fiber laser welding (using Yb fiber based lasers) is continuing at a rapid pace with ever increasing laser welding powers of up to 100 kW, challenging electron beam welding for single pass welding of material thicknesses in the 10–15 mm material thickness range.

The investment in additive manufacturing techniques by major aerospace companies continues at a rapid pace, and is likely to lead to increased use of the technology in Class C and B aerospace hardware in the future. Cost effective qualification of AM components must be addressed however, before we see the increased use of AM as predicted by many in the industry.

References

[1] O. Masefield, Eclipse Aviation Inc 'The VLJ vision becomes a reality—Development & Certification of the Eclipse 500', in: Meeting of the Royal Aeronautical Society at CAA, Gatwick 11 January 2006, 2006.

[2] O. Masefield, Eclipse Aviation Inc 'An update on the production of the Eclipse 500 aircraft', in: Presentation at TWI Annual Dinner in London 6 November 2007, 2008.

[3] Palm F (2008) EADS Innovation Works 'Can welding be an option in future fuselage structures—Lessons learnt and new concepts derived from 10 years in laser beam welding research' TWI-EWI seminar on Joining of Aerospace Materials, Toulouse 1–2 October 2008.

[4] S.M. Beech, D. Clark, J. Allen, Rolls-Royce, The repair of turbine seal segments using blown powder direct laser deposition, in: TWI-EWI seminar on joining of aerospace materials, Toulouse 1–2 October 2008, 2008.

[5] T. Khaled, FAA, An Outsider Looks at Friction Stir Welding Report #: ANM-112N-05-06, 2005.

[6] F. Delany, S.W. Kallee, M.J. Russell, TWI, Friction stir welding of aluminum ships, in: International forum on welding technologies in the shipping industry, held at Beijing Essen Welding and Cutting Fair in Shanghai 16–19 June 2007, 2007.

[7] G. Verhaeghe, TWI, Low porosity laser welding of thick section aluminum, in: TWI-EWI seminar on Joining of Aerospace Materials, Toulouse 1–2 October 2008, 2008.

[8] C. Punshon, TWI, Reduced pressure electron beam welding—development of a prototype local vacuum system, in: TWI Core Research Report 898/2008 available to TWI Industrial members via secure password at, 2008. http://www.twi.co.uk/content/mr898.pdf.

[9] B.G.I. Dance, A.L. Buxton, TWI, An introduction to Surfi-Sculpt ® technology - new opportunities, new challenges, in: Paper presented at 7th International Conference on Beam Technology, Halle, Germany, 17–19 April 2007, 2007.
[10] Chujoya (Daido Steel), The Development of the G-Coat™ Titanium Alloy Welding Wire for GMAW, 2004. Daido Steel Co Ltd, March 2004.
[11] A. Kostrivas, A. Plewka, G.B. Melton, L.S.S. Smith, TWI, Pulsed MIG welding of titanium with a novel wire, in: TWI Core Research Report 900/2008 available to TWI Industrial members via secure password at, 2008. http://www.twi.co.uk/content/mr900.pdf.

Inertia friction welding (IFW) for aerospace applications

2

M.M. Attallah[a] and M. Preuss[b,c]
[a]University of Birmingham, Birmingham, United Kingdom, [b]University of Manchester, Manchester, United Kingdom, [c]Monash University, Melbourne, VIC, Australia

2.1 Introduction

The need for high-quality joints, combined with the inherent difficulty in welding most aerospace materials, has fostered the use of solid-state friction-based welding techniques within the past decade in the aerospace industry, such as: friction stir welding (FSW), linear friction welding and rotary friction welding (RFW) with its two variants; continuous-drive friction welding (CDFW) and inertia friction welding (IFW) [1]. Except for FSW, friction-based welding processes can be described as self-cleaning as a result of the ejection of the plasticized material in the form of flash at the end of the welding cycle, which carries alongside any surface contamination or oxides, making it unnecessary to use shielding gas during welding. Among the friction-based welding processes, the use of inertia welding in the aerospace industry has been steadily increased in the past 2 decades, especially in joining nickel-based superalloys, titanium alloys and steel aero engine cylindrical components, owing to the significant improvements it provides in the joint quality, compared with the use of fusion welding. This chapter contains six themes. The first theme introduces the process, with respect to its operation, parameters, differences from other friction welding techniques and equipment. The second theme explains the application of the technique and the selection of the process parameters, and the different mathematical, analytical and numerical approaches that are used to model the thermal fields and residual stress development. The third and fourth themes discuss the microstructural and mechanical property development respectively, in inertia friction-welded Ni-based superalloys, titanium alloys, steels and other alloys. The impact of the process on the residual stress development is discussed in the fifth theme, focusing on the application of neutron and synchrotron X-ray diffraction in measuring the residual stress development. Finally, the chapter is concluded with a section outlining the future trends and possible developments in IFW.

2.1.1 Process development

Although it is believed that the interest in using rotary frictional heating for joining dates back to a late 19th-century US patent, further developments in the first half of the 20th century resulted in the development of commercially applicable RFW

Welding and Joining of Aerospace Materials. https://doi.org/10.1016/B978-0-12-819140-8.00002-X
Copyright © 2012 Woodhead Publishing Ltd. All rights reserved.

techniques [2]. Concurrent yet separate efforts by Russian and American engineers resulted in the development of the two variants of RFW in the second half of the 20th century. Around 1954–1957, Russian engineers Chudikov and Vill were first to suggest an RFW technique for joining cylindrical sections mounted on a modified lathe, which was later termed direct or continuous-drive friction welding [3, 4]. Upon successful commercialization of this technique in Russia, the concept of RFW became familiar with American and British engineers. Pioneering work at the Caterpillar Tractor Company led to the development of inertia (flywheel) friction welding (IFW), which was US patented in 1965 [2, 4, 5]. Because of this, IFW remains more commonly used in the USA until today, while CDFW is mostly used in Europe and Japan.

2.1.2 Inertia friction welding (IFW) process description

In IFW, the kinetic energy stored in a rotating flywheel is conserved into frictional thermal energy to mostly join two components of cylindrical geometry; one component is clamped to the flywheel, while the other component is clamped in a nonrotating chuck connected to a hydraulic ram. During welding, once the flywheel is brought to a certain rotation speed, the motor is disengaged, and a forging pressure is applied to the hydraulic ram to bring the two components to contact. Following the initial contact, the flywheel speed starts to decelerate owing to the conservation of the stored energy into thermal energy, causing the temperature to increase sharply at the interface owing to the generated friction. Ultimately, a plasticized layer forms between the two components, where consolidation occurs. The application of pressure causes the plasticized material to flow outside the joint line forming a flash, which dissipates some of the weld energy causing the interface region to cool slightly even before the rotating part has come to a halt [2, 6].

IFW thus differs from CDFW in the braking mechanism. In CDFW, braking is performed by declutching the spindle from the hydraulic or electric motor that "continuously" drives the rotating component, followed by applying the brakes upon the application of the forging force for a certain time [6]. In IFW, braking occurs upon the dissipation of the energy stored in the flywheel, which occurs gradually during IFW, with the maximum energy transfer occurring upon the first touch between the two interfaces [1]. This difference affects the application of the power input to the weld throughout the process; where the power input in IFW changes to supply the required power to first plasticize the interface and then to forge the components, while the power input in CDFW is limited by the power rating of the motor. The main differences between the IFW and CDFW systems are shown in Fig. 2.1.

2.1.3 IFW process parameters

Both IFW and CDFW processes differ in the parameters that control the process [6]. IFW is controlled by two main parameters, which are the welding energy (rotation speed and flywheel inertia) and the forging pressure, while CDFW is controlled by the rotation speed and the time pressure cycle to be used, including the braking time (Fig. 2.2). It is important to mention that the spindle speed (and hence the power input)

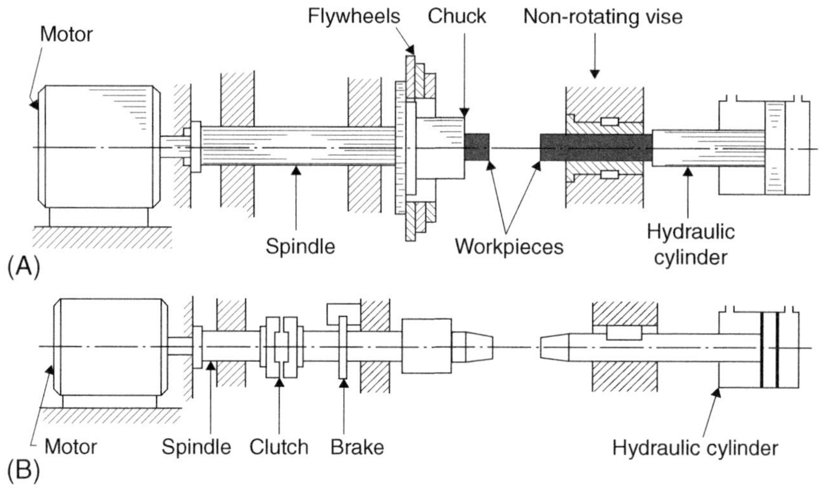

Fig. 2.1 Schematic diagrams for the set-ups of the welding systems for (A) inertia welding and (B) CDFW [6].

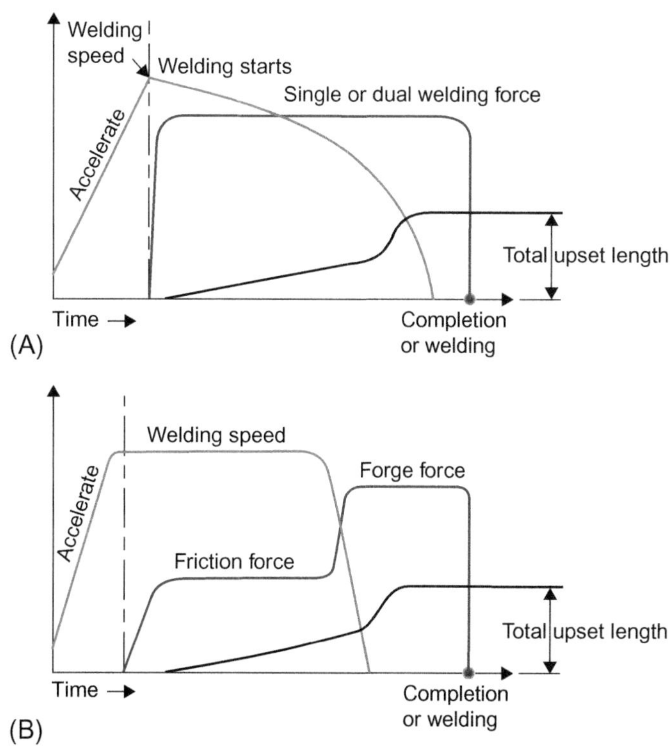

Fig. 2.2 A comparison between the welding cycle for (A) inertia welding and (B) direct drive friction welding (DDFW); CDFW.
Courtesy of Manufacturing Technology, Inc.

gradually decreases from the maximum (set) value following contact in IFW, whereas the spindle speed is mostly constant in CDFW [4]. For further details on CDFW (e.g. process parameters, machine specifications and applications), the reader is directed to other references that fully discuss CDFW [6–8].

2.1.4 IFW process stages

A three-stage model is generally used to define the IFW stages depending on the fluctuation in the frictional torque owing to the contact between the rotating components [9], or four stages if an initial stage is added during which the flywheel reaches the desired rotation speed [10]. According to the three-stage model (Fig. 2.3), the three stages are described as follows:

- Stage I (*initial contact*): The two components are brought in contact, which results in a rapid increase in the frictional torque, a deceleration in the rotation speed and accordingly a sizeable dissipation of the stored energy owing to the dry friction between the interfaces. The friction also leads to the removal of any surface irregularities and asperities, similar to what happens during wear, until perfect contact is reached. With the increase in temperature, a decrease in the torque occurs at the end of this stage, owing to the softening of the material and adhesion of asperities at the interface, forming a plasticized layer (Fig. 2.4A). The high torque that is experienced and the high rotation speed at start mean that the maximum power input is achieved within this stage.

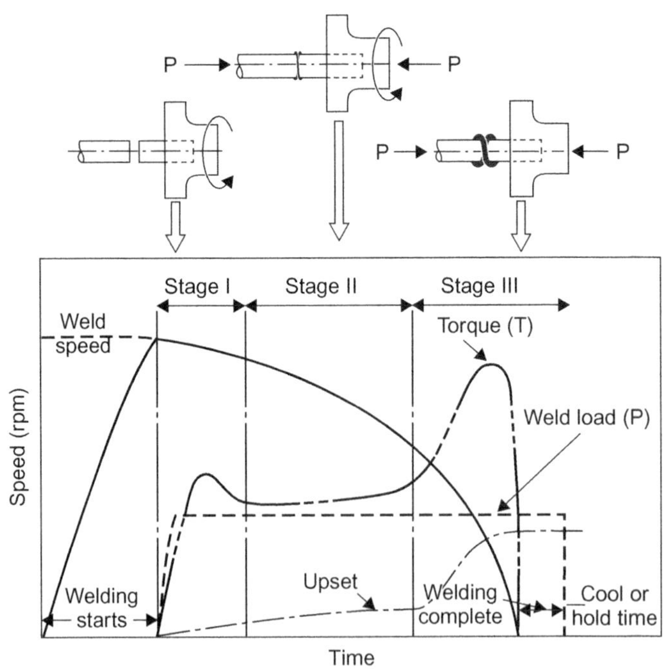

Fig. 2.3 The stages of inertia welding, showing the variation in process variables [6, 9].

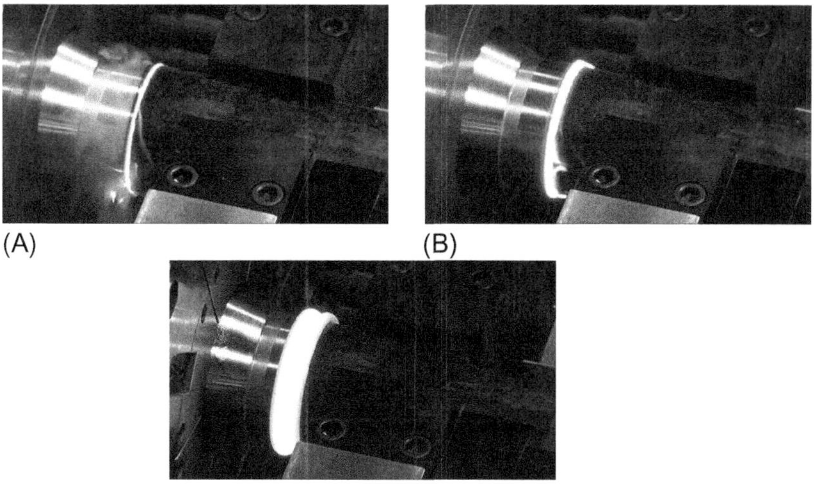

(A) (B) (C)

Fig. 2.4 The stages of inertia welding in a drill-pipe weld (A) initial contact, (B) transitional stage, and (C) forging.
Courtesy of Manufacturing Technology, Inc.

- Stage II (*transition stage*): The friction-induced thermomechanical deformation makes the material at the thin interface fully plasticized (visco-plastic). Thus, the process reaches a transitional steady-state condition, where the strain hardening is overcome by frictional heating. This is manifested in a roughly constant torque, and a gradually decreasing rotation speed (and power). A gradual increase in the upset also occurs owing to burn-off and the initiation of flash formation at the interface. However, continuous frictional heating leads to the widening of the plasticized region (Fig. 2.4B).
- Stage III (*forging stage*): With the decrease in rotation speed while the forging pressure is still being applied, the torque increases to another peak to overcome the cooling and hardening. This increase in torque is believed to refine the joint microstructure at this final stage, as well as the ejection of the flash that carries along any oxides or inclusions [2]. The upset increases and more flash is formed, leading to further cooling of the interface (Fig. 2.4C). After reaching the maximum upset, the forging load is kept applied until the weld cools.

2.1.5 IFW production machines

IFW machines are generally classified according to the forge force capacity, which ranges from a fraction of a ton to 4500 ton lb. (~8896 kN). The choice of the machine depends on the application geometry and material, which accordingly controls the required process parameters. Typical machines and products are shown in Figs. 2.5 and 2.6. For relatively small aerospace components (e.g. pistons, pipes, turbine wheels and shafts), machines with forging force capacity ~1–50 ton lb. (~2–250 kN) are normally used (Fig. 2.5). For larger aero engine assemblies (e.g. compressor-rotor, disk-to-cone, drum drive compressor and disk-to-shaft assemblies), machines with large capacities are used to provide the required energy [11] (Fig. 2.6).

Major suppliers of IFW machines include Manufacturing Technology Inc. (MTI), Blacks equipment, Thompson friction welding, Swanson industries, AI Welders, Kuka and the Welding Institute (TWI). Most IFW machines are horizontal and driven with hydraulic systems, although there are some machines (mostly with small forging force capacity) that run with DC or AC motors [6]. In the hydraulic-driven machines,

(A)

(B)

Fig. 2.5 Typical IFW machines and products (A) 15 ton welder, (B) 40 ton welder,

(Continued)

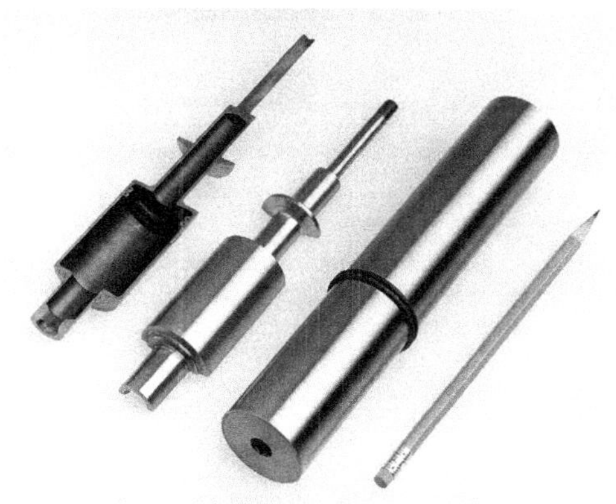

(C)

(D)

Fig. 2.5, Cont'd (C) light-weight piston for aircraft pump produced by the 15 ton welder, and (D) turbine wheels produced by the 40 ton welder.
Courtesy of Manufacturing Technology, Inc.

the spindle-flywheel assembly is operated using a hydraulic pump. The pump itself is operated with an electric motor, which switches off once the desired rotation speed is reached. In the electric motor-driven machines, the motor is directly connected to the spindle, and is either declutched or switched off when the required speed is reached.

2.1.6 Advantages and disadvantages of IFW

The use of IFW provides several advantages, either from a manufacturing viewpoint or with respect to the weld structural integrity. Compared with fusion welding, the process is fully automated and repeatable, and does not require the use of a filler material, shielding gases or vacuum owing to its self-cleaning mechanism (i.e. flash formation) [1]. The process control and optimization is also simple as it is controlled by

Fig. 2.6 (A) A 2000 ton forge force inertia welder (model 2000B) and (B) a titanium low-pressure rotor assembly for an aero engine produced using the 2000 ton welder.
Courtesy of Manufacturing Technology, Inc.

only two variables (weld energy and forging pressure) [6], or three variables if the weld energy is separated into the flywheel inertia and rotation speed. Similar, dissimilar and components of different geometries are also weldable using IFW. In addition, the process results in a more efficient material utilization, weight reduction and a longer component life compared with using bolted joints [12]. Moreover, the solid-state nature means that any solidification defects are avoided. The joint possesses several unique characteristics, with respect to its microstructure and mechanical properties, as will be discussed later. Finally, IFW is a safe and environmentally friendly technique as it does not produce any harmful gases, fumes, etc.

Nonetheless, IFW requires a remarkable capital investment in the machinery and tooling, although the capability of the IFW machine can be tailored according to the geometry of the application required [13]. Still, the introduction of new applications can be expensive and requires a long lead time. In addition, there is a shortage in qualified welders as it is a very specialized process. Thus, if only a limited number of IFW joints are required, acquiring an IFW machine might not be economically sustainable.

2.2 Process parameters, heat generation and modeling

2.2.1 Process parameters and joint design

The selection of the welding parameters in IFW is dependent upon the component material and geometry, which is then used to determine the required welding energy. Manufacturers of IFW machines have typically produced charts that determine the required energy based on the section diameter for solid cylinders, or the wall thickness for tubular sections. Owing to the origin of IFW being in the USA, most IFW variables, parametric tables and charts are available in imperial units, although modern inertia welders follow the metric system.

The stored energy (E, lb./ft^2) in the flywheel of inertia (Wk^2, lb./ft^2) and rotation speed (N, rpm) is given as:

$$\frac{E = \left(Wk^2 \times N^2\right)}{5873} \tag{2.1a}$$

In SI units, the above equation becomes:

$$E = \frac{I \times N^2}{182.38} \tag{2.1b}$$

where I is the inertia (kg m^2).

The linear speed of the outer diameter is presented as surface feet per minute (s.f.p.m.), which is calculated using:

$$\text{s.f.p.m.} = 2\pi N \times r \tag{2.2}$$

where r is the outer radius in feet.

It is usually required to calculate the energy per unit area (E/A) and load applied per unit area (L/A), which is performed by dividing over the contact area. The following example illustrates the methods for the calculation of the IFW process parameters. The material data and parameters are based on the information supplied in the manual of the M120 inertia welder, Manufacturing Technology Inc. [14].

2.2.1.1 Example

Most machine manuals include charts similar to the one shown in Fig. 2.7, which can be used to determine the energy and load required for welding tubular sections or bars of mild steel. For any material other than mild steel and any geometry other than bar-to-bar or tube-to-tube, material factors (e.g. for Al, Ti, Ni alloys) and geometry factors (e.g. bar-to-tubes, bar-to-plate, tube-to-plate) are used to calculate the required energy and load, whereby:

$$E = \text{energy for mild steel} \times \text{material factor} \times \text{geometry factor} \tag{2.3a}$$

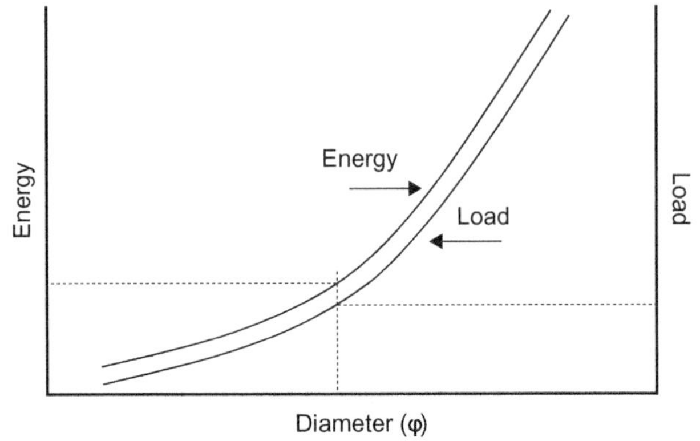

Fig. 2.7 A typical "process-parameters" determination chart.

$$L = \text{load for mild steel} \times \text{material factor} \times \text{geometry factor} \tag{2.3b}$$

To weld a 3/4 in. (19.05 mm) mild-steel bar, a weld energy of $E = 15{,}000$ ft lbs. (20.3 kJ) and a load of $L = 7250$ lbs. (32.25 kN) are required. For the same alloy, a speed (s.f.p.m.) range of ~1200–1800 ft./min (365–548 m/min) can be possibly used. As the used chart was for bars of mild steel, both the material and geometry factors are reduced to unity.

To calculate the inertia and rotation speed required based on the energy required, machine manufacturers suggest two approaches. In the first approach, an average value of the typical rotation speed range to weld a specific material is used to calculate the inertia required using Eq. (2.1). As the flywheels available for each machine have specific inertias that most likely do not match the theoretical inertia, the nearest flywheel assembly is used. Finally, the rotation speed is recalculated based on the inertia of the available flywheel. It is important to point out that the energy is more sensitive to changes in the rotation speed than the inertia (Eq. 2.1). In the second approach, a chart for the total energy plotted against the spindle speed is used for each of the flywheels or flywheel combinations available as shown in Fig. 2.8. This approach eliminates the need to recalculate the rotation speed.

To ensure reliable forging between the two parts, machines are equipped with displacement transducers to control the amount of upset, as a minimum upset is required to ensure that any inclusions or oxides are fully ejected out of the weld in the form of flash. The following formulae give an approximate estimate for the target upset to yield an acceptable weld [13]:

$$\text{Target upset} = 0.15''(3.81\ \text{mm}) + 0.2 \times \text{tube thickness} \tag{2.4a}$$

$$= 0.05''(1.27\ \text{mm}) + 0.2 \times \text{bar diameter} \tag{2.4b}$$

Knowing that there is a reasonable range of welding parameters that can be used for different materials, the increase in a certain parameter can change the extent of upset

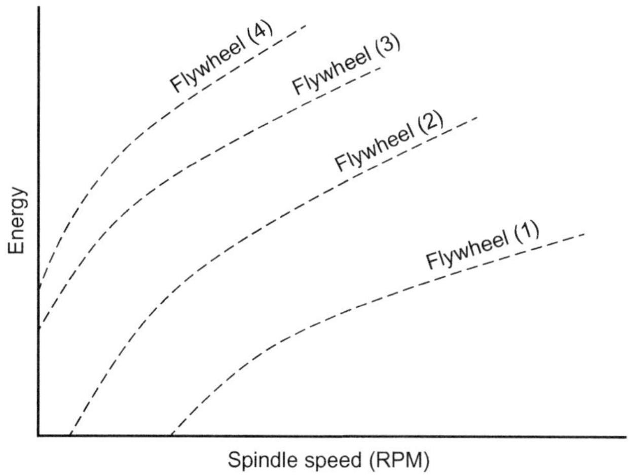

Fig. 2.8 An alternative chart for the determination of the rotation speed and inertia.

and the morphology of the welding region. Upon increasing the flywheel energy, the amount of flash and upset are known to increase. For the peripheral speed, the morphology of the weld region develops from being concave to convex on increasing the speed, with the weld performed at low speed having poor quality at the center. This is also the case with the welds performed with a high forging pressure [15].

2.2.2 Heat generation

An important issue in modeling friction welding is the mathematical representation for heat generation. Early analytical models utilized the actual power measured by the welding machine after subtracting the idle power [16], until Cheng [17] suggested that the power input can be described using the instantaneous measured value of the torque (M_t) and rotation speed (N_t) such that:

$$q = \frac{2\pi}{K} M_t N_t \tag{2.5}$$

where q is the power input and K is a constant.

On modeling IFW, Wang and Nagappan [18] suggested a friction-induced heat generation model, where the power input at a certain radial position, r, can be represented by:

$$q = K_I \mu p r N_t \tag{2.6}$$

where K_I is a constant, μ is the coefficient of friction, and p is the forging pressure. The total heat generation for an annular element or thickness dr can be represented as:

$$Q = K_{II} \int_0^T \int_0^R \mu(r,t) \cdot p(r,t) \cdot N_t \cdot r^2 \cdot dr \cdot dt \tag{2.7}$$

where K_{II} is a constant. The friction coefficient was also suggested to be varying with position and time as given by:

$$\mu(r,t)=\frac{K}{(N_t r)^2} \tag{2.8}$$

This formulation, although analytically correct, results in a complicated computation to estimate the heat input. Thus, it was suggested that the product μp is constant throughout the welding process, which is calculated from the average value of the heat generation. Eq. (2.7) can be integrated after substituting N_t with a second-order polynomial function for the variation of N with time.

It is apparent that the spatial variation in the friction coefficient, as well as with temperature, resulted in making the measured power input-based approaches more popular than the friction-based approach. Other power input [19] models were suggested for IFW, where the actual power input was modeled in two functions (stages). The first stage represents the initial contact stage, followed by the decrease in power in the second stage, as shown in Fig. 2.9:

$$\text{Stage I}: q_I = q_{\max}\ \sin\,\omega t \tag{2.9a}$$

$$\text{Stage II}: q(t)=\left(\frac{k\rho c}{\pi}\right)^{1/2} T_S t^{-1/2} \tag{2.9b}$$

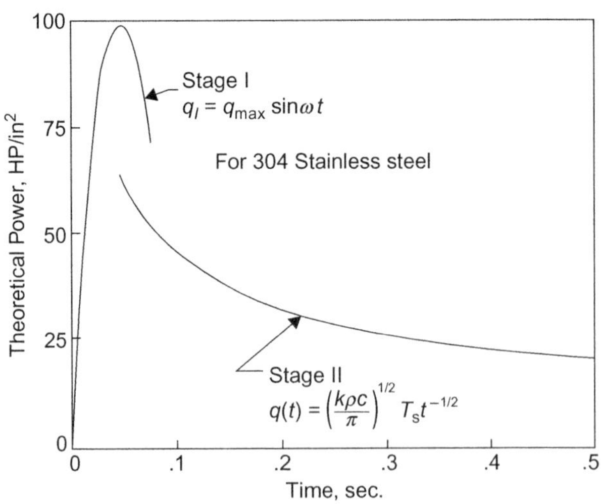

Fig. 2.9 Power input-based modeling for inertia welding according to Johnson et al. model [9]. From K. K. Wang, W. Lin, Flywheel friction welding research, Weld. J. Res. Suppl. 53, 1974, 233S–241S is being republished with permission of the American Welding Society (AWS).

where $q_{\max}$ is the maximum power input, T_s is the maximum interface temperature, c is the specific heat, ρ is the density, t is time, and k is the thermal conductivity. A review of other models that rely on the measured power input is available elsewhere [20].

2.2.3 Analytical and numerical (finite-difference) modeling

Early IFW modeling efforts focused on developing thermal analytical models, prior to the wider application of the finite element method for thermal and thermomechanical modeling in recent years. A review of the early analytical models for friction welding, which were mostly generated in the former Soviet Union, is available elsewhere [20]. These models rely on obtaining closed-form solutions for the two-dimensional heat-transfer differential equation:

$$\frac{\partial^2 T}{\partial r^2} + \frac{1}{r}\frac{\partial T}{\partial r} + \frac{\partial^2 T}{\partial z^2} = \frac{1}{\alpha}\frac{\partial T}{\partial t} \tag{2.10}$$

This approach uses some assumptions to simplify obtaining the mathematical solution (e.g. assuming a semi-infinite solid, constant thermophysical properties and constant heat flux). The resulting solutions are given by equations in the form of:

$$T = \left(\frac{2q_O}{k}\right) \cdot \sqrt{\alpha t} \cdot ierfc\left(\frac{x}{2\sqrt{\alpha t}}\right); \qquad 0 \le t \le t_h \tag{2.11}$$

where T is the temperature, q_o is the surface heat flux, k is the thermal conductivity, α is the thermal diffusivity, x is the distance from the weld line, is the total time for the heat-flux application. It is not t is the time and t_h is the total time for the heat-flux application. It is not clear though how the heat flux is calculated based on the welding parameters. Further details on this approach can also be found elsewhere [21].

Later, Cheng was the first to develop numerical (finite-difference) solutions for a one-dimensional heat-transfer equation, including in the model a melt (moving) boundary at the interface [16, 17]. The calculated thermal fields were compared with thermocouple measurements during CDFW that showed that the model under-estimated the temperatures in the initial heating phase, prior to decreasing the under-shoot by the end of the heating cycle. Cheng also investigated the influence of using variable and constant thermophysical properties [16], and compared the numerical results with the heat-balance (closed-form) integral method. The concept of the melting interface was later disproved by performing thermocouple measurements across a mild-steel inertia friction weld [18] and using analytical flow modeling [20]. This also becomes clearer by investigating the weld microstructure as will be discussed later. Most of the aforementioned modeling attempts were mainly for CDFW, which was the more familiar RFW process in the 1960s. Thus, it was not until the 1970s when Cheng's numerical approach was applied to model IFW [9, 18]. A two-dimensional numerical formulation was used, with temperature-dependent thermophysical properties. Nonetheless, the models also showed a deviation from the thermocouple measurements, which was more noticeable toward the center of the welded section.

2.2.4 Thermal and thermomechanical modeling

With the advance in using the finite element (FE) method, several models were developed to model the thermal fields [10, 22–29], deformation stresses [10, 23, 27, 28] and residual stress development [10, 22, 24, 29, 30]. The main factor in assessing the quality of an FE model is the performance of the validation using the measured process variables (e.g. welding time, temperatures, upset and weld (flash) morphology), as well as the weld properties (e.g. microstructural development and residual stresses). Nonetheless, early models did not sufficiently characterize the microstructure or the residual stress characterization.

Among the early models, Moal and co-workers established a two-dimensional FE code (INWELD) to model the thermal and strain-field development, owing to IFW in the Astroloy powder metallurgy Ni-based superalloy [27, 28]. Their model was complemented with in-depth mechanical (torsion) testing and microstructural characterization to model the high-temperature deformation, dissolution/precipitation and rapid-heating kinetics of the γ' dissolution process [28]. The mechanical characterization was used to construct the rheological thermomechanical constitutive equations for the material using Norton-Hoff law. The microstructural studies were used to predict the thermal fields owing to IFW based on the γ' precipitates development, which suggested that the temperature at the weld center approached ~1280 °C (above the γ' solvus but well below melting temperature). The heat generation in the model was friction-induced using Coulomb's friction law developing to viscous flow at high temperatures. This required the calculation of the rotational velocity at each step, with the temperature and stress/strain fields being computed simultaneously. Validation was performed using the total upset, temperature (pyrometer) measurements and welding time. Their later work [27] included further validation using the rotational speed. In spite of the model potential, it was not further tested nor validated. Later, D'Alvise et al. developed a two-dimensional coupled thermomechanical IFW model using FORGE2®, which is capable of predicting the temperature, stress/strain fields and residual stresses in similar and dissimilar welds [10]. The model used the same heat generation scheme as the previous model. However, the model was only validated using temperature measurements, the weld upset, flash morphology and the rotational speed, with reasonable agreement between the model's prediction and experimental data. Nonetheless, some features of the reported model were not discussed or validated (e.g. residual stresses and plastic strains), while the microstructural development was not even considered.

Following the earlier models, several researchers utilized available FE packages (e.g. MSC Marc and DEFORM 2D) to model the thermal and stress/strain fields due to IFW, yet model validation was always limited (mostly thermal, rotation speed and weld morphology) without considering the microstructure nor the residual stress development [23, 25, 26]. Yet, with the advance in electron microscopy and in residual stress characterization using neutron diffraction, it became possible to further validate the FE models using the residual stress predictions with measured stress profiles [24, 29, 30].

The work by Wang et al. focused on modeling the residual stress development in RR1000 (a high γ' Ni-based superalloy) welds using DEFORM, and comparing it with residual stress data obtained using neutron diffraction [29, 30]. Their energy-input FE models were two-dimensional axisymmetric models, with a visco-plastic material model and using a frictional heat flux (q), which is given by:

$$q(t) = \frac{1}{2}\frac{\eta}{A}\frac{dE}{dt} = \frac{1}{2}\frac{\eta}{A}\frac{\frac{d(I\omega^2(t))}{2}}{dt} = \frac{1}{2}\frac{\eta}{A}I\omega(t)\frac{d(\omega(t))}{dt} \tag{2.12}$$

where η is the efficiency, A is contact interface area, I is the moment of inertia, ω is the rotational velocity and t is the time. The model used the calculated thermal fields to perform the residual stress analysis by performing creep (elastic) analysis during cooling. Generally, the predicted trends matched the measured ones, whereby a bending moment was predicted in the axial direction, while the hoop direction showed very high stress levels that approached the yield strength, compared with the radial direction that showed minimal stresses. Nonetheless, the predictions were ~15–35% higher than the measured stresses, which was attributed to the lack of high-temperature creep data (>750 °C) and the influence of the flash machining [29]. Further validation was also performed through microstructural characterization of the γ' precipitates development in three different welds, and comparing it with the model thermal predictions. The work of Wang et al. was further improved by Grant et al. [24], using a larger material property database (up to 1150 °C), better representation of the heat generation model and the forging force application, and using an elasto-plastic analysis of the stress development. Validation of the model was performed using microstructural characterization of base metal specimens that were thermally cycled using the model predictions, and the residual stress measurements. The predicted residual stresses were both spatially and quantitatively similar to the measured stresses, especially in the hoop direction, although the model failed to predict the axial stresses with the same accuracy (Fig. 2.10). The radial stresses were generally negligible, especially considering the accuracy in the stress analysis using neutron diffraction (±70 MPa).

Grant et al. also used their model to perform parametric simulations investigating the influence of the forging pressure on the thermal fields. It was found that the increase in the forging pressure leads to a slight increase in the maximum temperature, resulting in steep thermal gradients (Fig. 2.11A). Nonetheless, by calculating the residual stress development, it was evident that the increase in pressure only affected the location of the maximum hoop stress resulting in a narrower heat-affected zone (HAZ), but did not affect the maximum stress quantity (Fig. 2.11B).

Recent IFW modeling reports used new numerical approaches (e.g. Eigenstrain FE modeling [31]), but still focus on Ni-based superalloys welds. There are only a limited number of reports on IFW models when joining other materials than Ni-based superalloys. In ferritic steels, a phase transformation occurs during cooling after welding

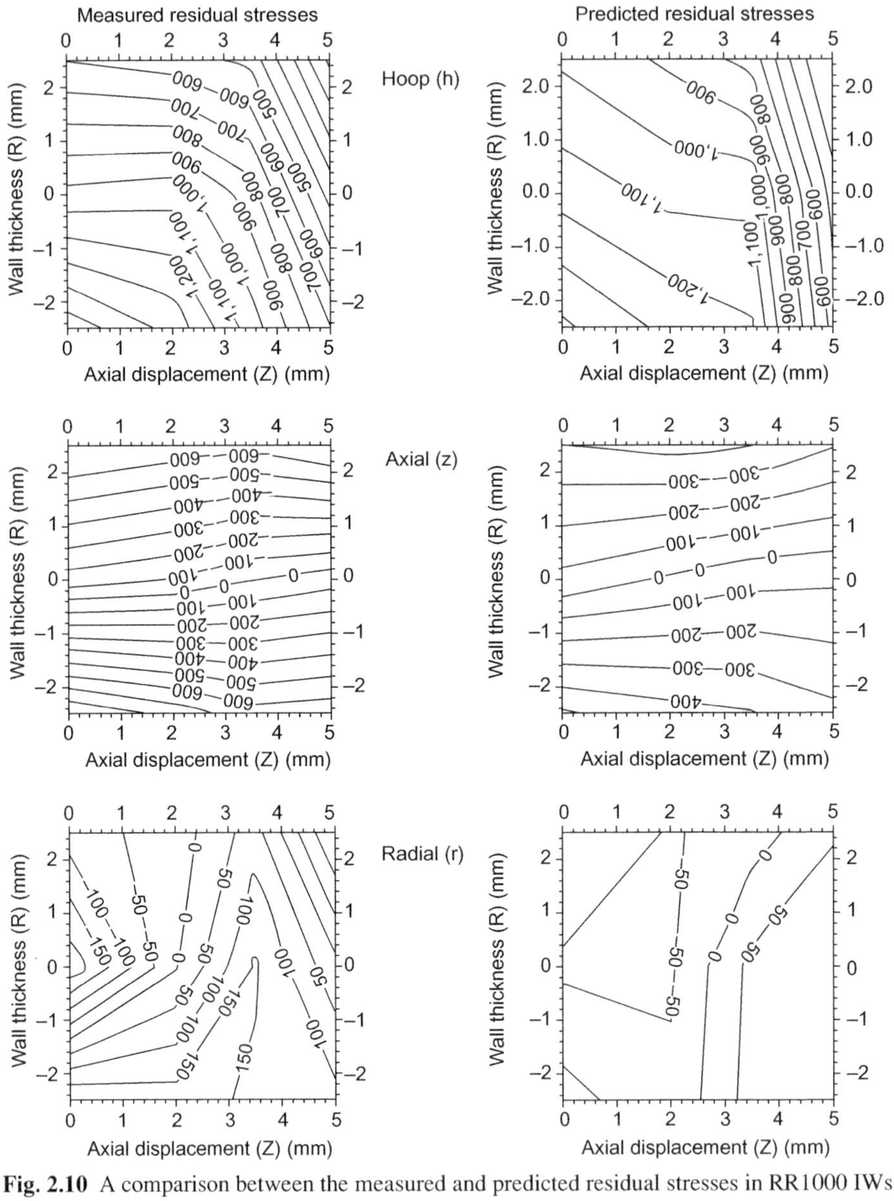

Fig. 2.10 A comparison between the measured and predicted residual stresses in RR1000 IWs using Grant et al. model (2009).

(e.g. martensitic or bainitic transformation), resulting in a volumetric change affecting the residual stress evolution in this region [32]. In their model, Bennett et al. studied the residual stress development in SCMV steel welds using a DEFORM 2D model [22], validated using the rotation speed, upset and thermocouple measurements. At the end of the IFW cycle, the residual stress development was traced throughout the martensitic transformation, where the fraction transformed and a volumetric

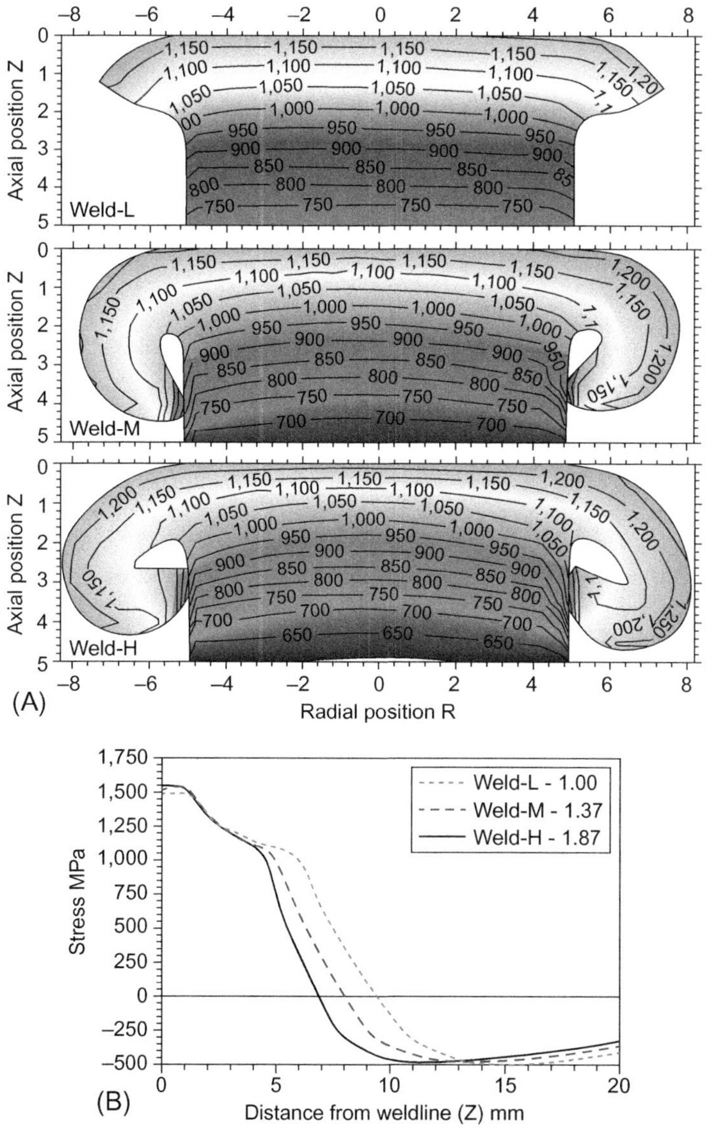

Fig. 2.11 Influence of the forging pressure on (A) the thermal fields and (B) hoop stress (forging pressure low (L), medium (M) and high (H)) [24].

change parameter were calculated, which was then used to investigate the influence of the transformation on the residual stress development. Their findings showed that the occurrence of the transformation resulted in a significant drop in the residual stresses owing to the volumetric increase during cooling (Fig. 2.12). Although this model is the first to investigate the influence of the phase transformation on the residual stress development, the residual stress capability was not validated.

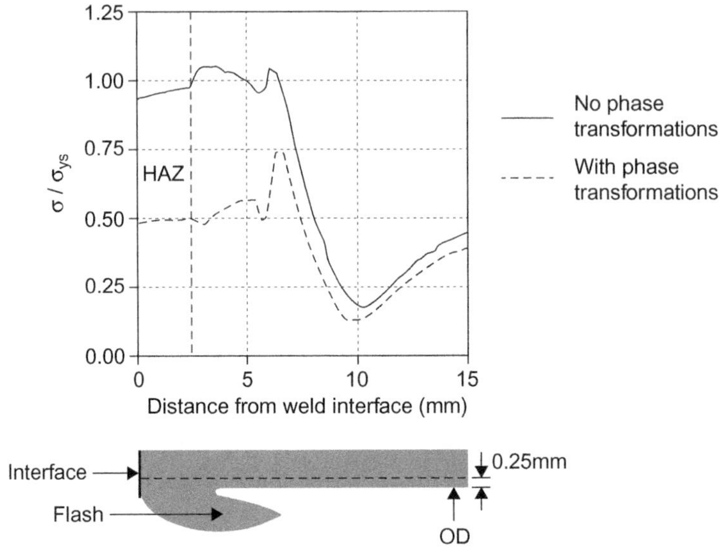

Fig. 2.12 Influence of the phase transformation on the von Mises residual stress distribution [22].

2.3 Microstructural development

Because of the nature of IFW, the resulting joint demonstrates a unique yet localized microstructural residual stress, and mechanical properties develop across the joint. Although it is generally believed that CDFW and IFW result in a similar microstructural development [1], early IFW literature suggested that the use of the flywheel results in circumferential flow lines at the weld plane, compared with radial flow lines in non-flywheel welds [2]. It was argued that such a metallurgical difference would result in better fatigue properties of IFW welds owing to the spiraling of the flow lines that become almost tangential to the surface, compared with being normal to the surface in CDFW. Nonetheless, no comparative studies have ever been performed to quantitatively assess the metallurgical differences between IFW and CDFW to date. In the following sections, the influence of IFW on the microstructural development in Ni-based superalloys, titanium alloys, steels and aluminum alloys will be discussed.

2.3.1 Nickel-based superalloys

The majority of the microstructural studies on Ni-based superalloy inertia friction welds focused on the development of the precipitate structure in commonly used aerospace alloys, especially RR1000, IN718, U720Li, Astroloy, and Waspaloy (Table 2.1).

Table 2.1 Chemical compositions (wt.%) of selected aerospace Ni-based superalloys.

Alloy	Ni	Cr	Co	Mo	W	Nb	Ti	Al	Fe	C	Other
RR1000	52.3	15.0	18.5	5.0	–	–	3.6	3.0	–	0.027	0.015B, 2Ta, 0.06Zr, 0.5Hf
Astroloy	56.5	15.0	15.0	5.25	–	–	3.5	4.4	<0.3	0.06	0.03B, 0.06Zr
U 720Li	57.0	16.0	15.0	3.0	1.25	–	5	2.5	–	0.025	0.03Zr
IN718	52.5	19.0	–	3.0	–	5.1	0.9	0.5	18.5	0.08	0.15 max Cu
Waspaloy	57.0	19.5	13.5	4.3	–	–	3.0	1.4	2.0	0.07	0.06B, 0.09Zr

Source: M.J. Donachie, S. J. Donachie, 2002. Superalloys: A Technical Guide. ASM International, Ohio.

According to Adam [34], inertia friction welds of Ni-based superalloys can be classified into five regions from the center outwards (based on Astroloy):

- *Fine grain region*: this is the weld line where dynamic recrystallization occurs.
- *Deformed and rotated grain region*: where the intense plastic deformation leads to the rotation and re-orientation of the grains.
- *Deformed grain region*: where plastic deformation has occurred, without rotating the grain structure.
- *HAZ*: where the material is only thermally affected.
- *Base (unaffected) metal*: where the material is thermally and mechanically unaffected.

Conversely, Preuss et al. [35] found in RR1000 IFWs that weld line microstructure is actually not heavily deformed, showing coarse recrystallized grain structure, followed by fine grains within 0.5 mm of the weld line. Beyond the central region, the material was plastically deformed, with the temperature insufficient to lead to recrystallization.

Owing to the nature of the precipitation strengthening mechanism in Ni-based superalloys, several IFW studies investigated the development of the γ' precipitate [28, 29, 35–39], and γ'' precipitate [37, 39–41] structures using scanning electron microscopy (SEM), transmission electron microscopy (TEM) and synchrotron X-ray diffraction. Limited studies investigated the development of grain boundary (GB) carbides/borides [36, 42] and microtexture [35]. Qualitatively, Huang et al. investigated the precipitate development at the vicinity of the weld region in a IN718-U U720Li weld using SEM and TEM [37]. In the IN718 side (Fig. 2.13), it was clear that the thermal fields experienced during welding led to complete dissolution of the γ'', γ', or δ precipitates as far as 2.1 mm from the weld center, and also created a low dislocation density structure. This led to a hardness drop observed in the IN718 side of the weld. At 4 mm from the weld center evidence of the presence γ'' precipitate was found. Similar observations were reported by Roder et al., who investigated the extent of dissolution of the γ'' and δ phase (Ni_3Nb phase for grain-size control), [41]. It was reported that whereas the γ'' phase was fully dissolved

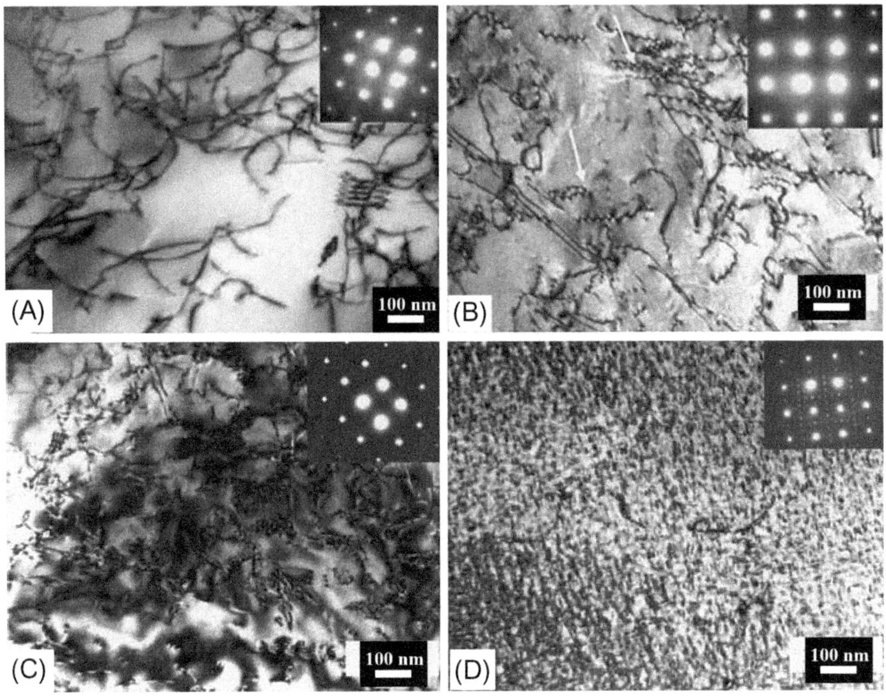

Fig. 2.13 The γ'' precipitate development in IN718 (AW) at (A) 0.2 mm, (B) 0.7 mm, (C) 2.1 mm, and (D) 4 mm from the weld line [37].

up to ~500 μm from the weld center, the δ phase was dissolved up to only ~300 μm from the weld center.

In highly γ'-strengthened alloys (Astroloy [28], RR1000 [35], and U720Li [37]) a different development in the weld center was observed owing to the fast precipitation kinetics of γ' and the high levels of γ'' stabilizers (this excludes Waspaloy). For instance, at the weld center U720Li IFW, a high fraction of reprecipitated γ' was observed (Fig. 2.14A), which has decreased dramatically only 1 mm from the weld line (Fig. 2.14C) [38].

Preuss and co-workers used both synchrotron X-ray diffraction and image analysis to map the γ' and γ'' fraction development in RR1000, IN718, and U720Li IFWs across the weld line [35, 37–39]. It was evident that IFW led to the dissolution of the strengthening precipitates in the weld region, which ranges from full dissolution of γ'' in the case of IN718, to dissolution and reprecipitation of γ' in RR1000 and U720Li (Fig. 2.15). These measurements show that IFW disturbs the precipitate volume fraction within a region of ±3 mm from the weld line, resulting in hardness peaks caused by reprecipitation during cooling in the presence of a strong driving force for reprecipitation in RR1000 and U720Li, or hardness troughs caused by partial or full dissolution [35]. By correlating the γ' distribution with the microhardness distribution in RR1000 [35] and U720Li [38], it becomes evident that the hardness variation is controlled by a combination of strengthening mechanisms (precipitate strengthening,

Fig. 2.14 The γ' precipitate development in U720Li (AW) (A) at the weld line, (B) 0.5 mm, and (C) 1 mm from the weld line [38].

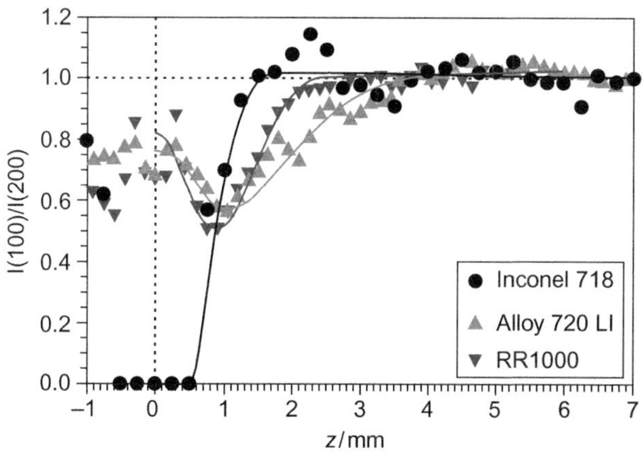

Fig. 2.15 The normalized integrated intensity ($\pm$0.04 accuracy) of the γ' superlattice reflection as a function of the distance from the weld line (z) in the AW condition of IN718, 720Li and RR1000 IWs [39].

GB strengthening and work hardening), as correlated through measurements of the grain size, tertiary γ' volume fraction and γ/γ' misfit, although the contributions of these effects have not been quantified [35]. In inertia friction-welded Astroloy, Soucail et al. [28] found a high dislocation density, with tangled dislocations, Orowan loops and sheared particles with varying extents depending on the distance from the weld center. Electron backscatter diffraction (EBSD) maps also showed regions of high stored energy about 1 mm from the weld line in inertia friction-welded RR1000 [35]. This demonstrates that a region exists adjacent to the recrystallized region where material has been deformed significantly, but the stored energy was not sufficient to result in dynamic recrystallization.

In addition to the microstructural investigations of the precipitate structure, other microstructural features of interest include the development of the GB carbides [36, 42]. The high temperature experienced during welding (~1200 °C) is sufficient to dissolve (fully or partially) any existing carbides, which subsequently reprecipitate at the grain boundaries during cooling or post-weld heat treatment (PWHT). The morphology of the GB carbides has a strong influence on the ductility and fatigue properties, where discrete or globular carbide particles are beneficial as they act as obstacles for GB motion, while continuous carbide films are harmful [43]. Li et al. detected the formation of continuous films of $M_{23}C_6$ carbides with a relatively high Cr content close to the weld region within the primary γ'-free zone in the U720Li side of a IN718-U720Li IFW following PWHT [43] (Fig. 2.16). As it is known that the high Cr content of the carbides results in a localized depletion of Cr in their vicinity, this is expected to result in poor oxidation damage resistance in these welds. The formation of the carbides was reported to have a negative influence on the fatigue-crack propagation (FCP) properties of Ni-superalloy welds as will be discussed later [43, 44].

It was suggested that performing a solution-plus-aging treatment to redistribute the carbides could solve the problem of the continuous GB carbides. Following this approach, Ferte compared three PWHT conditions in N18 IFWs, where it was found that on increasing the PWHT time from 4 to 8 h at 800 °C, grain boundaries became increasingly decorated with Cr-Mo carbides [36]. However, on performing a solution treatment plus aging PWHT, a bimodal γ' precipitate structure was obtained in the weld center, whose sizes are similar to those of the parent metal. Although this latter treatment led to a significant improvement in the FCP resistance, it also led to large grain growth in the HAZ, which can be attributed to the large stored energy present in some parts of the HAZ. It is also worth noticing that a solution heat treatment of large welded components is economically not viable.

2.3.2 Steels

The majority of the microstructural studies on steel IFW dealt with ferritic steels. During IFW of ferritic steels, the temperature of the material at the interface exceeds the austenization temperature, which, alongside the thermomechanical deformation, results in a very unique microstructure and texture. The temperature decreases gradually from the weld center toward the HAZ and the parent metal. Upon cooling, the microstructure shows varying amounts of ferrite, martensite, bainite or retained austenite across the weld regions, depending on the cooling rates and the stresses.

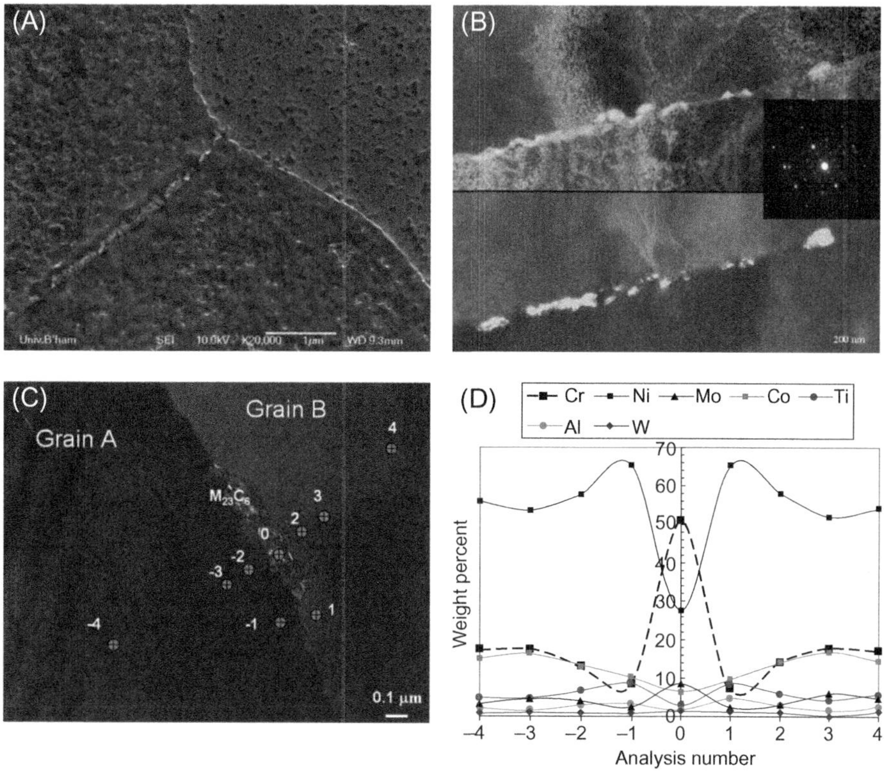

Fig. 2.16 GB carbides ($M_{23}C_6$) precipitation in 720 Li in the primary γ' free zone following PWHT (A) SEM, (B) bright and dark-field TEM with SAD pattern, (C) and (D) chemical analysis across a carbide particle [43].

Among the most notable IFW applications in the aerospace industry is the dual drive shaft (dissimilar SCMV-AerMet100 IFW joint) [32, 45, 46], which received the most thorough characterization among the ferrous welds. SCMV (0.3% C, 3.15% Cr, 1.6% Mo, 0.1% Va, 0.6% Si) is used at the turbine (high temperature) end of the aero engine drive shaft, while Aermet 100 (0.2% C, 2.5% Cr, 10.1% Ni, 12.7% Co, 1.37% Mo, 3.26% Nb, 0.01% Mn) is used at the compressor end of the shaft, where a high-strength alloy is required to withstand the high torque [46]. Moat et al. [45] divided the weld into four regions (Fig. 2.17):

- *the weld (bond) line (interface)*: where a slight banding between the two alloys was observed, yet without full mixing
- the (partially or fully) deformed zone or the thermomechanically affected zone (TMAZ): where varying thermomechanical effects result in the deformation of the parent microstructure, but not to annihilate it completely
- *the HAZ*: where the microstructure experiences only thermal effects, resulting in the formation of a recrystallized equiaxed grain structure, with some coarsening [46]
- the unaffected parent metal.

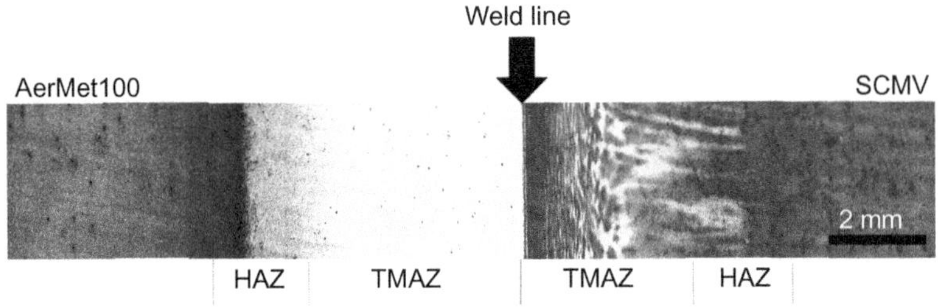

Fig. 2.17 Microstructural zones in the AerMet100-SCMV IW.
Courtesy of R. J. Moat.

The width of the regions is dependent on the process parameters with the forging pressure having a major impact on HAZ (high pressure narrows HAZ) and TMAZ (high pressure increases extent of deformation). Conversely, the increase in the rotation speed (also weld energy) widens the HAZ accompanied with a relatively coarse grain morphology. As a result, a two-stage pressure can be occasionally used to refine the grain structure, by applying a higher pressure following the stop of the spindle [46].

It is known that IFW creates a heterogeneous microstructure, with varying quantities of the phases present (e.g. martensite, bainite or retained austenite) [47]. Thus, it is necessary to quantify the extent of the microstructural heterogeneity, especially the presence of retained austenite and martensite, owing to their influence on the mechanical properties. Laboratory and synchrotron X-ray diffraction was used by Moat et al. [45] to map the austenite fraction variation across the weld in the as-welded (AW) and PWHT conditions of before-mentioned SMCV/AerMet100 IFWs (Fig. 2.18).

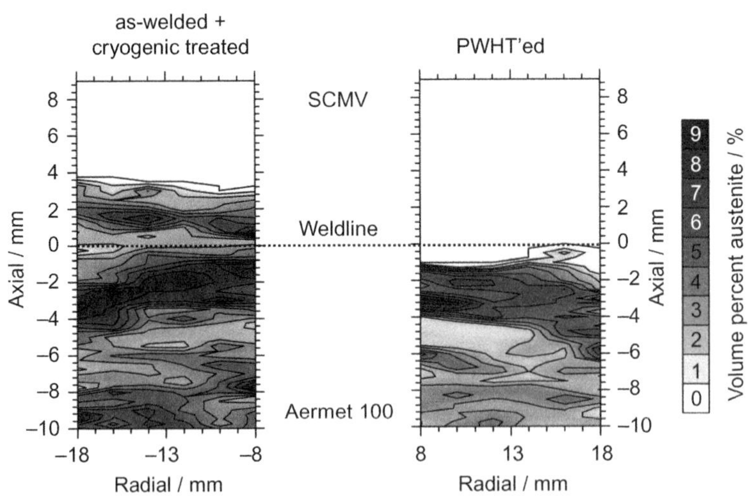

Fig. 2.18 A two-dimensional map for the austenite fraction in SCMVAerMet100 IWs (AW and PWHT conditions) as measured by high energy synchrotron X-ray diffraction [32].

A significant variation in the retained austenite was found especially toward the AerMet100, whereby the volume fraction decreased from ~7% to 9% in the parent AerMet100 to ~2–4% within the HAZ, prior to reaching 8–9% within the TMAZ. In the SCMV side, there is a pronounced localized increase in the austenite fraction (~8%) within the TMAZ, while keeping the ferrite-cementite structure within the parent metal and HAZ. The high austenite fraction in the TMAZ of both sides was attributed to the rapid cooling, resulting in the retention of some austenite, with varying amounts because of the difference between both alloys in the austenite stabilizers content. In the HAZ, the temperature was insufficient to fully austenite the SCMV side, which resulted in the absence of austenite. Following PWHT, the SCMV side regained the ferrite-cementite structure in all its regions. In the AerMet100 side, the overall austenite content decreased owing to PWHT, yet without changing the local variation across the different regions. Further microstructural evidence was obtained using electron microscopy, which showed that the SCMV TMAZ AW microstructure was predominantly martensitic with some retained austenite, as manifested in the laths and the absence of carbides. In the AerMet100, the thermal exposure led to some precipitation of the M_2C carbides, with varying degrees. These complicated developments resulted in a unique microhardness development as will be discussed in Section 2.4.1.1.

2.3.3 Titanium alloys

The microstructural investigations of inertia-friction-welded aerospace titanium alloy are very limited, and do not provide sufficient information on the several classes of titanium alloys (α, $\alpha + \beta$ and β alloys). Among the earliest reports on IFW of $\alpha + \beta$ titanium alloys, Nessler et al. performed preliminary microstructural investigations of Ti-8Al-1V-1Mo and Ti-6Al-4V [48]. No evidence of melting was found, although a narrow alpha-case region was observed at the center of the flash. Generally, the weld region was narrow (~3 mm) and mostly composed of a highly deformed structure with fine grains. At the weld center, martensitic α' was observed, which apparently resulted from rapid cooling from above the β-transus temperature (~1000 °C). In a different investigation, the HAZ microstructure in Ti-6Al-4V IFW was also found to contain Widmanstätten α (transformed β grains) [49]. Similarly, inertia-friction-welded Ti-6Al-2Sn-4-Zr-2Sn also displayed a refined lamellar α structure in the weld region [50].

Baeslack III et al. studied an IFW of a rapidly solidified titanium alloy (Ti-6.4Al-3.2Sn-3.0Zr-2.7Er-2.5Hf-1Nb-0.33Ge-0.3Ru-0.15Si) [51]. The weld regions were classified into an inner heat and deformation zone (HDZ), which is ~25 μm wide with dynamically recrystallized fine β grains with fine Er-rich dispersoid (100–350 nm) precipitates along the grain boundaries and an outer HDZ (~500 μm wide) that contained transformed β grains with α colonies. The unique composition of this experimental alloy makes it difficult to generalize the findings of this study.

A single report by Roder et al. [52] is available on the microstructural development of inertia-friction-welded Ti-6Al-2Sn-4Zr-6Mo, an $\alpha + \beta$ Ti alloy that is more heavily

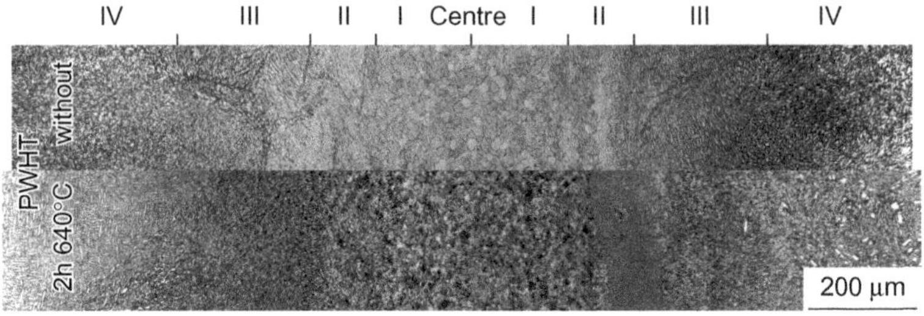

Fig. 2.19 Microstructural zones in a Ti-6246 IW (AW and PWHT) [52].

β-stabilized than Ti-6Al-4V. Upon welding, four unique microstructural zones were observed (Fig. 2.19):

- region I (center to 175 μm): showing recrystallized β grains (~20 μm in size)
- region II (175–275 μm): showing recrystallized β grains (~10 μm in size), and deformed β grains (>20 μm in size), which are elongated, in the radial direction
- region III (275–525 μm): showing deformed β grains similar to those in region II
- region IV (525–1500 μm): showing a deformed parent microstructure.

An interesting feature in Ti-6246 IFWs is the observation of the so-called ghost α plates in region III, which are α_P plates from the base metal that was heated up to the β-phase field, without allowing sufficient time for diffusion to homogenize the chemical composition. Following PWHT, fine α needles precipitated within the β grains in all the regions (I–IV).

It is important to note that the publically available investigations do not provide detailed information on the microstructure development in inertia-friction-welded titanium alloy. Several alloys need to be further investigated (e.g. Ti-64 and Ti-6246) and the feasibility of welding alloys such as Ti-5553 need to be pursued to increase the opportunities of using titanium IFW in the aerospace industry.

2.3.4 Other alloys

There are few preliminary reports in the literature on the weldability of other alloys using IFW, including B2 aluminides [53], aluminum-based alloys [54, 55], metal-matrix composites [56] and dissimilar welds [25, 57]. The interest in iron and nickel aluminides as potential high-temperature materials in the mid-1980s led to an interest in performing investigations of their weldability. Whittenberger et al. [53] investigated a dissimilar FeAl and NiAl IFW. In contrast to fusion welding, IFW of these alloys did not lead to cracking or porosity formation, but instead it was composed of fine equiaxed grains. Nonetheless, upon performing post-weld annealing at 1027 °C for 16 h, recrystallization occurred leading to grain coarsening, which was detrimental for high-temperature creep strength. This initial report was never followed by any other investigations for the metal aluminides.

Preliminary studies were performed in the early 1990s to investigate the inertia friction weldability of rapidly solidified alloys, Al-Fe-Mo-V [54] and Al-Fe-V-Si [55], which involved microstructural characterization of the weld regions using electron microscopy and mechanical testing. This work was further extended by studying SiC-reinforced Al-Fe-V-Si alloy, which showed that composites are also weldable using IFW. Nonetheless, beyond these studies, the interest of IFW of Al alloys has not been revived again for over a decade.

There has been some recent interest in investigating the weldability of dissimilar superalloy-steel IFWs. Jeon et al. performed a preliminary microstructural and mechanical property characterization of a Nimonic 80A/SNCrW IFW [25]. Although there is a lack of extensive mixing at the weld interface, the welds had mechanical properties (strength and fatigue) better than the properties of SNCrW. In another report, Zhu et al. [57] investigated a Nimonic 80A/$4Cr_{10}Si_2Mo$ dissimilar weld using electron microscopy. Chemical analysis by X-ray spectroscopy was used to characterize the extent of mixing at the weld interface that showed that there is a central mixing zone (equivalent to the weld line) of ~100 μm width. The weld line was composed of austenite grains of 3–5 μm size, with fine carbides (50 nm in size). Beyond the mixing region, a high dislocation density region was found, similar to that described earlier for inertia-friction-welded Ni-based superalloys called, in this case, the pure shearing region.

As it was discussed, the studies on alloys other than Ni-based superalloys, steel and titanium are very limited. Although IFW is known to have been applied to other aerospace materials, the majority of the welds produced are mostly for Ni-based superalloys, and only recently on steel and titanium. Further work is needed to investigate the utility of IFW in welding new aerospace materials and particularly the ability to join dissimilar metals and alloys.

2.4 Development of mechanical properties

2.4.1 *Ni-based superalloys*

2.4.1.1 *Microhardness development*

Owing to the nature and the volume fraction of the strengthening phases, differences in the microhardness distribution were observed between inertia friction welds of advanced Ni-based alloys with a high γ' volume fraction (e.g. RR1000, U720Li) and welds of superalloys with a volume fraction of about 20–25% precipitates, such as Waspaloy and the $\gamma' + \gamma''$-strengthened IN718. In general, the latter class of superalloys tends to display a significant hardness trough in the weld region in the AW condition, towing to either a complete absence or at least significantly reduced volume fraction of strengthening precipitates (Waspaloy Fig. 2.20A) [34, 41]. However, in alloys with γ'-volume fractions around 40–50% (e.g. RR1000 [35, 39], N18 [36], U700 [34], U720Li [38, 39] and Astroloy [36], with ~50% γ'), both the region close to the weld center and the HAZ show hardness peaks even in the AW condition (RR1000, Fig. 2.20B). This behavior was attributed to the higher levels of γ'-stabilizing alloying elements such as Al and Ti found in such alloys compared with, for example, Waspaloy providing a high driving force of γ'

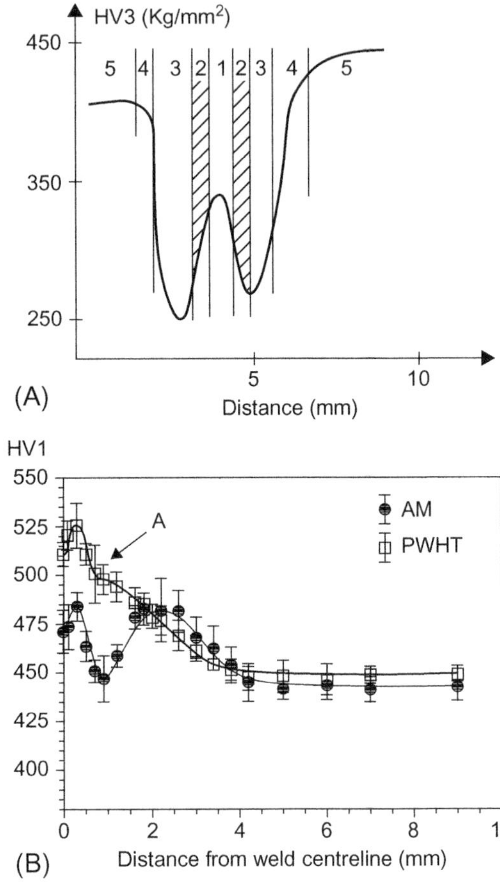

Fig. 2.20 The effect of the γ' content on the microhardness development (mid-thickness) in IFWs of Ni-based superalloys: (A) low γ' Waspaloy [34] and (B) high γ' RR1000 [35].

reprecipitation during rapid cooling following welding [35]. However, a small hardness trough, as shown in Fig. 2.20B, can still be observed attributed to the partial γ' dissolution during welding not providing a sufficiently high driving force for reprecipitation during cooling. Following PWHT in γ'-rich alloy inertia friction welds, additional γ' precipitation occurs resulting in an overall increase in hardness, with the maximum hardness occurring toward the weld center (tertiary γ'), and decreasing toward the base metal with a change in the slope in the previously observed hardness trough in the AW condition (Fig. 2.20B) [35]. Similar PWHT hardness trends were observed in other γ'-rich alloys inertia friction welds (e.g. U700 [34], U720Li [38, 39], and N18 [36]).

In the $\gamma' + \gamma''$-strengthened alloy IN718 following a PWHT, the hardness of the weld region is either slightly or fully regained depending on whether a low-temperature anneal or a complex multistage PWHT was undertaken (Fig. 2.21)

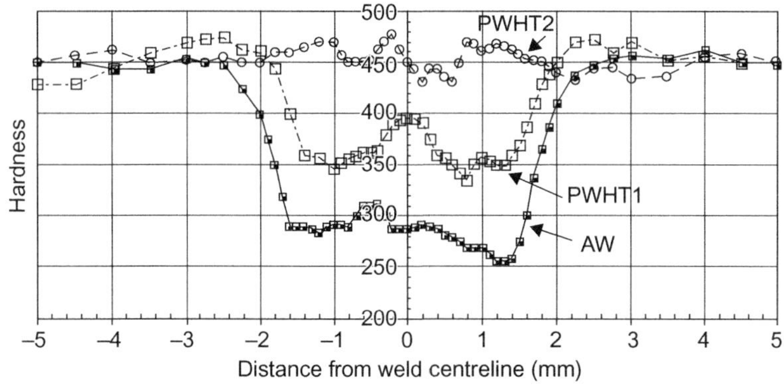

Fig. 2.21 Microhardness distribution (mid-thickness) in $\gamma' + \gamma''$-strengthened IN718 [41]. Adapted from O. Roder, D. Helm, S. Neft, J. Albrecht, G. Lutjering, 2005. Mixed INCONEL alloy 718 inertia welds for rotating applications—microstructures and mechanical properties, in: E.A. Loria, (Ed.), Proceedings of the International Symposium on Superalloys and Various Derivatives, Pittsburgh, PA with permission of The Minerals, Metals & Materials Society.

[41]. In fact, the latter PWHT led to a slight hardness increase in the weld region compared with the base metal, which was attributed to superposition of precipitate and grain-size strengthening.

In dissimilar inertia friction welds, Daus et al. [44] studied the influence of the process parameters on the hardness development in RR1000-IN718 welds. Three welds were investigated, representing different parameter combinations (Fig. 2.22A). It was evident that the hardness traces created a combination of the typical trends in the welds of γ' and $\gamma' + \gamma''$-strengthened alloys as previously discussed, with the RR1000 side showing two peaks with a small trough in between, while the IN718 side showed a significant hardness trough in the weld region. The dissimilar hardness trend following PWHT was observed in U720Li-IN718 welds [37] (Fig. 2.22B). Following PWHT, the hardness of the weld region ($\pm$6 mm from the weld centerline) generally increased to a maximum of ~550H_V in the U720Li side and ~490H_V in the IN718 side, with the weld region in each side showing hardness values above its respective base metal. It was also found that the increase in PWHT time led to the parent IN718 being overaged, while the hardness of the U720Li side remained roughly unchanged. It was generally recommended that PWHT of IN718 is performed either by full annealing at 955 °C to avoid heterogeneous aging effects [33], or below 732 °C to maximize the weld strength. However, U720Li requires significantly higher temperatures between 760 °C [37] and 1135 °C [33] to relieve the high stresses that result from the welding process. This highlights one of the fundamental issues when joining Ni-base superalloys with different temperature capability. While a low-temperature PWHT might avoid overaging of the alloy with the comparatively low-temperature capability, such heat treatment is unlikely to be effective in terms of relieving residual stresses in the alloy with high-temperature capability. In contrast, if the high-temperature alloy guides the

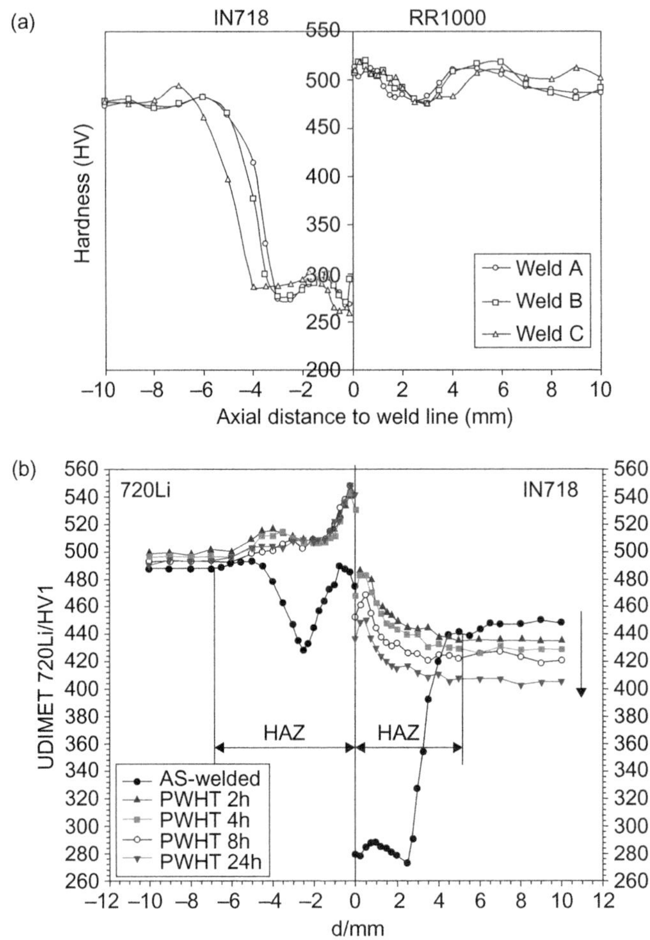

Fig. 2.22 Hardness distribution in dissimilar IN718-X IFWs. (A) IN718-RR1000 (AW) [44] and (B) U720Li-IN718 [37].

PWHT procedure, it is likely that the "low"-temperature alloy will get overaged resulting in low strength. This demonstrates the importance of considering microstructure/mechanical property development together with residual stress mitigation particularly in dissimilar welds when identifying appropriate PWHTs. Further dissimilar microhardness traces for IN718-X (X: U720Li, Incoloy909, René88) including the influence of different PWHT, can be found in the work of Roder and coworkers [40, 41].

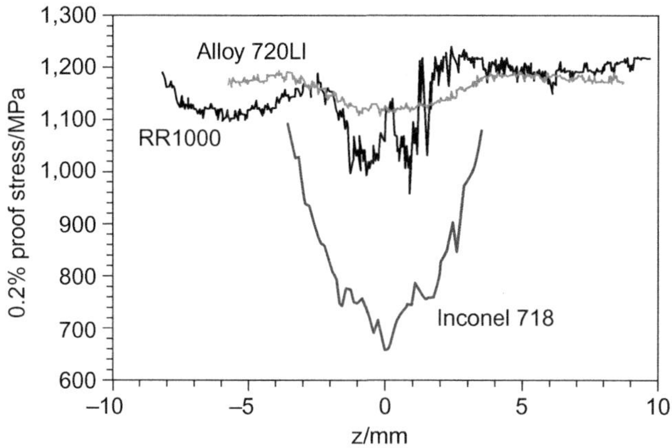

Fig. 2.23 Proof-stress (0.2%) distribution in AW IFWs of IN718, RR1000, and 720Li (average accuracy ±10–50 MPa) [39].

2.4.1.2 Tensile properties

To measure the cross-weld mechanical properties, Preuss et al. used electron speckle-pattern interferometry (ESPI) to measure the proof-stress distribution across the weld in IN718, RR1000 and U720Li alloys IFWs [39] (Fig. 2.23). The distribution shows a soft-weld region in IN718, with a yield strength of ~650 MPa. Although the proof-stress traces agree qualitatively with the microhardness profiles, they provide further quantitative information on the local tensile properties of the weld regions. To investigate the effect of the test temperature, Roder and co-workers measured the PWHT tensile strength and ductility in similar and dissimilar IN718 welds [40, 41]. It was found that failure generally occurred in the IN718 side of the weld, except for Incoloy909 alloy. A drop in tensile strength (200–300 MPa from strength at 20 °C) was observed for IN718 welds tested at high temperature (650 °C), yet strength was generally within the typical range for IN718. However, in IN718-U720Li welds failure occurred at strength levels below the typical levels for IN718 and U720Li regardless of the PWHT parameters.

2.4.1.3 Fatigue-crack propagation (FCP)

Limited work is available on the FCP measurements in similar Ni-superalloy inertia friction welds, with a single report on N18 welds [36]. The majority of FCP measurements in Ni-superalloys inertia friction welds were performed on dissimilar IN718-X welds (X: RR1000 or U720Li) [43, 44]. The FCP studies investigated the influence of the PWHT parameters [36], test temperature [42, 44] and the local FCP rates in different weld regions [42]. For more general information on FCP in Ni-based superalloys (especially RR1000), the reader is directed to other references [58–60]. Ferte [36] showed that the PWHT parameters significantly affect the FCP rates in N18 IFWs. By increasing the PWHT duration from 4 to 8 h at 800 °C, the FCP rates decreased owing to the precipitation of (Cr, Mo)-rich carbides along the grain boundaries. By introducing a post-weld solution stage at 1165 °C for 4 h in the process, roughly full retention of the parent FCP rates was observed in the welded sections (Fig. 2.24A), which was attributed to the formation of a similar microstructure to that of the parent metal. To consider the influence of the test temperature, Li et al. found that the FCP rates increased with the increase in temperature in RR1000-IN718 PWHTed welds [43] (Fig. 2.24B). The observed irregularity in the rates at 600 and 650 °C was attributed to the crack deviation from RR1000 to

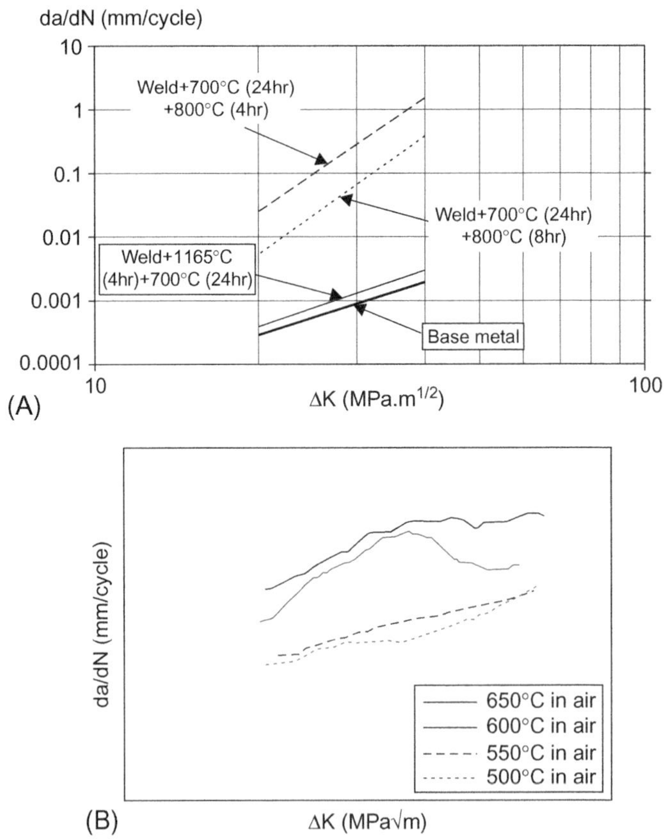

Fig. 2.24 FCP in Ni-superalloy inertia friction welds, showing the influence of (A) PWHT tested at room temperature and (B) the test temperature (following PWHT). (A) N18 [36] and (B) RR1000-IN718 [43].

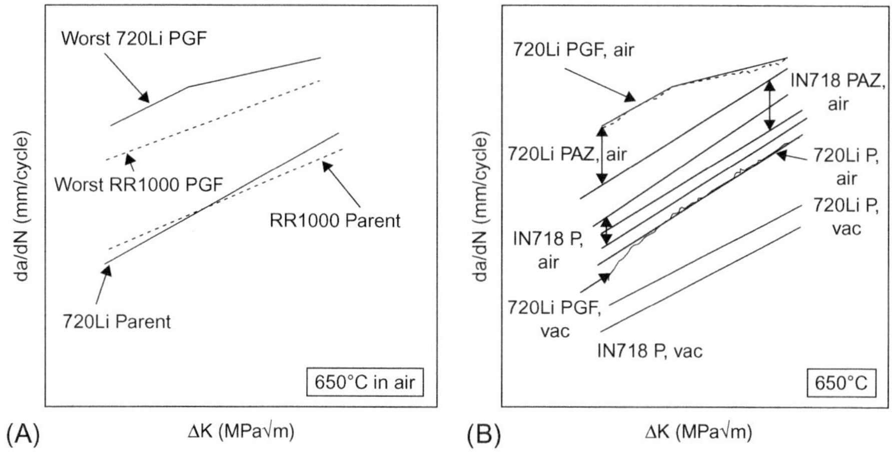

Fig. 2.25 FCP in Ni-based superalloy inertia PWHT welds [43], showing the influence of (A) the parent material (IN718/RR1000 or 720Li welds) and (B) the crack location (IN718–720Li).

IN718, as discussed elsewhere [44]. Moreover, the increase in test temperature was also associated with the dominance of intergranular fracture, compared with fully transgranular failure for the test performed at 500 °C. The transition temperature (~600 °C) between the transgranular and intergranular fracture supported the suggestion that an oxidation damage mechanism exists in Ni-based superalloys within this temperature range [61]. To identify the causes for the poor FCP at 650 °C, Li et al. further investigated the influence of the starting parent metal on the FCP rate, as well as the influence of the sample location with respect to the weld. When testing welds of alloys of similar γ' fraction (RR1000 and U720Li), it was found that the worst FCP condition in the U720Li weld half, corresponding to the primary γ'-free zone (PGF), had a faster FCP rate compared with the similar condition in RR1000 weld (Fig. 2.25A). Moreover, the FCP rates were generally lower when tested in vacuum, with the highest rates occurring in the PGF zone, followed by the TMAZ (Fig. 2.25B).

2.4.2 Steels

2.4.2.1 Microhardness development

Depending on the type of the alloy, the microhardness development in steel IFWs showed several trends. In a low-carbon steel solid inertia friction weld, the hardness was generally highest at the weld center, and decreased until it reached the base metal hardness [9] (Fig. 2.26). The hardness decreases from the center of the weld interface outwards owing to the concavity of the weld region.

In dissimilar steel inertia friction welds (e.g. SCMV-AerMet100 [45, 46]), the joining process results in a more complex microhardness distribution (Fig. 2.27). In an

Fig. 2.26 Microhardness development in steel inertia welds. A quarter of a 3/8″ (9.5 mm) diameter solid low-carbon steel weld section, showing the local variations across the weld line and the depth [9].

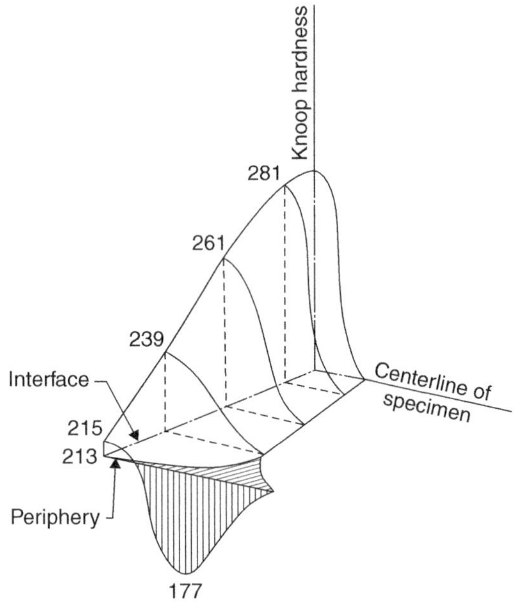

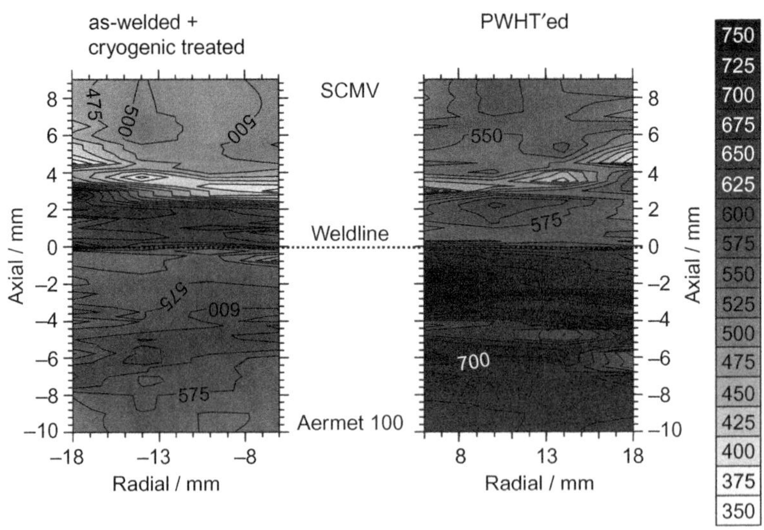

Fig. 2.27 Microhardness distribution in the AW and PWHT conditions for a SCMV-AerMet100 [32].

SCMV-AerMet100 IFW the three typical distinct zones, TMAZ, HAZ and base material, were identified on each weld side. While in SCMV the hardness of the HAZ was lower than the base metal, in AerMet100 the hardness was actually higher than the base metal. Conversely, in the TMAZ of SCMV the hardness was higher than the base metal, while in AerMet100 it was lower. Following the PWHT, the extent of heterogeneity of the microhardness distribution decreased, with each side retaining a roughly uniform hardness distribution, except for a localized drop in hardness in the HAZ of both sides. These differences were attributed to the variation in the fractions of the phases present (e.g. martensite, austenite, ferrite and cementite) across the weld regions, depending on the alloy, cooling rates and temperatures experienced, as well as the heterogeneous plastic deformation at the TMAZ and aging of the M_2C carbides in the AerMet100 side.

2.4.3 Titanium alloys

2.4.3.1 Tensile properties

Early work by Nessler et al. [48] studied the influence of IFW on the tensile, low-cycle fatigue (LCF) and reverse high-frequency bending properties of titanium alloy inertia friction welds. It was shown that certain welding parameters (welding speed, forging pressure, surface roughness, joint squareness and concentricity) did not have a statistically significant influence on the mechanical properties. Nonetheless, on investigating the influence of the weld upset on the tensile properties, it became clear that both the tensile strength and ductility improved with the increase in upset (Fig. 2.28). The properties were generally reproducible in Ti-6Al-4V and Ti-6Al-2Sn-4Zr-2Mo IFWs once they were welded with the parameters that produced an upset of ~4 mm. Roder et al. [62] also investigated the tensile strength of Ti-6246 IFWs at 20 and 300 °C. The base metal tensile strength was achieved in all of the investigated welds, although the elongation was relatively lower than the base metal. The PWHT specimens generally failed in the base metal, while the AW specimens failed close to the weld center.

2.4.3.2 Fatigue properties

Nessler et al. also investigated the LCF and high-cycle fatigue (HCF) properties of titanium inertia friction welds [48]. As shown in Fig. 2.29A, the Sonntag LCF shows that the majority of the PWHT welds displayed a superior behavior than the base metal and the AW conditions. >Similarly, the HCF tests (Fig. 2.29B) also show acceptable properties that occasionally surpass the base metal, with the failure occurring in the base metal. Other LCF investigations for Ti-6Al-2Sn-4Zr-6Mo [62] and Ti-17 [50] echo the findings of this study. It was found that even at relatively high temperatures (~300–400 °C), the welds of Ti-17 and Ti-6242S have LCF and HCF properties that are comparable to the properties of the base metals of these alloys.

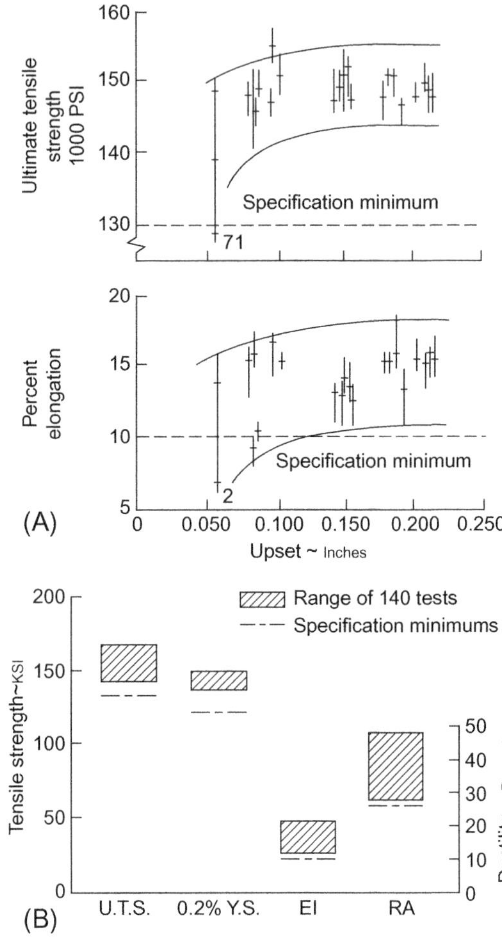

Fig. 2.28 Tensile properties of titanium inertia welds. (A) Influence of the upset on the strength and elongation in Ti-6Al-4V welds and (B) reproducibility of Ti-6A-2Sn-4Zr-2Mo IWs [48].
From C.G. Nessler, D.A. Rutz, R.D. Eng, P.A. Vozzella, 1971. Friction welding of titanium alloys, Weld. J. Res. Suppl. 50, 379S–385S is being republished with permission of the American Welding Society (AWS).

2.5 Residual stress development

Within the past decade, neutron and synchrotron X-ray diffraction have been extensively used to characterize the residual stress development in inertia friction welds owing to its capability of obtaining in-depth measurements, as opposed to hole drilling, which is only capable of performing surface measurements. Residual stress characterization by diffraction involves measuring the change in the lattice parameters or interplanar spacing by measuring peak shift, and comparing it to the unaffected metal. The strains are then calculated using:

$$\varepsilon = \frac{d - d_0}{d_0} = \frac{a - a_0}{a_0} \tag{2.13}$$

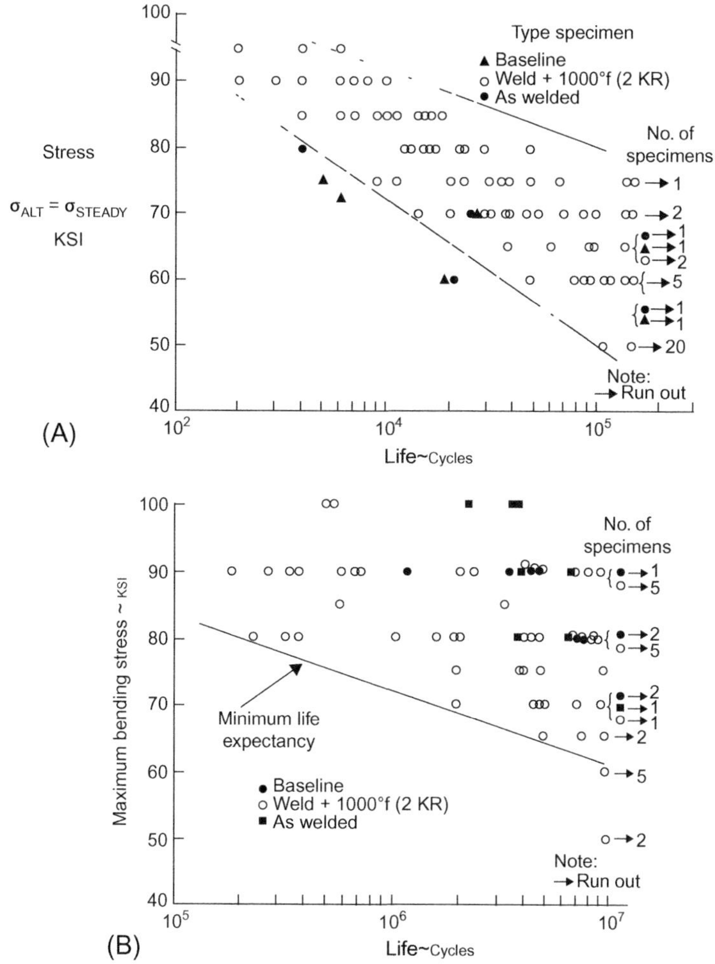

Fig. 2.29 Fatigue properties of Ti-6Al-2Sn-4Zr-2Mo IWs. (A) Sonntag LCF and (B) reverse-bending high-frequency fatigue strength [48].
From C.G. Nessler, D.A. Rutz, R.D. Eng, P.A. Vozzella, 1971. Friction welding of titanium alloys, Weld. J. Res. Suppl. 50, 379S–385S is being republished with permission of the American Welding Society (AWS).

where *d* and *a* are the *d*-spacing and the lattice parameter, respectively, with the subscript "0" indicating the strain-free values. Following the measurement and calculation of the strain in all three directions (axial, radial and hoop), a stress component can be calculated using:

$$\sigma_{\text{Hoop}} = \frac{E}{(1+\upsilon)(1-2\upsilon)}\left[(1-\upsilon)\varepsilon_{\text{Hoop}} + \upsilon(\varepsilon_{\text{Radial}} + \varepsilon_{\text{Axial}})\right] \tag{2.14}$$

When undertaking such measurements great care has to be taken when determining the strain-free value d_0. Microstructural changes across the weld mean that the chemical compositions of the phases present in a multiphase engineering alloy have changed too. Consequently, d_0 is unlikely to be constant across the weld region requiring separate measurements of the d_0 variation or using stress-balance conditions [63]. Further details on residual stress characterization and d_0 correction can be found elsewhere [64–67].

Studies on the residual stress characterization in tubular Ni-based superalloy inertia friction welds generally agree that the highest stresses in the weld occur in the hoop direction (Fig. 2.11), being tensile at the weld center especially close to the inner diameter, and being limited by the yield strength of the alloy [39]. In the axial direction, a lower stress intensity than in the hoop direction was measured but generated a bending moment. This bending moment was attributed to either the difference in temperature between the outer and inner diameters [30], a tourniquet effect from the hoop stress owing to its cross thickness variation [39] or the influence of the clamping forces of the tooling (Fig. 2.30) [68]. It was also suggested that the influence of the tooling on the residual stress development could possibly be larger than the effect of the process parameters. Finally, minimal stresses occurred in the radial direction of these IFWs, which had a wall thickness of 8–11 mm.

Although different Ni-based superalloys generally have roughly similar spatial stress distribution and relative stress levels, differences in the magnitude of the stresses are normally found that are caused by the nature of the alloy microstructure and chemistry. Preuss et al. compared the residual stress development in RR1000, IN718, and U720Li (Fig. 2.31). The three welds had low radial stresses (±100 MPa), while the axial stresses all showed the before-mentioned bending profile with the maximum tensile and compressive stresses being ±350, ±450, and ±600 MPa in IN718, U720Li, and RR1000, respectively. In the hoop direction,

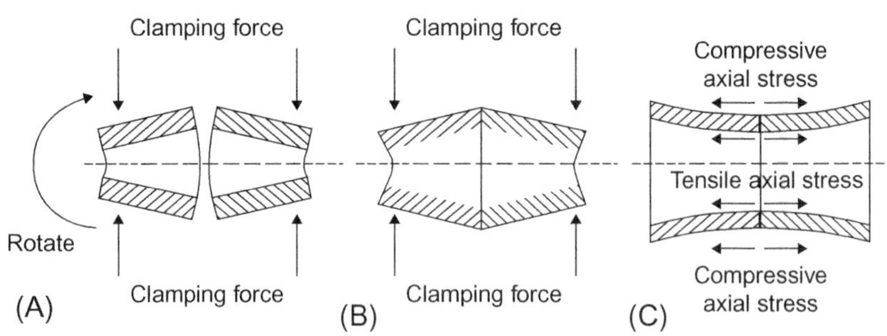

Fig. 2.30 A schematic illustration for the development of the bending stresses in the axial direction. (A) Clamping leads to bending of the tube sections (exaggerated) prior to welding, (B) clamping is still applied during welding, and (C) upon the removal of the clamping forces, the ends are released, resulting in the formation of the axial stresses [68].

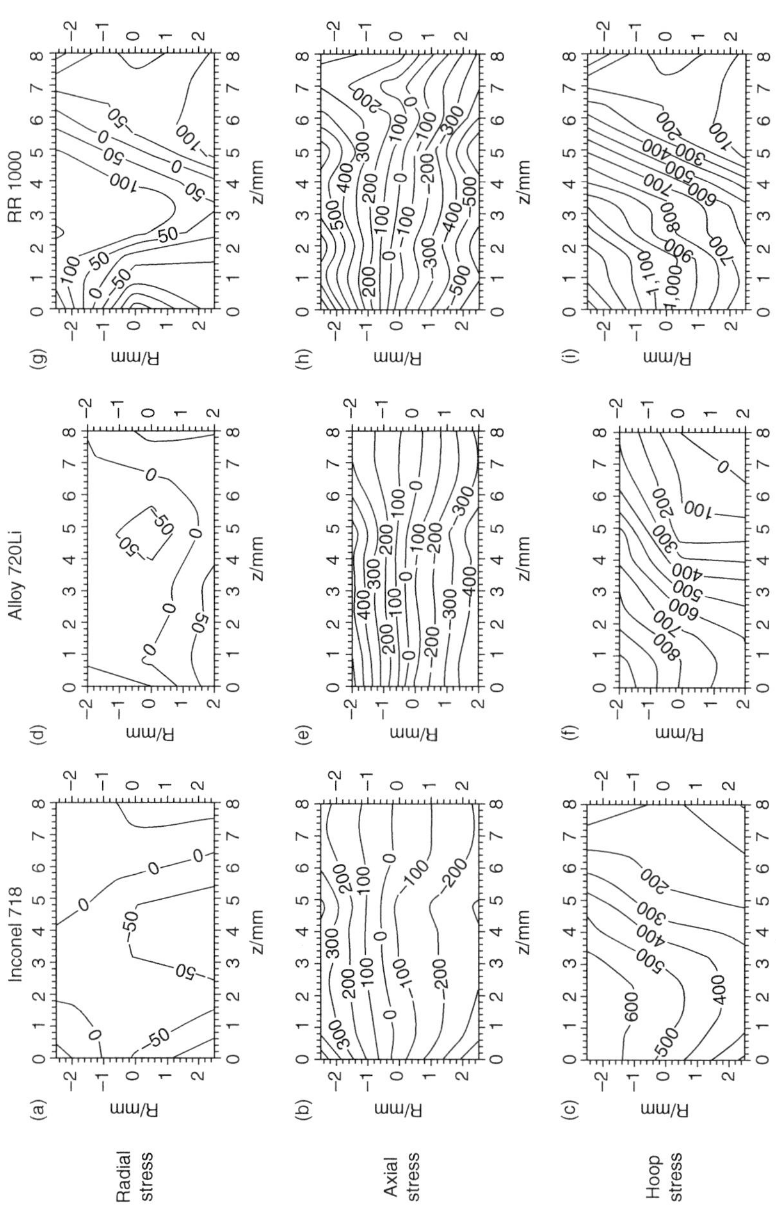

Fig. 2.31 Residual stress contours in IN718 (A–C), U720Li (D–F) and RR1000 (G–I) [39].

maximum tensile stresses of 700, 1000, and 1500 MPa were calculated in IN718, U720Li, and RR1000, respectively. By calculating an equivalent stress using Von Mises equation, it was found that the equivalent stress in RR1000 weld center exceeded the yield stress, suggesting that the stresses close to the weld line are limited by the local yield stress. In IN718 and U720Li, the weld stresses were generally lower than the yield stress ($\sim 0.8\sigma_{ys}$). This difference in the magnitude of stresses was attributed to the relatively poor short-term creep properties in IN718 and U720Li, allowing the misfit strains in these alloys to be reduced more during cooling than in RR1000. Upon performing PWHT, the major part of the stresses observed gets annihilated, although the hoop stresses remain generally high (~300–500 MPa) [39], while the axial and radial stresses decrease to $\sim\pm100$ MPa (the experimental error in these measurements is ±60 MPa).

In contrast, residual stress characterization studies of inertia friction welds produced of other materials (e.g. steels, titanium and aluminum alloys) are very limited, with a single report only available on dissimilar SCMV-AerMet100 steel welds ([32]; section 11.3.2). Ferritic steels differ from Ni-based superalloys in that IFW results in a phase transformation during heating (ferrite to austentite), followed by the martensitic or bainitic phase transformation upon rapid cooling depending on the cooling rate and is affected by the wall thickness. As the formation of martensite is associated with a volume increase, this alters the stresses to become compressive near the weld line instead of the typical tensile stresses observed in this region. This can be viewed in the SCMV-AerMet100 hoop stress (Fig. 2.32), where the weld region (±2 mm) was predominantly compressive, as opposed to the Ni-based superalloys, which was tensile. Small stresses (100–300 MPa) were observed in both the radial and axial directions. Following PWHT, the stresses in AerMet100 generally dropped to acceptable levels, while almost no change was observed in the SCMV. These differences are because of the nature of the selected PWHT, and the differences between the two alloys in their thermal stability.

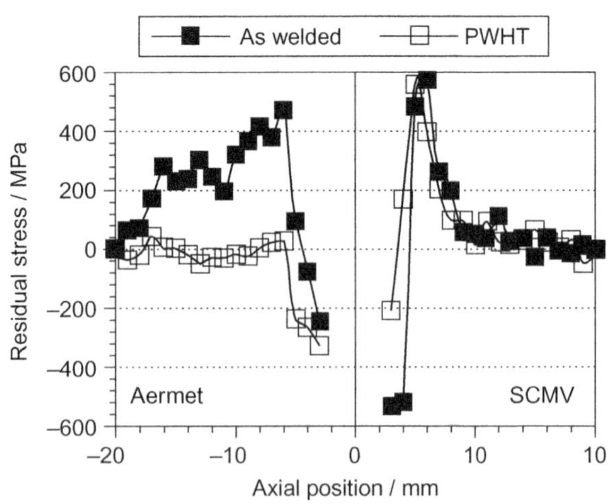

Fig. 2.32 AW and PWHT hoop residual stress distribution in SCMVAerMet100 [32].

2.6 Future trends

Considering that IFW has been around for over half a century, it is clear that it can be classified as a mature technique, especially with respect to IFW of Ni-based superalloy aero engine parts. Nonetheless, there is a need for more extensive studies on IFW of steels, titanium and aluminum alloys owing to their potential applications in the aerospace industry. By browsing through the recent patents that are associated with IFW, several trends can be identified as potential future applications. There is an interest in using IFW to weld cylindrical components to thin non-cylindrical components, and dissimilar materials. The PWHT processes in several alloy inertia friction welds seem to be a concern, as it is crucial to use a PWHT that results in the retention of the mechanical properties, especially the fatigue-resistance properties. Several patents also suggested new applications, including aircraft actuator pistons, gas turbine parts and crankshafts.

2.7 Source of further information and advice

For further information on IFW, especially machines and products, the reader is directed to the following web sites:

- Interface Welding: http://www.interfacewelding.com/
- Manufacturing Technology, Inc.: http://www.mtiwelding.com
- MTU Aeroengines: http://www.mtu.de/en/technologies/manufacturing_processes/inertia_friction_welding/index.html
- Swanson Industries: http://www.swansonindustries.com/inertiawelding.php
- The Welding Institute (TWI Ltd.): http://www.twi.co.uk
- Thompson Friction Welding: http://www.thompson-friction-welding.co.uk/

References

[1] S.W. Kallee, E.D. Nicholas, M.J. Russell, Friction welding of aero engine components, in: 10th World Conference on Titanium Ti-2003. Hamburg, Germany, 2003.

[2] T.L. Oberle, C.D. Loyd, M.R. Calton, Caterpillar's inertia weld process, SAE Trans. 75 (660470) (1967) 28–35.

[3] Chudikov, A. I. (1956) Friction welding. Russian Patent, RU106270.

[4] P.T. Houldcroft, Welding Process Technology, Cambridge University Press, Cambridge, 1977.

[5] T.L. Oberle, Applications of inertia welding, SAE Trans. 680047 (1968) 144–149.

[6] Anon, Inertia welding: simple in principle and application, Weld. Met. Fabr. 10 (1979) 585–589.

[7] C.R.G. Ellis, Continuous drive friction welding of mild steel, Weld J. 51 (1972) 183S–197S.

[8] M.B. Hollander, C.J. Cheng, J.A. Quimby, Friction welding—a natural for production, SAE Trans. 640800 (1964) 242–254.

[9] K.K. Wang, W. Lin, Flywheel friction welding research, Weld J. 53 (1974) 233S–241S.

[10] L. D'Alvise, E. Massoni, S.J. Walloe, Finite element modelling of the inertia friction welding process between dissimilar materials, J. Mater. Process. Technol. 125–126 (2002) 387–391.

[11] MTI, Inertia Friction Welding, Manufacturing Technology, Inc., 2009. http://www.mti-welding.com/inertia-friction-welding-process.cfm.

[12] J.M. Heberling, Building strong bonds with inertia welding, Mach. Des. 62 (1990) 119–122.

[13] B. Benn, Inertia Welding, 2000. unpublished.

[14] MTI, M120 Inertia Welder Parameter Book, MTI, IN, USA, 1974.

[15] J.W. Elmer, D.D. Kautz, Fundamentals of friction welding—volume 6: welding, brazing and soldering, in: ASM Handbook, ASM International, Ohio, 1993.

[16] C.J. Cheng, Transient temperature distribution during friction welding of two similar materials in tubular form, Weld J. 41 (1962) 542S–550S.

[17] C.J. Cheng, Transient temperature distribution during friction welding of two dissimilar materials in tubular form, Weld J. 42 (1963) 233S–240S.

[18] K.K. Wang, P. Nagappan, Transient temperature distribution in inertia welding of steels, Weld J. 49 (1970) 419S–426S.

[19] P.C. Johnson, B.A. Stein, R.S. Davis, Inertia welding, Unpublished Technical Report, Caterpillar Tractor Co, 1966.

[20] V.R. Davé, M.J. Cola, G.N.A. Hussen, Heat generation in the inertia welding of dissimilar tubes, Weld J. 80 (2001) 246–252.

[21] Ø. Grong, Metallurgical Modelling of Welding, The Institute of Materials, London, 1997.

[22] C.J. Bennett, T.H. Hyde, E.J. Williams, Modelling and simulation of the inertia friction welding of shafts, Proc. Inst. Mech. Eng. Pt. L J. Mater. Des. Appl. 221 (2007) 275–284.

[23] L. Fu, L.Y. Duan, S.G. Du, Numerical simulation of inertia friction welding process by finite element method, Weld J. 82 (2003) 65–70.

[24] B. Grant, M. Preuss, P.J. Withers, G. Baxter, M. Rowlson, Finite element process modelling of inertia friction welding advanced nickel-based superalloy, Mater. Sci. Eng. A 513–514 (2009) 366–375.

[25] H.S. Jeong, J.R. Cho, H.C. Park, Development of dissimilar inertia welding process of large superalloy spindle, Key Eng. Mater. 345–346 (II) (2007) 1429–1432.

[26] Z. Liwen, L. Chengdong, Q. Shaoan, Y. Yongsi, Z. Wenhui, Q. Shen, W. Jinghe, Numerical simulation of inertia friction welding process of GH4169 alloy, J. Phys. IV 120 (2004) 681–687.

[27] A. Moal, E. Massoni, Finite element simulation of the inertia welding of two similar parts, Eng. Comput. 12 (1995) 497–512.

[28] M. Soucail, A. Moal, L. Naze, Microstructural study and numerical simulation of inertia friction welding of astroloy, in: S.D.R.W.S. Antolovich, R.A. Mackay, D.L. Anton, T. Khan, R.D. Kissinger, D.L. Klarstrom (Eds.), Superalloys 1992, TMS, Champion, PA, 1992.

[29] L. Wang, M. Preuss, P.J. Withers, G. Baxter, P. Wilson, Energy-input-based finite-element process modeling of inertia welding, Metall. Mater. Trans. B Process Metall. Mater. Process. Sci. 36 (2005) 513–523.

[30] L. Wang, M. Preuss, P.J. Withers, G.J. Baxter, P. Wilson, Residual stress prediction for the inertia welding process, J. Neutron Res. 12 (2004) 21–25.

[31] A.M. Korsunsky, G.M. Regino, D. Nowell, M. Karadge, B. Grant, P.J. Withers, M. Preuss, G. Baster, Inertia friction welds between nickel superalloy components: analysis of residual stress by eigenstrain distributions, J. Strain Anal. 44 (2009) 159–170.

[32] R.J. Moat, D.J. Hughes, A. Steuwer, N. Iqbal, M. Preuss, S.E. Bray, M. Rawson, Residual stress in inertia-friction-welded dissimilar high-strength steels, Metall. Mater. Trans. A Phys. Metall. Mater. Sci. 40 (A) (2009) 2098–2108.

[33] M.J. Donachie, S.J. Donachie, Superalloys: A Technical Guide, ASM International, Ohio, 2002.

[34] P. Adam, Material response during inertia welding of superalloys, Z. Werkstofftech. 13 (1982) 258–262.

[35] M. Preuss, J.W.L. Pang, P.J. Withers, G.J. Baxter, Inertia welding nickel-based superalloy: part I'. Metallurgical characterization, Metall. Mater. Trans. A Phys. Metall. Mater. Sci. 33 (2002) 3215–3225.

[36] J.P. Ferte, Assemblage metallurgique des moteurs. Soudage par friction inertielle de disques de compresseur HP en superalliage de nickel. Metallurgical assembly of engines. Inertial friction welding of nickel-base superalloy HP compressor disks, J. Phys. 3 (1993) 1019–1028.

[37] Z.W. Huang, H.Y. Li, M. Preuss, M. Karadge, P. Bowen, S. Bray, G. Baxter, Inertia friction welding dissimilar nickel-based superalloys alloy 720Li to IN718, Metall. Mater. Trans. A Phys. Metall. Mater. Sci. 38 (A) (2007) 1608–1620.

[38] M. Preuss, J.Q. Da Fonseca, I. Kyriakoglou, P.J. Withers, G.J. Baxter, Characterisation of γ' across inertia friction welded Alloy 720Li, in: K.A. Green, T.M. Pollock, H. Harada, T. E. Howson, R.C. Reed, J.J. Schirra, S. Walston (Eds.), Proceedings of the International Symposium on Superalloys. Champion, PA, 2004.

[39] M. Preuss, P.J. Withers, G.J. Baxter, A comparison of inertia friction welds in three nickel base superalloys, Mater. Sci. Eng. A 437 (2006) 38–45.

[40] O. Roder, J. Albrecht, G. Lutjering, Microstructure and mechanical properties of an inertia friction welded INCOLOY alloy 909—INCONEL alloy 718 joint for rotating applications, Mater. High Temp. 23 (2006) 171–177.

[41] O. Roder, D. Helm, S. Neft, J. Albrecht, G. Lutjering, Mixed INCONEL alloy 718 inertia welds for rotating applications—microstructures and mechanical properties, in: E.A. LORIA (Ed.), Proceedings of the International Symposium on Superalloys and Various Derivatives. Pittsburgh, PA, 2005.

[42] G. Montay, M. Francois, M. Tourneix, B. Guelorget, C. Vial-Edwards, I. Lira, Strain and strain rate measurement during the bulge test by electronic speckle pattern interferometry, J. Mater. Process. Technol. 184 (2007) 428.

[43] H.Y. Li, Z.W. Huang, S. Bray, G. Baxter, P. Bowen, High temperature fatigue of friction welded joints in dissimilar nickel based superalloys, Mater. Sci. Technol. 23 (2007) 1408–1418.

[44] F. Daus, H.Y. Li, G. Baxter, S. Bray, P. Bowen, Mechanical and micro-structural assessments of RRiooo to IN718 inertia welds—effects of welding parameters, Mater. Sci. Technol. 23 (2007) 1424–1432.

[45] R. Moat, M. Karadge, M. Preuss, S. Bray, M. Rawson, Phase transformations across high strength dissimilar steel inertia friction weld, J. Mater. Process. Technol. 204 (2008) 48–58.

[46] W.S. Robotham, T.H. Hyde, E.J. Williams, P. Brown, I.R. Mccoll, C.J. Kong, Mechanical testing of dual alloy inertia friction welded shafts, in: Applied Mechanics and

Materials. Fourth International Conference on Advances in Experimental Mechanics. Southampton, 2005.

[47] M.D. Tumuluru, Parametric study of inertia friction welding for low alloy steel pipes, Weld J. 63 (1984) 289–296.

[48] C.G. Nessler, D.A. Rutz, R.D. Eng, P.A. Vozzella, Friction welding of titanium alloys, Weld J. 50 (1971) 379S–385S.

[49] C.L. English, Inertia welding of jet engine components: transition from electron welding to inertia welding for a titanium alloy fan disk, in: Fabtech International '95. Society of Manufacturing Engineers, Chicago, IL, USA, 1995.

[50] A. Barussaud, A. Prieur, Structure and properties of inertia welded assemblies of Ti-based alloy disks, in: Titanium '95—Science and Technology; the 8th World Conference on Titanium. Birmingham, United Kingdom, 1996.

[51] W.A. Baeslack III, D. Phillips, C. English, A.P. Woodfield, Inertia-friction welding of an advanced rapidly solidified titanium alloy, J. Mater. Sci. Lett. 10 (1991) 1401–1408.

[52] O. Roder, D. Helm, G. Lutjering, Microstructure and mechanical properties of inertia and electron beam welded Ti-6246, in: G. Lutjering (Ed.), Tenth World Conference on Titanium, Wiley VCH, Hamburg, Germany, 2003.

[53] J.D. Whittenberger, T.J. Moore, D.L. Kuruzar, Preliminary investigation of inertia friction welding B2 aluminides, J. Mater. Sci. Lett. 6 (1987) 1016–1018.

[54] K.H. Hou, W.A. Baeslack III, Electron microscopy of inertia-friction weldments in a rapidly solidified Al-Fe-Mo-V alloy, J. Mater. Sci. 25 (1990) 2642–2653.

[55] H.H. Koo, S. Krishnaswamy, W.A. Baeslack III, Solid-Phase Welding of a Rapidly-Solidified Dispersion-Strengthened Al-Fe-V-Si Alloy—FVS1212, Minerals, Metals & Materials Society (TMS), Detroit, MI, USA, 1991.

[56] T.J. Lienert, W.A. Baeslack III, J. Ringnalda, H.L. Fraser, Inertia-friction welding of SiC-reinforced 8009 aluminium, J. Mater. Sci. 31 (1996) 2149–2157.

[57] Y. Zhu, Z. Zhu, Z. Xiang, Z. Yin, Z. Wu, W. Yan, Microstructural evolution in 4Cr10Si2Mo at the 4Cr10Si2Mo/Nimonic 80A weld joint by inertia friction welding, J. Alloys Compd. 476 (2009) 341–347.

[58] S. Everitt, M.J. Starink, H.T. Pang, I.M. Wilcock, M.B. Henderson, P.A.S. Reed, A comparison of high temperature fatigue crack propagation in various subsolvus heat treated turbine disc alloys, Mater. Sci. Technol. 23 (2007) 1419–1423.

[59] D.M. Knowles, D.W. Hunt, The influence of microstructure and environment on the crack growth behavior of powder metallurgy nickel superalloy RR1000, Metall. Mater. Trans. A Phys. Metall. Mater. Sci. 33 (A) (2002) 3165–3172.

[60] M.J. Starink, P.A.S. Reed, Thermal activation of fatigue crack growth: analysing the mechanisms of fatigue crack propagation in superalloys, Mater. Sci. Eng. A 491 (2008) 279–289.

[61] B.A. Pint, J.R. Distefano, I.G. Wright, Oxidation resistance: one barrier to moving beyond Ni-base superalloys, Mater. Sci. Eng. A 415 (2006) 255–263.

[62] O. Roder, D. Helm, G. Lutjering, Microstructure and mechanical properties of inertia and electron beam welded Ti-6246, in: Tenth World Conference on Titanium, Wiley VCH, Hamburg, Germany, 1999.

[63] M. Preuss, J.W.L. Pang, P.J. Withers, G.J. Baxter, Inertia welding nickel-based superalloy: Part II. Residual stress characterization, Metall. Mater. Trans. A Phys. Metall. Mater. Sci. 33 (2002) 3227–3234.

[64] P.J. Withers, Residual stresses: measurement by diffraction, in: K.H.J. Buschow (Ed.), Encyclopedia of Materials: Science & Technology, Pergamon, Oxford, 2001, pp. 8158–8169.
[65] P.J. Withers, H.K.D.H. Bhadeshia, Residual stress. Part 1—Measurement techniques, Mater. Sci. Technol. 17 (2001) 355–365.
[66] P.J. Withers, H.K.D.H. Bhadeshia, Residual stress. Part 2—Nature and origins, Mater. Sci. Technol. 17 (2001) 366–375.
[67] P.J. Withers, M. Preuss, A. Steuwer, J.W.L. Pang, Methods for obtaining the strain-free lattice parameter when using diffraction to determine residual stress, J. Appl. Cryst. 40 (2007) 891–904.
[68] J.W.L. Pang, M. Preuss, P.J. Withers, G.J. Baxter, C. Small, Effects of tooling on the residual stress distribution in an inertia weld, Mater. Sci. Eng. A 356 (2003) 405–413.

Laser welding of metals for aerospace and other applications

3

J. Blackburn
TWI Ltd., Cambridge, United Kingdom

3.1 Introduction

The term laser (an acronym for Light Amplification by Stimulated Emission of Radiation) refers to the mechanism for generating electromagnetic radiation, normally between the ultraviolet and infra-red frequencies of the electromagnetic spectrum, by the process of stimulated emission. Gould [1] coined the term in a paper published shortly before the first successful demonstration of a working laser source by Maiman [2], a flash-lamp-pumped ruby crystal (Cr^{3+}:Al_2O_3) emitting at wavelengths of 694.3 nm and 692.9 nm, advancing the theoretical and practical research performed by Einstein [3], Ladenburg [4], Schawlow and Townes [5], Gould [1], Javan [6] and numerous others. In the same decade the development of several other laser sources were reported, including the semi-conductor (GaAs) laser [7], the Nd:YAG laser [8] and the CO_2 laser [9].

The emitted electromagnetic radiation from these laser sources is commonly referred to as a laser beam, which in contrast with other light sources is highly monochromatic, coherent and of very low divergence. The potential for using focused laser beams as a tool for materials processing was reported (for example, [10, 11]) shortly after Maiman's demonstration and the laser materials processing industry was born. In 2008 the total revenue generated world-wide by the sale of laser systems for industrial materials processing was approximately \$6.1 billion [12]—with welding applications (excluding microprocessing activities) accounting for an estimated 12% of total unit sales [13]. Laser welding has been reported [14] as a potential manufacturing technique for aerospace components since the 1960s, although it has received considerably more attention in recent years, as a result of advances in high-power laser sources making deep penetrations possible.

This chapter provides an introduction into laser technology and laser welding, and reviews the most relevant published literature concerned with the laser welding of titanium alloys. An overview of laser technology and the key characteristics of laser light are detailed at the beginning of this chapter, followed by a summary of the fundamental laser-beam interactions with metallic materials that occur. The utilization of laser beams for welding applications is then discussed, and the key concepts behind welding in the conduction-limited and keyhole modes are examined. A subchapter then follows, centered on laser welding of titanium alloys. A short section on expected future trends follows, with the objective of emphasizing the exciting potential of the

Welding and Joining of Aerospace Materials. https://doi.org/10.1016/B978-0-12-819140-8.00003-1
Copyright © 2012 Woodhead Publishing Ltd. All rights reserved.

new generation of solid-state laser sources for joining metallic aerospace materials. It is acknowledged that this chapter is not sufficiently detailed to cover a subject as broad as laser welding, and therefore suggested sources of further information on this subject are given at the end of this chapter. In particular, literature concerned with laser welding of other metallic aerospace materials is emphasized.

3.2 Operating principles and components of laser sources—An overview

It is known from quantum mechanics that electrons can only exist in discrete orbits, with each orbit having a specific energy level. Atoms in an excited state may spontaneously decay into a lower energy level whereby an electron makes the transition from one discrete energy level, for example E_3, to a lower energy level, for example E_2. The energy difference between the two levels (E_3–E_2) is simultaneously emitted as a quantum of electromagnetic radiation, a photon, in a random direction. This process is known as spontaneous emission and is one of three possible photon-related electron transitions. The wavelength (λ), and therefore frequency (f), of the emitted electromagnetic radiation is directly related to the photon's energy (E_3–E_2), and can be calculated knowing Planck's constant (h) and the speed of light (c). Photons can also be absorbed by an atom, thereby raising the energy level of the atom, as long as its energy matches that of a possible electron transition in the atom (e.g. from E_1 to E_2). The absorption and emission of photons is also possible in molecules, where their discrete energy levels are associated with molecular vibration or rotation.

$$E_3 - E_2 = hf = hc/\lambda \tag{3.1}$$

Laser sources operate on the principle of a different type of photon-related electron transition, stimulated emission; a concept proposed by Einstein [3] and experimentally confirmed by Ladenburg [4]. Stimulated emission occurs when an atom or molecule in an energy level above the ground state interacts with a photon that has energy equal to that between the atom or molecule's current energy level and a lower energy level. This results in a photon being emitted as the atom or molecule makes the transition from, for example, E_3 to E_2, (as is shown in Fig. 3.1) which is of the same energy (i.e. frequency and wavelength), direction of travel and phase as the photon that induced the transition. The materials used in an industrial laser source that have energy levels conducive to the stimulated emission process, and are therefore capable of carrying out the lasing process, are referred to as the gain medium (also called the active medium or lasing medium). Table 3.1 details the gain mediums and their associated photon wavelength for industrial laser sources ordinarily used for welding operations.

However, electromagnetic radiation incident on a gain medium in thermo-dynamic equilibrium will result in a net absorption, rather than stimulated emission, of radiation. This is because the number of atoms or molecules with a lower energy level (for example N_1) is much greater than those with a higher energy level (for example N_2) for

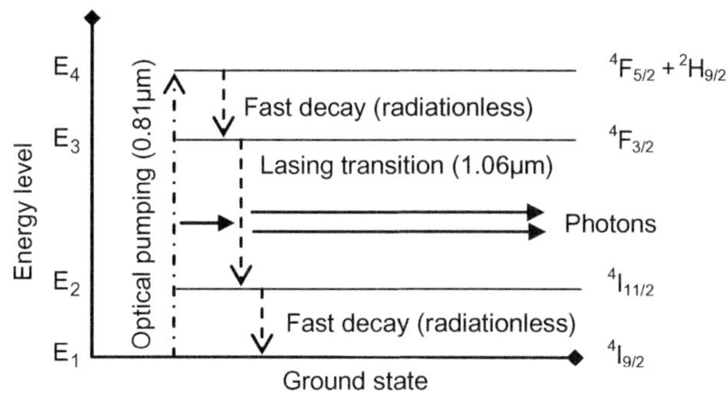

Fig. 3.1 Simplified energy-level diagram for a typical Nd:YAG laser (Nd^{3+} ion in YAG) showing optical pumping at 0.81 µm and stimulated emission of radiation at 1.06 µm. Values from A.A. Kaminskii, Laser Crystals, Springer, Berlin, 1981.

Table 3.1 Typical laser sources utilized for high-power welding applications.

	Wavelength (µm)	**Maximum commercially available power (kW)**	***M*² value**	**Wall-plug efficiency (%)**
CO_2	~10	20	≥1.1 (increasing with power)	~10
Nd:YAG	~1	5	≥35 (increasing with power)	~4
Yb-fiber	~1	50	≥1.1 (increasing with power)	~30
Yb:YAG disc	~1	16	≥6 (increasing with power)	~20

a material in thermodynamic equilibrium. In order to achieve a net stimulated emission of radiation, N_2 must exceed N_1 by a threshold amount. This situation is known as a population inversion and is reached by supplying energy for atoms or molecules to excite into higher energy levels. The mechanism by which the energy is supplied depends upon the gain medium, but optical and electrical (either through electron collisions with the molecules of the gain medium, or through electron collisions with other molecules that subsequently collide with the gain medium) pumping methods are frequently used. For Nd:YAG lasers, and most other solid-state lasers, optical pumping (as schematically shown in Fig. 3.2) was traditionally performed with a

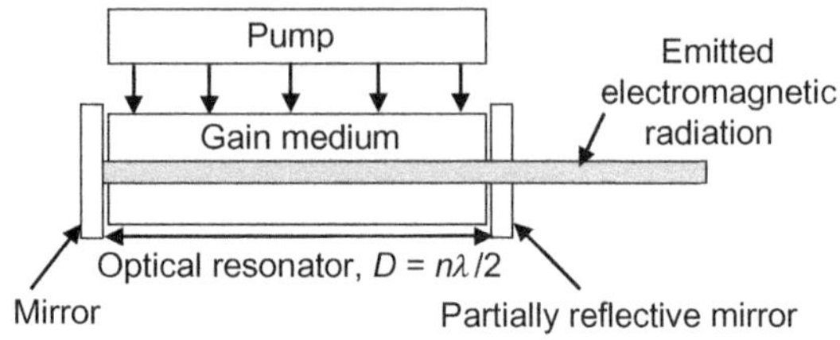

Fig. 3.2 Schematic representation of a laser source. The emitted electromagnetic radiation would be delivered to the workpiece through a series of optics and/or optical fibers.

high-intensity flash lamp, although diodes are now commonly used to achieve increased wall-plug efficiency. A population inversion cannot be efficiently achieved in a system that has only two discrete energy levels, as the photons interacting with the atoms may either cause stimulated emission or be absorbed. Systems must be used that have three or four energy levels (with one of the upper states being meta-stable) for pulsed and continuous-wave (i.e. operating at constant power) lasers, respectively.

In order to achieve amplification of the electromagnetic radiation the gain medium is contained within an optical resonator, which in its simplest form consists of a mirror at each end of the gain medium. Photons traveling perpendicular to the axis of the gain medium are reflected between the mirrors and stimulate the emission of further photons from the pumped gain medium. If the photons are to be in phase with each other, the length of the laser cavity must conform to Eq. (3.2); the length of the cavity (D) should be equal to an integral number (n) of half wavelengths of the electromagnetic radiation.

$$D = n\lambda/2 \tag{3.2}$$

One end of the resonator is only partially reflective and this serves as the aperture of the laser source, whereby a shutter controls the release of the laser beam. For a continuous-wave laser a steady-state is reached in the optical resonator whereby the number of photons being transmitted out of the resonator is nominally identical to the number being generated by stimulated emission from the gain medium, thereby ensuring a constant output power. The emitted photons, or laser beam, may then be delivered to the workpiece through a series of optics and/or optical fibers (wavelength dependent). Laser radiation emitted from Nd:YAG, Yb-fiber and Yb:YAG disc laser sources are all approximately 1 μm in wavelength, and can therefore be transmitted down optical fibers tens of meters long, to a process head containing a series of optics that focus the beam for materials processing applications. This allows the laser source to be situated away from the production environment, and the flexible optical fiber enables the process head to be mounted on industrial jointed-arm robots capable of navigating particularly complex component geometries. Also possible is the switching of the laser beam between different delivery fibers, which allows the laser source to be time-shared between different processes and/or provide redundancy in the system. In comparison, CO_2 laser sources emit radiation of $\sim$10 μm wavelength that cannot be focused down optical fibers, and must therefore be directed toward the workpiece using a series of optical components. In practice, this limits the component geometries that can be processed with CO_2 laser sources unless additional component-handling manipulators are introduced.

3.3 Key characteristics of laser light

Laser light has a number of key characteristics; it is highly monochromatic, has a low beam divergence and is highly coherent, characteristics that are conducive to using it as a materials processing tool. The design, manufacture and integration of the gain medium, population inversion pump and optical resonator will all determine the properties of the emitted electromagnetic radiation (laser light). When deciding on a laser source for a particular materials processing application the emission wavelength, temporal and spatial operating modes, available output power and beam quality are among the key factors that should be considered. In addition, the chosen optics used to guide the laser light to the workpiece will determine crucial parameters such as the beam waist and the depth of focus.

The light emitted from the majority of laser sources is highly monochromatic, in that the total spectrum of the light has a very narrow spectral linewidth—the exact spectrum is dependent upon the bandwidth available in the gain medium and the longitudinal modes present in the optical resonator. Laser sources that operate in the continuous-wave mode are capable of generating a constant power to the workpiece, although this may be modulated and/or ramped up/down if required, so long as it does not exceed the maximum-rated output power of the laser source. Conversely, pulsed laser sources are capable of generating very high peak powers for a short duration, either through Q-switching or pulsed pumping.

The standing longitudinal electromagnetic waves established in the optical resonator may be separated by varying angles—related to the design of the resonator. Constructive and destructive interference between these longitudinal standing waves give rise to the formation of an electromagnetic radiation field pattern transverse to the longitudinal waves. This is referred to as the transverse electromagnetic mode (TEM_{mn}, where the integers m and n indicate the number of zero fields in a particular direction) structure of the laser beam, and determines the intensity distributions perpendicular to the direction of the laser-beam propagation. A complete description of the potential TEM modes is outside the scope of this chapter and the reader should refer to the further information section for recommended reading concerning this subject. Nevertheless, it can be summarized that laser beams with a higher TEM mode are more difficult to focus than a laser beam with a low TEM mode [15]. The TEM_{00} mode, or fundamental mode, is the simplest mode, and its intensity distribution, I, as a function of radius, r, from the central axis can be theoretically described by a Gaussian function.

$$I(r) = I_0 \exp\left(-2r^2/{w_0}^2\right) \tag{3.3}$$

where I_0 is the axial irradiance of the laser beam, and w_0 is the beam waist.

A Gaussian beam radius is usually defined as the radius where its irradiance is $1/e^2$ of the axial irradiance. The beam waist is the point in propagation direction where the laser-beam diameter converges to a minimum, and the beam radius at this point is referred to as w_0. A focused laser beam propagating in free space will converge to a minimum, the beam waist, before diverging. The angle at which the laser beam diverges is termed the beam divergence angle, 2θ. The half-angle divergence, θ, is shown in Fig. 3.3. Knowing the half-angle divergence, the beam waist of a Gaussian laser beam can be calculated according to Eq. (3.4).

$$\theta = \lambda / \pi w_0 \tag{3.4}$$

The product of the beam waist and the half-angle divergence is a constant and known as the beam parameter product (BPP), stated in mm mrad. Therefore, the BPP of a Gaussian laser beam will be λ/π, which is the theoretical optimum. However, the emitted outputs of actual laser sources are not truly Gaussian, although single-mode Yb-fiber and Yb:YAG disc lasers are very near, and are characterized by measures of their beam quality. Perhaps the most commonly used measure of beam quality is the M^2 value of the laser beam, which compares the BPP of an actual laser beam to that of a Gaussian laser beam of identical wavelength. ISO 11146–2:2005 defines the M^2 value of a laser beam as its BPP divided by λ/π [16]. For laser sources emitting beams of approximately 1 μm in length, the BPP is often used as a measure of beam quality. Nonetheless, both these beam-quality values approximate the propagation of actual laser beams with an expansion of Gaussian beam analysis. Eq. (3.5) [15] can be used to calculate the beam waist, w, of a real laser beam.

$$w = 4M^2 \lambda f / \pi R \tag{3.5}$$

where f is the lens focal length, and R is the radius of the beam at the focusing lens.

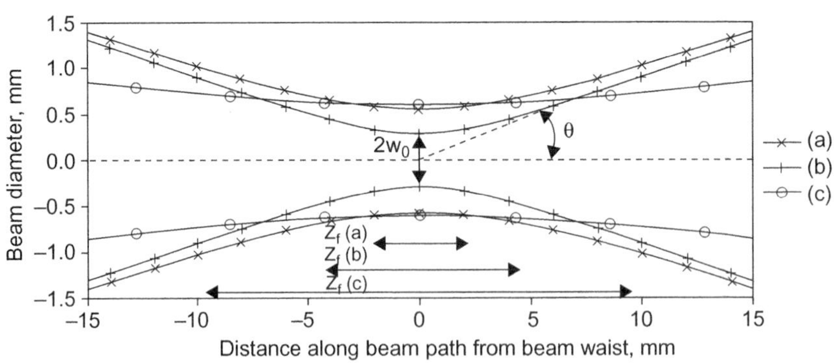

Fig. 3.3 Calculated two-dimensional profiles of three different laser beams of ~1-μm wavelength. (A) BPP = 12 mm mrad, 300 mm focal-length focusing lens. (B) BPP = 6 mm mrad, 300 mm focal-length focusing lens. (C) BPP = 6 mm mrad, 640 mm focal-length focusing lens.

Eq. (3.5) indicates that laser beams with a smaller value of M^2, or BPP, can be focused into smaller diameter spots than those with higher M^2 or BPP values. Another important factor when determining the characteristics of a laser beam used for a particular materials processing application is its depth of focus, Z_f. The depth of focus is equal to the distance traveled in either direction from the beam waist over which the intensity remains about the same, which, in practical terms corresponds to approximately a 5% increase in the beam diameter. Materials processing applications that use laser beams with a long depth of focus are less susceptible to shifts in the focal plane position. Eq. (3.6) [17] details the calculation of the depth of focus for a 5% increase in beam diameter. However, there is no exact definition of this and other authors define Z_f in different ways.

$$Z_f = w^2/\lambda M^2 \tag{3.6}$$

Fig. 3.3 details the profiles of three different laser beams, all of ~1 μm wavelength. It illustrates that laser beams with lower BPPs may be focused into smaller beam waists using the same optical system, thereby maintaining an acceptable stand-off distance (distance between the focusing optic and the workpiece). Conversely, laser beams with lower BPPs may be focused into a similar beam waist to that produced with a laser beam of higher BPP, but have a greater depth of focus and stand-off distance.

3.4 Basic phenomena of laser light interaction with metals

3.4.1 Absorption

The previous section dealt with the key characteristics of laser light. If this light is incident on the surface of a solid metal, it may be absorbed provided the metal has a quantized energy level (electronic, atomic or molecular) that matches that of the incident electromagnetic radiation, according to Eq. (3.1). Absorption of the laser light is dependent upon the substrate properties, as well as the characteristics of the incident laser light. Fig. 3.4 details the reflectivity of solid Al, Fe, Ni and Ti, for a normal angle of incidence and at room temperature, over a range of wavelengths.

For the metals shown in Fig. 3.4, there is a considerable difference in their reflectivity between wavelengths of ~1 μm (Nd:YAG, Yb:YAG disc, Yb-fiber) and ~10 μm (CO_2). The value of reflectivity, R_f, used in this chapter may fall within a range of 0–1, where 1 indicates that all the incident electromagnetic radiation is reflected. For opaque materials with a smooth surface, $R_f = 1 - A$, where A is the absorptivity. The absorption of the incident laser radiation is also dependent upon its angle of incidence with the metal's surface and the light's polarization. Fig. 3.5 details the effect of the angle of incidence, Ø, on the absorption of the substrate for Ti and Al at wavelengths of 1 μm and 10 μm. The maximum absorption of parallel polarized light by a metal occurs at the Brewster angle, which is wavelength and

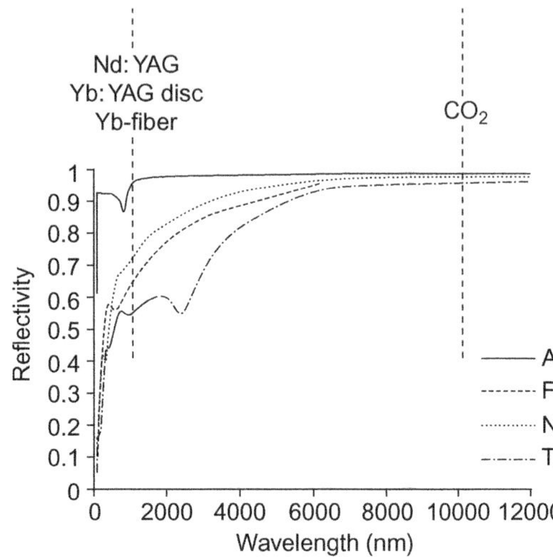

Fig. 3.4 Reflectivity of Al, Fe, Ni and Ti, for a normal angle of incidence and at room temperature, over a range of wavelengths.
Values from D.R. Lide (Ed.), Handbook of Chemistry and Physics, 78th ed., CRC, Boca Raton, FL, 1997.

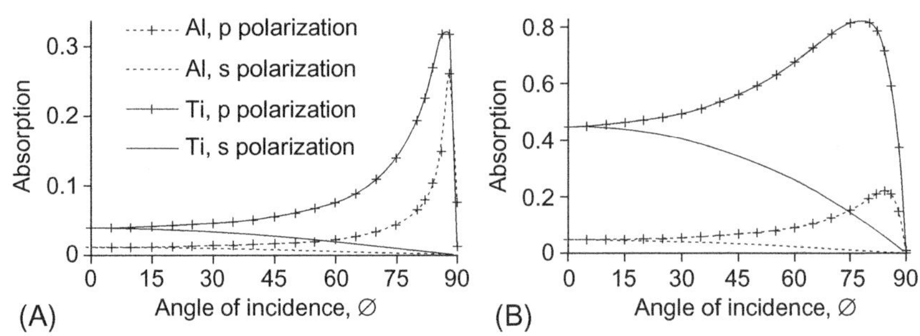

Fig. 3.5 Absorption of Al and Ti at room temperature for parallel (p) and perpendicular (s) polarization as a function of the angle of incidence, Ø and for (A) 10 μm wavelength, and (B) 1 μm wavelength. Note different *y*-axis scales.
Values from D.R. Lide (Ed.), Handbook of Chemistry and Physics, 78th ed., CRC, Boca Raton, FL, 1997.

material dependent. This may be significant when welding with laser sources whose output is polarized in a certain direction, as indicated in research by Sato et al. [18].

The proportion of electromagnetic radiation absorbed by the substrate will change during the interaction time because the absorption is also dependent upon the temperature; as shown for a laser beam of 10.6 μm wave-length in Fig. 3.6 for Al, Cu and Sn. As the temperature of the solid metal rises there is a steady increase in absorptivity. Subsequently, the absorptivity rises appreciably at the melting point of the material. In keyhole laser welding, the absorption of the incident electromagnetic radiation will increase extensively either though multiple Fresnel absorptions at the keyhole walls, or through inverse Bremsstrahlung absorption by the metallic vapor in the keyhole.

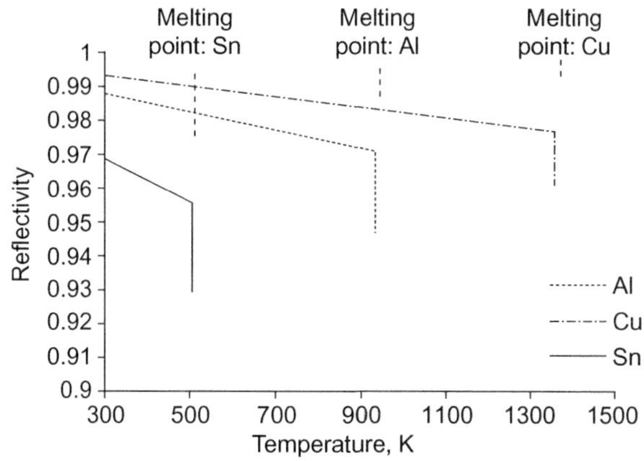

Fig. 3.6 Change in reflectivity of Al, Cu and Sn during interaction with a 10.6-μm laser at a normal angle of incidence as a function of temperature.
Values from M. Brückner, J.H. Schäfer, J. Uhlenbusch, Ellipsometric measurement of the optical constants of solid and molten aluminum and copper at =10.6 μm, J. Appl. Phys. 66 (1989) 1326–1332; M. Brückner, J.H. Schäfer, J. Uhlenbusch, Measurements of the optical constants of solid and molten gold and tin at =10.6 μm, J. Appl. Phys. 70 (1991) 1642–1647.

Eq. (3.7) [19] indicates the depth of absorption, d_α, over which the absorbed intensity reduces by $1/e^2$. The absorption depth of metals is typically less than the wavelength of the incident electromagnetic radiation, since the k value of the refractive index is greater than 1, and therefore for metallic substrates the laser beam can be initially treated as a surface heat source.

$$d_\alpha = \lambda/4\pi k \tag{3.7}$$

3.4.2 Conduction and melting

At very low values of applied laser intensity, there will be insufficient energy deposited at the surface of the substrate for a transition into the liquid phase to take place. The rate at which this thermal energy diffuses through the substrate is characterized by the material's thermal diffusivity, k_d, a property that is related to the thermal conductivity, k_c, the specific heat, c_p and the density, ρ, of the material. If the intensity of the laser radiation incident on the substrate is increased sufficiently ($\sim 10^2$ Wmm^{-2} for most metals), surface melting will begin to occur and a pool of molten material will form. The depth of the melt pool, X, can be approximated by Eq. (3.8) [20] under the assumption that the thermal conductivity and diffusivity in the solid and liquid phases are nominally identical.

$$X(t) \approx 0.16\, P(t - t_m)/\rho L \tag{3.8}$$

where t is the time the laser radiation is applied for, t_m is the time required to reach the melting temperature, T_m, at the surface of the workpiece, P is the absorbed laser power density and L is the latent heat of fusion.

The time taken for the surface of the substrate to reach its melting temperature can be approximated by Eq. (3.9) [20].

$$t_m = \pi k_{cs}{}^2 T_m{}^2 / 4K_{ds} I^2 \tag{3.9}$$

where k_{cs} and k_{ds} are the thermal conductivity and thermal diffusivity of the solid phase.

3.4.3 Vaporization and plasma formation

Vaporization of metallic substrates can occur at applied intensities as low as 10^2 Wmm^{-2} if the interaction time is sufficient. However, applied laser intensities exceeding 10^4 Wmm^{-2} are often used to achieve vaporization of the surface for materials processing applications such as keyhole laser welding, laser cutting and laser drilling. Eq. (3.10) [21] estimates the time taken, T_v, to reach the vaporization temperature T_v, of the substrate.

$$t_v = \pi k_c \rho c_p (T_v - T_0)^2 / 4P \tag{3.10}$$

where T_0 is the ambient temperature.

This initial vaporization of the substrate creates a depression in the molten pool through the recoil pressure exerted by the vapor, thereby forcing the molten metal to the peripheries of the interaction area. It is a crucial stage in the formation of a vapor cavity, or keyhole, in the substrate as it leads to a significant increase in absorption. Efficient coupling of the laser beam into the substrate is then achieved through multiple Fresnel absorptions by the molten material for 1-μm wavelength laser sources. For wavelengths of the order of 10 μm, the absorption method is more complex. Absorption of the incident laser energy by free-free electrons (inverse Bremsstrahlung absorption) is possible at this wavelength, as is Fresnel absorption. The proportion absorbed by each mechanism is a function of the welding parameters [22]. Fig. 3.7 details the formation of a keyhole through a series of high-speed photographs taken when Nd:YAG laser welding of Ti-6Al-4V.

Above it was discussed that the formation of a keyhole in a substrate with a laser beam is dependent on the vaporization of the substrate. For laser beams of ~10 μm wavelength the vapor becomes ionized and the radiation is efficiently absorbed via the inverse Bremsstrahlung process. The proportion of incident radiation absorbed by the plasma, a gas that is constituted of both electrons and ions, depends upon the ratio of the electron, n_e and ion density, n_i, to the density of the vapor atoms, n_o. This can be calculated by Eq. (3.11) [23], which was derived from earlier work by Saha [24], if the gas is assumed to be in local thermodynamic equilibrium.

$$\frac{(n_e n_i)}{n_0} = \left(\frac{g_i g_e}{g_0}\right)\left(\frac{(2\pi m_e k_b T_e)^{(3/2)}}{h^3 \exp^{(E_i/k_b T_e)}}\right) \tag{3.11}$$

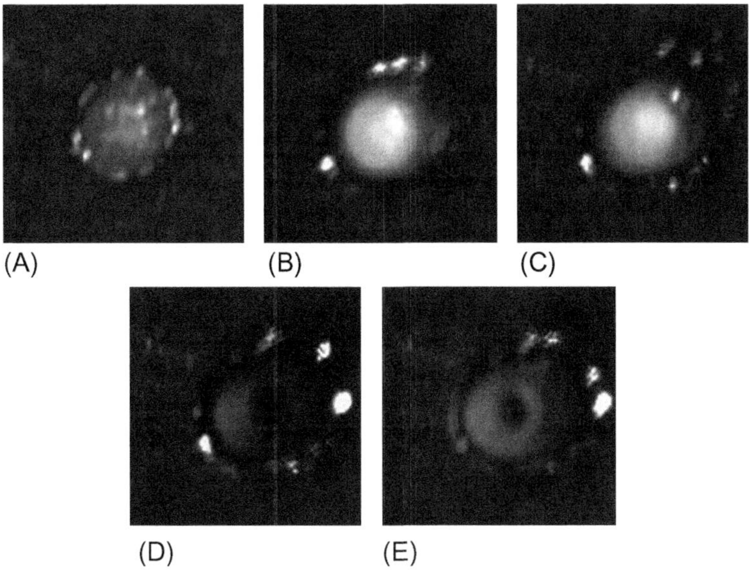

Fig. 3.7 Formation of a vapor cavity in Ti-6Al-4V using a Nd:YAG laser, (A) surface melting, (B, C) vaporization of the substrate occurs, molten metal is pushed to the peripheries and absorption of the laser beam significantly increases, and (D, E) this leads to the formation of a high aspect-ratio vapor cavity.

where, k_b is the Boltzmann constant, m_e is the electron mass, E_i is the ionization energy of the gas, $g_{e,i,o}$ are the degeneracy factors of the electrons, ions and neutral atoms respectively, and T_e is the electron temperature.

It is possible that a plasma may form above the keyhole and attenuate the incident laser radiation, either through inverse Bremsstrahlung absorption or defocusing as a result of a gradient electron density, and lead to a reduction in depth of the keyhole [25]. For laser beams of $\sim$1 μm wavelength, inverse Bremsstrahlung absorption of the laser beam is not a concern. However, it has been shown that when welding with an Nd:YAG laser the beam can be attenuated and defocused by a vapor plume of nano-scale particles, causing a reduction in penetration depth or keyhole instabilities [26].

3.5 Laser welding fundamentals

The absorption of laser light by a metallic substrate and the conduction of the resultant thermal energy leading to possible phase changes in the substrate were detailed in the previous section. These potential phase changes can be used to distinguish between the two fundamental modes of laser welding; conduction-limited and keyhole welding. Only solid-liquid and liquid-solid phase changes occur when laser welding in the conduction-limited mode, whereas during keyhole laser welding the gaseous phase is also present. Furthermore, there is the potential for plasma to be present when keyhole laser welding. Both conduction-limited and keyhole laser welding are capable of joining a wide variety of metallic materials including, but not limited to, aluminum, carbon steels, copper, galvanized steel, nickel, stainless steels and titanium. Several combinations of dissimilar metals can also be joined.

This chapter is primarily concerned with keyhole laser welding, because, if the process is optimized, it is more advantageous to laser weld typical grades and thicknesses of metallic aerospace materials in the keyhole mode than in the conduction-limited mode. Principally, this is a result of the lower heat input and increased processing speeds possible when keyhole laser welding. Table 3.2 summarizes the characteristics and subsequent advantages of using keyhole laser welding in an industrial environment. The advantages associated with the flexibility and repeatability of the process are also valid for conduction-limited laser welding. Disadvantages associated with laser welding include:

- Equipment cost—the cost of laser sources can be high compared with arc-welding equipment, although, the initial cost of equipment is ordinarily offset by an increase in

Table 3.2 Advantages of keyhole laser welding.

Characteristic	Reasons why characteristic occurs	Industrial significance/ advantage
High processing speed	– Laser beam can be narrowly focused – Very high-power laser sources available – Efficient energy transfer into workpiece	– High productivity, potential for cost savings – Possibility for longer weld seams, increasing component stiffness
Low heat input	– Laser beam can be narrowly focused – High-intensity heat source making high processing speeds possible	– High aspect ratio (width: depth) welds – Narrow HAZ – Minimal thermal distortion – Possibility for simpler clamping
Flexible process	– Can operate at atmospheric pressure – Noncontact process – Autogenous process or with filler material – Fiber-optic delivery of laser beam (λ dependent) – Easy robotic automation	– Few/no component size limitations – Complex welding geometries possible – Variety of joint configurations possible (butt, lap, t-butt, etc.)
Repeatability	– Easy robotic automation – Excellent equipment reliability – Laser beam is not affected by magnetic fields	– Accurate reliable welding process

productivity. Recent advances in solid-state laser technology is driving down initial and operating costs, as well as increasing wall-plug efficiency.
- Safety—absorption of laser radiation by the skin, and in particular the retinal hazard region, is of great concern. The exact safety measures are dependent on wavelength and power. BS EN 60825-1:2007 or ANSI Z136.1-2007 should be referred to for exact safety requirements [27, 28].
- Joint fit-up requirements—narrowly focused laser beams may stray from the required joint line through workpiece misalignment or thermal distortion. This tolerance can be increased by using a seam tracking system (potentially including adaptive control) and/or hybrid laser-arc welding.

3.5.1 Conduction-limited laser welding

Conduction-limited laser welding involves only the solid and liquid phases of the substrate and consequently the energy from the incident laser radiation is only absorbed by the surface of the substrate. Subsequently, this thermal energy is transferred from the surface into the bulk of the substrate via thermal conduction, melting occurs and a weld is made when the molten material solidifies. Conduction-limited laser welding can be used to produce spot or seam (either through overlapping spots or a continuous process) welds.

Since the process relies solely on conduction, the weld depths possible are limited by the thermal conductivity of the substrate. A power density of approximately 10^2–10^4 Wmm^{-2} is ordinarily sufficient for conduction-limited welding to be performed. The resulting fusion zone has a hemispherical weld profile, with a width exceeding the depth by a factor of $\sim$2. A significantly larger heat-affected zone (HAZ) compared with welds made using keyhole laser welding also occurs. For conduction-limited spot welds Eq. (3.8) can be used to approximate the depth of penetration. This equation is subject to certain boundary conditions so that only parameters that cause melting are chosen. It has been reported [29] that fully penetrating welds in Al (2000 series) at least 6.35 mm in thickness can be produced by conduction-limited laser welding if the focused intensity is optimized, such that the surface temperature of the weld pool is just below the vaporization temperature. In comparison with keyhole laser welding, conduction-limited laser welding is an inherently stable process. High-integrity welds with few, or no, defects can therefore be more easily produced when conduction-limited laser welding. Shows a partial-penetration conduction-limited laser weld in 8 mm thickness 2024 aluminum alloy, produced with a Nd:YAG laser.

Accurate control of the melt-pool temperature, and hence the heat input, is required to ensure the penetration depth remains constant. Figs. 3.4 and 3.5 show that the absorption of the laser beam by a particular substrate is dependent upon the wavelength of the beam and its angle of incidence with the workpiece. A result of this is that laser beams of wavelengths $\leq$1 μm are more suited to conduction-limited welding of metallic substrates, as they are more readily absorbed, than longer wavelength laser beams. It is also known that the proportion of the incident laser light absorbed is a function of the temperature of the substrate (Fig. 3.6). An accurate knowledge of the temperature-absorptivity relationship, as well as the temperature-dependent values of thermal conductivity and effective viscosity would allow an approximate heat input to be calculated [30]. Furthermore, it is crucial that the temperature of the substrate does not exceed its vaporization temperature, resulting in a significant increase in absorption and the formation of a vapor cavity. However, in practice these relationships are not known and it is not easy to determine them. Real-time monitoring and feedback of the weld-pool temperature is a possible approach to controlling penetration depth when conduction-limited laser welding [31].

3.5.2 Keyhole laser welding

Keyhole laser welding (also referred to as deep-penetration laser welding) is similar in concept to electron-beam welding, in that a vapor cavity is formed in the substrate and subsequently traversed across it. A liquid sheath surrounds the vapor cavity, or keyhole, which is in turn surrounded by the solid substrate. The keyhole is primarily maintained by the pressure of the vapor within it. A portion of this vapor is ejected from the keyhole and therefore a steady-state cannot be achieved with a stationary keyhole, as ultimately it will fully penetrate the substrate and the vaporized material that is ejected cannot be replenished to sustain the vapor pressure.

However, a quasi steady-state can be considered for a moving keyhole. As the keyhole is traversed through the substrate, the sheath of molten material surrounding it is continuously transported from the region in front of the keyhole to the trailing weld pool. The dominant transportation process is the flow of molten material around the keyhole, although a proportion of the molten material is vaporized and transported across the keyhole maintaining the vapor pressure and potentially producing a quasi steady-state. Thermal conduction in the direction of travel ensures the continuous replenishment of the molten material. Fig. 3.8 shows the formation of the keyhole

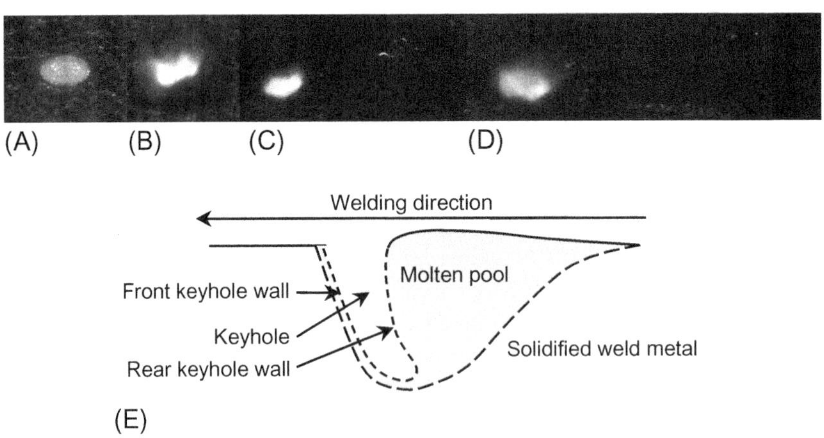

Fig. 3.8 Formation of keyhole laser-welding process in C–Mn steel, (A) surface melting, (B) vaporization of substrates occurs, (C) keyhole traverses across the workpiece and weld pool begins to form, and (D) the weld-pool length increases and stabilizes; (E) schematic of the side view of a keyhole.

laser-welding process in C–Mn steel, and a schematic diagram of a keyhole and molten-pool geometries. The keyhole is not cylindrical in shape and has a characteristic curve to it, which is determined by the absorption mechanism, travel speed and the thermal conductivity of the substrate. Analogous to conduction-limited laser welding, keyhole laser welding can be used for either spot or seam welding. Most typically a continuous process is employed to maximize the potential advantages of the process, although pulsed laser sources are particularly suited to spot-welding applications. Characteristic profiles of keyhole laser welds produced in Ti-6Al-4V with a 1 μm laser source are shown.

Power intensities of $>10^4$ $\mathrm{Wmm^{-2}}$ are ordinarily sufficient for a keyhole to be initiated in the substrate, although vaporization is possible at lower power densities if the values of welding speed and thermal conductivity of the substrate are conducive. Eq. (3.12) [32] can be used to approximate the critical power required for keyhole laser welding with a Gaussian beam, as a function of welding speed, v.

$$P = \rho\pi(2k_d v)^{0.5} {w_0}^{1.5} E_v \tag{3.12}$$

where E_v is the energy required per kilogram to vaporize the material.

This relationship should be treated as an approximation only, since it is known from empirical evidence that oscillations in the keyhole behavior may occur even with a constant set of parameters [33], making a quasi steady-state particularly difficult to achieve. Variations in the keyhole and weld-pool behavior may lead to certain weld defects, such as intermittent penetration and sub-surface porosity. Consequently, the formation and subsequent dynamic behavior of a keyhole has been the subject of intense theoretical and practical investigations (for example [33–37]). Thorough details of the keyhole and weld-pool dynamics is particularly extensive and challenging, and the published investigations of the aforementioned researchers should be referred to for an excellent understanding of this area. Fundamentally, the transient behavior of the keyhole is dependent upon the forces acting to maintain it, and those tending to close it. Kroos et al. [38] summarized that the forces maintaining the keyhole are the evaporative and radiative pressures, while those acting to close it are the hydrostatic and hydrodynamic pressures of the surrounding molten material and its surface tension. Optimization of the process parameters is crucial if high-quality welds are to be produced with the keyhole-laser-welding process. Table 3.3 specifies the crucial process parameters that should be considered when keyhole laser welding, either autogenously or with filler material. Further process parameters will emerge if more complicated processes are chosen, such as hybrid laser-arc welding and dual-focus keyhole laser welding.

Theoretical research by Kroos et al. [38] indicated that keyhole oscillations may arise owing to a pressure imbalance in the keyhole. It was previously discussed that a portion of the incident laser radiation may be attenuated by a plasma or a vapor plume, therefore providing one possible mechanism for keyhole instability and the formation of weld defects. Keyhole laser welding performed under vacuum has shown that plasma formation can be easily suppressed [39]. Unfortunately, this limits the inherent flexibility of laser welding since, unlike electron-beam welding, it can be

Table 3.3 Potential process parameters for keyhole laser welding.

	Parameters	
Laser source/ focused beam	Wavelength Power Focused spot size Beam quality Stand-off distance Pulse time (pulsed laser output) Pulse frequency (pulsed laser output) Modulation frequency (modulated output)	Polarization Welding speed Focal plane position Depth of focus Pulse energy (pulsed laser output) Pulse shape (pulsed laser output) Modulation waveform (modulated output) Modulation duty cycle (modulated output)
Workpiece	Surface preparation Joint geometry Jigging/fixturing Shielding gas type Shielding gas arrangements	
Filler material	Chemical composition Type (wire or powder) Gauge/diameter Feed rate Position with respect to welding process	

performed at atmospheric pressure. When keyhole laser welding with a CO_2 laser beam, helium gas (>99.995% purity) is ordinarily used to shield the welding process, as its ionization potential and thermal conductivity are higher than other shielding gases (such as argon and nitrogen), which will inhibit the formation of plasma outside the keyhole. However, keyhole fluctuations may still occur when using helium as a shielding gas, resulting in weld defects [40]. Miyamoto et al. [41] reported that a jet of inert gas, directed toward the laser-material interaction point, could be used to further prevent the plasma formation above the keyhole when CO_2 laser welding, and produce weld beads with few defects. The mechanisms for beam attenuation when welding with 1 μm wavelength lasers are not related to ionization, and, therefore, shielding gases with particularly high ionization potentials are not required. Nevertheless, a directed jet of gas has also been reported to reduce keyhole fluctuations and improve weld quality when welding with 1 μm wavelength laser beams [42, 43].

Attenuation of the incident laser radiation by a plasma or a vapor plume is not the only mechanism for defects to occur when keyhole laser welding. A theoretical study by Matsunawa and Semak [44] indicated that a small discontinuity of the front-keyhole-wall angle may cause localized absorption, as a result of the increased angle of incidence. Consequently, humps of molten material may be formed on the front keyhole wall that will be driven down the wall to the root of the keyhole causing inconsistent penetration. If the velocity of the humps is particularly high then weld-pool volumetric oscillations may be generated, raising the potential for more defects. Furthermore, the localized evaporation will cause a large dynamic pressure locally exerted on the rear keyhole wall that may cause keyhole instability and generate further defects in the weld. This has been observed using an X-ray transmission imaging system when laser-keyhole welding A5083, and also in C–Mn and stainless steels [37].

Despite the reported mechanisms for introducing keyhole instability and hence weld defects, techniques have been developed to suppress key-hole instabilities that cannot be controlled through optimization of process parameters, and produce high-integrity welds in a variety of metallic materials. In addition to the methods using shielding gases of high ionization potential and/or directed gas jets, power modulation and dual-focused laser beams have been reported to reduce the occurrence of weld defects.

Geometrical defects in the weld profile, such as undercut and concavity at the weld face or root, are particularly undesirable for components subject to dynamic loading. The defects may act as stress concentrators and subsequently be initiation sites for fatigue cracks. Fundamentally, the defects can either be attributed to the laser-welding process or to the joint configuration/restraint. For instance, joint misalignment, where the laser beam wanders from the joint line resulting in a portion of the weld seam not being welded, is an example of a defect that is not related to the welding process. It may occur when the laser beam is not correctly aligned with the joint line or when the component is not correctly clamped allowing movement of the joint during welding. Consequently, this defect is more commonly observed in keyhole laser welding than in conduction-limited laser welding, as the finely focused beams are more susceptible to missing the joint line. Incorrect joint configuration/restraint, errors in gap fit-up or inadequate machining of the abutting edges may lead to other geometrical defects in the weld profile; specifically [19] burn through, drop out or loss of penetration.

The above defects can be eliminated through the adoption of adequate weld clamping and workpiece-preparation procedures. Geometrical weld-profile defects associated with the laser-welding process, such as humping, undercut and concavity, may occur as a consequence of incorrect welding parameters. Often, a balance of process parameters is found to achieve the required weld penetration and minimize weld-profile defects. Defects in the weld profile that cannot be eliminated through alteration of process parameters may be removed by further processing and/or machining. Filler material, either powder or wire, may be added during the process, or a low-power cosmetic pass utilized to re-shape the top bead. Workpieces that are thicker at the joint could be used, which would allow postweld machining to eliminate the defects without having an undersized weld.

3.6 Laser weldability of titanium alloys

The properties of titanium and its alloys, such as high specific strength and corrosion resistance, are ideal for high-performance aerospace applications, where the relatively high material cost can be justified. They are often used in applications requiring high specific strength, or where metallurgical stability and high strength are required at elevated operating temperatures. Keyhole laser welding of titanium alloys should be particularly advantageous as its low thermal conductivity (~20 $Wm^{-1} K^{-1}$) and high absorption of infra-red light will result in increased weld penetrations than other metallic materials.

A result of the high-focused power density combined with high processing speeds when keyhole laser welding, is that the weld metal undergoes a particularly fast thermal cycle compared with other fusion welding processes. Cooling rates exceeding 410 °C s^{-1} are easily achievable, which for Ti-6Al-4V results in a completely martensitic microstructure [45]. Compared with arc-welding processes, keyhole laser welding of Ti-6Al-4V produces finer martensite needles with a smaller content of alloying elements [46]. Yunlian et al. [47] compared the weld microstructures produced when keyhole laser welding 0.5 mm thickness commercially pure titanium with a CO_2 laser, to those from electron-beam and Tungsten inert gas (TIG) welds. The grain size produced when keyhole laser welding, a fine acicular α structure was observed, was smaller than those produced when electron-beam or TIG welding. This is a direct result of the increased cooling rates possible when keyhole laser welding. Li et al. [48] investigated the resultant microstructures when Yb-fiber laser welding of commercially pure titanium. At increased welding speeds, and hence increased cooling rates, the content of fine-grained acicular α present at the weld centerline increased. The potential also exists for a number of defects to occur, including weld-bead embrittlement, cracking, geometrical defects in the profile and weld-metal porosity.

3.6.1 Embrittlement

Titanium has an elevated affinity for light elements (such as hydrogen, nitrogen and oxygen) at temperatures exceeding 500 °C, which may result in embrittlement of the weld metal if they are absorbed. The discoloration of the weld metal can be used as an indicator of the shielding adequacy, whereby the weld metal follows the color sequence [49]: silver (indicating no discoloration), light straw, dark straw, bronze, brown, violet, green, blue, grey and white (indicating heavy discoloration and embrittlement). This discoloration is directly related to the degree of weld-metal embrittlement and hardness of the weld metal. However, it is imperative that the weld-bead discoloration not be utilized as an inspection tool for shielding adequacy, as the discoloration sequence will repeat as the oxidation thickness increases [50].

To avoid embrittlement titanium alloys are ordinarily shielded with a high-purity inert gas when they are laser welded; argon and helium are the only shielding gases that should be considered when laser welding titanium alloys. As discussed previously, helium is often used when CO_2 laser welding because of its high ionization potential, and argon can be used when welding titanium alloys with a Nd:YAG laser, as 1 μm wavelength laser radiation does not easily ionize the shielding gas. The shielding gas is often delivered through a trailing shield that covers the weld face. This is particularly well documented in the available literature and numerous different designs have been reported to provide adequate coverage (for example [51, 52]).

A typical trailing shield is shown. Protection of the weld root is usually performed with an efflux channel, also supplied with an appropriate flow rate of a suitably pure inert gas. If the weld speed is sufficiently low, shielding gas delivered through a co-axial or lateral nozzle may provide adequate shielding of the weld pool and weld metal. Such an approach is often used when welding titanium alloys with a pulsed laser output (e.g. [53]).

The effects of oxygen contamination (0.001–10%) in argon shielding gas was studied by Li et al. [54] when welding commercially pure titanium with a 1-μm wavelength laser beam. At increased oxygen contents the weld-bead discoloration followed the sequence outlined in AWS D17.1, which also corresponded to a change in the surface hardness. Fig. 3.9 details the hardness data reported by Li et al. [54]. A mixed coarse-grained serrated α and a small amount of fine-grained acicular α were observed in the weld microstructure when welding with high-purity argon shielding gas. The microstructure became dominated by acicular and platelet α as the oxygen content in the argon shielding gas increased above 2%; which had the effect of decreasing the tensile strength of the weld metal as a result of brittle fracture caused by the formation of acicular and platelet α.

3.6.2 Cracking

Titanium alloys are, in general, not considered susceptible to solidification cracking since they contain low concentrations of impurities. However, research has suggested [55] that titanium alloys may be vulnerable to solidification cracking if they are incorrectly restrained during welding. It is likely that the degree of vulnerability is related to the amount of back diffusion of solute-elements in the solid [55]. Of more concern when welding titanium alloys is contamination cracking, which may occur if the weld

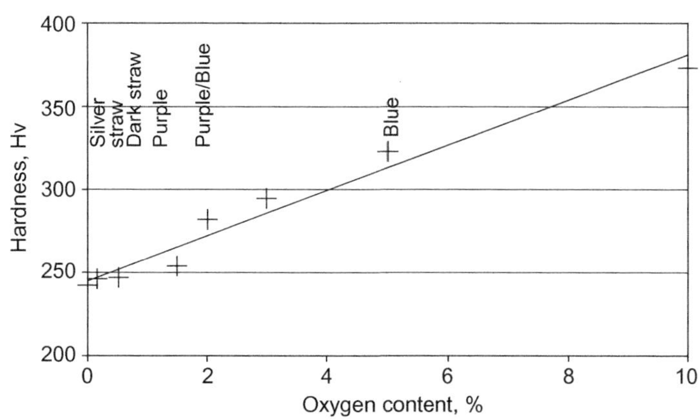

Fig. 3.9 Surface color and hardness of Nd:YAG laser welds made in 0.5 mm thickness commercially pure titanium sheets with varied oxygen content in the argon shielding gas. Data from X. Li, J. Xie, Y. Zhou, Effects of oxygen contamination in the argon shielding gas in laser welding of commercially pure titanium thin sheet, J. Mater. Sci. 40 (2005) 3437–3443.

metal is exposed to either light elements at temperatures exceeding 500 °C or iron particles during welding. When absorbed, the light elements, such as hydrogen, nitrogen and oxygen, will migrate to interstitial sites and may cause cracking as a result of the welding stresses. Particularly high levels of these elements are required in the welding atmosphere, for example 3000 ppm oxygen in the weld metal may cause transverse cracking [56]. As previously discussed, the presence of light elements in the vicinity of the welding process and cooling weld bead may be reduced by using an effective trailing shield and a high-purity inert shielding gas with a low dew point. Particles of iron on the workpiece may be dissolved into the melt pool and, potentially, causing embrittlement of the weld metal [56]. An appropriate materials preparation and handling procedure will minimize the risk of iron particle contamination.

3.6.3 Hydrogen porosity

Lakomski and Kalinyuk [57] determined the hydrogen solubility in titanium as a function of temperature. A sharp decrease in solubility was reported at the solidification point, which promotes the rejection of small diameter (<0.1 mm) hydrogen gas bubbles close to the fusion zone/HAZ boundary as the weld solidifies. Typically, a hydrogen content >210 mL/100 g would be required for hydrogen bubbles to be rejected. Modern titanium alloys have hydrogen contents much less than this specified value and therefore other potential hydrogen sources are of greater concern, specifically: the hygroscopic titanium oxide layers that may be present on the workpiece surfaces, the shielding gas and filler materials.

3.6.3.1 Workpiece preparation

Titanium oxide is hygroscopic and will form on the surface of titanium alloys if sufficient levels of oxygen are present in the media surrounding it. Removal of hydrated layers prior to welding is crucial in minimizing the potential hydrogen content of the melt pool. Consequently, the effectiveness of the method used to remove the hydrated layer and other surface contaminants may have a large influence on the formation of weld-metal porosity.

Mechanical cleaning methods are often employed because they are relatively straightforward processes compared with chemical pickling. Mueller et al. [52] produced welds with small amounts of weld-metal porosity when laser welding Ti-6Al-4V, if the joint was cleaned with a stainless-steel brush prior to welding. It should be noted there is a slight risk of iron pick-up with stainless-steel brushes, and titanium brushes should be used for critical applications [58]. Autogenous melt runs and butt welds with very low levels of porosity were also produced in titanium alloys up to 9.3 mm in thickness by Hilton et al. [59]. In this work the surfaces were cleaned with an abrasive pad and acetone degreased prior to welding, and abutting joint edges were dry machined. However, it should be noted that in both of the above publications [52, 59] a directed gas jet was also used to achieve the low levels of weld-metal porosity. Chemical pickling of titanium alloys prior to welding is usually performed with an aqueous solution of hydrofluoric and nitric acid [58]. In comparison with mechanical cleaning methods, a more uniform surface finish is possible and there is less dependency on the operator. A comparison of different pickling solutions was performed by Gong et al. [51], who concluded that 10%HF–30%HNO_3–60%H_2O was the most suitable formula for removing the oxide layer of the titanium alloy BT20.

It is difficult to directly compare the effectiveness of the different surface preparation techniques from the above, since there are a number of other process variables that must be accounted for. Nevertheless, it appears from the published research that mechanical cleaning with an abrasive pad, scratch brushing with a stainless steel or titanium brush and chemical pickling with an appropriate aqueous solution of hydrofluoric and nitric acid are all appropriate methods of removing hydrated layers. The time between welding and workpiece preparation should be minimized, as the oxide layer will continue to adsorb moisture.

3.6.3.2 Shielding gas

As mentioned previously, titanium has a high affinity for light elements at temperatures of ~500 °C and above. Consequently, rigorous inert-gas shielding is used when welding titanium alloys to prevent discoloration and possible embrittlement of the weld metal. However, the inert shielding gases are a potential source of hydrogen, which may be absorbed by the melt pool as it cools prior to solidification. Therefore, an increased amount of hydrogen present in the shielding gases may increase the possibility of hydrogen porosity present in the solidified weld metal. Shielding gases with a low dew point should be used when welding titanium alloys to reduce the possibility of hydrogen porosity. Adequate pre- and postweld purging times should be established to minimize the amount of hydrogen present in the shielding shoe.

3.6.3.3 Filler material

In certain instances, wire addition may be required to correct geometric defects present in the weld profile, such as undercut at the weld face and/or root. Investigations performed by Gorshkov and Tret'Yakov [60] when arc welding titanium alloys have shown that increased hydrogen content in the filler metals tended to increase weld-metal porosity. Many welding consumables are now extra-low-interstitial grade with nominal hydrogen contents. As a result, the hydrogen content of welding consumables should not be of concern provided that a suitable grade is chosen. However, as with the parent material, a hydrated layer can form on the surface of titanium welding wires, which may need to be removed prior to welding.

3.6.4 Processing porosity and its prevention

Despite minimizing potential sources of hydrogen, porosity can still form in the weld metal when keyhole laser welding. In particular, larger pores have been observed that cannot be attributed to hydrogen precipitation as the cooling rate is too high [61]. It has been reported [62] that keyhole instability can lead to metal vapor and/or inert shielding gases being trapped in the weld metal. Keyhole instabilities occur when the forces tending to keep the keyhole open are not in equilibrium with those trying to close it. The majority of the work that has studied keyhole instabilities has been performed using C steels and aluminum alloys. The same keyhole instability mechanisms that were discussed in Section 3.5 are most likely valid for titanium alloys. For instance, Caiazzo et al. [63] reported that shielding of the weld pool with helium when CO_2 laser welding results in a higher weld quality than when using argon as a shielding gas. As discussed previously, this is a result of the lower ionization potential of argon, which promotes inverse Bremsstrahlung absorption at lower irradiances than when helium shielding is used. The use of helium shielding gas can therefore reduce the variation in the vaporization pressure and increase keyhole stability/internal weld quality. Other reported methods of successfully preventing porosity formation when keyhole laser welding titanium alloys include: a directed inert gas jet, laser power modulation and using a dual-focus beam configuration.

3.6.4.1 Directed inert-gas jet

Both Denney and Metzbower [64] and Li et al. [65] have reported the use of a directed inert-gas jet to control the welding process when keyhole laser welding titanium alloys. More recently Hilton et al. [59] and Mueller et al. [66] have quantified the effects of using a directed gas jet when keyhole laser welding titanium alloys up to 9.3 mm in thickness. Hilton et al. [59] reported that butt welds in Ti-6Al-4V in thicknesses up to 9.3 mm, with internal porosity contents lower than specified in company specific aero engine weld criteria and significantly lower than detailed in AWS D17.1:2001 Class A, could be produced with a 7 kW Yb-fiber laser, if a directed jet argon was accurately positioned toward the laser-material interaction point. An extension of the work performed by Hilton et al. [59] has indicated, through a parametric study, that the position and flow rate of this inert gas side jet must be controlled within a stringent tolerance range, if the weld quality is to be consistently reproduced [67].

3.6.4.2 Laser power modulation

Modulation of the output power is an accepted method of reducing weld-metal porosity when laser welding metallic materials other than titanium alloys. Kuo and Jeng [68] reported that modulating the output power of an Nd:YAG laser reduced the resultant porosity levels when welding 3.0 mm thickness SUS 304L and Inconel 690, compared with a continuous-wave output power. The same effect has also been reported when CO_2 laser welding A5083 [69]. They attributed the reduction in porosity as owing to pores entrapped in the weld pool generated by one laser pulse having a second chance to escape, when a part of the weld was remelted by the subsequent pulse. A comparison of modulation amplitudes by Kawaguchi et al. [70] indicated that larger modulation amplitudes were more effective in reducing the occurrence of porosity when CO_2 laser welding SM490C. The potential for producing high-quality keyhole Nd:YAG laser welds in titanium alloys by modulating the output power has been evaluated by Blackburn et al. [71]. Butt welds with an excellent internal quality were produced when a square waveform was used with a modulation frequency $\geq$125 Hz and a duty cycle of 50%. Analysis of the keyhole and emitted vapor plume behaviors suggested that they both exhibited the same periodic tendencies, and with the correct parameters an oscillating wave can be set up in the weld pool that appears to manipulate the vapor plume behavior and reduce unintended keyhole instabilities. Furthermore, the oscillations in the molten metal will act to agitate the molten metal and may encourage the trapped bubbles to float to the surface, reducing porosity.

3.6.4.3 Dual-focus laser-beam configuration

The keyhole behavior and weld pool shape may be manipulated if a dualfocus (i.e. two focused laser beams) focusing arrangement is adopted. A number of researchers have reported that welding with a dual-focus laser beam is an effective method of reducing weld-metal porosity in metallic materials (for example [72–74]). Direct observation of the plasma behavior when CO_2 laser welding 5052 aluminum alloy has indicated that there was less variation in its behavior when a dual-focus technique was used, compared with a conventional single-focused laser beam [74]. Therefore transient variations in the proportion of incident laser radiation attenuated will be reduced and a more stable keyhole is possible. Hayashi et al. [73] researched the effect of an in-line dual-focus configuration for CO_2 laser welding of austenitic stainless steel SUS 304. Observation of the key-hole using X-rays showed that the keyholes coalesced to form one large keyhole and it was suggested that this was more resistant to changes in keyhole forces and subsequently reduces the generation of bubbles and hence the formation of porosity. The feasibility of reducing the formation of porosity when Nd:YAG laser welding titanium alloys by using a dualfocus technique was reported by Blackburn et al. [75]. The effects of the foci orientation, foci separation, welding speed and power distribution ratio on the resulting porosity and weld profile were examined. It was found that both transverse and in-line foci orientations, with controlled foci separations and welding speeds, can be used to establish a stable keyhole and vapor-plume regime, promoting low-porosity welds. The weld-metal porosity levels were reduced to within levels stipulated in the stringent aerospace weld quality criteria.

3.7 Future trends

A significant proportion of the literature referenced in this chapter is concerned with the utilization of CO_2 or Nd:YAG rod laser sources. In comparison with Nd:YAG rod laser sources, superior beam qualities at higher output powers are possible from CO_2 laser sources. Thermal lensing effects, caused by temperature-induced changes in the refractive index of the gain medium, are difficult to overcome in Nd:YAG rod lasers. This limits the maximum available output power that can achieved while still maintaining a useable beam quality. In recent years, significant advances in solid-state laser technologies have been achieved through the development of the Yb:YAG disc and Yb-fiber laser sources. These laser sources are capable of emitting laser beams with a beam quality comparable to CO_2 laser sources and at powers exceeding 10 kW. Fundamentally, both sources have increased the aspect ratio of the gain medium (in disc, diameter >> length; in fiber length >> diameter) to achieve efficient cooling and reduce thermal-lensing effects. Furthermore, the wall-plug efficiencies of these laser sources is >20%, achieved through efficient cooling and diode laser pumping of the gain medium, and claimed service intervals are particularly long. As with Nd:YAG rod laser sources, the ~1-μm wavelength emitted radiation can be focused down an optical fiber, providing increased flexibility at the workpiece and enabling the laser source to be located many meters away from the production line. The excellent beam quality offered by Yb:YAG disc and Yb-fiber lasers are allowing increased processing speeds to be achieved. Verhaeghe and Dance [76] reported that the high-quality beams from Yb-fiber and Yb:YAG laser sources are now capable of producing welds with an aspect ratio that only previously could have been produced with in-vacuum electron-beam welding. From market data [12] it is evident that these laser sources, along with direct diode lasers, are attracting an increasing share of the laser market for metal processing.

3.8 Sources of further information and advice

In writing this chapter, the author has sourced material from numerous journals and conference proceedings. These have been referenced accordingly and are excellent sources of further detailed information concerning laser welding of metallic materials. A comprehensive list of published literature on the subject of laser welding of other metallic aerospace materials would be particularly lengthy and cannot be reproduced here. A sparse few are detailed below, which can be used to gain an understanding of how the laser-welding process differs for different metals, and act as a starting point for locating other important research.

For literature concerned with laser welding of aluminum alloys, the reviews by Cao et al. [77, 78] are a good starting point. Research performed by Brenner et al. [79] and Dittrich et al. [80] provide information on the application of laser welding to produce aluminum fuselage panels, and published studies by Allen et al. [81] and Verhaeghe et al. [82] give details regarding the effects of certain process parameters on weld quality when laser welding aluminum aerospace alloys. Cao et al. [83] gives a summary of laser welding techniques for magnesium alloys.

The books in the Further Reading section should be consulted for further details concerning laser sources and laser welding.

References

[1] R.G. Gould, The LASER, light amplification by stimulated emission of radiation, in: The Ann Arbor Conf. on Opti. Pumping, 1959, p. 128.
[2] T.H. Maiman, Stimulated optical radiation in ruby, Nature 187 (1960) 493–494.
[3] A. Einstein, Strahlungs-emission und -absorption nach der quantentheorie, Verhandl. Dtsch. Phys. Ges. 18 (1916) 318–323.

[4] R.W. Ladenburg, Anomalous dispersion in excited gases. I. Proof of the dispersion formula in excited neon, Z. Phys. 48 (1928) 15–25.
[5] A.L. Schawlow, C.H. Townes, Infrared and optical masers, Phys. Rev. 112 (1958) 1940–1949.
[6] A. Javan, Possibility of production of negative temperature in gas discharges, Phys. Rev. Lett. 3 (1959) 87–89.
[7] R.N. Hall, G.E. Fenner, J.D. Kingsley, T.J. Soltys, R.O. Carlson, Coherent light emission from GaAs junctions, Phys. Rev. Lett. 9 (1962) 366–368.
[8] J.E. Geusic, H.M. Marcos, L.G. Van Vitert, Laser oscillations in Nd doped yttrium aluminium, yttrium gallium and gadolinium garnets, Appl. Phys. Lett. 4 (1964) 182–184.
[9] C.K.N. Patel, Continuous-wave laser action on vibrational-rotational transitions of CO_2, Phys. Rev. Lett. 136 (1964) 1187–1193.
[10] C.J. Bahun, R.D. Eng-Quist, Metallurgical applications of lasers, Proc. Nat. Elecron. Conf. 18 (1962) 607–619.
[11] H. Schwarz, A.J. DeMaria, Electron, ion, and light beams as present and future material working tools, in: Proc. of the Nat. Electron. Conf, 1962, pp. 351–365.
[12] D. Belforte, Economic review and forecast—the worst is over, Ind. Laser Solut. 25 (2010) 4–10.
[13] D. Belforte, Economic review and forecast—keeping the economy in perspective, Ind. Laser Solut. 24 (2009) 4–11.
[14] L.P. Earvolino, J.R. Kennedy, Laser welding of aerospace structural alloys, Weld. J. 44 (3) (1966) 127S–134S.
[15] W.M. Steen, Laser Material Processing, second ed., Springer-Verlag, London, 1998.
[16] ISO 11146–2:2005, Lasers and Laser-Related Equipment—Test Methods for Laser Beam Widths, Divergence Angles and Beam Propagation Ratios—Part 2: General Astigmatic Beams, 2005.
[17] D. Havrilla, Process Fundamentals of Industrial Laser Welding and Cutting, Rofin-Sinar Inc, Plymouths, MI, 2002.
[18] S. Sato, K. Takahashi, B. Mehmetli, Polarization effects of a high power CO_2 laser beam on aluminum alloy weldability, J. Appl. Phys. 79 (1996) 8917–8919.
[19] W.W. Duley, Laser Welding, Wiley, New York, 1999.
[20] M.I. Cohen, Melting of half-space subjected to constant heat input, J. Frankl. Inst. 283 (1967) 271–285.
[21] J.F. Ready, Industrial Applications of Lasers, second ed., Academic Press, San Diego, 1997.
[22] P. Solana, G. Negro, A study of the effect of multiple reflections on the shape of the keyhole in the laser processing of materials, J. Phys. D Appl. Phys. 30 (1997) 3216–3222.
[23] A.R. Hochstim, Kinetic Processes in Gases and Plasmas, Academic Press, New York, 1969.
[24] M.N. Saha, Ionisation in the solar chromospheres, Philos. Mag. 40 (1920) 472–488.
[25] A. Poueyo-Verwaerde, R. Fabbro, G. Deshors, A.M. de Fruttos, J.M. Orza, Experimental study of laser induced plasma in welding conditions with continuous CO_2 laser, J. Appl. Phys. 74 (1993) 5773–5780.
[26] J. Greses, Plasma Effects in CO_2 and Nd:YAG Laser Welding, Ph.D. Thesis, University of Cambridge, Cambridge, 2003.
[27] ANSI Z136.1-2007, American National Standard for Safe Use of Lasers, 2007.
[28] BS EN 60825-1:2007, Safety of Laser Products – Part 1: Equipment Classification and Requirements, 2007.
[29] S.W. Williams, G. Scott, N.J. Calder, Direct diode laser welding of aerospace alloys, LaserOpto 33 (4) (2001) 50–54.

[30] A. De, T. DebRoy, Improving reliability of heat and fluid flow calculation during conduction mode laser spot welding by multivariable optimisation, Sci. Technol. Weld. Join. 11 (2006) 143–153.
[31] F. Bardin, S. Morgan, S. Williams, R. McBride, A.J. Moore, D.C. Jones, D.P. Hand, Process control of laser conduction welding by thermal imaging measurement with a color camera, Appl. Opt. 44 (2005) 6841–6848.
[32] G. Qin, X. Qi, S. Lin, Critical power of keyhole formation in CW Nd:YAG laser, China Weld. 16 (2007) 36–40.
[33] Y. Arata, N. Abe, T. Oda, Beam hole behaviour during laser beam welding, LIA 38 (1984) 59–66.
[34] J.M. Dowden, P. Kapadia, R. Ducharme, Analytical model of deep penetration welding of metals with a continuous CO_2 laser, Int. J. Join. Mater. 7 (1995) 54–61.
[35] R. Fabbro, K. Chouf, Dynamical description of the keyhole in deep penetration laser welding, J. Laser Appl. 12 (2000) 142–148.
[36] A. Kaplan, Model of deep penetration laser welding based on calculation of the keyhole profile, J. Phys. D Appl. Phys. 27 (1994) 1805–1814.
[37] A. Matsunawa, J.D. Kim, N. Seto, M. Mizutani, S. Katayama, Dynamics of keyhole and molten pool in laser welding, J. Laser Appl. 10 (1998) 247–254.
[38] J. Kroos, U. Gratzke, G. Simon, Towards a self-consistent model of the keyhole in penetration laser beam welding, J. Phys. D Appl. Phys. 26 (1993) 474–480.
[39] Y. Arata, N. Abe, T. Oda, N. Tsujii, Fundamental phenomena during vacuum laser welding, LIA 44 (1985) 1–7.
[40] N. Seto, S. Katayama, A. Matsunawa, High-speed simultaneous observation of plasma and keyhole behaviour during highs power CO_2 laser welding—effect of shielding gas on porosity formation, in: Proc. 19th ICALEO, USA, E, 1999, pp. 19–27.
[41] I. Miyamoto, H. Maruo, Y. Arata, Role of assist gas in CO_2 laser welding, in: Proc. Int. Conf of Appl. of Lasers and Electro-Opt., ICALEO, Boston, USA, 1984, pp. 68–75.
[42] R. Fabbro, S. Slimani, I. Doudet, F. Coste, F. Briand, Experimental study of the dynamical coupling between the induced vapour plume and the melt pool for Nd-yag CW laser welding, J. Phys. D Appl. Phys. 39 (2006) 394–400.
[43] K. Kamimuki, T. Inoue, K. Yasuda, M. Muro, T. Nakabayashi, A. Matsunawa, Prevention of welding defect by side gas flow and its monitoring method in continuous wave Nd:YAG laser welding, J. Laser Appl. 14 (2002) 136–145.
[44] A. Matsunawa, V. Semak, The simulation of front keyhole wall dynamics during laser welding, J. Phys. D Appl. Phys. 30 (5) (1997) 798–809.
[45] T. Ahmed, H.J. Rack, Phase transformation during cooling in $\alpha+\beta$ titanium alloys, Mater. Sci. Eng. A 243 (1998) 206–211.
[46] A. Costa, R. Miranda, L. Quintino, D. Yapp, Analysis of beam material interaction in welding of titanium with fiber lasers, Mater. Manuf. Process. 22 (2007) 798–803.
[47] Q. Yunlian, D. Hu, H. Quan, Z. Liying, Electron beam welding, laser beam welding and gas tungsten arc welding of titanium sheet, Mater. Sci. Eng. A 280 (2000) 177–181.
[48] C. Li, K. Muneharua, S. Takao, H. Kouji, Fiber laser-GMA hybrid welding of commercially pure titanium, Mater. Des. 30 (2009) 109–114.
[49] AWS D17.1:2001, Specification for Fusion Welding for Aerospace Applications, 2001.
[50] J. Talkington, D. Harwig, H. Castner, G. Mitchell, Development of titanium weld color inspection tools, Weld. J. 79 (2000) 35–39.
[51] S.L. Gong, L. Chen, W. Yao, Y. Tang, Research on laser welding of BT20 titanium alloy, in: Proc. of 22nd ICALEO, Jacksonville, USA, 2003, pp. 545–558.

[52] S. Mueller, E. Stiles, R. Dienemann, Study of porosity formation during laser welding of Ti6Al4V, in: Proc. 26th ICALEO, Scottsdale, USA, 2006, pp. 133–138.
[53] K. Richter, U. Reisgen, W. Behr, R. Holtz, Welding of titanium and other refractory metals with a free pulse shaping laser, J. Laser Appl. 19 (2007) 116–123.
[54] X. Li, J. Xie, Y. Zhou, Effects of oxygen contamination in the argon shielding gas in laser welding of commercially pure titanium thin sheet, J. Mater. Sci. 40 (2005) 3437–3443.
[55] H. Inoue, T. Ogawa, Weld cracking and solidification behavior of titanium alloys, Weld. J. 74 (1995) 21S–27S.
[56] M.J. Donachie, Titanium: A Technical Guide, second ed., ASM International, Ohio, 2000.
[57] V.I. Lakomski, N.N. Kalinyuk, The solubility of hydrogen in liquid titanium, Autom. Weld. 16 (1963) 28–32.
[58] L.S. Smith, P. Threadgill, M. Gittos, Welding Titanium—A Designers and Users Handbook, Titanium Information Group, Cambridge, UK, 1999.
[59] P. Hilton, J. Blackburn, P. Chong, Welding of Ti-6Al-4V with fibre delivered laser beams, in: Proc. of the Laser Mater. Proc. 26th ICALEO, Orlando, USA, 2007, pp. 887–895.
[60] A.I. Gorshkov, F.E. Tret'Yakov, The effects of hydrogen and alloying elements on the formation of pores in welded joints in titanium, Autom. Weld. 16 (1963) 32–37.
[61] M. Pastor, H. Zhao, R.P. Martukanitz, T. DebRoy, Porosity, underfill and magnesium loss during continuous wave Nd:YAG laser welding of thin plates of aluminium alloys 5182 and 5754, Weld. J. 78 (1999) 207S–216S.
[62] A. Matsunawa, Problems and solutions in deep penetration laser welding, Sci. Technol. Weld. Join. 6 (2001) 351–354.
[63] F. Caiazzo, F. Curcio, G. Daurelio, F.C.M. Minutolo, Ti6Al4V sheets lap and butt joints carried out by CO_2 laser: mechanical and morphological characterization, J. Mater. Process. Technol. 149 (2004) 546–552.
[64] P.E. Denney, E.A. Metzbower, Laser beam welding of titanium, Weld. J. 68 (1989) 342–346.
[65] Z. Li, S.L. Gobbi, I. Norris, S. Zolotovsky, K.H. Richter, Laser welding techniques for titanium alloy sheet, J. Mater. Process. Technol. 65 (1997) 203–208.
[66] S. Mueller, C. Bratt, P. Mueller, J. Cuddy, K. Shankar, Laser beam welding of titanium—a comparison of CO_2 and fiber laser for potential aerospace applications, in: Proc. of 27th ICALEO, Temecula, USA, 2008, pp. 846–909.
[67] J.E. Blackburn, C.M. Allen, P.A. Hilton, L. Li, Nd:YAG laser welding of titanium alloys using a directed gas jet, J. Laser Appl. 22 (2010) 71–78.
[68] T.Y. Kuo, S.L. Jeng, Porosity reduction in Nd-YAG laser welding of stainless steel and inconel alloy by using a pulsed wave, J. Phys. D Appl. Phys. 38 (2005) 722–728.
[69] A. Matsunawa, M. Mizutani, S. Katayama, N. Seto, Porosity formation mechanism and its prevention in laser welding, Weld. Int. 17 (2003) 431–437.
[70] I. Kawaguchi, S. Tsukamoto, G. Arakane, H. Honda, Characteristics of high power CO_2 laser welding and porosity suppression mechanism with nitrogen shielding—study of high power laser welding phenomena, Weld. Int. 20 (2006) 100–105.
[71] J.E. Blackburn, C.M. Allen, P.A. Hilton, L. Li, M.I. Hoque, A. Khan, Modulated Nd:YAG Laser Welding of Ti-6Al-4V, Sci. Technol. Weld. Join. 15 (2010) 433–439.
[72] A. Haboudou, P. Peyre, A.B. Vannes, G. Peix, Reduction of porosity content generated during Nd:YAG laser welding of A356 and AA5083 aluminium alloys, Mater. Sci. Eng. A 363 (2003) 40–52.
[73] T. Hayashi, K. Matsubayashi, S. Katayama, S. Abe, A. Matsunawa, A. Ohmori, Reduction mechanism of porosity in tandem twin-spot laser welding of stainless steel, Weld. Int. 17 (2003) 12–19.

[74] J. Xie, Dual beam laser welding, Weld. J. 10 (2002) 283S–290S.
[75] J.E. Blackburn, C.M. Allen, P.A. Hilton, L. Li, Dual focus Nd:YAG laser welding of titanium alloys, in: Proc. of 36th MATADOR Conf., Manchester, England, 2010, pp. 279–282.
[76] G. Verhaeghe, B. Dance, An assessment of the welding performance of high-brightness lasers and a comparison with in-vacuum electron beams, in: Proc. Int. Conf of Appl. of Lasers and Electro-Opt, ICALEO, Temecula, USA, 2008, pp. 406–414.
[77] X. Cao, W. Wallace, C. Poon, J.P. Immarigeon, Research and progress in laser welding of wrought aluminium alloys. I. Laser welding processes, Mater. Manuf. Process. 18 (2003) 1–22.
[78] X. Cao, W. Wallace, J.P. Immarigeon, C. Poon, Research and progress in laser welding of wrought aluminium alloys ii, metallurgical and micro-structures, defects, and mechanical properties, Mater. Manuf. Process. 18 (2003) 23–49.
[79] B. Brenner, J. Standfuss, D. Dittrich, J. Liebscher, J. Hackius, Laser beam welding of aircraft fuselage panel, in: Proc. of 27th ICALEO, Temecula, USA, 2008.
[80] D. Dittrich, B. Brenner, B. Winderlich, E. Beyer, J. Hackius, Progress in laser beam welding of aircraft fuselage panel, in: Proc. of 27th ICALEO, Temecula, USA, 2008.
[81] C.M. Allen, G. Verhaeghe, P.A. Hilton, C.P. Heason, P.B. Prangnell, Laser and hybrid laser-MIG welding of 6.35mm and 12.7mm thick aluminium aerospace alloy, in: Proc. of ICAA-10, Vancouver, Canada, 2006, pp. 1139–1144.
[82] G. Verhaeghe, C. Allen, P. Hilton, Achieving low-porosity laser welds in 12.7mm thickness aerospace aluminium using a Yb-fibre laser, in: Proc. 4th Lasers In Manufacturing Conf., Munich, Germany, 2007, pp. 17–23.
[83] X. Cao, M. Jahazi, J.P. Immarigeon, W. Wallace, A review of laser welding techniques for magnesium alloys, J. Mater. Process. Technol. 171 (2006) 188–204.

Laser sources

J. Hecht, The Laser Guidebook, second ed., TAB Books, Blue Ridge Summit, PA, 1992.

Properties of laser light

D. Schuöcker (Ed.), Handbook of the EuroLaser Academy, vol. 1, Chapman & Hall, London, 1998.

Laser materials processing

J. Ion, Laser Processing of Engineering Materials: Principles, Procedure and Industrial Application, Elsevier, Oxford, 2005.

Linear friction welding in aerospace engineering

4

I. Bhamji[a], A.C. Addison[a], P.L. Threadgill[a,#], and M. Preuss[b,c]
[a]TWI Ltd., Cambridge, United Kingdom, [b]University of Manchester, Manchester, United Kingdom, [c]Monash University, Melbourne, VIC, Australia

4.1 Introduction to linear friction welding

Linear friction welding (LFW) is a solid-state joining process in which a stationary part is forced against a part that is reciprocating in a linear manner in order to generate frictional heat (Fig. 4.1) [1–3]. The heat, along with the force applied perpendicular to the weld interface, causes the material at the interface to deform and plasticize. Much of this plasticized material is removed from the weld as flash, because of the combined action of the applied force and part movement. Surface-oxides and other impurities are removed, along with the plasticized material, and this allows metal-to-metal contact between parts and allows a joint to form. To date, LFW is best known for joining titanium-alloy blades to discs forming so called blisk components used in the aero-engine industry. It is also now considered for creating complex-shaped titanium structures for airframes.

4.2 History and major applications of linear friction welding

LFW was first patented in 1929 [4]; however, the description of the process was vague. Some discussion of the concept was then recorded in the 1960s, but it was described as "very doubtful" because of the difficulty in generating linear reciprocation [3]. The Caterpillar Tractor Company [5] was the next to mention the process in a patent. However, the patent primarily focused on the machine that generates the linear reciprocation and not the actual welding process. Indeed a patent search has shown that no currently valid patents exist that protect the fundamentals of the LFW process. However many patents protect certain aspects of LFW, such as particular applications, welding methods or tooling concepts (for example see Slattery [6] and Trask et al. [7]).

The first structured research into LFW was undertaken at TWI Ltd. from 1989 onward on a prototype LFW machine that is thought to be the first of its kind. The process was an evolution from orbital friction welding, which uses a type of motion where the center point of one of the parts is rotated around the center point of the other. Orbital friction welding was in turn an evolution from rotary friction welding, which is one of the oldest and most well-established friction welding processes. Rotary friction

Retired.

Welding and Joining of Aerospace Materials. https://doi.org/10.1016/B978-0-12-819140-8.00004-3
Copyright © 2012 Woodhead Publishing Ltd. All rights reserved.

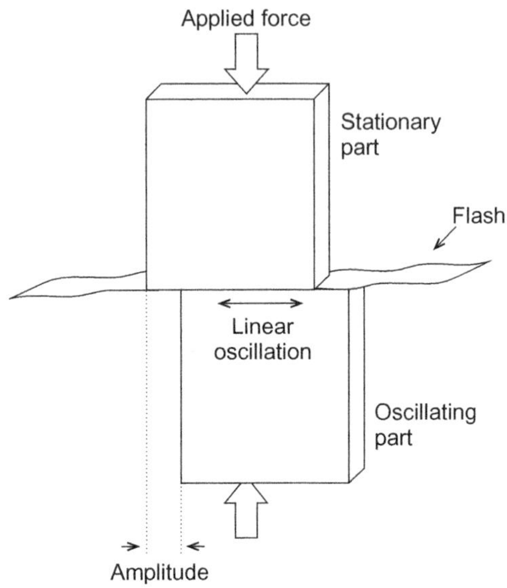

Fig. 4.1 Schematic diagram of the LFW process.

welding involves rotating an axi-symmetric part and forcing it against a stationary part to form a weld. Its key limitation is that one of the parts must be axi-symmetric for the process to work. This is not the case for LFW, and parts with irregular cross-section and complex shapes can be welded.

The linear motion of LFW naturally lends itself to the repair and manufacture of aero-engine compressor blades and this was the initial driver for the development of the process. Indeed, to date, the only major commercial use of LFW is for the joining of aero-engine compressor blades to compressor discs, to form blisks (Fig. 4.2) [8], although the process is also being seriously considered for the near-net-shape manufacture of components in high-value materials (for example see Slattery [9]). LFW offers many advantages for the manufacture of components for the aerospace industry, these include:

- The replacement of mechanical blades to disc attachments in aero engines with a welded structure can enable overall weight reduction by 20–30%, as blisk weight reductions are further reflected in the design of shafts and related parts of the engine. This weight reduction should have a significant impact on the fuel consumption and efficiency of aircraft utilizing linear friction welded blisks.
- The process has been proven to provide good reliability and it has been reported by certain aero-engine manufacturers that tens of thousands of welds have been produced without a single failure.
- Different disc and blade alloys or materials can be welded together when producing blisks so that an optimum welded structure can be produced. The process is also considered to be good for producing welds more generally between some dissimilar materials and alloys.
- Fine-grained blade alloys, which are more suited to high-cycle fatigue endurance, can be joined to coarse-grained disc alloys, which are more suited to low-cycle fatigue endurance.
- Damaged turbine blades can be easily replaced by removing the damaged blade and linear friction welding another in its place.

Fig. 4.2 A linear friction welded blisk assembly.
Courtesy of MTU Aero Engines M.E. Nunn, Aero engine improvements through linear friction welding, in: Proc. 1st Int. Conf. on 'Innovation and Integration in Aerospace Sciences', Paper no 2005–0040, Queen's University Belfast, 4–5 August 2005, CEIAT, Queen's University Belfast, 2005.

- Only limited machining is needed after weld production (mainly to remove flash), and adaptive machining can be used to accommodate potential variations in blade location owing to distortion [10]. However, positional accuracy of the process when welding aerofoils is thought to be good, with tolerances quoted to being less than 0.2 mm and 0.2° in all dimensions [8].
- The flexibility of the process and the fact that in can produce low distortion parts allows the process to be effective for near-net-shape manufacture (Fig. 4.3). When the process is used for this a number of linear friction welds can be used to build-up a shape that is close to that of the final component. This provides significant cost savings over machining from solid, and is therefore economically beneficial, especially in high-value materials.
- The process is fairly quick in terms of total weld time. A typical cycle time for welding Ti-6Al-4V (Ti-64) is less than 10 s.
- The process allows the welding of irregular cross-section and complex geometry parts.
- A defining feature of LFW, and indeed all friction welding processes, is that it takes place in the solid state and involves no melting of the parts to be welded. This means the process offers advantages over fusion welding when joining metals that exhibit solidification problems (e.g. porosity, hot cracking, segregation, etc.).
- The severe deformation, high-temperature and high-cooling rates that are experienced close to the weld line in linear friction welds can allow a refined microstructure to form. This can provide improved strength at the weld line relative to the parent material [11].
- Oxide formation is also reduced, because of the close contact between parts [12], which means shielding gases are only rarely needed.

One of the main disadvantages of the process is the high capital cost of the equipment. The high cost of both the equipment and tooling means the LFW process can only be justifiably used for producing high-value-added components [11]. This has generally confined the process to niche applications such as producing bladed discs for aero engines, however the business case for using LFW to produce blisks and to

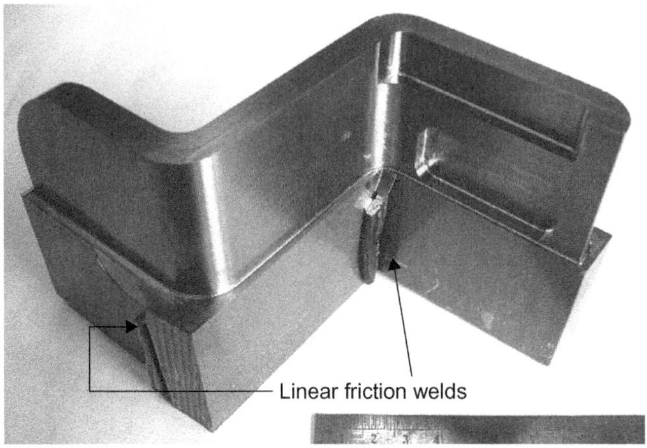

Fig. 4.3 Near-net-shape manufacture of a Ti-64 component. Welded structure can be seen at the bottom of the image and the final machined component at the top. Work was carried out by TWI in conjunction with Thompson Friction Welding.
Picture courtesy of TWI.

near-net-shape manufacture parts in high-value materials seems very sound. Machines based on the principle of stored energy rather than direct drive have also significantly lowered costs, and this may mean the LFW of lower-value-added components can be justified [13]. A further disadvantage of the process is that it can be very noisy, but this can be mitigated to some extent by sound proofing.

4.2.1 Quality control systems for linear friction welding

Although LFW is an established process for welding aero-engine blisks, there are currently no public domain guidelines on the production and testing of linear friction welds. Procedures for producing and testing rotary friction welds have been disclosed however (BS EN 15620:2000 [14]), and as rotary friction welding and LFW are similar processes these guidelines should be equally applicable to LFW. For most applications a combination of statistical process control (monitoring welding parameters as well as the final loss of length, or upset) and periodic mechanical, metallographic and nondestructive testing should be adequate.

4.3 Linear friction welding machines

LFW machines at a fundamental level involve a mechanism that allows one of the work pieces to oscillate, and a further mechanism that allows the remaining work piece to be forced against the oscillating work piece. The force application is always by means of a hydraulic ram, whereas oscillation can be achieved either through hydraulic or mechanical means. An example of each of these types of machine is shown in Fig. 4.4.

(A)

(B)

Fig. 4.4 Examples of (A) electromechanical and (B) hydraulic LFW machines. Both of these machines are located at TWI, Cambridge, UK.
Pictures courtesy of TWI.

Fig. 4.5 shows a basic schematic diagram of the oscillation mechanism of a hydraulic-type LFW machine. The components of the system are a pump that pumps high-pressure hydraulic fluid into a stack of accumulators. A valve then allows the high-pressure fluid from the accumulators, to be alternately transferred into each end of a hydraulic cylinder and piston assembly creating the reciprocating motion.

For the mechanical variant, a common assembly involves a drive motor, which rotates two crankshafts that are linked to one another (Fig. 4.6). The crankshaft nearest the drive motor (crankshaft A) is solid, whereas the other (crankshaft B) is hollow and has a mechanism that allows it to be phase shifted with respect to crankshaft A. To each of the crankshafts a crank is attached, and a whipple beam is in turn attached to the cranks. The maximum stroke of the cranks, and the maximum amplitude achievable by the machine is reached when the two crankshafts are rotated in phase. When

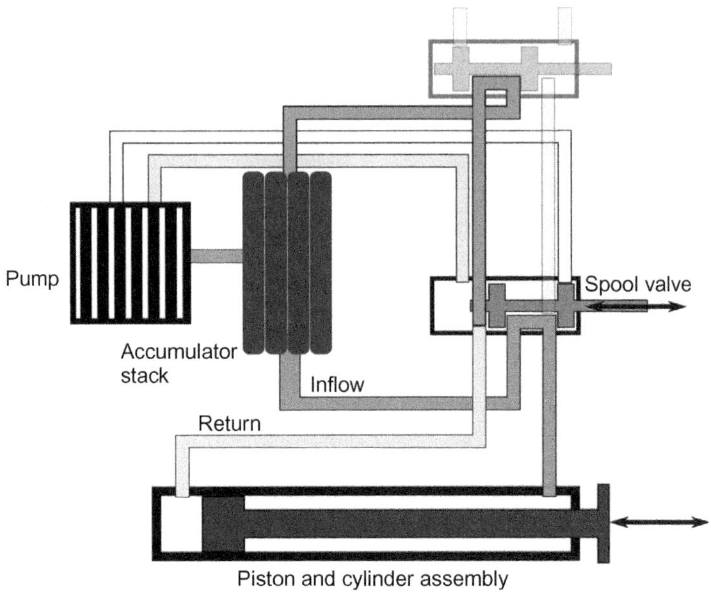

Fig. 4.5 Schematic diagram of the oscillation mechanism of a hydraulic-type LFW machine. The spool valve in the figure oscillates in and out, which allows hydraulic fluid to be alternately forced into the top and bottom of the cylinder and piston assembly, which in turn creates the reciprocating motion. The tooling to hold the parts to be welded is attached to the piston.

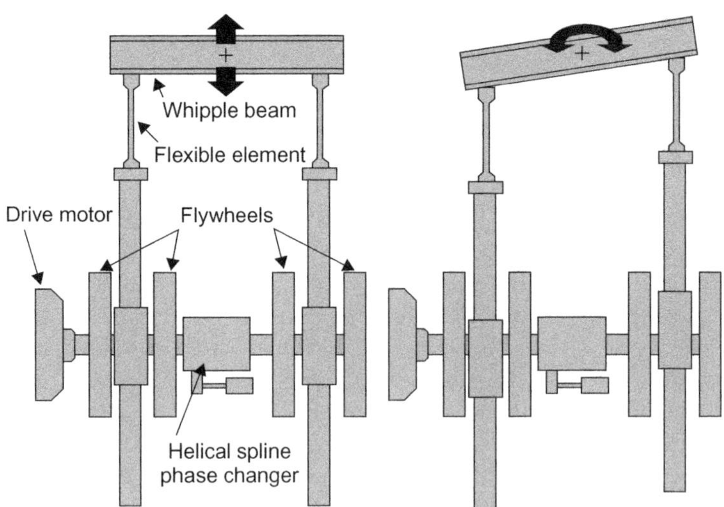

Fig. 4.6 Schematic diagram of the oscillating mechanism of a mechanical type LFW machine.

the crankshafts are rotated 180° out of phase the whipple beam rocks so that its center remains stationary. This effectively gives an amplitude of zero. To achieve amplitudes between zero and the maximum the phase shift between the two cranks must be less than 180°, or greater than 0°. In order to change the oscillation frequency the rotational speed of the crankshaft is changed.

The cranks are connected to the whipple beam by way of flexible elements and the tooling to hold the reciprocating part is attached to the whipple beam. The flexible elements are weighted with balance weights to help stabilize the reciprocating motion. It is necessary that parts used in the machine are either machined from solid or are forged as they are subjected to cyclic loads, and these loads can be high when the machine is used at high frequencies and amplitudes.

Fig. 4.6 shows a schematic diagram of the oscillating mechanism of a mechanical-type LFW machine. Tooling for the reciprocating part is attached to the center of the whipple beam. The left-hand image shows the crankshafts moving in phase so that the center of the whipple beam reciprocates at the maximum amplitude. The right-hand image shows the crankshafts moving 180° out of phase so that the center of the whipple beams remains stationary, giving a zero amplitude. The assembly also has four flywheels to stabilize the motion of the crankshafts.

4.3.1 Linear friction welding machine operation

The machine operation of the LFW process can be broken down into six separate stages:

Part clamping: The parts are held using tooling designed to withstand the forces experienced during the process. Specimen and tooling preparation is crucial to the process, with accurate sides and edges needed on the specimen, and a tight fit needed between the specimen and the tooling. This generally means that the tooling is custom built to fit particular specimen geometries.

Datum and retract: The clamped parts are brought together under a small compressive force in order to determine the location of the parts and set the machine datum to zero. The parts are then retracted to leave a small separation distance between the work pieces.

Conditioning phase: Oscillation of one of the parts is increased and stabilized over a set period (usually very quickly), and the parts are brought together under a small force for a predetermined time (Fig. 4.7).

Frictional phase: The compressive force (friction force) is increased to a set level and heat is generated at the interface. The material at the interface becomes plastic and flows out of the weld as flash because of the shearing motion between the two parts and the applied force. This loss of material from the weld causes the parts to shorten (or burn-off). This phase usually ends, and the next is triggered when a predetermined loss of length, or burn-off distance, is reached. However, the next phase can also be triggered after the frictional phase has continued for a predetermined time (burn-off time) or number of oscillation cycles (burn-off cycles).

The LFW process is always carried out under load control, but other parameters also play a role in controlling the welding process. For example when using a

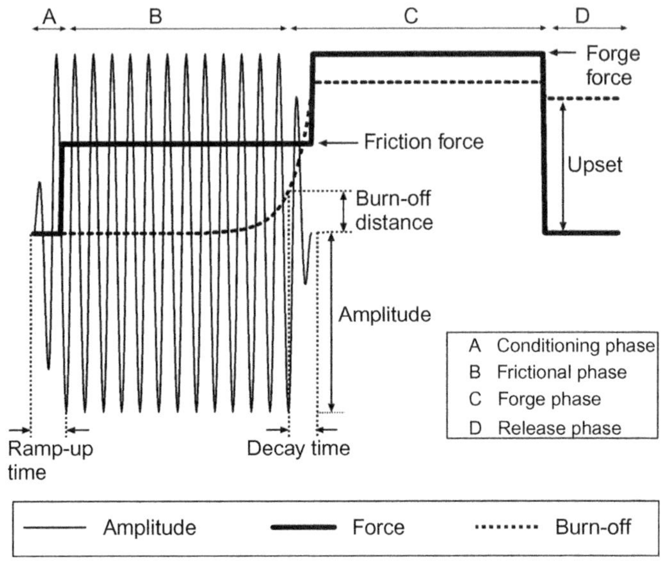

Fig. 4.7 Schematic diagram of the parameter traces that are obtained during the LFW process. A number of input variables are defined in the diagram. Burn-off is defined as the loss of length occurring as the process continues, whereas upset is the total loss of length measured after the weld has been produced.

burn-off distance the load is controlled throughout the welding process, however the burn-off is also monitored (although not controlled), and at a set burn-off distance the next phase (forge phase) is triggered. Similarly with burn-off time or cycles the load is controlled throughout welding and the amount of time or cycles determines the transition to the next phase.

Forge phase: The amplitude is decayed to zero over a predetermined time to ensure good alignment (usually very quickly), and a forge force is rapidly applied and held for a set time to consolidate the joint. The forge force can either be the same as or higher than (more common) the friction force.

Release phase: The welded parts are released from the clamps and removed from the machine.

The frictional phase can be further broken down into three separate phases that are determined by material behavior (Fig. 4.8) [15]:

Initial phase: The parts are forced together and heat is generated through friction. The rubbing together of parts causes asperities to wear down and flatten so that the true area of contact increases toward 100%. A small amount of part shortening occurs owing to wear particle expulsion. It is critical for the rest of the process that enough frictional heat is generated in this phase of the process.

Transition phase: The temperature at the interface rises and therefore the strength of the material at the interface decreases. The applied load can then cause this low-strength material to soften and plastically deform.

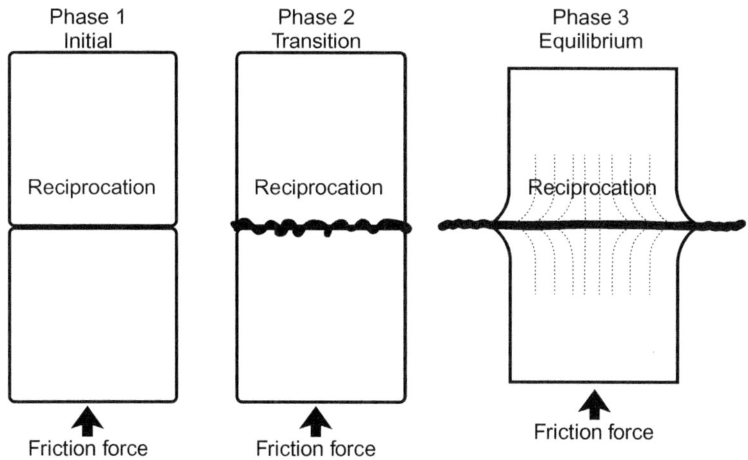

Fig. 4.8 Phases of the LFW process, determined from consideration of material behavior. Redrawn from A. Vairis, M. Frost, On the extrusion stage of linear friction welding of Ti6Al4V, Mater. Sci. Eng. A Struct. Mater. Prop. Microstruct. Process 271 (1–2) (1999) 477–484.

Equilibrium phase: The hot plasticized material at the interface is expelled as flash, with the help of the oscillatory motion and applied pressure. The main input variables during the process are:

- *Frequency*: Number of oscillatory cycles per second
- *Ramp-up time*: Time taken to increase the welding parameters to the required steady-state level (Fig. 4.7)
- *Amplitude*: Maximum displacement of the oscillating sample from its equilibrium position
- *Friction pressure*: Pressure applied during the frictional phase of the process; pressure is calculated by using the nominal area of contact at zero amplitude
- *Burn-off distance, time or cycles*: Possible factors that trigger the start of the forge phase
- *Decay time*: Time taken to reduce the amplitude to zero at the beginning of the forge phase
- *Forge pressure*: Pressure applied during the forge phase of the process
- *Forge time*: The amount of time for which the forge pressure is applied

There are also other variables that are of importance, but which are a consequence of the main variables and cannot be easily and precisely changed through the variation of the main variables. These consequent parameters are:

- *Upset*: The total loss of length (shortening) owing to the process (the upset will always exceed the burn-off distance mainly because of the loss of length occurring during the forge phase)
- *Shear (or in-plane) force*: Force parallel to the oscillatory movement
- *Burn-off rate*: Rate of shortening (i.e. gradient of the burn-off curve, Fig. 4.7)
- *Welding time*: Total time taken to weld a specimen

4.4 Macroscopic features of and defects in linear friction welds

The macroscopic features of a linear friction weld can be seen in Fig. 4.9. The most striking feature is the flash, which consists of expelled plasticized material coming from the weld interface. This flash can be in the form of a single wing-like structure or be bifurcated (i.e. the flash from each weld half forms two separate collars). The type of flash is usually determined by the material being welded. Titanium alloys usually produce flash in a wing-like structure, whereas other materials tend to form flash that is bifurcated.

The extremities of the weld can be common areas for defects in linear friction welds. This is not surprising as these areas experience greater heat loss owing to convection relative to the central region, and are exposed to the atmosphere when the oscillating part is reciprocated. The pressure distribution is also thought to be elliptical so that particularly the corners of the weld experience a lower axial pressure relative to the central region.

Common defects include a lack of bonding running from the flash and into the parent material geometry. This can be difficult to avoid in welds where the flash is bifurcated, but this lack of bonding usually only penetrates a small distance into the parent material geometry and this region is usually within machining tolerances. In extreme cases, however, particularly when brittle materials are welded, the bifurcation of the flash can allow a lack of bonding to penetrate large distances (Fig. 4.10A). Tearing in the heat-affected zone (HAZ) (see Section 4.5) can also occur in some materials if the amplitude is decayed too quickly at the end of the welding cycle (Fig. 4.10B). This is again uncommon and usually only occurs in very brittle materials. Corner defects that are caused by a lack of flash extrusion from the corners of the welds can also occur. However these defects can be avoided by using chamfered, as opposed to sharp, corners.

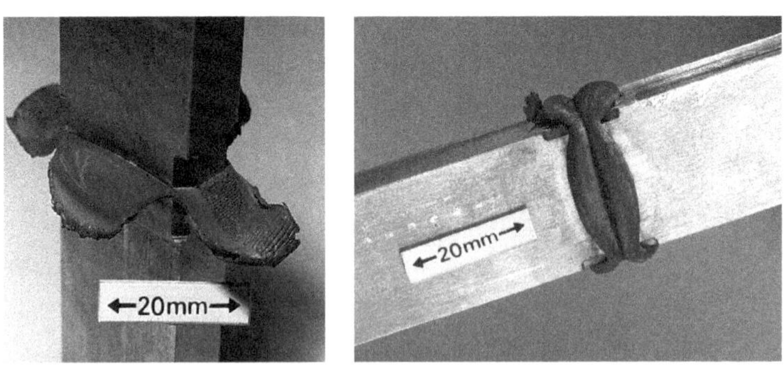

(A) (B)

Fig. 4.9 Linear friction welds in (A) titanium showing flash in the form of a single wing-like structure and (B) Waspalloy showing bifurcated flash.

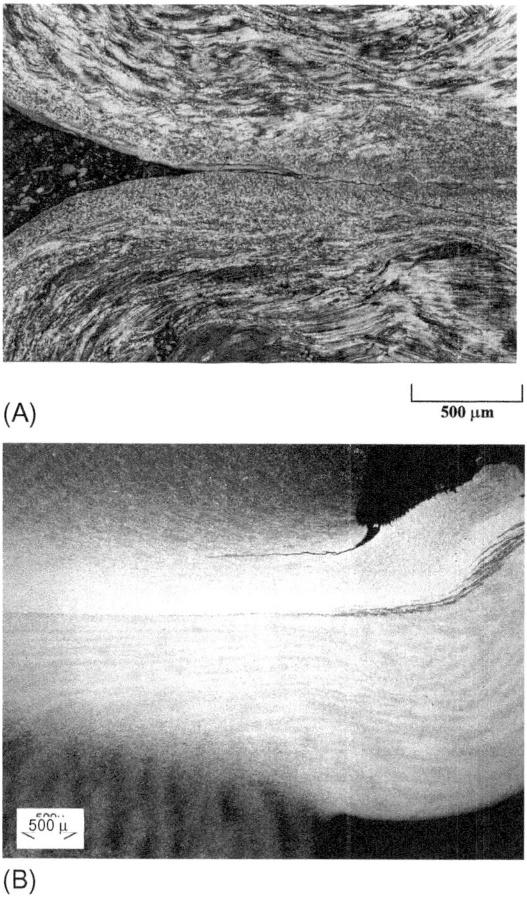

Fig. 4.10 Defects in linear friction welds. (A) Weld in Ti-48Al-2Cr-2Nb showing a crack penetrating into the parent material geometry from the flash; (B) weld between a martensitic (top) and stainless (bottom) steel showing cracking in the HAZ of the martensitic steel as a result of rapid amplitude decay. Pictures courtesy of TWI.

4.5 Microscopic features of linear friction welds

Although the interface temperature during LFW is not expected to exceed the melting temperature of the material being welded, the peak interface temperature is nevertheless very high, and could be close to the solidus temperature of the material being welded [16]. In all reported cases, this high temperature, along with the applied pressure, causes significant microstructural variation close to the weld interface.

No widely accepted nomenclature exists for microstructural regions in linear friction welds, but there would be a clear advantage in establishing such a system as it would avoid much confusion in the literature. The following proposal is based on a nomenclature that has received widespread acceptance for friction stir welding [17]. It is proposed that the weld is divided into four regions, each defined as follows:

- Parent material (PM): This is material, some distance from the weld line, where no change in microstructure, mechanical properties or other properties can be detected.

- Heat-affected zone (HAZ): In this region the microstructure and/or other properties have been changed by heat from the weld, but there is no optically visible plastic deformation. Changes could, for example, include one or more of grain growth, change in precipitate morphology, changes in mechanical or physical properties.
- Thermo-mechanically affected zone (TMAZ): In this region the material has been subjected to more heat than in the HAZ and shows clear evidence of plastic deformation (see Fig. 4.11D for a typical TMAZ microstructure in Ti-64). Phase transformations could also take place in some materials in this zone. Changes in this region would be expected to be more apparent than in the HAZ.
- Weld zone (WZ): In many materials there will be a region close to the weld line where the microstructure is very different to that in other parts of the weld. This is usually because of recrystallization, which produces a region consisting of very fine equiaxed grains (see Fig. 4.11C for a typical WZ microstructure in Ti-64), and/or phase transformations. As this area has been subjected to heat and plastic flow, it is a subgroup of the TMAZ.

An identical approach to this could also be used for rotary and inertia friction welds.

4.6 Linear friction welding of titanium alloys

Thus far all published work on the linear friction welding of titanium has concentrated on welding $\alpha + \beta$ alloys, although attempts have been made to weld β and near-β alloys. Therefore, this section of the report will concentrate solely on the welding of $\alpha + \beta$ alloys.

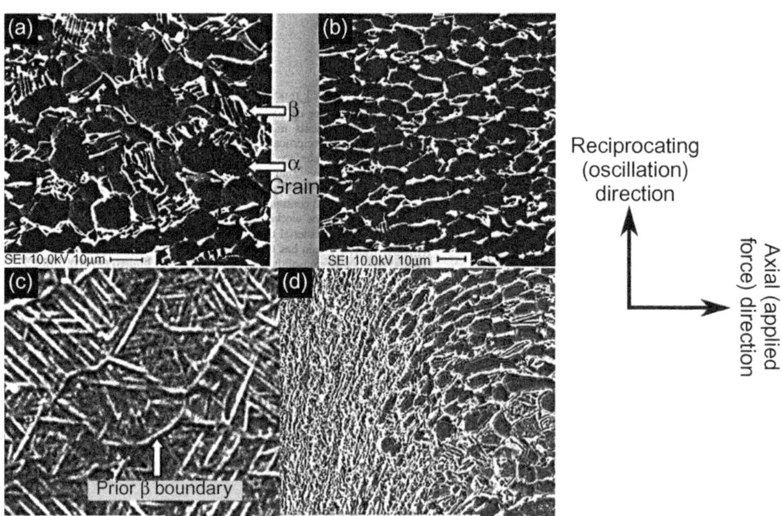

Fig. 4.11 Microstructure of a Ti-64 linear friction weld [11]. (A) and (B) Parent material (consisted of alternating layers of the two types of microstructure); (C) Widmanstatten microstructure in the WZ (that had a width of ~180 μm); (D) Nonrecrystallized region of the TMAZ (that had a width of ~290 μm on one side of the weld), showing deformation of the parent microstructure [11].

4.6.1 Mechanical properties

The mechanical properties of welds in titanium alloys have generally been very good, with the yield and ultimate tensile strengths of defect-free Ti-64 welds usually surpassing the strength of the parent materials [11, 18]. The weld properties were also found to be very tolerant to changes in welding parameters, with poor welds only being produced at very extreme welding parameters [19]. The impact toughness of the WZ in Ti-64 linear friction welds was found to be higher than that of the parent material and a notch, centered at the weld line, formed during testing a crack that quickly progressed through the TMAZ and into the parent material [20]. These good mechanical properties are a result of grain refinement and work hardening in regions close to the weld interface [20].

The effects of producing successive linear friction welds (i.e. one weld on top of the other) have been reported [21]. The aim was to simulate the effects of blisk repair, and involved the removal of a weld (welds were produced in Ti-6Al-2Sn-4Zr-6Mo) by cutting parallel to the weld plane, 3 mm from the weld line, and welding onto this cut surface. The results from this study, which simulated up to two repairs (i.e. up to three successive welds), showed that tensile failure always occurred away from the weld lines in the parent material. High-cycle fatigue tests also produced failure in the parent material, but not enough samples were examined to determine if the presence of the welds had a definite impact on fatigue properties.

4.6.2 Weldability

Most of the publicly available work on weldability and welding parameters has been conducted on Ti-64, however similar concepts should be equally applicable to other materials. Vairis and Frost [22] showed that for sound linear friction welds to be formed in Ti-64, a specific power input parameter must be exceeded. It was shown that frequency, amplitude and pressure have an effect on this parameter, which was defined as:

$$w = \frac{\alpha f P}{2\pi A} \tag{4.1}$$

with α being the amplitude; f the frequency; P the pressure; and A the weld area. From this equation it can be seen that the power input can be increased by increasing the frequency, amplitude or pressure. A similar critical power input has also been suggested to exist for linear friction welds in 316L stainless steel [23].

Wanjara and Jahazi [11] showed that another parameter, the upset, was also important in forming sound welds and demonstrated that a minimum level of upset was necessary to consolidate the weld. Therefore the power input parameter devised by Vairis and Frost [22] cannot be used as an exclusive criterion for obtaining sound welds.

4.6.3 Microstructure

The microstructure of linear friction welds in bimodal (α and β structure, Fig. 4.11A and B) titanium alloys have been studied by a number of researchers [11, 24, 25]. From this work it is evident that during LFW the material close to the weld line exceeds the β-transus temperature. A very fine-grained structure is typical of microstructures in this region, which is related to the material being exposed to high temperature and strain resulting in recrystallization. The fully β-transformed microstructure is rapidly cooled, after the frictional phase of the process, which avoids β-grain coarsening and leaves a Widmanstatten microstructure of α and β plates delineated by prior β grain boundaries (Fig. 4.11C). It has also been suggested that the cooling rates experienced by the weld could be high enough to produce martensite (α') (with some retained metastable β) [25], suggesting that cooling rates greater than 410 °C s^{-1} can be achieved in titanium LFW [11, 24].

Slightly further away from the weld line the nonrecrystallized region of the TMAZ is present. In this region grains are heavily deformed and are reoriented (in terms of morphology) so that their long dimension is perpendicular to the applied force (Fig. 4.11D) [11]. Titanium welds do not display a prominent HAZ (although subtle HAZs may exist that are very hard to detect), and there is possibly a direct transition from the TMAZ to the parent material [11].

4.6.4 Effects of welding parameters

The influences of various welding parameters on the size of recrystallized β grains in the WZ of Ti-64 welds have been reported [11]. This work gives a good insight as to how the interface temperature varies as welding parameters are changed, as the β-grain growth will be dependent on it. It was shown that an increase in frequency or pressure increases the size of the prior β grains (Fig. 4.12). This is thought to be because an increase in power input, associated with an increase in frequency or pressure, causes the temperature at the interface to be greater. However, the increased prior β-grain size could also be related to slow cooling rates when high parameters were used [11].

More recent work, however, has shown that the prior β-grain size actually decreases at high pressures, which has been interpreted as a reduction in the peak welding temperature [26], although changes in the overall weld thermal cycle may also have contributed to producing these effects. This proposed decrease in temperature was attributed to a large amount of flash expulsion at high pressures, which caused large heat rejection and short welding times. The conclusion that the welding temperature decreases when using higher pressures is further supported by Romero et al. [27] The results from Romero et al. [27] will be discussed in more detail in Sections 4.6.5 and 4.6.6.

The reasons for the discrepancies between studies are not clear. However, there are differences in the materials being welded (Ti-64 [11, 27], Ti-6Al-2Sn-4Zr-6Mo [26]) and the machine, parameters and specimen geometry being used, which could have caused the differing findings. It was also shown in Bhamji et al. [23] that the final weld

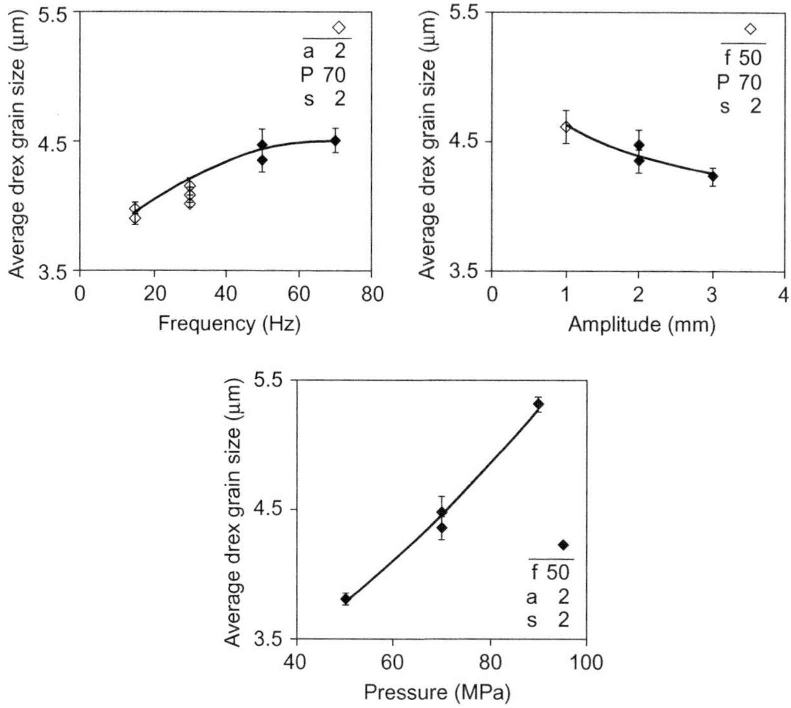

Fig. 4.12 Effects of frequency, amplitude and pressure on average prior β-grain size in the WZ of Ti-64 linear friction welds [11]. a, amplitude; P, pressure; f, frequency; s, burn-off distance. Open markers indicate poor welds.

microstructure of 316L linear friction welds was sometimes dependent on a number of different welding parameters working in combination to produce a particular effect. A similar occurrence in the work with titanium alloys may have contributed to the discrepancies, but as welding parameters were not disclosed by Attallah et al. [26] it is difficult to be sure that this was the case. Results from Bhamji et al. [23] are discussed in more detail in section 8.1.

It was also demonstrated, in Wanjara and Jahazi [11], that an increase in oscillation amplitude results in a reduction of the prior β-grain size even though the power input (according to Eq. (4.1)) increases (Fig. 4.12). It was reasoned that large oscillation amplitudes expose a considerable amount of material to the surrounding atmosphere, resulting in increased convective heat dissipation. Therefore, an increase in power input would not create the expected temperature increase. The greater exposure of material at high amplitudes may also have allowed a greater amount of oxygen penetration and oxidation, which could have affected weld microstructures.

These parameter studies clearly demonstrate that the welding temperature, in titanium linear friction welds, can be controlled to some extent through the control of welding parameters. Minimizing the peak welding temperature is an important objective, particularly when attempting to join highly dissimilar materials (e.g. aluminum to copper) by LFW. In such cases, the formation of intermetallics can be a major obstacle. Therefore, minimizing the welding temperature aims to avoid the formation of detrimental intermetallic phases at the weld line. A similar control of welding temperature by controlling welding parameters has also been demonstrated in 316 L stainless-steel linear friction welds [23].

4.6.5 Residual stresses

Measurements of residual stresses, deep inside the weldments, have been carried out on linear friction welds in a number of different titanium alloys [27–31]. Each of these studies has been carried out by using either high-energy synchrotron X-ray or neutron diffraction, and each rely on calculating strains and stresses from changes in lattice parameter [32]. The studies are in broad agreement, and most of them show tensile stresses in all three directions at central locations (i.e. mid-width and mid-thickness) near the weld line (Fig. 4.13). The stresses were highest in the transverse direction (in the weld plane but perpendicular to the direction of oscillation, which is the reciprocating direction) and lowest in the axial direction (direction of force application). In each of the studies stresses dropped sharply on either side of the weld line, with stresses becoming compressive in the adjacent region before finally approaching zero.

Different studies reported significantly different values for the highest stress even if the same materials were welded (comparing maximum stresses reported by Romero et al. [27] with those reported by Karadge et al. [30]; Ti-64 was welded in both of these studies but welding parameters were not disclosed in either). A particular issue with residual stress calculations is the accurate determination of the strain-free lattice spacing (d_0), which is used as a reference value when determining residual strains and stresses. The d_0 value will change across the weld line because of microstructural changes and changes in the phase-specific chemical composition. When these changes in d_0 are taken into account much lower stress values are reported (as by Romero et al. [27]), relative to without d_0 correction (as by Karadge et al. [30]). As well as this, the stresses in the axial direction become negligible when d_0 variations are taken into account, suggesting that the residual stress state is predominantly biaxial in the plane of weld interface [27]. Whereas the d_0 variation is clearly important for an accurate understanding of residual stresses in linear friction welds, it is important to note that some uncertainty in d_0 values occurs at locations very close to the weld line (<0.5 mm) [29]. This also means that there is some uncertainty associated with the residual stress results directly at the weld line, and more work is needed to accurately determine the stresses in this region.

Although the magnitude of residual stress in weldments can vary with direction as has been discussed in the preceding paragraphs, it is important to note that this variation is related to sample geometry rather than the oscillation and applied force directions defined by the welding process. Residual stresses in linear friction welds are a result of the thermal mismatch created by different cooling rates at various positions across the weld (material at the surface of the weld will cool more quickly than that in the center). As the weld geometry will affect the thermal profile within a weld, it is clear that it will also have a substantial effect on residual stresses for a given material.

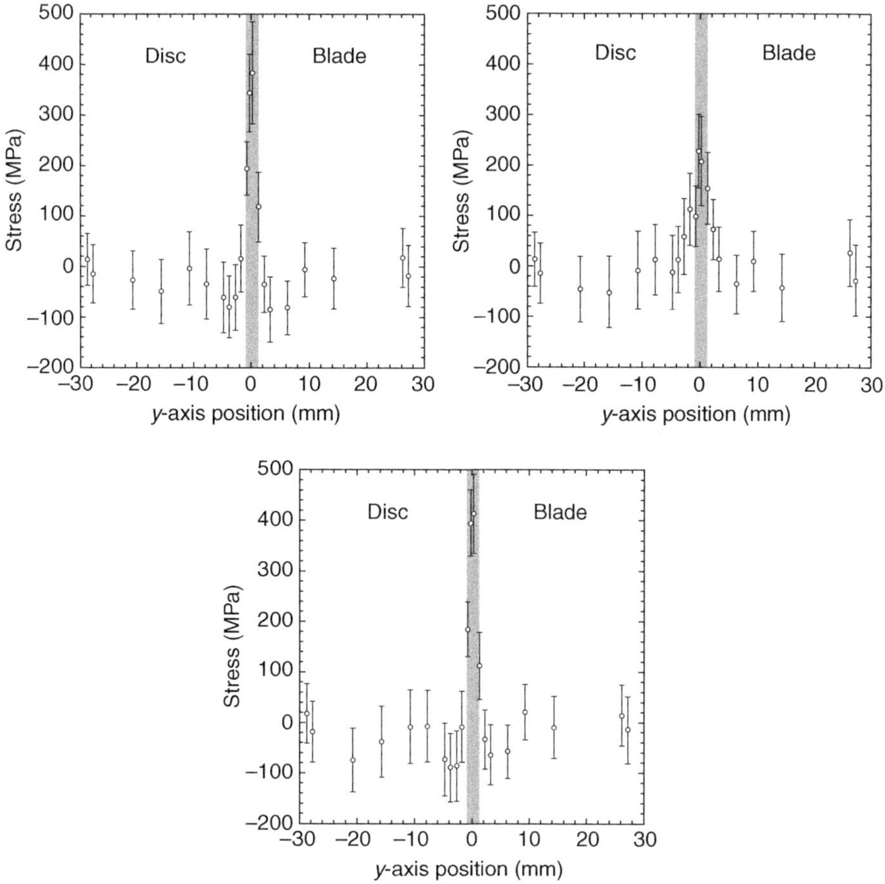

Fig. 4.13 Stress in the three spatial directions as a function of position along the axial direction (*y*-axis) [28]. (A) Reciprocating direction; (B) axial direction; (C) transverse direction (study conducted on Ti-64).

In almost all of the welds studied for residual stress the sample was reciprocated in the small dimension, with a larger dimension in the transverse direction. As the distance between the weld center and the surface is longer in the transverse direction relative to that in the reciprocating, it is likely that there will be a larger thermal gradient and hence thermal mismatch in this direction. This explains why generally larger tensile stresses are observed in the transverse direction compared with the reciprocating direction.

The low or negligible stresses in the axial direction observed particularly when high welding pressures were used and appropriate d_0 corrections were applied across the weld line [27], demonstrate that only small thermal mismatches are generated in this direction. The precise reasons for this are however not clear at present.

The results described in this section show that very significant stresses develop during LFW, which can have a detrimental effect on the long-term performance of the welded component. Consequently, the development of an appropriate post-weld heat treatment (PWHT) that relieves the stresses sufficiently without compromising microstructure and mechanical properties is of major importance. A few studies have investigated the influence of PWHT on residual stresses, and reductions in stresses ranging between 75% and 90% have been reported for various titanium alloys [29, 31]. An important dependence of residual stress relief on specimen size has been reported in Ti-64 [30]. In a small, lab-scale, linear friction welded sample, negligible stress levels existed after PWHT, whereas in a full-scale blisk significant tensile stresses still existed after an identical PWHT. Both welds displayed similar residual-stress profiles in the as-welded condition [30].

It has also been suggested that residual stress development can be minimized by optimizing welding parameters with low stress levels being observed when high applied pressures were used during LFW of Ti-64 [27] and Ti-6Al-2Sn-4Zr-6Mo [33]. These low stresses are thought to result from a low peak welding temperature at high pressures, which was also suggested by Attallah et al. [26] (see section 6.4). This result clearly demonstrates that stresses can be minimized by the choice of appropriate welding parameters, which is particularly important if sufficient residual-stress mitigation by a subsequent PWHT is difficult to obtain (for instance in dissimilar material welds).

4.6.6 Texture

Textures in Ti-64 linear friction welds of different sizes (lab-scale and full-scale specimens) have been investigated [25]. With both specimen geometries a strong transverse texture in the α phase, $\{10\overline{1}0\}\langle 11\overline{2}0\rangle$, was seen close to the weld line. In simple terms, the c-axis of the hexagonal close-packed α crystallites are predominantly orientated parallel to the transverse direction (Fig. 4.14). This strong texture was present across a region of approximately ± 100 μm from the weld line in the lab-scale specimen and $\pm$ 50 μm from the weld line in the full-scale specimen [25]. The TMAZ in the lab-scale specimen spanned a region of ± 250 μm from the weld line with a WZ of ± 80 μm, whereas the TMAZ in the full-scale specimen spanned ± 500 μm with a WZ of ± 150 μm [25]. It has been suggested that this strong α texture is the result of a $\{111\}\langle 111\rangle$ β-deformation texture generated during LFW. On subsequent cooling this β texture, in combination with strong variant selection during β-to-α phase transformation (12 different α variants can form from a single β grain, but here only a single variant was activated), resulted in the strong transverse texture observed at the weld line [25]. Although these strong textures have been reported in Ti-64 linear friction welds, more research is needed to determine the significance of these textures on the mechanical and physical properties of the weld.

Although textures close to the weld line in lab- and full-scale specimens were similar, they were quite different slightly further away from it [25]. In the full-scale specimens alternating bands of transverse and $\{11\overline{2}2\}\langle 11\overline{2}3\rangle$-type textures (c-axis of the

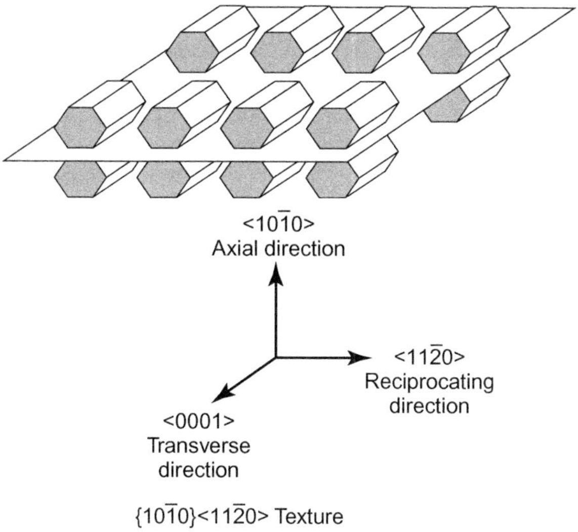

Fig. 4.14 Crystallite orientation at the weld line of lab- and full-scale Ti-64 linear friction welds [25]. The region of strong transverse texture spans ±100 μm from the weld line in the lab-scale specimens and ±50 μm in the full-scale specimens.

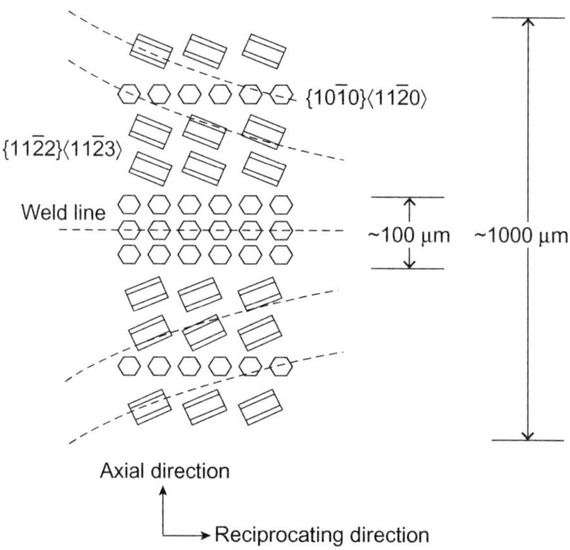

Fig. 4.15 α-crystallite orientation at and near the weld line in a full scale Ti-64 linear friction welded blisk [25].

α crystallites aligned almost parallel to the reciprocating direction) were observed a small distance away from the weld line (Fig. 4.15). It appears, therefore, that two (of the 12 possible) types of variant were selected on β-to-α transformation in full-scale welds. At present, this difference in variant selection between lab-scale and full-scale welds has not been explained.

The influence of welding parameters on texture in Ti-64 has been studied by a few authors. No trends could be identified between welding parameters and texture by

Dalgard et al. [34], but a strong influence of weld pressure on texture was seen by Romero et al. [27] (along with an influence on residual stress, as described in section 6.5). Trends may have been more apparent for Romero et al. [27] as a larger range of welding pressures was investigated. Results from Romero et al. [27] showed that when welding with a low applied pressure the before-mentioned strong transverse α texture is observed at the weld line. However, an almost random α texture developed at the weld line when high pressures were used. To date, the mechanisms by which welding pressure can minimize the α texture at the weld line is not entirely clear, but it seems obvious that variant selection is less active when using high applied pressures to weld Ti-64.

4.6.7 Modeling

There have been a few attempts at modeling the LFW of different titanium alloys, although similar concepts can of course be used to model welds in other materials. A heat-input equation for the equilibrium stage of the process was developed by Vairis and Frost [15], whereas results from finite element modeling have been reported by Jun et al., Vairis and Frost, Sorina-Muller et al. and Li et al. [35–38] Each of these finite element modeling studies used a power input at the weld interface that was dependent on the welding parameters and the coefficient of friction at the weld interface, and each of the studies were predictive rather than retrospective. Temperature-dependent material properties and coefficients of friction were used in a number of cases [37, 38].

Results from these studies are quite promising and generally matched experimental observations well. Jun et al. [35] found the discrepancy between the predicted and measured temperature profiles was smaller than 12%, whereas the predicted and measured upset differed by less than 16%. Vairis and Frost [36] found temperature measurements corresponded very well to the model in the initial stages of the welding process, but there were large discrepancies between predicted and measured temperatures during the later stages of LFW. Microstructural features and the burn-off was used to validate models by Sorina-Muller et al. [37], with a reasonable match between model and experimental validation shown.

An area where modeling will be particularly useful is in the identification and confirmation of trends in welding temperature with different welding parameters. Results described in this chapter clearly show that weld microstructures and properties can be changed significantly by changing welding parameters, and changes in welding temperature are thought to be responsible for these changes in weld properties. A model to better understand these trends would be very useful and could help to optimize welding parameters and achieve a particular set of weld properties. Unfortunately, no modeling studies, predictive or retrospective, have thus far been undertaken to analyze welds produced using a very wide range of welding parameters.

The energy required to produce welds and the efficiency of the process with different welding parameters has been studied by Ofem et al. [39] It was found that the process efficiency when welding medium carbon steel could be optimized by optimizing the frequency and amplitude welding parameters. This observation may be important for minimizing the energy usage of LFW machines in a production environment.

4.7 Linear friction welding of nickel-based superalloys

4.7.1 Mechanical properties and weldability

There have been a number of reports on the LFW of nickel-based super-alloys, but details of the mechanical properties of these welds is scarce. Karadge et al. [40] studied welds between a polycrystalline nickel-based super-alloy and a single-crystal alloy of different grade. Under ideal conditions the joints met design properties, although the particulars of these design properties were not mentioned. It was also shown that there was a relationship between weldability and the orientation of the single crystal. The materials were easiest to weld when the primary slip system of the single crystal was favorably orientated to give a high Schmid factor. When the single-crystal material was not orientated for easy activation of primary slip, the welding process was unsuccessful.

Crack-free linear friction welds have been reported in IN738 [41, 42] and the single crystal CMSX-486 [43], which are generally considered to be difficult to weld because of susceptibility to liquation and solidification cracking. The authors [41–43] suggested that although liquation does occur when these materials are linear friction welded, the materials are less prone to cracking because of the rapid re-solidification of the liquated phases. The compressive axial stresses imposed on the material during the frictional phase of the process, where the interface temperature is still rising, is thought to increase diffusion rates and allow solute diffusion from the liquid films into the solid matrix material. This is further thought to allow the rapid isothermal solidification of the previously liquid regions. This solid material is then less prone to tearing during cooling of the weld, when predominantly tensile stresses develop because of differences in cooling rates at different regions in the sample (see Section 4.6.5 on residual stresses).

An issue with welding nickel-based superalloys appears to be oxide formation at the weld line. Oxides have been reported in a number of different studies [40, 44, 45], and Chamanfar et al. [45] found these oxides were present at the extremities of the weld, and failure during tensile testing was initiated at these sites. It has been suggested that these materials be welded under vacuum or in an inert-gas environment, so that these oxides do not form, or large burn-off distances be used to allow any oxides to be expelled [45]. The amplitude used during welding could also be reduced in order to reduce the amount of material exposed to the atmosphere during part reciprocation and therefore reduce the likelihood of oxidation. Good fatigue properties have been reported for some welds between nickel-based superalloys, with welds between 720Li and IN718 having a resistance to fatigue-crack propagation at least comparable with the parent IN718 [46].

4.7.2 Microstructure

A significant loss in strength and hardness is seen at the weld line of nickel superalloys owing to the dissolution of strengthening precipitates [16, 44, 45]. Fig. 4.16 shows the variation in γ' and hardness in Waspalloy linear friction welds plotted against distance from the weld line. It shows that to a large extent the hardness drop matches the drop in γ'. There is a slight increase in hardness between 1.5 and 0.9 mm from the interface, which is thought to be caused by an increase in metal carbides. It was suggested that the dissolution of γ' could have provided Ti that could have reacted with carbon that was either dissolved in the alloy matrix, or was remaining after the dissolution of low-solvus-temperature metal carbides.

4.8 Linear friction welds in other materials

Aside from the materials already mentioned, LFW has been demonstrated to be capable of producing welds in [8]:

- aluminum alloys
- steels and stainless steels
- metal matrix composites
- intermetallic alloys
- tungsten-based alloys
- cobalt-based superalloys

Details of welds in some of these materials will be detailed in this section.

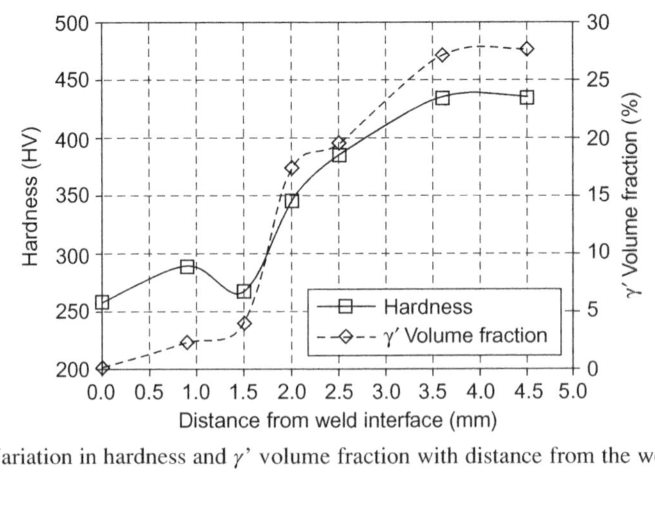

Fig. 4.16 Variation in hardness and γ' volume fraction with distance from the weld line [45].

4.8.1 Steels

Welds in a number of different steels have been attempted [23, 47–50]. Good mechanical properties have been reported in ferritic [49, 50] as well as austenitic stainless steels [23], with the strengths of these welds suggested to be superior than those of the respective parent materials. Indeed welds in a C–Mn steel [50] and in AISI316L austenitic stainless steel [23] demonstrated very good tensile strength unless very extreme welding parameters were used. This suggests that when certain steels are linear friction welded there is a large parameter window where sound welds are achievable. It would be beneficial to have a large welding-parameter window in a production environment.

Bhamji et al. [23] describe a detailed analysis that was undertaken into the effects of welding parameters on weld microstructure. In particular the variation of weld-line delta-ferrite (a high temperature phase in steels) fraction was observed with different welding parameters, and it was shown that at higher burn-off rates (rate at which plasticized material is ejected from the weld as flash) a lesser amount of delta-ferrite was formed at the weld interface. This was attributed to the quicker expulsion of hot plasticized material from the weld, which did not allow enough time at temperature for large amounts of delta-ferrite to form. Conversely in low burn-off rate welds a slow-moving layer of hot plasticized material was present that did allow time for delta-ferrite formation and therefore much greater amounts than in the parent material were present at the weld line. The highest burn-off rates occurred at mid-range frequencies and amplitudes and high forces.

4.8.2 Aluminum alloys and metal matrix composites

Linear friction welds in Al-Fe-V-Si 8009 [51] aluminum alloy showed a decrease in strength at the weld line relative to the parent material. The overaging or dissolution of hardening particles during the welding process may have been significant in reducing strength at the weld line in this material. In contrast welds in AA5083–0 (non-age-hardening and fully annealed) showed increased strength at the weld line relative to the parent material [19]. The LFW of aluminum matrix composites have also been reported [52–55], and results showed the welds to have good tensile and fatigue properties [56]. The welding of metal matrix composites may be a future application for LFW as these materials can be difficult to weld with fusion techniques [56].

4.8.3 Titanium and nickel aluminides

There has been some degree of success in LFW α_2 [57] and γ [58] titanium aluminides, and the welding of dissimilar titanium aluminide alloys have also been attempted [59]. When welding α_2 titanium aluminides it is important that cooling rates are kept to a minimum to avoid undesirable microstructures [57]. At high cooling rates the relatively slow β-to-α transformation causes a microstructure of retained β, which has low notch toughness, or α_2 martensite, which is very brittle. In order to produce a more desirable microstructure of a mixture of α_2 and β, very slow cooling rates are needed.

It has been suggested that this could be achieved by optimizing welding parameters, with a regime of low forces and amplitudes and high frequencies proposed [58]. Despite these difficulties crack-free welds have been produced in α_2 titanium aluminide.

Crack-free welds were also produced in a γ titanium aluminide [58]. A fine lamellar α_2/γ microstructure was formed in this material by using welding parameters that gave long welding times and therefore relatively slow heating and cooling rates.

There has also been some success in the welding of nickel aluminide, Ni_3Al [60, 61], although the mechanical properties of the welds have not been fully characterized. These results clearly show that LFW is a promising process for welding brittle intermetallic materials. Although α_2 titanium aluminide is now obsolete, γ titanium aluminide and Ni_3Al are successful and the welding of these materials may be a potential future application for LFW.

4.9 Conclusion

A review of published information on LFW has been conducted, and from this the following general conclusions can be drawn.

- LFW is a process that has grown to assume considerable importance in the past few decades. As understanding of the process continues, further important applications, both in aerospace and in other industries, are emerging.
- LFW can be used to join a number of different materials and should be appropriate for a number of different applications both within and outside of the aerospace industry. Welds in various titanium alloys, nickel-based superalloys, steels and aluminum alloys have been demonstrated. Good results have also been seen when welding aluminum metal matrix composites, as well as titanium and nickel aluminides.
- Welding parameters appear to have a significant influence on the welding temperature, and this property of the process may be important when welding challenging materials (e.g. when welding dissimilar materials). The modification and optimization of weld microstructure and properties just by modifying welding parameters may be one of the key benefits of the process.
- The peak interface temperature during LFW is very high, although it is not expected to exceed the melting points of the materials being welded. As a result of such high temperatures near the weld line, along with the applied force and plastic deformation taking place during welding, significant microstructural changes have been reported in linear friction welds.
- The microstructure of a linear friction weld is usually characterized by a TMAZ close to the weld line, a HAZ further from the weld line and the unaffected parent material microstructure further still. The TMAZ usually displays a region at the weld line that is recrystallized and contains fine grains (the WZ), and a region further away from the weld line that is heavily deformed but not recrystallized. The HAZ is affected by the transferred heat during welding, but is not plastically deformed (this region is not prominent in some materials, e.g. Ti-64).
- Residual stresses have been studied in a small number of linear friction welded systems, and in all cases significant stresses have been reported in the material close to the weld interface. In the case of titanium it has been demonstrated that these stresses can be relieved by more

than 75% when applying alloy-typical annealing procedures during PWHT. In addition, a significant reduction in stress generation was found when using high weld pressures.
- Results have shown that the modeling of LFW is feasible, although a large amount of work has not been completed on this aspect at present. Even though such a model might not be fully predictive at the first stage owing to the complexity of LFW, it is clear that it could help to significantly improve understanding of the process, and in this way provide strategies to further improve welding parameters for LFW.
- The use of LFW for the production of titanium aero-engine blisks is now established, and there are a number of advantages to the use of the process for this application.
- Use of the process for near-net-shape manufacture in high-value materials is being developed and is being seriously considered by the aerospace industry.

References

[1] I. Bhamji, M. Preuss, P.L. Threadgill, A.C. Addison, Solid state joining of metals by linear friction welding: a literature review, Mater. Sci. Technol. 27 (1) (2011) 2–12.

[2] A.W.E. Nentwig, Untersuchungen zum linear-reibsscheissen von metallen, Schweissen und Schneiden 47 (8) (1995) 648–653.

[3] V.I. Vill, Friction Welding of Metals, American Welding Society, New York, 1962.

[4] W. Richter, Herbeifuehrung einer haftverbindung zwischen plaettchen aus werkzeugstahl und deren traegern nach art einer schweissung oder loetung, Patent: DE477084, 1929.

[5] R. Maurya, J. Kauzlarich, Bonding apparatus—friction welding by reciprocal motion, Patent: Caterpillar Tractor Co, US3420428-A; DE1552871-A; CA844858-A, 1969.

[6] K.T. Slattery, Structural assembly's e.g. wing, preform for e.g. aircraft, has set of structural members connected by linear friction weld joint to base member, where each structural member defines connection surfaces, Patent: Boeing Co, US2007186507-A1, 2007.

[7] R.D. Trask, S.H. Goetschius, S.A. Hilton, Process for linear friction welding—comprises providing two members to be joined to one another at an interface by linear friction welding, where one of the two members is stub, etc, Patent: United Technologies Corp, EP850718-A; EP850718-A1; JP10193141-A; US5865364-A; EP850718-B1; DE69734232-E; DE69734232-T2, 1998.

[8] M.E. Nunn, Aero engine improvements through linear friction welding, in: Proc. 1st Int. Conf. on 'Innovation and Integration in Aerospace Sciences', Paper no 2005–0040, Queen's University Belfast, 4–5 August 2005, CEIAT, Queen's University Belfast, 2005.

[9] K.T. Slattery, Preform forming method for structural assembly, involves moving pair of structural units relative to two base units to form plasticized region of material, and urging structural units against surface of base units, Patent: Boeing Co, US2005127140-A1, US7225967-B2, 2005.

[10] D.H. Zhang, Y. Zhang, B.H. Wu, Research on the adaptive machining technology of blisk, in: J.L. Yuan, S.M. Ji, D.H. Wen, M. Chen (Eds.), Ultra-Precision Machining Technologies, Trans Tech Publications Ltd, Stafa-Zurich, 2009, pp. 446–450.

[11] P. Wanjara, M. Jahazi, Linear friction welding of Ti-6Al-4V: processing, microstructure, and mechanical-property inter-relationships, Metall. Mater. Trans. A 36A (8) (2005) 2149–2164.

[12] W.M. Thomas, Friction stir welding and related friction process characteristics, in: M.H. Ogle, P.L. Threadgill, S.J. Maddox (Eds.), Proc. 7th Int. Conf. on 'Joints in Aluminium – INALCO '98', Woodhead Publishing, 1999, pp. 157–174. TWI Ltd., Cambridge, UK, 15–17 April 1998.

[13] S. Kallee, Development of low-cost linear friction machine, Assembly Autom. 21 (2) (2001).

[14] BS EN 15620:2000, Welding—Friction welding of metallic materials, British Standards Institution, 2000.

[15] A. Vairis, M. Frost, On the extrusion stage of linear friction welding of Ti6Al4V, Mater. Sci. Eng. A-Struct. Mater. Prop. Microstruct. Process. 271 (1–2) (1999) 477–484.

[16] C. Mary, M. Jahazi, Linear friction welding of IN-718: process optimization and microstructure evolution, Adv. Mater. Res. 15–17 (2007) 357–362.

[17] P.L. Threadgill, Terminology in friction stir welding, Sci. Technol. Weld. Join. 12 (2007) 357–360.

[18] M. Corzo, O. Casals, J. Alcala, A. Mateo, M. Anglada, Mechanical evaluation of linear friction welds in titanium alloys through indentation experiments, Revista De Metal. 41 (6) (2005) 403–409.

[19] A.C. Addison, Linear Friction Welding of Engineering Metals, TWI Ltd. Core research program report, 2008.

[20] T.J. Ma, W.Y. Li, S.Y. Yang, Impact toughness and fracture analysis of linear friction welded Ti-6Al-4V alloy joints, Mater. Des. 30 (6) (2009) 2128–2132.

[21] A. Mateo, M. Corzo, M. Angada, J. Mendez, P. Villechaise, J.P. Ferte, O. Roder, Welding repair by linear friction in titanium alloys, Mater. Sci. Technol. 25 (7) (2009) 905–913.

[22] A. Vairis, M. Frost, High frequency linear friction welding of a titanium alloy, Wear 217 (1) (1998) 117–131.

[23] I. Bhamji, M. Preuss, P.L. Threadgill, R.J. Moat, A.C. Addison, M.J. Peel, Linear friction welding of AISI 316L stainless steel, Mater. Sci. Eng. A-Struct. Mater. Prop. Microstruct. Process. 528 (2) (2010) 680–690.

[24] D. Helm, G. Lutjering, Microstructure and properties of friction-welds in titanium alloys, in: I.V. Gorynin, S.S. Ushkov (Eds.), Proc. 9th Int. Conf. 'Titanium '99', CRISM, 1999, pp. 1726–1733. St Petersburg, Russia, 7–11 June 1999.

[25] M. Karadge, M. Preuss, C. Lovell, P.J. Withers, S. Bray, Texture development in Ti-6Al-4V linear friction welds, Mater. Sci. Eng. A 459 (1–2) (2007) 182–191.

[26] M.M. Attallah, M. Preuss, S. Bray, Microstructural development during linear friction welding of titanium alloys, in: S.A. David (Ed.), Proc. 8th Int. Conf. on 'Trends in Welding Research', ASM, 2008. Pine Mountain, Georgia, 1–6 June 2008.

[27] J. Romero, M.M. Attallah, M. Preuss, M. Karadge, S.E. Bray, Effect of the forging pressure on the microstructure and residual stress development in Ti6-Al-4V linear friction welds, Acta Mater. 57 (18) (2009) 5582–5592.

[28] M.R. Daymond, N.W. Bonner, Measurement of strain in a titanium linear friction weld by neutron diffraction, Phys. B-Condens. Matter 325 (1–4) (2003) 130–137.

[29] P. Frankel, M. Preuss, A. Steuwer, P.J. Withers, S. Bray, Comparison of residual stresses in Ti-6Al-4V and Ti-6Al-2Sn-4Zr-2Mo linear friction welds, Mater. Sci. Technol. 25 (5) (2009) 640–650.

[30] M. Karadge, P. Frankel, A. Steuwer, C. Lovell, S. Bray, P.J. Withers, M. Preuss, Microstructure, texture, local tensile properties and residual stress relief in Ti6-Al-4V linear friction welds, in: Proc. of 'Materials Science & Technology, Association for Iron and Steel Technology, Cincinnati, Ohio, 2006, pp. 35–46.

[31] M. Preuss, J. Quinta da Fonseca, A. Steuwer, L. Wang, P.J. Withers, S. Bray, Residual stresses in linear friction welded IMI550, J. Neutron Res. 12 (1) (2004) 165–173.

[32] P.J. Withers, Residual stress and its role in failure, Rep. Prog. Phys. 70 (12) (2007) 2211–2264.

[33] M.M. Attallah, M. Preuss, P.J. Withers, S. Bray, Influence of linear friction welding parameters on the residual stress development in Ti-6246, in: S.A. David, T. DebRoy, J.N. DuPont, T. Koseki, H.B. Smartt (Eds.), Trends in Welding Research, Asm International, Materials Park, 2009, pp. 751–757.

[34] E. Dalgard, M. Jahazi, J.J. Jonas, Texture and microstructural evolution during linear friction welding of Ti-6Al-4V, in: R.S. Mishra, M.W. Mahoney, T.J. Lienert (Eds.), Friction Stir Welding and Processing V, Warrendale, Minerals, Metals & Materials Soc., 2009, pp. 55–64.

[35] T. Jun, Z. Tiancang, L. Pengtao, L. Jing, M. Yu, Numerical computation of a linear friction welding process, Phys. Numer. Simulat. Mat. Process. 575–578 (Pts 1 and 2) (2008) 811–815.

[36] A. Vairis, M. Frost, Modelling the linear friction welding of titanium blocks, Mater. Sci. Eng. A 292 (1) (2000) 8–17.

[37] J. Sorina-Muller, M. Rettenmayr, D. Schneefeld, O. Roder, W. Fried, FEM simulation of the linear friction welding of titanium alloys, Comput. Mater. Sci. 48 (4) (2010) 749–758.

[38] W.-Y. Li, T. Ma, J. Li, Numerical simulation of linear friction welding of titanium alloy: effects of processing parameters, Mater. Des. 31 (3) (2010) 1497–1507.

[39] U.U. Ofem, P.A. Colegrove, A. Addison, M.J. Russell, Energy and force analysis of linear friction welds in medium carbon steel, Sci. Technol. Weld. Join. 15 (6) (2010) 479–485.

[40] M. Karadge, M. Preuss, P.J. Withers, S. Bray, Importance of crystal orientation in linear friction joining of single crystal to polycrystalline nickel-based superalloys, Mater. Sci. Eng. A 491 (1–2) (2008) 446–453.

[41] O.T. Ola, O.A. Ojo, P. Wanjara, M.C. Chaturvedi, Enhanced resistance to weld cracking by strain-induced rapid solidification during linear friction welding, Philos. Mag. Lett. 91 (2) (2011) 140–149.

[42] O.T. Ola, O.A. Ojo, P. Wanjara, M.C. Chaturvedi, Analysis of microstructural changes induced by linear friction welding in a nickel-base superalloy, Metall. Mater. Trans. A 42 (12) (2011) 3761.

[43] O.T. Ola, O.A. Ojo, P. Wanjara, M.C. Chaturvedi, Linear friction welding of single crystal nickel-base CMSX-486 superalloy, in: 16th International Conference & Exhibition on the Joining of Materials Tisvildeleje, Denmark, Institute for the Joining of Materials, 2011.

[44] C. Mary, M. Jahazi, Multi-scale analysis of IN-718 microstructure evolution during linear friction welding, Adv. Eng. Mater. 10 (6) (2008) 573–578.

[45] A. Chamanfar, M. Jahazi, J. Gholipour, P. Wanjara, S. Yue, Mechanical property and microstructure of linear friction welded WASPALOY, Metall. Mater. Trans. A-Phys. Metall. Mater. Sci. 42A (3) (2011) 729–744.

[46] H.Y. Li, Z.W. Huang, S. Bray, G. Baxter, P. Bowen, High temperature fatigue of friction welded joints in dissimilar nickel based superalloys, Mater. Sci. Technol. 23 (12) (2007) 1408–1418.

[47] B.J. Ginn, T.G. Gooch, Toughness of 12%Cr ferritic/martensitic steel welds produced by non-arc welding processes, Weld. J.l 77 (8) (1998) 341S–349S.

[48] W.Y. Li, T.J. Ma, S.Q. Yang, Q.Z. Xu, Y. Zhang, J.L. Li, H.L. Liao, Effect of friction time on flash shape and axial shortening of linear friction welded 45 steel, Mater. Lett. 62 (2) (2008) 293–296.

[49] T.J. Ma, W.Y. Li, Q.Z. Xu, Y. Zhang, J.L. Li, S.Q. Yang, H.L. Liao, Microstructure evolution and mechanical properties of linear friction welded 45 steel joint, Adv. Eng. Mater. 9 (8) (2007) 703–707.

[50] A.C. Addison, P.L. Threadgill, Initial studies of linear friction welding of a C—Mn steel, Weld. Cut. 9 (6) (2010) 364–370.

[51] H.H. Koo, W.A. Baeslack, Structure, properties, and fracture of linear friction welded Al-Fe-V-Si alloy 8009, Mater. Charact. 28 (2) (1992) 157–164.

[52] L. Ceschini, A. Morri, F. Rotundo, A. Korsunsky, T.S. Jun, Linear friction welding (LFW) of metal matrix composites, Metall. Ital. 3 (2010) 23–30.

[53] R.J. Harvey, M. Strangwood, M.B.D. Ellis, Bond-line structures in friction welded Al2O3 particle reinforced aluminium alloy metal matrix composites (MMC's), in: H.B. Smartt, J. A. Johnson, S.A. David (Eds.), Proc. 4th Int. Conf. on 'Trends in Welding Research, ASM, Gatlingburg, Tennessee, 1995, pp. 803–808. 5–8 June 1995.

[54] T.S. Jun, A.M. Korsunsky, Eigenstrain reconstruction method in linear friction welded aluminium alloy and MMC plates, Int. J. Numer. Methods Eng. 84 (8) (2010) 989–1008.

[55] T.S. Jun, F. Rotundo, L. Ceschini, A.M. Korsunsky, A study of residual stresses in Al/SiCp linear friction weldment by energy-dispersive neutron diffraction, Adv. Fract. Dam. Mech. VII 385–387 (2008) 517–520.

[56] F. Rotundo, L. Ceschini, A. Morri, T.S. Jun, A.M. Korsunsky, Mechanical and microstructural characterization of 2124Al/25 vol%SiCp joints obtained by linear friction welding (LFW), Compos. Part A: Appl. Sci. Manufact. 41 (9) (2010) 1028–1037.

[57] P.L. Threadgill, The prospects for joining titanium aluminides, Mater. Sci. Eng. A-Struct. Mater. Prop. Microstruct. Process. 193 (1995) 640–646.

[58] W.A. Baeslack, P.L. Threadgill, E.D. Nicholas, T.F. Broderick, Linear friction welding of Ti-48Al-2Cr-2Nb (at. pct) titanium aluminide, in: P.A. Blenkinsop, W.J. Evans, H.M. Flower (Eds.), Proc. of 'Titanium '95 – Science and Technology', Institute of Materials, Birmingham, UK, 1996, pp. 424–431. 22–26 October 1995.

[59] W.A. Baeslack, T.F. Broderick, M. Juhas, H.L. Fraser, Characterization of solid-phase welds between TI-6Al-2Sn-4Zr-2Mo-0.1Si and Ti-13.5Al-21.5Nb titanium aluminide, Mater. Charact. 33 (4) (1994) 357–367.

[60] P.L. Threadgill, Joining of a nickel aluminide alloy, in: H.B. Smartt, J.A. Johnson, S.A. David (Eds.), Proc. 4th Int. Conf. on 'Trends in Welding Research', ASM, Gatlingburg, Tennessee, 1995. 5–8 June 1995.

[61] P.L. Threadgill, Solid state joining of nickel aluminides: Preliminary study, in: F.H. Froes, T. Khan, C.M. Ward-Close (Eds.), Proc. 3rd Int. Conf. on 'Synthesis, Processing and Modelling of Advanced Materials', ASM, Paris, France, 1997, pp. 153–158. 25–27 June 1997.

Hybrid laser-arc welding in aerospace engineering

5

J. Zhou[a], T.T. Zhang[b], H.L. Tsai[c], and P.C. Wang[d]
[a]Pennsylvania State University, The Behrend College, Erie, PA, United States, [b]Taiyuan University of Technology, Taiyuan, China, [c]Missouri University of Science and Technology, Rolla, MO, United States, [d]General Motors Corporation, Alpharetta, GA, United States

5.1 Introduction

In the last three decades, hybrid laser-arc welding has been increasingly applied to various applications in engineering [1–3]. As shown in Fig. 5.1, in hybrid laser-arc welding, the laser and the arc are integrated to provide primary and secondary heating sources for the joining process. Due to the synergic action of the laser beam and welding arc, hybrid welding offers many advantages over laser welding and arc welding alone [4–7], such as higher welding speed [6], deeper penetration [8], better weld quality with reduced susceptibility to pores and cracks [7, 9–22], good gap bridging ability [23–32], as well as process stability, as shown in Fig. 5.2 [6].

The development of the hybrid laser-arc welding technique can be divided into three stages [1]. The concept of hybrid laser welding was first proposed by Steen et al. [33–35] in the late 70s. In their studies, a CO_2 laser was combined with a tungsten inert gas (TIG) arc for welding and cutting applications. Their tests showed clear benefits of combining an arc and a laser beam, such as a stabilized arc behavior under the influence of laser radiation; a dramatic increase in the speed of welding of thin sheets; and an increase in penetration depth compared with laser welding. Japanese researchers continued Steen's effort and developed various methods and corresponding devices for laser-arc welding, cutting, and surface treatment. However, these efforts did not result in the introduction of this joining technique into engineering applications particularly because laser welding itself was not yet an economic and viable joining technique at that time [19]. In the second stage of the development of the hybrid laser welding technique, the observed influencing of the arc column behavior by laser radiation was used to improve the efficiency of arc welding processes, which leads to the laser enhanced arc welding technology [1]. A characteristic feature of this technology was that only a low-intensity laser beam was needed, i.e., the required laser power was small compared to the arc power. For TIG welding, Cui and Decker [37–40] demonstrated that a low-energy CO_2 laser beam with a power of merely 100 W could facilitate arc ignition; enhance arc stability; improve weld quality; and increase welding speed due to a reduced arc size and higher arc amperages. However, despite such reported improvements of the arc welding process by laser support, there were neither subsequent extensive

Welding and Joining of Aerospace Materials. https://doi.org/10.1016/B978-0-12-819140-8.00005-5
Copyright © 2021 Elsevier Ltd. All rights reserved.

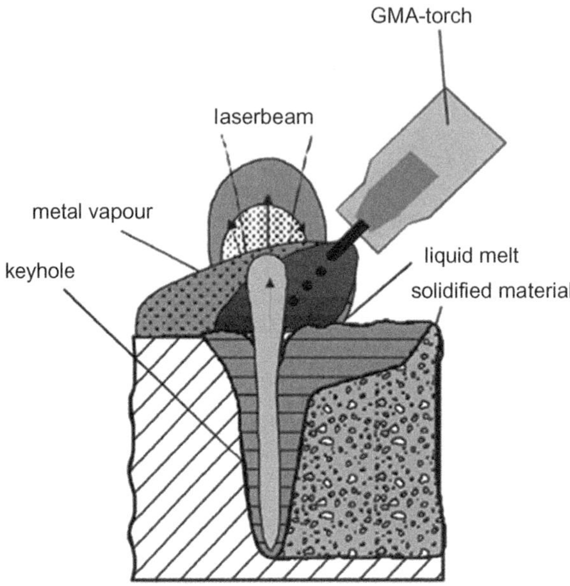

Fig. 5.1 Schematic diagram of a hybrid laser-GMA welding process.

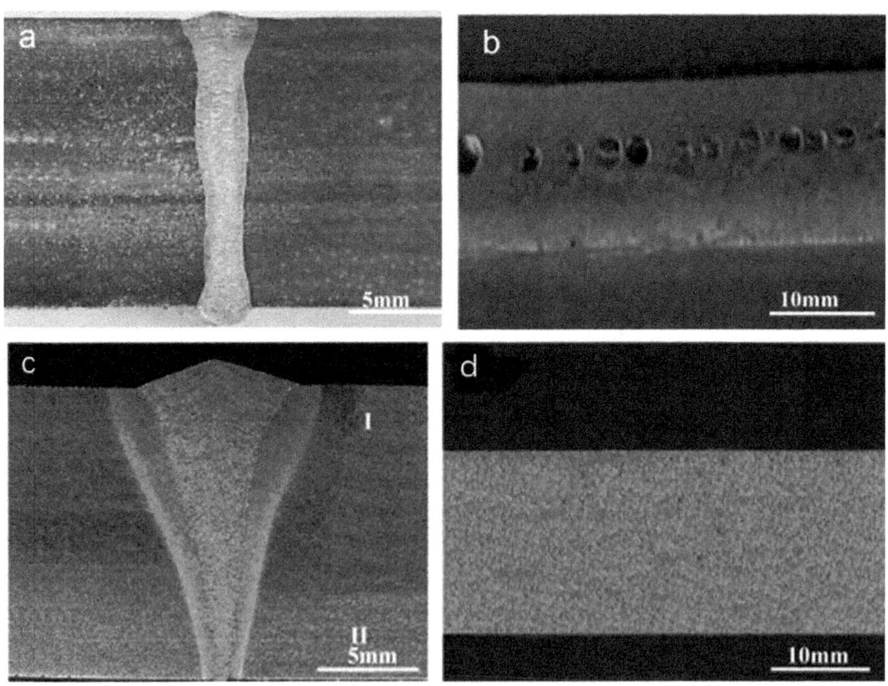

Fig. 5.2 Comparison between laser welding and hybrid welding: (A) weld section and (B) weld defect of laser welding; (C) weld section and (D) weld defect of hybrid welding.

investigations of this subject nor known industrial applications of the laser-enhanced arc welding technology. The third stage of hybrid welding technology started in the early 1990s with the development of combined welding processes using a high-power laser beam as the primary and an additional electric arc as the secondary heating source [41–48]. At that time, although the continuous wave CO_2 laser welding process was already well established in industry, it had some known disadvantages, e.g., high requirements of edge preparation and clamping; high solidification rates leading to material-dependent pores and cracks; as well as the high investment and operating costs for the laser equipment. Additionally, some welding applications of highly practical interest could not be solved satisfactorily by the laser welding process alone, e.g., joining of tailored blanks in the automotive engineering; welding of heavy plate under the conditions of the shipbuilding industry; as well as high speed welding of crack-susceptible materials. In searching for suitable solutions, a hybrid welding technique was developed into a viable joining technique with significant industrial acceptance during the last two decades.

According to the combination of various heating sources used, hybrid welding can be generally categorized as: (1) laser-gas tungsten arc (GTA) welding; (2) laser-gas metal arc (GMA) welding; and (3) laser-plasma welding [36]. Since laser welding offers deep penetration, primary heating sources commonly used in hybrid welding are CO_2, Nd: YAG, and fiber lasers. The first two types of lasers are well established in practice and used for various hybrid welding process developments. While the fiber laser is still in development for industrial applications, it seems to be a future primary heating source for hybrid welding due to its high beam quality. The secondary heating sources used in hybrid welding are mainly electric arcs. Dedicated processes can be divided into GMA welding with consumable electrodes and GTA welding with non-consumable tungsten electrodes. In GMA welding, the arc is burning between a mechanically supplied wire electrode and the workpiece. The shielding gas used in GMA welding was found to have significant effects on arc shape and metal transfer [49, 50]. Hence, GMA welding can be subdivided into metal inert-gas (MIG) and metal active-gas (MAG) welding according to the type of shielding gas used. In GTA welding, a chemically inert gas, such as argon or helium, is often used. A special form of this is the plasma arc welding (PAW), which produces a squeezed arc due to a special torch design and results in a more concentrated arc spot.

In hybrid welding, the laser and arc are arranged preferably in a way that they can compensate and benefit from each other during the welding process, which implies the creation of a common interaction zone with changed characteristics in comparison to laser welding and arc welding alone. In contrast to this is the arrangement in which the laser and arc are serving as two separate heating sources during the welding process. Several configurations have been proposed. In a parallel arrangement, there is a distance in either the vertical or horizontal direction along the path between both heating sources. In a serial arrangement, the primary and secondary heating sources are moved along the same welding path with a certain working distance, and the secondary heating source can either lead or follow the primary heating source [1]. The first one enables a preheating of the region to be welded. It can increase the efficiency of the laser welding process because the material to be welded is locally preheated and energy losses by heat conduction are reduced. In comparison, the second one often acts like a short-time post-heat treatment of the weld that can change the weld microstructure favorably. It is worth considering that there is an essential difference between parallel and serial process arrangement. In a serial arrangement, additional energy is dissipated within the weld seam region, whereas the parallel arrangement only reduces the heat flow across the weld seam. The option to move the working area temporally enables flexibility in influencing the cooling rates in order to avoid defects.

Hybrid laser-arc welding has been used to weld a wide variety of metals from ferrous metals like steels, stainless steels to the non-ferrous metals like aluminum, magnesium, nickel, copper, titanium and their alloys, etc. Current understanding of hybrid laser-arc welding is primarily based on experimental observations. In order to expand the applications of this joining technique in some special areas and optimize the process of its current applications, knowledge of the mechanisms governing the physical processes in hybrid laser-arc welding must be better understood. In the following, some fundamental physics involved in hybrid laser-arc welding will be discussed.

5.2 Fundamentals of hybrid laser-arc welding

Since hybrid laser-arc welding involves laser welding, arc welding and their interactions as well, complicated physical processes like metal melting and solidification; melt flow; keyhole plasma formation; arc plasma formation and convection are typically involved, which results in very complex transport phenomena in this welding process [51]. As known, transport phenomena in welding, such as heat transfer; melt flow; and plasma flow, can strongly affect both metallurgical structure and mechanical properties of the weld [52–56]. In the following, transport phenomena in hybrid welding will be discussed and particular attention is given to (1) arc plasma formation and its effect on metal transfer and weld pool dynamics; (2) laser-induced plasma formation and laser-plasma interaction; (3) recoil pressure and other possible mechanisms contributing to keyhole formation and dynamics; (4) the interplay among various process parameters; and (5) plasma-filler metal-weld pool interactions.

Due to the different nature of heat and mass transfer mechanisms in metal and plasma, separate models are developed for the studies of fundamental physics in hybrid laser-arc welding. One is for the metal region containing base metal, electrode, droplets, and arc plasma. The other is for the keyhole region containing laser-induced plasma. There is a free surface (liquid/vapor interface) separating these two regions. For the metal region, continuum formation is used to calculate the energy and momentum transport [51]. For the keyhole plasma region, laser-plasma interaction and the laser energy absorption mechanism will be discussed. These two regions are coupled together and the VOF technique is used to track the interface between these two regions [51].

5.2.1 Transport phenomena in metal (electrode, droplets, and workpiece) and arc plasma

Differential equations governing the conservation of mass, momentum and energy based on continuum formulation are given below [57]:

5.2.1.1 Conservation of mass

$$\frac{\partial}{\partial t}(\rho) + \nabla \cdot (\rho \boldsymbol{V}) = 0 \tag{5.1}$$

where t is the time, ρ is the density, and $\boldsymbol{V}$ is the velocity vector.

5.2.1.2 Conservation of momentum

$$\frac{\partial}{\partial t}(\rho u) + \nabla \cdot (\rho \boldsymbol{V} u) = \nabla \cdot \left(\mu_l \frac{\rho}{\rho_l} \nabla u\right) - \frac{\partial p}{\partial x} - \frac{u_l}{K} \frac{\rho}{\rho_l}(u - u_s) - \frac{C\rho^2}{K^{0.5}\rho_l}|u - u_s|(u - u_s)$$
$$- \nabla \cdot (\rho f_s f_l \boldsymbol{V_r} u_r) + \nabla \cdot \left(\mu_s u \nabla \left(\frac{\rho}{\rho_l}\right)\right) + \boldsymbol{J} \times \boldsymbol{B}|_x \tag{5.2}$$

$$\frac{\partial}{\partial t}(\rho v) + \nabla \cdot (\rho \boldsymbol{V} v) = \nabla \cdot \left(\mu_l \frac{\rho}{\rho_l} \nabla v\right) - \frac{\partial p}{\partial y} - \frac{u_l}{K} \frac{\rho}{\rho_l}(v - v_s) - \frac{C\rho^2}{K^{0.5}\rho_l}|v - v_s|(v - v_s)$$
$$- \nabla \cdot (\rho f_s f_l \boldsymbol{V_r} v_r) + \nabla \cdot \left(\mu_s v \nabla \left(\frac{\rho}{\rho_l}\right)\right) + \boldsymbol{J} \times \boldsymbol{B}|_y \tag{5.3}$$

$$\frac{\partial}{\partial t}(\rho w) + \nabla \cdot (\rho \boldsymbol{V} w) = \rho g + \nabla \cdot \left(\mu_l \frac{\rho}{\rho_l} \nabla w\right) - \frac{\partial p}{\partial z} - \frac{u_l}{K} \frac{\rho}{\rho_l}(w - w_s) - \frac{C\rho^2}{K^{0.5}\rho_l}|w - w_s|(w - w_s)$$
$$- \nabla \cdot (\rho f_s f_l \boldsymbol{V_r} w_r) + \nabla \cdot \left(\mu_s w \nabla \left(\frac{\rho}{\rho_l}\right)\right) + \rho g \beta_T (T - T_0) + F_{drag} + \boldsymbol{J} \times \boldsymbol{B}|_z \tag{5.4}$$

where u, v and w are the velocities in the x-, y- and z-directions, respectively, and $\boldsymbol{V_r}$ is the relative velocity vector between the liquid phase and the solid phase. $\boldsymbol{J}$ is the current field vector and $\boldsymbol{B}$ is the magnetic field vector. The subscripts s and l refer to the solid and liquid phases, respectively; Subscript 0 represents the reference conditions; p is the pressure; μ is the viscosity; f is the mass fraction; K, the permeability, is a measure of the ease with which fluid passes through the porous mushy zone; C is the inertial coefficient; β_T is the thermal expansion coefficient; g is the gravitational acceleration; and T is the temperature.

5.2.1.3 Conservation of energy

$$\frac{\partial}{\partial t}(\rho h)+\nabla\cdot(\rho \boldsymbol{V} h)=\nabla\cdot\left(\frac{k}{c_p}\nabla h\right)-\nabla\cdot\left(\frac{k}{c_p}\nabla(h_s-h)\right)$$
$$-\nabla\cdot(\rho(\boldsymbol{V}-\boldsymbol{V}_s)(h_l-h))+\frac{|\boldsymbol{J}|}{\sigma_e}-S_R+\frac{5k_b}{2e}\boldsymbol{J}\cdot\frac{\nabla h}{c_p} \tag{5.5}$$

where h is the enthalpy, k is the thermal conductivity, and c_p is the specific heat. The first two terms on the right-hand side of Eq. (5.5) represent the net Fourier diffusion flux. The third term represents the energy flux associated with the relative phase motion. σ_e is the electrical conductivity; S_R is the radiation heat loss; k_b is the Stefan-Boltzmann constant; and e is the electronic charge.

The third and fourth terms on the right-hand side of Eqs. (5.2)–(5.4) represent the first and second order drag forces of the flow in the mushy zone. The fifth term represents an interaction between the solid and the liquid phases due to the relative velocity. The second term on the right hand side of Eq. (5.5) represents the net Fourier diffusion flux. The third term represents the energy flux associated with the relative phase motion. All these aforementioned terms in this paragraph are zero except in the mushy zone. In addition, the solid phase is assumed to be stationary ($\mathbf{V}_S=0$).

5.2.1.4 Conservation of species

$$\frac{\partial}{\partial t}(\rho f^\alpha)+\nabla\cdot(\rho \boldsymbol{V} f^\alpha)=\nabla\cdot(\rho D\nabla f^\alpha)-\nabla\cdot\left(\rho D\nabla\left(f_l^\alpha-f^\alpha\right)\right)$$
$$-\nabla\cdot\left(\rho(\boldsymbol{V}-\boldsymbol{V}_s)\left(f_l^\alpha-f^\alpha\right)\right) \tag{5.6}$$

where D is a mass diffusivity and f^α is a mass fraction of constitute. Subscript, l and s, represents liquid and solid phase, respectively.

5.2.2 Transport phenomena in laser induced plasma

The vapor inside the keyhole is modeled as a compressible, inviscid ideal gas. No vapor flow is assumed in the keyhole and the energy equation is given in the following form [58]:

$$\frac{\partial}{\partial t}(\rho_v h_v)=\nabla\cdot\left(\frac{k_v}{c_v}\nabla h_v\right)+\nabla\cdot(-\mathrm{q}_\mathrm{r})+k_{pl}\cdot I_{laser}\cdot(1-\alpha_{iB,1})$$
$$+\sum_{mr=1}^{n}k_{pl}I_{laser}\cdot(1-\alpha_{iB,1})\cdot(1-\alpha_{Fr})\cdot(1-\alpha_{iB,mr}) \tag{5.7}$$

where h_v and ρ_v represent the enthalpy and density of the plasma; k_v and c_v represent the thermal conductivity and specific heat of the plasma. The first term on the right-hand side of Eq. (5.7) represents the heat conduction term. The second term represents the radiation heat term and $\boldsymbol{q}_r$ stands for the radiation heat flux vector. The fourth term represents energy input from the original laser beam. The last term represents the energy input from multiple reflections of the laser beam inside the keyhole.

5.2.3 Electrical potential and magnetic field

Arc plasma from GMA welding will not only provide heat to the base metal, but will also exert magnetic force on the weld pool. The electromagnetic force can be calculated as follows [59]:

5.2.3.1 Conservation of current

$$\nabla \cdot (\sigma_e \nabla \varphi) = 0 \tag{5.8}$$

$$\boldsymbol{J} = -\sigma_e \nabla \varphi \tag{5.9}$$

where φ is the electrical potential. According to Ohm's law, the self-induced magnetic field B_θ is calculated by the following Ampere's law:

$$B_\theta = \frac{\mu_0}{r} \int_o^r j_z r dr \tag{5.10}$$

where $\mu_0 = 4\pi \times 10^{-7}$H m^{-1} is the magnetic permeability of free space. Finally, three components of the electromagnetic force in Eqs. (5.2)–(5.4) are calculated via

$$\boldsymbol{J} \times \boldsymbol{B}|_x = -B_\theta j_z \frac{x - x_a}{r} \tag{5.11}$$

$$\boldsymbol{J} \times \boldsymbol{B}|_y = -B_\theta j_z \frac{y}{r} \tag{5.12}$$

$$\boldsymbol{J} \times \boldsymbol{B}|_z = -B_\theta j_r \tag{5.13}$$

5.2.4 Arc plasma and its interaction with metal zone (electrode, droplets, and weld pool)

In welding, shielding gas is ionized and forms a plasma arc between the electrode and workpiece. In the arc region, the plasma is assumed to be in local thermodynamic equilibrium (LTE) [60], implying the electron and the heavy particle temperatures are equal. On this basis, the plasma properties, including enthalpy, specific heat, density, viscosity, thermal conductivity and electrical conductivity, are determined from an equilibrium composition calculation [60]. It is noted that the metal vaporized from the metal surface may influence plasma material properties, but this effect is omitted in the present study. It is also assumed that the plasma is optically thin, thus the radiation may be modeled in an approximate manner by defining a radiation heat loss per unit volume [48]. The transport phenomena in the arc plasma and the metal are calculated separately in the corresponding arc domain and metal domain, and the two domains are coupled through interfacial boundary conditions in each time step.

5.2.4.1 Heat transfer

At the plasma-electrode interface, there exists an anode sheath region [60]. In this region, the mixture of plasma and metal vapor departs from LTE, thus it no longer complies with the model presented above. Since the sheath region is very thin, it is treated as a special interface to take into account the thermal effects on the electrode. The energy balance equation at the surface of the anode is modified to include an additional source term, S_a [61, 62], for the metal region.

$$S_a = \frac{k_{eff}(T_{arc} - T_a)}{\delta} + J_a\varphi_w - q_{rad} - q_{evap} \tag{5.14}$$

The first term on the right-hand side of Eq. (5.14) is the contribution due to heat conduction from the plasma to the anode. The symbol k_{eff} represents the thermal conductivity taken as the harmonic mean of the thermal conductivities of the arc plasma and the anode material. δ is the length of the anode sheath region. T_{arc} is the arc temperature and T_a is the temperature of the anode. The second term represents the electron heating associated with the work function of the anode material. J_a is the current density at the anode and φ_w is the work function of the anode material. The third term q_{rad} is the black body radiation loss from the anode surface. The final term q_{evap} is the heat loss due to the evaporation of electrode materials.

Similar to the anode region, there exists a cathode sheath region between the plasma and the cathode. However, the physics of the cathode sheath and the energy balance at the nonthermionic cathode for GMA welding are not well understood [61–67]. The thermal effect due to the cathode sheath has been omitted in many models and reasonable results were obtained [61–65]. Thus, the energy balance equation at the cathode surface will only have the conduction, radiation, and evaporation terms.

$$S_a = \frac{k_{eff}(T_{arc} - T_c)}{\delta} - q_{rad} - q_{evap} \tag{5.15}$$

where k_{eff} is the effective thermal conductivity at the arc-cathode surface taken as the harmonic mean of the thermal conductivities of the arc plasma and the cathode material. δ is the length of the cathode sheath. T_c is the cathode surface temperature.

5.2.4.2 Force balance

The molten part of the metal is subjected to body forces, such as gravity and electromagnetic force. It is also subjected to surface forces, such as surface tension due to surface curvature, Marangoni shear stress due to temperature difference, and arc plasma shear stress and arc pressure at the interface of arc plasma and metal. For cells containing a free surface, surface tension pressure normal to the free surface can be expressed as [68].

$$p_s = \gamma\kappa \tag{5.16}$$

where γ is the surface tension coefficient and κ is the free surface curvature.

The temperature-dependent Marangoni shear stress at the free surface in a direction tangential to the local free surface is given by [69].

$$\tau_{Ms} = \frac{\partial\gamma}{\partial T}\frac{\partial T}{\partial \mathbf{s}} \tag{5.17}$$

where $\mathbf{s}$ is a vector tangential to the local free surface.

The arc plasma shear stress is calculated at the free surface from the velocities of arc plasma cells immediately adjacent the metal cells.

$$\tau_{ps} = \mu\frac{\partial \mathbf{V}}{\partial T} \tag{5.18}$$

where μ is the viscosity of arc plasma.

The arc pressure at the metal surface is obtained from the computational result in the arc region. The surface forces are included by adding source terms to the momentum equations according to the CSF (continuum surface force) model [68]. Using F of the VOF function as the characteristic function, surface tension pressure, Marangoni shear stress, arc plasma shear stress, and arc pressure are all transformed to the localized body forces and added to the momentum transport equations as source terms for the boundary cells. Based on these assumptions, Hu et al. has successfully simulated the droplet formation and impingement and arc formation in a GMA welding process, as shown in Figs. 5.3 and 5.4 [70].

5.2.5 Laser-induced recoil pressure and keyhole dynamics

In the laser welding process, the laser beam is directed to the metal surface, which melts the material forming a small molten pool. The liquid metal is heated to high temperatures resulting in large evaporation rates. The rapid evaporation creates a large recoil pressure on the surface of the molten layer depressing it downward. Then, a keyhole is formed. Many investigators believe that this recoil pressure balanced with surface tension force will determine the shape of the keyhole. So the consideration of this recoil pressure is critical for determining the laser welding process. This recoil pressure results from the rapid evaporation from the surface that has been heated to high temperatures. When the liquid-vapor interface temperature reaches the boiling point, evaporation begins to occur. The flow field is shown in Fig. 5.5. There is a Knudsen layer adjacent to the liquid surface where the vapor escaping from the liquid surface is in a state of thermodynamic non-equilibrium, i.e., the vapor molecules do

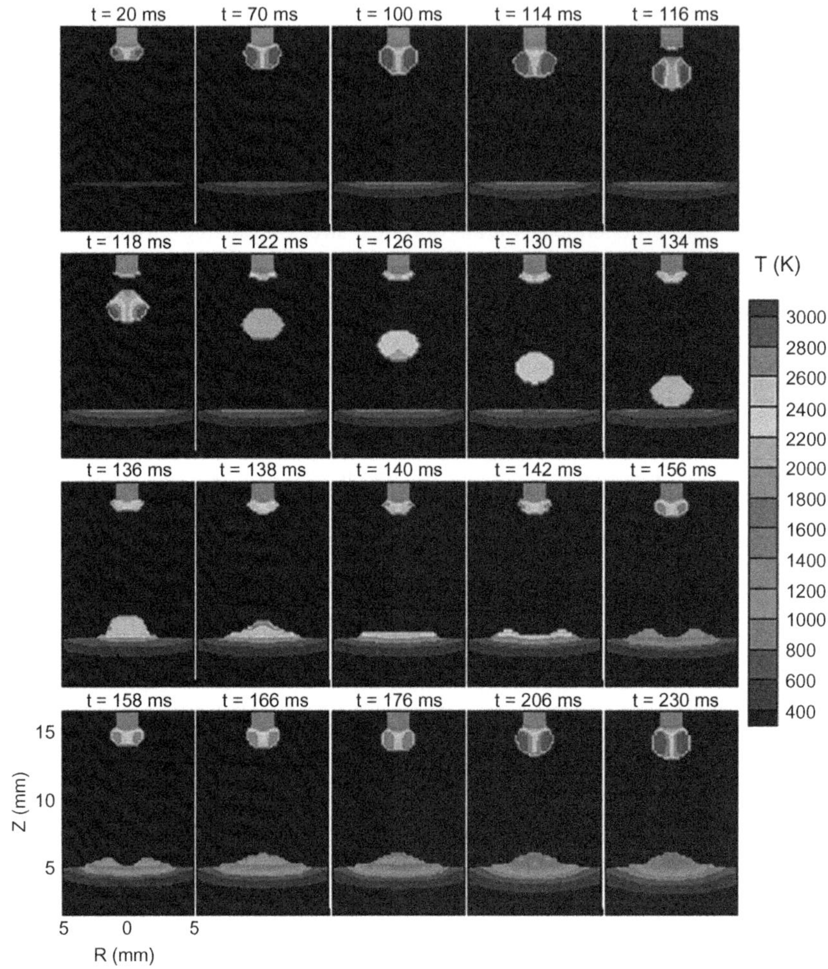

Fig. 5.3 Droplet formation and impingement in a GMA welding process.

not have a Maxwellian velocity distribution. This occurs when the equilibrium vapor pressure (i.e., the saturation pressure) corresponding to the surface temperature is large compared to the ambient partial pressure of the vapor. Under these conditions the vapor adjacent to the surface is dominated by recently evaporated material that has not yet experienced the molecular collisions necessary to establish a Maxwellian velocity distribution. The Knudsen layer is estimated to be a few molecular mean free paths thick in order to allow for the molecular collisions to occur that bring the molecules into a state of translational equilibrium at the outer edge of the Knudsen layer.

Anisimov [71] and Knight [72] did the early investigations on the Knudsen layer. Here a kinetic theory approach [73] is used in the present study. The analysis proceeds by constructing an approximate molecular velocity distribution adjacent to the liquid surface. Equations describing the conservation of mass, momentum and energy across

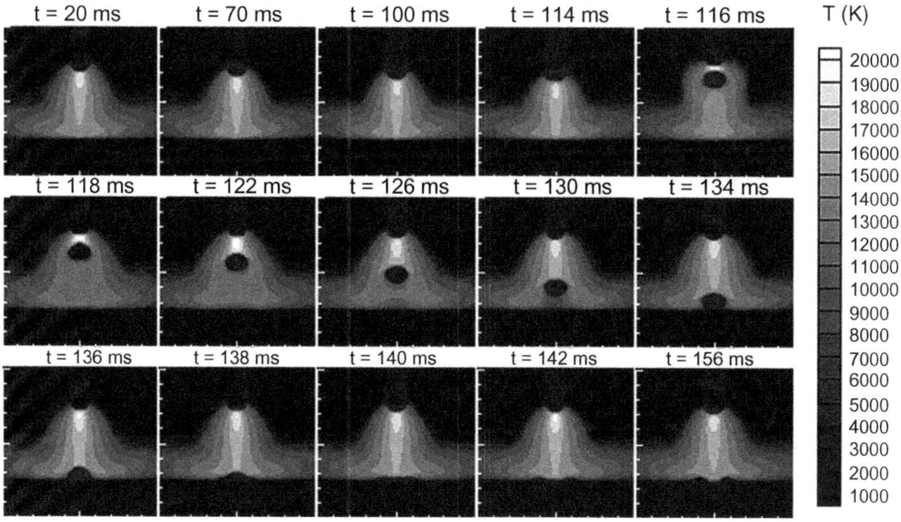

Fig. 5.4 Arc formation in a GMA welding process.

the Knudsen layer are developed in terms of this velocity distribution. This gives Eqs. (5.19) and (5.20), as given below, for gas temperature, T_K, and density, ρ_K, outside of the Knudsen layer as functions of the liquid surface temperature and the corresponding saturation density, ρ_{sat}.

$$\frac{T_K}{T_L} = \left[\sqrt{1+\pi\left(\frac{\gamma-1}{\gamma+1}\frac{m}{2}\right)^2} - \sqrt{\pi}\frac{\gamma-1}{\gamma+1}\frac{m}{2}\right]^2 \tag{5.19}$$

$$\frac{\rho_K}{\rho_{sat}} = \sqrt{\frac{T_L}{T_K}}\left[\left(m^2+\frac{1}{2}\right)e^{m^2}\operatorname{erfc}(m) - \frac{m}{\sqrt{\pi}}\right] + \frac{1}{2}\frac{T_L}{T_K}\left[1-\sqrt{\pi}me^{m^2}\operatorname{erfc}(m)\right] \tag{5.20}$$

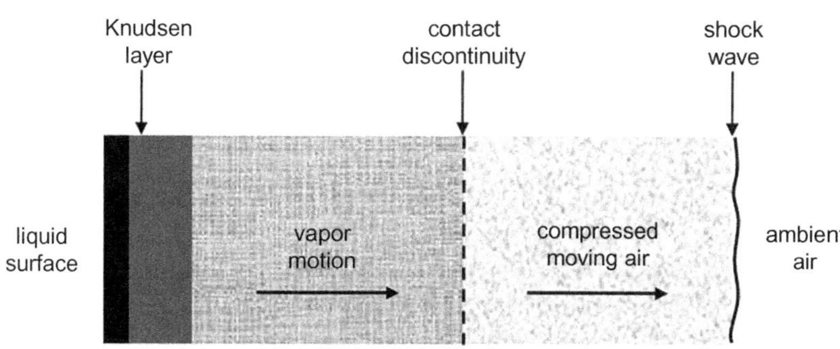

Fig. 5.5 A schematic of the gas dynamic of vapor and air away from a liquid surface at elevated temperature.

The quantity, m, is closely related to the Mach number at the outer edge of the Knudsen layer, M_K, and is defined as, $m = u_K/\sqrt{2R_V T_K} = M_K\sqrt{2/\gamma_V}$, where γ_V and R_V are the ratio of specific heats and the gas constant for the vapor, respectively. The value of m depends on the gas dynamics of the vapor flow away from the surface. The gas temperature, pressure and density throughout the vapor region (outside of the Knudsen layer) are uniform. The contact discontinuity, that is, the boundary between vapor and air, is an idealization that results due to the neglect of mass diffusion and heat conduction. The velocity and pressure are equal in these regions, $u_K = u_S$ and $P_K = P_S$, where the subscript, S, denotes properties behind the shock wave. Note that, in general, $T_K \neq T_S$ and $\rho_K \neq \rho_S$.

The thermodynamic state and velocity of the air on each side of the shock wave are related by the Rankine-Hugoniot relations, where the most convenient forms to this application are given by Eqs. (5.21) and (5.22). M_K is the Mach number in the vapor, $M_K = u_K/\sqrt{2\gamma_V R_V T_K}$.

$$\frac{P_S}{P_\infty} = 1 + \gamma M_K\sqrt{\frac{\gamma_V R_V T_K}{\gamma_\infty R_\infty T_\infty}}\left[\frac{\gamma_\infty+1}{4}M_K\sqrt{\frac{\gamma_V R_V T_K}{\gamma_\infty R_\infty T_\infty}} + \sqrt{1+\left(\frac{\gamma_\infty+1}{4}M_K\sqrt{\frac{\gamma_V R_V T_K}{\gamma_\infty R_\infty T_\infty}}\right)^2}\right] \quad (5.21)$$

$$\frac{T_S}{T_\infty} = \frac{P_S}{P_\infty}\left(1+\frac{\gamma+1}{\gamma-1}\frac{P_S}{P_\infty}\right)\Big/\left(\frac{\gamma+1}{\gamma-1}+\frac{P_S}{P_\infty}\right) \quad (5.22)$$

The saturation pressure, P_{sat}, is obtained from Eq. (5.23), where A, B and C are constants which depend on the material. This is used to obtain the saturation density, $\rho_{sat} = P_{sat}/(R_V T_L)$, assuming an ideal gas.

$$\log(P_{sat}) = -\frac{A}{T_L} - B\log(T_L) + C \quad (5.23)$$

Eqs. (5.20)–(5.23) are solved as a function of T_L using an iterative solution method. The vapor was assumed to be iron in the form of a monatomic gas with a molecular weight of 56, and $\gamma_V = 1.67$. Quantities of particular interest are the recoil pressure, P_r, and rate of energy loss due to evaporation, q_e, and they are given below.

$$P_r = P_K + \rho_K u_K^2, \quad q_e = H_V \rho_K u_K \quad (5.24)$$

5.2.6 Laser-plasma interaction and multiple reflections of laser beam in keyhole

In the keyhole, the laser beam is reflected several times by the keyhole wall. Also, each time the laser beam travels inside the keyhole, it will interact with the keyhole plasma. Multiple reflection of the laser beam and absorption mechanism is critical in determining the energy distribution and is discussed below.

5.2.6.1 Inverse bremsstrahlung (IB) absorption

With the continuous heating of the laser beam, the temperature of the metal vapor inside the keyhole can reach much higher than the metal evaporation temperature, resulting in strong ionization, which produces keyhole plasma. The resulting plasma absorbs laser power by the effect of Inverse Bremsstrahlung (IB) absorption. Eqs. (5.25) and (5.26) define the IB absorption fraction of laser beam energy in plasma by considering multiple reflection effects [74]:

$$\alpha_{iB,1} = 1 - \exp\left(-\int_0^{s_0} k_{pl} ds\right) \tag{5.25}$$

$$\alpha_{iB,mr} = 1 - \exp\left(-\int_0^{s_m} k_{pl} ds\right) \tag{5.26}$$

here, $\alpha_{iB,\,1}$ is the absorption fraction in plasma due to the original laser beam; $\alpha_{iB,mr}$ is the absorption fraction due to the reflected laser beam. $\int_0^{s_0} k_{pl} ds$ and $\int_0^{s_m} k_{pl} ds$ are, respectively, the optical thickness of the laser transportation path for the first incident and multiple reflections, and k_{pl} is the plasma absorption coefficient due to inverse Bremsstrahlung absorption [75]:

$$k_{pl} = \frac{n_e n_i Z^2 e^6 2\pi}{6\sqrt{3m}\varepsilon_0^3 c h \omega^3 m_e^2}\left(\frac{m_e}{2\pi k T_e}\right)^{0.5}\left[1 - \exp\left(-\frac{\omega}{kT_e}\right)\right]\overline{g} \tag{5.27}$$

where Z is the average ionic charge in the plasma, $\omega$is the angular frequency of the laser radiation, ε_0 is the dielectric constant, k is the Boltzmann's constant, n_e and n_i are particle densities of electrons and ions, h is Planck's constant, m_e is the electron mass, T_e is the excitation temperature, c is the speed of light, and $\overline{g}$ is the quantum mechanical Gaunt factor. For the weakly ionized plasma in the keyhole, the Saha equation [75] can be used to calculate the densities of plasma species:

$$\frac{n_e n_i}{n_0} = \frac{g_e g_i}{g_0}\frac{(2\pi m_e k T_e)^{1.5}}{h^3}\exp\left(-\frac{E_i}{kT_e}\right) \tag{5.28}$$

5.2.6.2 Fresnel absorption

As discussed before, part of the laser energy will be absorbed by keyhole plasma and part of the laser energy can reach the keyhole wall directly. So, the energy input (q_{laser}) for the keyhole wall consists of two parts: (1) Fresnel absorption of the incident intensity directly from the laser beam ($I_{\alpha,Fr}$) and (2) Fresnel absorption due to multiple reflections of the beam inside the keyhole ($I_{\alpha,mr}$).

$$q_{laser} = I_{\alpha,Fr} + I_{\alpha,mr} \tag{5.29}$$

$$I_{\alpha,Fr} = I_{laser} \cdot (1 - \alpha_{iB,1}) \cdot \alpha_{Fr}(\varphi_1) \tag{5.30}$$

$$I_{\alpha,mr} = \sum_{mr=1}^{n} I_{laser} \cdot (1 - \alpha_{iB,1}) \cdot (1 - \alpha_{Fr}) \cdot (1 - \alpha_{iB,mr}) \cdot \alpha_{Fr}(\varphi_{mr}) \tag{5.31}$$

where I_{laser} is the incoming laser intensity. We assume the laser beam has in the simplest case a Gaussiam-like distribution:

$$I_{laser}(x, y, z)^{laser} = I_0 \left(\frac{r_f}{r_{f0}}\right)^2 \exp\left(-\frac{2r^2}{r_f{}^2}\right) \tag{5.32}$$

where r_f is the beam radius and r_{f0} is the beam radius at the focal position; I_0 is the peak intensity.α_{Fr} is the Fresnel absorption coefficient and can be defined it in the following formula [76]:

$$\alpha_{Fr}(\varphi) = 1 - \frac{1}{2}\left(\frac{1 + (1 - \varepsilon\cos\varphi)^2}{1 + (1 + \varepsilon\cos\varphi)^2} + \frac{\varepsilon^2 - 2\varepsilon\cos\varphi + 2\cos^2\varphi}{\varepsilon^2 + 2\varepsilon\cos\varphi + 2\cos^2\varphi}\right) \tag{5.33}$$

where φ is the angle of incident light with the normal of keyhole surface, n is the total number incident light from multiple reflections, $\vec{I}$ is the unit vector along the laser beam radiation direction and $\vec{n}$ is unit vector normal to the free surface. ε is a material-dependent coefficient.

5.2.7 Radiative heat transfer in laser-induced plasma

When an intense laser beam interacts with metal vapor, a significant amount of the laser radiation is absorbed by the ionized particles. The radiation absorption and emission by the vapor plume may strongly couple with the plume hydrodynamics. This coupling, shown on the right-hand side of Eq. (5.7), will affect the plasma laser light absorption and radiation cooling terms. The radiation source term $\nabla \cdot (-\mathbf{q_r})$ is defined via

$$\nabla \cdot \mathbf{q_r} = k_a \left(4\pi I_b - \int_{4\pi} I d\Omega\right) \tag{5.34}$$

where k_a, I_b and Ω denote the Planck mean absorption coefficient, blackbody emission intensity and solid angle, respectively. For the laser-induced plasma inside the keyhole, the scattering effect is not significant compared with the absorbing and emitting effect. So it will not lead to large errors to assume the plasma is an absorbing-emitting medium. The radiation transport equation (RTE) has to be solved for the total directional radiative intensity I [77]:

$$(\mathbf{s}\cdot\nabla)I(\mathbf{r},\mathbf{s})=k_a(I_b-I(\mathbf{r},\mathbf{s})) \tag{5.35}$$

where $\boldsymbol{s}$ and $\boldsymbol{r}$ denote a unit vector along the direction of the radiation intensity and the local position vector. The Planck mean absorption coefficient is defined in the following [77]:

$$k_a=\left(\frac{128}{27}k\right)^{0.5}\left(\frac{\pi}{m_e}\right)^{1.5}\frac{Z^2e^6\overline{g}}{h\sigma c^3}\frac{n_en_i}{T_v^{3.5}} \tag{5.36}$$

where n_i and n_e represent the particle density of ions and electrons, T_v is the temperature of the plasma, Z stands for the charge of ions, e is the proton charge and m_e is the mass of electrons.

5.2.8 Tracking of free surfaces

The algorithm of volume-of-fluid (VOF) is used to track the moving free surface [59]. The fluid configuration is defined by a volume of fluid function, $F(x,y,z,t)$, which is used to track the location of the free surface. This function represents the volume of fluid per unit volume and satisfies the following conservation equation:

$$\frac{dF}{dt}=\frac{\partial F}{\partial t}+(\boldsymbol{V}\cdot\nabla)F=0 \tag{5.37}$$

When averaged over the cells of a computing mesh, the average value of F in a cell is equal to the fractional volume of the cell occupied by the fluid. A unit value of F means a cell full of fluid and a zero value indicates a cell containing no fluid. Cells with F values between zero and one are partially filled with fluid and identified as surface cells.

Based on the aforementioned scientific principles governing the hybrid laser-arc welding process, Zhou et al. [46, 78, 79] have successfully developed mathematical models to simulate the heat and mass transfer and fluid flow phenomena in both pulsed and three-dimensional moving hybrid laser-MIG welding. As shown in Fig. 5.6 [46], hybrid laser-MIG welding can prevent pores that are easily found in laser welding. In the hybrid welding process, the mixing and heat transfer process in the weld pool are also found to be greatly affected by the droplet size, droplet frequency, etc. [80]. Hence, the microstructure and final weld quality can be improved. Fig. 5.7 shows the temperature distributions in a moving three-dimensional hybrid laser-MIG process [78]. The heat transfer process is greatly affected by laser-to-arc distance, welding speed, etc. More details have been given by Zhou et al. [78].

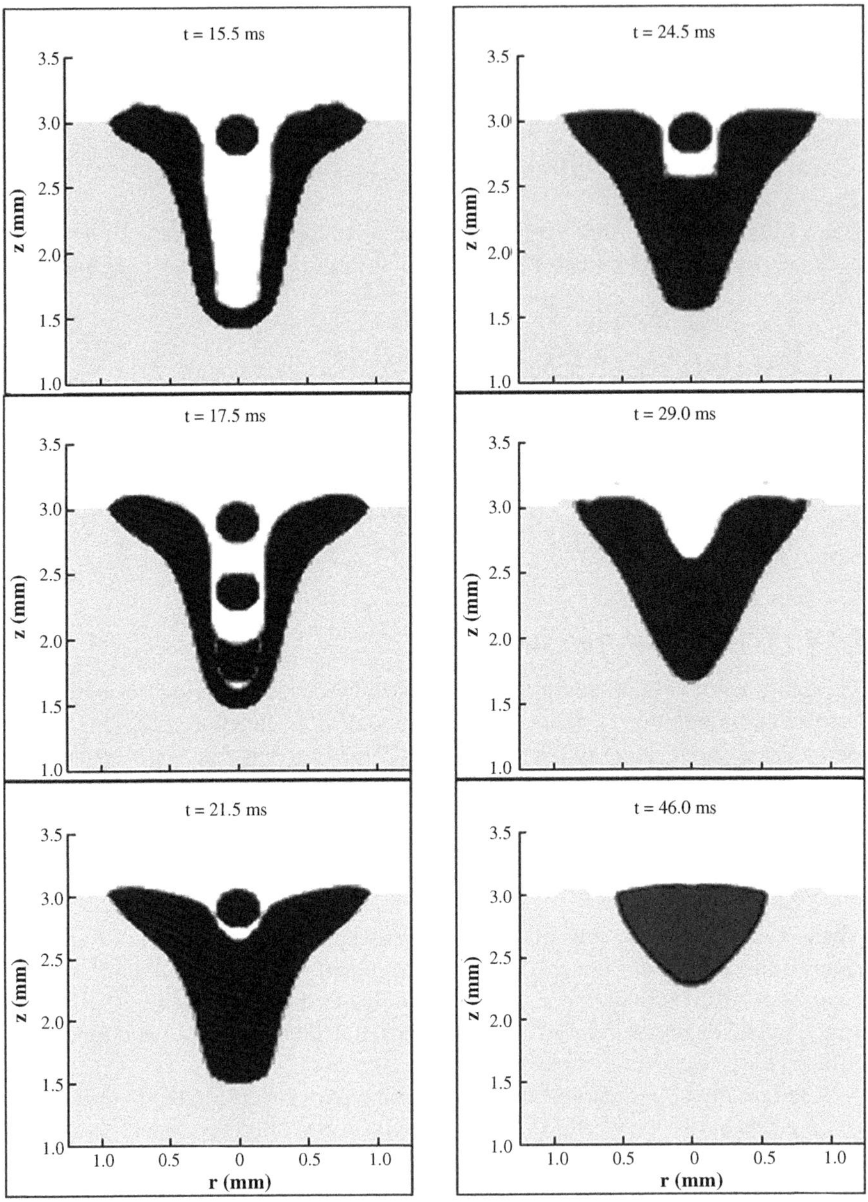

Fig. 5.6 A sequence of the keyhole collapse and solidification processes in hybrid laser-arc welding.

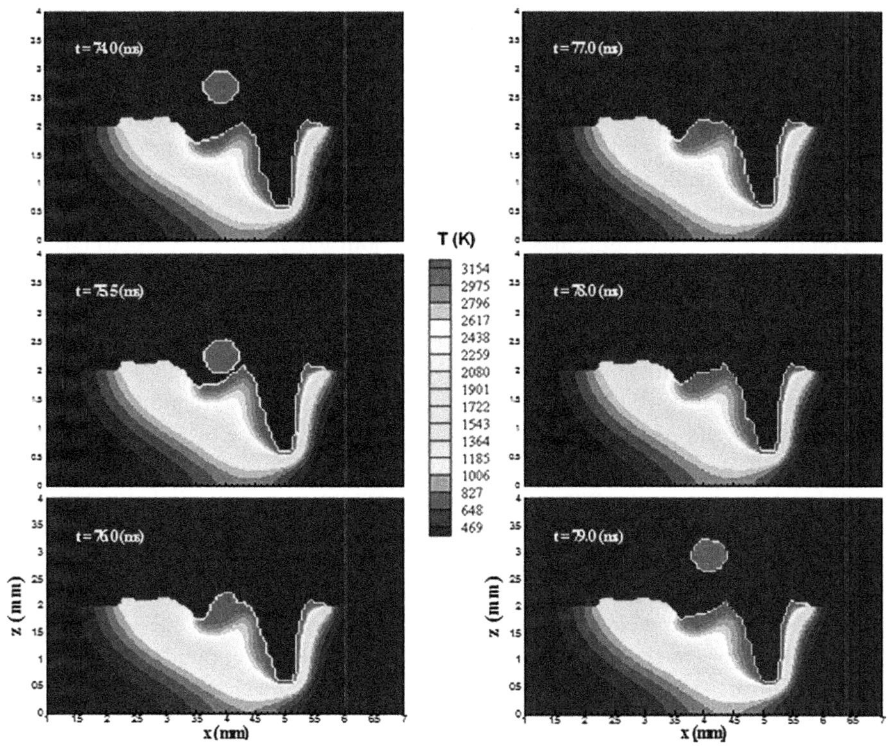

Fig. 5.7 A sequence of temperature evolution in three-dimensional moving hybrid laser-MIG welding.

5.3 Hybrid laser-arc welding of aeronautical materials

5.3.1 Hybrid laser-arc welding of magnesium and its alloys

Magnesium and its alloys have been widely used in aerospace, automotive and electronic industry due to their high specific strength, low density, excellent machinability, and recyclability [81]. They are considered as advanced materials for coping with energy conservation and environmental pollution regulations. However, since magnesium has a low melting point (650 °C), a low vaporization point (1100 °C) and surface tension, developing appropriate joining techniques is critical in expanding the applications of magnesium alloys in aerospace industries. Various welding techniques including the arc welding laser welding laser-arc hybrid welding, and friction stir welding have been applied to improve the weld ability of Mg alloys [82–84]. However, in arc welding, a wide fusion zone (HZ) and heat affected zone (HAZ) are formed, which are harmful to the mechanical properties of the welds. GTA welding has been commercially employed mainly for repairing cast magnesium parts. In MIG welding, the energy input must be controlled in such a way that the wire will only be melted but not vaporized, so special power sources are needed for the MIG welding of magnesium alloys. High energy beams, such as the laser and electron beam, have also been tried for the welding of magnesium alloys. Although high-energy-beam welding produces a very narrow HAZ, a large ratio of welding depth/width and high welding speeds, the gap bridging ability is very low. Among these welding techniques, hybrid laser-arc welding showed a great potential since more stable joining process, welds with better mechanical properties, and greater adaptability under severe conditions are achieved due to the synergic effects between the laser beam and arc column [81].

Hybrid laser-TIG welding has been used to join wrought (AZ31, AZ61) and cast magnesium-based alloys (AZ91) [85]. It showed that magnesium alloys could be easily welded by laser-TIG welding. The grain of FZ was finer than that in base metal, as shown in Fig. 5.8 [85]. The width of the HAZ was narrower than that of welded by TIG. With the increase of Al content in magnesium alloys, the width of HAZ increased, as well as the content of β phase ($Mg_{17}Al_{12}$). The hardness in FZ and HAZ of AZ61 and AZ91 has a large change compared to base metal due to the existence of β phase, while there is no change for AZ31. The corrosion behavior of hybrid laser-GTA welded magnesium alloy has also been investigated [70]. Corrosion resistance of the joint is predominantly influenced by grain refinement or interactions of grain refinement and continued net-shaped β phases and the welding mode as well.

Laser-arc hybrid welding has received significant attention in joining Mg alloys because of its higher welding efficiency and stronger ability to avoid weld metallurgical defects compared with pure laser welding [84]. For example, for a 5-mm-thick hybrid laser-arc weld, its tensile strength efficiency can reach up to 109% and grain size of the fusion zone and the width of PMZ both increase with additional heat input. In addition, Hall-Petch relationship between microhardness and grain size is obtained [84]. It was also found that high power laser–MIG hybrid welding might have greater potential in Mg industry since it can offer deeper penetration depth, fuller weld appearance and lower weld defects [84]. The width of the HAZ was narrower than that of welded by TIG. With the increase of Al content in magnesium alloys, the width of HAZ increased, as well as the content of β phase ($Mg_{17}Al_{12}$). The hardness in FZ and HAZ of AZ61 and AZ91 has a large change compared to base metal due to the

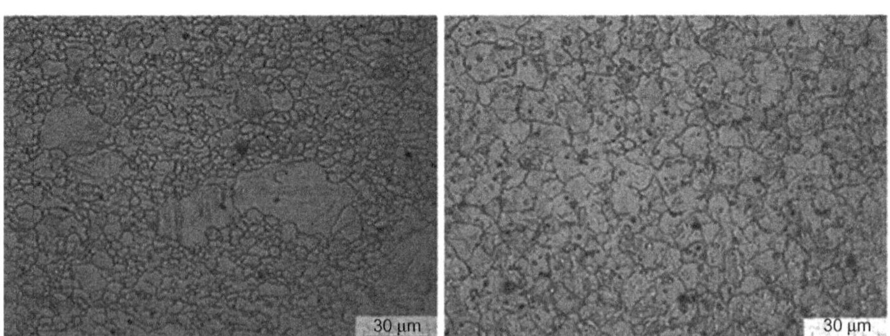

(A) (B)

Fig. 5.8 Microstructure of hybrid laser-TIG weld in (A) base metal and (B) fusion zone.

existence of β phase, while there is no change for AZ31. The corrosion behavior of hybrid laser–GTA welded magnesium alloy has also been investigated [85]. Corrosion resistance of the joint is predominantly influenced by grain refinement or interactions of grain refinement and continued net-shaped β phases and the welding mode as well.

One of the major concerns during high speed welding of magnesium alloys is the presence of in the weld that can deteriorate mechanical properties. Studies have been conducted to analyze the porosity formation during hybrid laser-TIG welding of magnesium alloy AZ31B15. It shows that when there is a lack of shielding gas for the laser beam, air is easily introduced in the molten metal, thus causing the formation of large pores. Hydrogen does not play a major role in the formation of large pores. When laser shielding gas is coaxial, it will disturb the arc stability strongly. However, a favorable weld without porosity can be obtained by appending lateral shielding gas for the laser beam. Evaporative loss of alloy elements in welding of magnesium alloys causes welding defects and reduces the mechanical properties of weld joints. Hybrid low-power laser-TIG welding with filler metal is proposed to resolve this problem [86].

Gao et al. analyzed the mechanical properties of welds of magnesium alloy AZ31B plates with thickness between 5 and 8 mm during hybrid laser-TIG welding [84]. As shown in Fig. 5.9, it was found that the arc plays an important role in porosity reduction by proving additional heat input and enhancing the melt flow in the molten pool. However, its impact on penetration depth is limited due to its low power density. The effects of arc and laser on blowhole overflow in weld pool varies according to the thickness of plate. When a thin sheet was welded, arc heat and pressure affect most of the molten pool. Arc plasma could be entrapped into the laser keyhole to stabilize the keyhole, thus reducing bubble formation in the molten pool. Meanwhile, the bubbles formed in laser-melted zone could be pushed into the wider arc-melted zone by the melt flow and subsequently escape into the air. However, when a thick (8 mm) plate welded, then arc-melted zone counts a smaller portion of the molten pool. The deeper portion of the laser-melted zone is hard to be affected by the arc and the molten pool is sensitive to bubble formation similar to that in laser welding. Thus, the porosity formation cannot be prevented only by varying welding speed when a thick plate is to be welded.

Studies [20] have been conducted to investigate the effect of the laser pulse on the arc in the positive and negative discharge periods when a pulsed laser-AC arc is applied for the welding of magnesium alloy. It was found that phase matching between the laser pulse and arc discharge is important. Under optimal welding conditions, a high quality weld joint is obtainable. However, the interaction mechanisms between the laser pulse and different (negative and positive) periods of arc discharge are not well understood and should be further studied. Due to the different heating effects, the welding penetration in hybrid welding with back-feeding of filler wire is found to be twice as deep as that with front-feeding of filler wire. The grain size of FZ and HAZ becomes uniform, and the tensile strength of the welding joint reaches about 95% of base metal. The results also show an expansion of arc plasma and a decrease of electron temperature in hybrid welding, but the electric conducting route concentrates and the energy density of arc plasma enhances, which improves the welding penetration significantly.

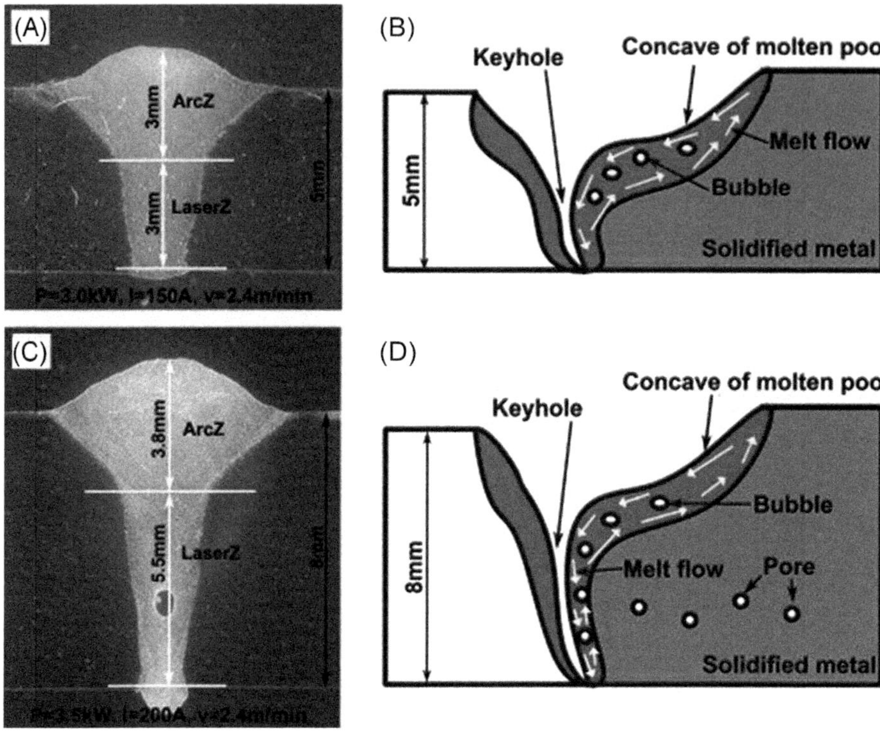

Fig. 5.9 Characteristic comparison of different thick joints: (A) weld cross-section of 5 mm-thick weld and (B) corresponding schematic molten pool; (C) weld cross-section of 8 mm-thick weld and (D) corresponding schematic molten pool of 8 mm-thick weld.

Hybrid laser-MIG welding was also found to be an efficient joining technique for magnesium alloys [86]. Stable arc, reliable droplet transfer and regular welds which are difficult to obtain in MIG welding can be obtained in hybrid welding by laser arc synergic effects. The ultimate tensile strength and elongation of the hybrid weld are far higher than those of the laser weld and reach 97% and 87.5% of the base metal, respectively. As shown in Fig. 5.10 [86], the microstructure of a hybrid weld in FZ is composed of equiaxed dendrites. Differences exist between the wide upper part (arc zone) and the narrow lower part (laser zone) of the weld. The arc zone has coarser grain size and a wider partially-melted zone than the laser zone due to the slow solidification rate there. The porosity reduction in hybrid laser-MIG welding of magnesium alloys is due to: faster bubble escaping velocity caused by arc pressure; shorter bubble escaping distance caused by the concave surface of molten pool; wider bubble escaping area caused by the wide and big molten pool; and longer bubble escaping time caused by slow solidifying speed.

Hybrid laser-arc welding also has been tested to join dissimilar AZ based magnesium alloys [85]. Excellent bead appearance and a compact microstructure without porosity or fracture defects were obtained. The microstructure of the dissimilar

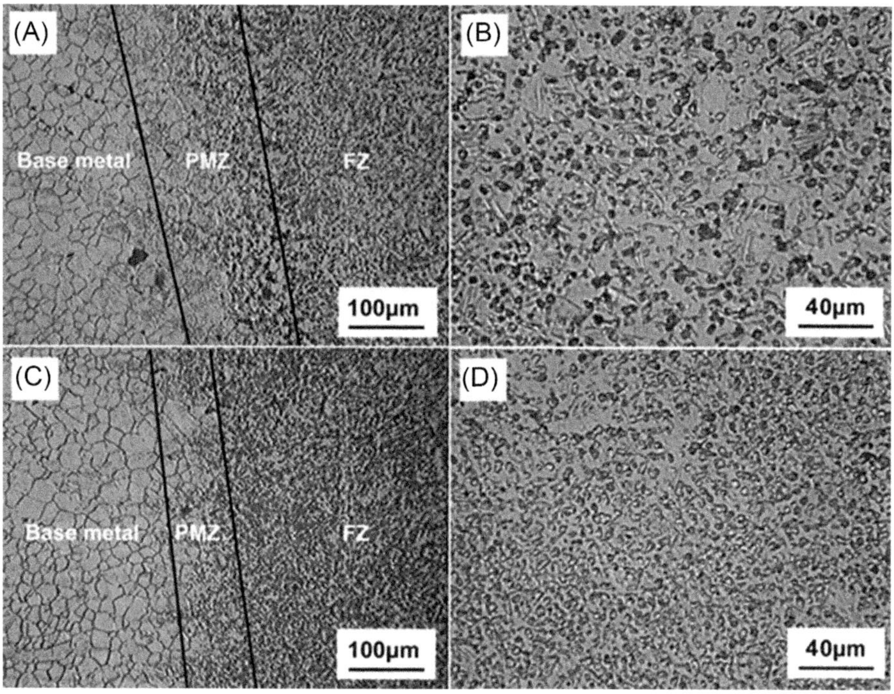

Fig. 5.10 Microstructure of hybrid laser-MIG weld of AZ31 magnesium alloy.

magnesium alloy joints is composed of primary α phase (Mg) and β phase. In addition, the tensile strengths of AZ31B–AZ61 and AZ31B–AZ91 joints are equal to that of AZ31B base metal. The presence of β phase was found to have a severe influence on the tensile strength and micro-hardness of dissimilar magnesium alloy joints. The grain size of the FZ was finer than that of both base metals. The width of HAZ on the AZ91 and AZ61 side was larger than that on the AZ31B side. The microstructure of FZ in dissimilar magnesium alloy joints was composed of primary Mg and $Mg_{17}Al_{12}$. The strength of dissimilar magnesium alloy joints was slightly higher than that of the AZ31B base metal. The hardness in FZ and HAZ near AZ61 and AZ91 sides showed a large change from that of base metal due to the presence of β phase, while there was no corresponding change for AZ31B. Interlayer, such as nickel and copper, has been used in hybrid laser-TIG welding of magnesium alloy with steel [87, 88]. With nickel as the interlayer, as shown in Fig. 5.11 [86], Mg_2Ni was generated at upper edges of molten pool distributed in a ribbon-shape along the interface of Mg alloy/Ni interlayer. The solid solution of Ni in Fe was detected along the side surface of molten pool and accumulated at the bottom, which indicates the formation of intermetallic compound Mg_2Ni. The solid solution of Ni in Fe at the interface altered the bonding mode of joints from mechanical bonding to semi-metallurgical joining. Thus, the tensile strength of the hybrid joint is greatly increased compared with that from direct joining of Mg alloy to steel. However, it is noticeable that the corrosion problems may affect the properties of the joint and further studies on corrosion are

Fig. 5.11 SEM image showing the flow pattern in upper end of molten pool along the Ni interlayer surface.

needed. Copper has also been used as an interlayer for hybrid laser-TIG welding of AZ31B with mild steel [74]. The intermetallic compounds Mg_2Cu with rod-like structure in the joint and equiaxed structure at the interface were found, which not only refine grains but also strengthen the grain boundaries. The transitional zone (TZ) consists of: mostly the remelted steel including a little solid solution of Cu in Fe distributed along the edge of the molten pool at the steel side; and the mixture of Cu and Fe on the upper margins of the pool. Such distribution of TZ ensures the contact of Cu with steel. The reason why FZ is attached to TZ tightly is that the addition of Cu improves the wettability and facilitates its nucleation on steel, which leads to an intimate connection. The maximum shear strength of the joint can achieve almost 100% to that of AZ31B, compared with that of a joint without any interlayer. Furthermore, hybrid laser-TIG welding technique was successfully used for lap welding of AZ31B magnesium alloy and Q235 steel using nickel and copper as the interlayer [89]. The shear strength of the joint obtained is higher than that of the base metal AZ31B magnesium alloy. The TZ is formed along the edge of the molten pool on the steel side in the joint with nickel and copper as the interlayer, which greatly improves the weld's shear strength. The strengthening effect of the intermediate phase Mg2Cu was found significant, but the strengthening effect of Mg2Ni was not obvious. Therefore, from strengthening point of view, Cu interlayer is more suitable for the lap joint of AZ31B magnesium alloy and Q235 steel. Ce has also been used as an interlayer in hybrid laser-TIG welding of aluminum-magnesium-silicon alloy 6061 and a magnesium-aluminum-zinc alloy AZ31 [90]. With the addition of Ce as an interlayer, cracks were not found in most critical regions, while did show up in the welds from direct joining of Mg and Al. The microstructure in the middle fused bath of the weld was the mixture of base materials and Mg-Al eutectic. Ce distributed uniformly in the critical regions where composition changed from Mg to Al and also made the microstructure of FZ uniform and the fracture zone finer.

5.3.2 Hybrid laser-arc welding of titanium and its alloys

Titanium attracts the interests of aerospace industry due to its high specific weight, excellent corrosion resistance, and high temperature mechanical properties [91–94]. However, the joining of titanium is facing challenges. Most research on welding of titanium has utilized GTA welding [95–98]. However, GTA welding in thinner materials is done at much slower speeds and thus low deposition rates. Because of the higher heat input, the welded parts are more likely to be distorted. The GMA welding can improve weld formation with filler wire and increasing heat input compared with laser beam welding at a high speed of 1.8 m/min and power of 8 kW since much more liquid metal can be added into the welding seam [99]. Friction stir welding (FSW) and EB welding have also been tried for joining titanium [100, 101]. However, EB welding involves the use of high vacuum and is difficult in seam-tracking exactly the required joint line. FSW of titanium has not yet been demonstrated as a viable production process, primarily due to excessive tool wear and lack of penetration and worm holes are frequently observed in the joint. There is an increasing interest on the laser welding of titanium alloys by using CO_2 and Nd: YAG lasers. Although, laser welding results in low welding stress and small distortion, it suffers from the insufficient gap bridging ability and requires precision in positioning. Furthermore, the wall plug efficiency of CO_2 and Nd: YAG lasers is low. Recently, hybrid fiber laser-GMA welding has been proposed to weld titanium and its alloys to overcome the drawbacks while maintaining the key advantages of the laser welding.

In hybrid fiber laser-GMA welding of commercially pure titanium, full-penetration welds with silver color were successfully obtained at welding speeds up to 9 m/min [102]. Despite the difference in microstructure of the two kinds of welds, there was a remarkable similarity in tensile strength and micro-hardness of the hybrid and laser welds. Both laser and hybrid laser-GMA welds showed higher micro-hardness and tensile strength than the base metal. However, the hybrid welds had a good combination of strength and ductility. Coarse columnar alpha and a smaller amount of fine-grained acicular alpha appeared in laser welds. The microstructure of hybrid laser-GMA welds consisted of acicular alpha, platelet alpha and twins, which made a good combination of strength and ductility for hybrid laser-GMA welds. The hybrid welding is about seven times faster than the GMA welding and a stable arc is observed in hybrid laser welding. The weld width was found to increase with the weld current. However, the influence will be weakened with continuous increase of arc current. The ultimate tensile stress of the welds is not obviously influenced by the welding current, laser-arc distance, and defocused distance. However, increasing the gap leads to the increase of the tensile strength and the elongation of the weld is influenced significantly by the process parameters.

Hybrid laser-MIG welding was tested for joining Ti-Al-Zr-Fe titanium alloys and good uniform welds with full penetration, no crack/porosity, and low distortion were achieved [99]. In general, a hybrid weld has a slightly protruding top surface and smooth transition from the weld metal to the parent metal compared with a weld made by laser welding. In hybrid welding, the low cooling speed and additional heat and mass input to the welding pool from melted filler wire help prevent the undercut defect formation commonly observed in laser welding. The width of the hybrid weld seam is much bigger than that of the laser weld seam since the arc has a relatively wide heating area. In hybrid welds, coarse columnar prior-β grains nucleated epitaxial from the base metal substrate and grew towards the weld centerline. The prior-β grain of fusion zone in hybrid laser-MIG welded joint is much bigger than that in the laser welded joint due to the large heat input from the MIG arc. Also, there are fine acicular α' solidification structures within the prior-β column grains in the FZ of the hybrid welds. The HAZ of the hybrid weld consists of martensitic α', acicular α, and primary β grains. Hence, a hybrid weld has better combination of strength and ductility than a laser weld.

Hybrid laser-TIG welding technique was also successfully applied to weld TA15 titanium alloys [103]. It was found that the hybrid weld was made of columnar and equiaxed crystals. The columnar crystals are at the upside and the exiguous and equiaxed crystals are at the bottom of the hybrid welds. The equiaxed grains in the HAZ provide particles for crystallization in the FZ. The grains grow up based on these particles, while the molten pool is cooling down, so bulky columnar crystals grow towards the top of the weld due to the low thermal conductivity of titanium alloys. The formation of the exiguous and equiaxed crystals at the bottom of the weld is mainly affected by the laser beam. Since the laser beam energy is highly concentrated and the FZ is narrow and a lot of particles are supplied for crystallization by these exiguous crystals during the cooling stage, these crystals intersect with each other before they grow up. The growth of the grains is restrained and equiaxed crystals are formed as a result. Both crystal grains and equiaxed crystals contain reticular structure, a typical structure of titanium alloys, forming for the stoppage of grain growth when the lamellar structures meet with each other. Reticular structure can improve tensile strength, fatigue resistance, and fracture toughness of the welds. The tensile strength of the hybrid welds can reach 98% of that of the base metal and the fracture of samples proves to be a gliding fracture. In cross-direction of the weld, the hardness of the FZ is higher than that of the base metal but lower than that of the HAZ. In vertical direction of the weld, the hardness is lower at the center than that at both sides. Also, major elements, such as Ti, AI, Mo and V, were found to distribute uniformly in the welds and there was no burning loss and accumulation of them in hybrid welding.

Laser-tungsten inert gas (TIG) hybrid welding has been successfully applied for the welding of TA15 titanium alloys [95]. The effects of the torch and the CO_2 content in the shielding gas on weld shape and reinforcement/width (H/W) radio were discussed in laser-TIG hybrid welding. It was found that the shielding gas composition had significant impact on the weld formation. Smooth transition of weld reinforcement can be obtained when the CO_2 content in the torch shielding gas exceeds 2%. As known, Ar is one of the most widely applicable shielding gases in TIG welding process to protect the joint against atmospheric oxygen, nitrogen and hydrogen. Thus, adding CO_2 to pure argon shielding gas can inevitably cause the oxidation of the welding pool surface. When the added CO_2 content is more than 3%, a thick oxide layer was formed on the surface of the joint. The heavy oxidation layer can protect the liquid pool from the shielding gas and plasma arc, making the oxygen in the weld remain nearly constant. It also prevents the molten pool from moving freely and weakens the Marangoni convection, thus affecting the formation of molten pool. In addition, it was found that adding appropriate amount of CO_2 does not affect the microstructure and microhardness of the weld, but the broadband in the heat affected zone was found wider.

5.3.3 Hybrid laser-arc welding of aluminum and its alloys

Heat treatable 7XXX, 2XXX, and 6XXX series aluminum alloys have wide applications in aerospace industry due to their high strength and toughness and excellent strength to weight ratio [5, 104–107]. Unfortunately, weldability of these aluminum alloys is very poor. For arc welding, problems such as degradation of properties in HAZ and FZ and hot cracking are major challenges. Material loss due to the vaporization of low-boiling-point alloy elements such as zinc and magnesium creates additional difficulties, particularly in laser welding. A solid state welding process like FSW offers an alternative joining method [107–110] to avoid many welding defects such as voids, hot cracks and large residual stress, etc., since the aluminum alloy is in a plastic state during the welding process. Laser beam welding allows high welding speed, easy joint accessibility, low residual distortion, restricted HAZ, and high quality of the weld, however some technological limits arise when the conventional autogenously welding of the aluminum alloys is used [5, 104].

Current efforts on hybrid welding of 7xxx series aluminum alloys are mainly focused on using the hybrid laser-GMA welding processes [5, 104]. The welds made by the hybrid laser-GMA welding have smooth top and lower surfaces and good reinforcements. The height of the reinforcement increases with the increase of the GMA wire feed rate. In comparison, laser welds have tough surfaces and severe upper and lower surface undercut in the FZ. Traverse solidification cracks occurred at high welding speeds and were found to be related to the elongated temperature distribution in the welding direction, which induced a transverse tensile strain in the FZ during the cooling stage. In hybrid welds, the FZ was surrounded by a zone about 100 μm showing evidence of partial melting. The welds contained equiaxed dendritic structures in the FZ center, and directional columnar dendritic solidification structures in the region close to the fusion line. A wide softening HAZ was found and no microstructure change in HAZ was visible in the hybrid welds. Alloying element micro-segregation was observed in the dendritic microstructure and solidification rate played an important role in alloying element micro-segregation. In general, the higher the solidification rate, the finer the dendrite arm spacing and second phase boundaries, and the less the micro-segregation.

Hybrid welding was reported to be able to bring in less residual stress and lower thermal distortion in the welds [111]. It also offers more freedom to structural designers to use bigger gaps and reduces porosity formation inside the weld bead if the off-set distance between the two heating sources are carefully controlled. For hybrid welds, significant strength recovery (80–85% to base metal) can be achieved via precipitation hardening, as shown in Fig. 5.12 [105]. The size and distribution of these precipitates are dictated primarily by the temperature cycles and alloying element additions in the FZ of the welds. The average grain size of the FZ for hybrid welding is larger than that for laser welding. The microstructure consists of columnar dendrites at the FZ boundary and fine equiaxed grains along the weld centerline.

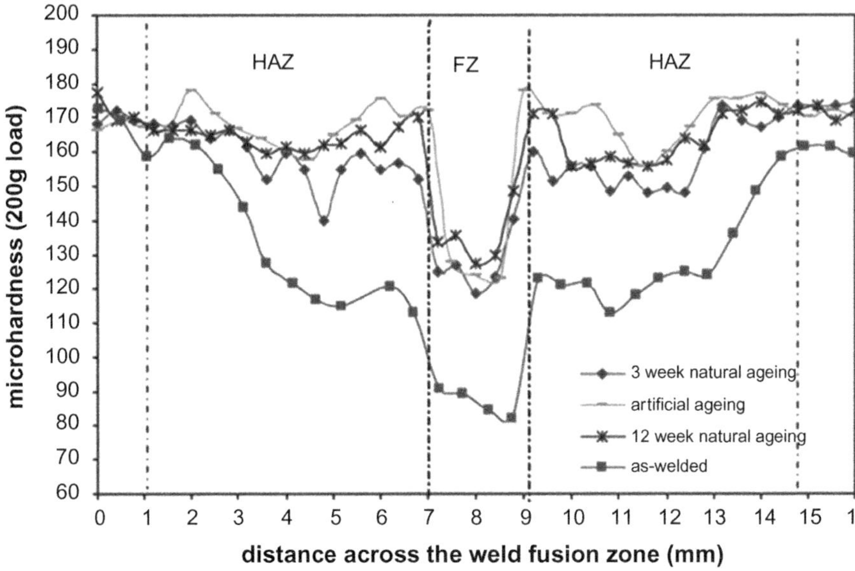

Fig. 5.12 Traverse micro-hardness measurements of AA7075-T6 hybrid welds with AA5754 filler wire at a welding speed of 80 mm/s.

During solidification, segregation of the alloying elements causes precipitates or eutectic films to form along the dendrite boundaries in the FZ, resulting in decreased ductility. Tensile testing of the hybrid welds shows that the FZ has lower average tensile strength and ductility than the base metal. The mechanical properties of the HAZ are significantly better than those of the FZ with higher strength and ductility. Since aging treatment cannot modify the alloy micro-segregation behavior, softening remains in the FZ of the welds and the welds can only tolerate limited deformation. However, it is found that a after a short duration solution heat treatment of the welds, a large proportion of the dendrite boundaries in FZ dissolve in the primary phase, which helps eliminate the alloying micro-segregation and improve the weld hardness.

For hybrid welds, it reported that the hybrid welding process can lead to the reduction in the level of the residual stress and cause lower thermal distortion [111]. This method also gives more freedom for the structural designers to use bigger gaps and have lower amount of porosity inside the weld bead, if the off-set distance between the heat sources are carefully controlled. Tensile testing of the hybrid welds shows that the FZ has lower average tensile strength and ductility than the base metal. The mechanical properties of the HAZ are significantly better than those of the FZ with higher strength and ductility. Since aging treatment cannot modify the alloy micro-segregation behavior, softening remains in the FZ of the welds and the welds can only tolerate limited deformation. However, it is found that a after a short duration solution heat treatment of the welds, a large proportion of the dendrite boundaries in FZ dissolve in the primary phase, which helps eliminate the alloying micro-segregation and improve the weld hardness.

In hybrid laser-arc welding of aluminum alloys, arc and laser power density were found to play an important role in affecting the weld quality [105, 106]. When high laser power is used, the stability of the process can be improved and the welds with good structural and geometrical properties can be achieved due to the formation of regular fused zone with low macro-porosity. Lowering the arc voltage and current can reduced the residual stress in the heat affected zone. In hybrid laser-arc welding of aluminum alloys, selection of filler wire was also found to play an important role in affecting the weld quality [112]. In hybrid laser-MIG welding of 7075-T6 aluminum alloy, it was found that the weld had a comparable strength to the base metal when an AA2319 filler wire was used, while a large amount of fine ductile type dimples and larger sized voids were found [5]. Fracture surfaces of tensile samples suggested that failure occurred trans-granularly at the dendrite boundaries. Use of novel filler additions can achieve grain refinement and prevent this defect. In hybrid laser-MIG welding of 2A12 aluminum alloy, full penetration welds without any defects were obtained when ER4043 and ER2319 filler wires were used [112]. SEM and XRD studies showed that silicon and copper were concentrated at the dendrite boundaries, and α-Al + Si + Al_2Cu + Mg_2Si eutectic was formed when ER4043 filler wire was used. However, only copper was concentrated at the dendrite boundaries and α-Al + u eutectic was formed when ER2319 filler wire was used. The tensile strength and elongation of welds decreased mainly due to the formation of eutectic phases in the FZ. The fracture test showed that the joint efficiency reached up to 78% and 69% by using ER2319 and ER4043 filler wires, respectively. The weld ability of AA6061-T6 aluminum alloys with a hybrid laser-GMA welding process using ER4043 filler wires was also examined [113]. It was found that dendritic structures were formed in the FZ of the welds and caused alloying element segregation in hybrid welding processes. The tensile fracture occurred in the FZ of the weld, while the face bend fracture was in the HAZ since the relatively coarse-grained microstructure of the HAZ might have less crack propagation resistance. The crack was found to propagate in a ductile mode. Severe plastic deformation occurred during this process, leaving a "troughlike" appearance of the surface surrounding the crack flanks.

Hybrid welding could also enhance corrosion susceptibility for AA6061-T6 aluminum alloy in nitric acid solution. The hybrid welding leads to the dendritic segregations of precipitate phase in the WFZ. The rate of anodic reaction of the hybrid welds is higher than that of the base alloy [114]. Severe pits and black porous surface films were observed in the FZ of the hybrid welds. The precipitation of intermetallic phases and the formation of galvanic corrosion couplings were found to be the major reasons for the corrosion of the hybrid laser welds. Plasma-MIG hybrid welded metal with as-cast structure presents a better corrosion resistance than the base metal with rolling structure [115]. Plasma-MIG hybrid welding can penetrate 5 mm Al5083 with only one welding procedure at a relatively fast speed of 4.75 mm/s. defect-free joint with good appearance can be achieved with optimized parameters. The localized and pitting corrosive morphologies and the EDS analysis of the tested samples confirms that the weld zone with as-cast structure has better corrosion resistance property than the base metal with the rolling structure.

5.4 Future trends

Although hybrid laser-arc welding has been gaining increasing acceptance in recent years, good understanding of the underlying physics remains a challenge. For example, whether the hybrid laser-arc welding technique can be extended for the joining of some advanced composites in aerospace applications are still unknown. The use of short and ultra-short pulse lasers in the hybrid laser welding needs to be explored. The comparative advantages or disadvantages during hybrid welding still need to be further studied. The interaction between the laser and the arc has been observed to enhance arc stability and push the arc towards the laser keyhole, resulting in a deeper penetration. However, the origin of this synergic interaction between the arc and laser plasma is not well understood. Measuring the distributions of electron temperatures and densities in the plasma can provide a better understanding of the laser-arc interaction [10]. Porosity formation is believed to be strongly related to the keyhole collapse process. Hence, better understanding of keyhole stability and dynamics through experimental and theoretical studies would be beneficial. Hybrid welding is known to produce welds with desirable widths and depths, but the maximum gap tolerance and weld penetration for various welding conditions have not been quantified. In the future, advanced mathematical modeling of the heat transfer and fluid flow will enable accurate predictions of weld profile and cooling rates in the welding process, which is critical in understanding the evolution of weld microstructures and residual stress formation in welds. Thus, the hybrid welding process can be optimized to obtain quality welds with no cracking, no brittle phase and less thermal distortion. Better sensing and process control of the hybrid welding process would also be helpful in expanding its applications.

In hybrid laser-arc welding of aeronautical materials, the benefit mainly originates from the ability of this process to adjust filler metal additions, heat input, and post heat treatment processes. Although some initial successes have been obtained to join aeronautical materials, like stainless steels, titanium, magnesium, aluminum, and their alloys, in-depth research, rigorous characterization, and cost analysis are still needed. Hybrid welding of these alloys often involves combinations of different filler and base metals, which have to be determined for various hybrid welding conditions in order to obtain optimum weld properties. Detailed characterization of structure and properties for each alloy remain a major task and rigorous studies need to be done to understand the correlation between welding conditions and the resulting weldment structure and properties for these important engineering alloys. Hence, the applications of hybrid laser-arc welding technique can be widened in aerospace engineering.

References

[1] J. Tusek, M. Suban, Hybrid welding with arc and laser beam, Sci. Technol. Weld. Join. 4 (5) (1999) 308–311.

[2] A. Mahrle, E. Beyer, Hybrid laser beam welding—classification, characteristics, and applications, J. Laser Appl. 18 (3) (2006) 169–180.

[3] B. Acherjee, Hybrid laser arc welding: state-of-art review, Opt. Laser Technol. 99 (2018) 60–71.
[4] G. Casalino, Statistical analysis of MIG-laser CO2 hybrid welding of Al-Mg alloy, J. Mater. Process. Technol. 191 (2007) 106–110.
[5] B. Hu, I.M. Richardson, Microstructure and mechanical properties of AA7075(T6) hybrid laser/GMA welds, Mater. Sci. Eng. A 459 (2007) 94–100.
[6] X. Zhan, Y. Li, W. Ou, F.Y. Yu, J. Chen, Y.H. Wei, Comparison between hybrid laser-MIG welding and MIG welding for the invar36 alloy, Opt. Laser Technol. 85 (2016) 75–84.
[7] L. Wang, X. Li, M. Gao, X. Zeng, Stabilization mechanism and weld morphological features of fiber laser-arc hybrid welding of pure copper, J. Manuf. Process 27 (2017) 207–213.
[8] Z.L. Feng, Processes and Mechanisms of Welding Residual Stress and Distortion, Woodhead Publishing Ltd, Cambridge, England, 2005.
[9] E.A. Metzbower, P.E. Denney, D.W. Moon, C.R. Feng, S.G. Lambrakos, Thermal analysis and microharness maping in hybrid laser welds in a structural steel, Mater. Sci. Forum 426–432 (2003) 4147–4152.
[10] B. Ribic, T.A. Palmer, T. Debroy, Problems and issues in laser-arc hybrid welding, Int. Mater. Rev. 54 (4) (2009) 223–244.
[11] A. Nonn, W. Dahl, W. Bleck, Numerical modelling of damage behaviour of laser-hybrid welds, Eng. Fract. Mech. 75 (2008) 3251–3263.
[12] J.X. Zhang, Y. Xue, S.L. Gong, Residual welding stresses in laser beam and tungsten inert gas weldments of titanium alloy, Sci. Technol. Weld. Joining 10 (2005) 643–646.
[13] L.M. Liu, G. Song, J.F. Wang, Mater. Sci. Forum 488–489 (2005) 361–364.
[14] L.M. Liu, G. Song, G.L. Liang, J.F. Wang, Pore formation during hybrid laser tungsten inert gas arc welding of magnesium alloy AZ31B-mechanism and remedy, Mat. Sci. Eng. A 390 (2005) 76–80.
[15] S. Katayama, S. Uchiumi, M. Mizutani, J. Wang, K. Fujii, Penetration and porosity prevention mechanism in YAG laser-MIG hybrid welding, Weld. Int. 21 (2007) 25–31.
[16] M. Mazar Atabaki, J. Ma, G. Yang, R. Kovacevic, Hybrid laser/arc welding of advanced high strength steel in different butt joint configurations, Mater. Des. 64 (2014) 573–587.
[17] P.M. Norman, J. Lamas, J. Frostevarg, J. Powell, A. Kaplan, A. Yanez, The effect of fit-up geometry on melt flow and weld quality in laser hybrid welding, J. Laser Appl. 25 (3) (2013) 7–13.
[18] Z.Y. Li, T.S. Srivatsan, L. Yan, W.Z. Zhang, Coupling of laser with plasma arc to facilitate hybrid welding of metallic materials: a review, J. Mater. Eng. Perform. 22 (2) (2013) 384–395.
[19] Z. Qi, J. Ping, X.Y. Shao, Z.M. Gao, L.C. Gao, C. Yue, X.B. Li, Optimization of process parameters of hybrid laser–arc welding onto 316L using ensemble of metamodels, Metall. Mater. Trans. B 47 (4) (2016) 2182–2196.
[20] M. Chen, L. Xin, Q. Zhou, L.J. He, F.F. Wu, Effect of laser pulse on alternative current arc discharge during laser-arc hybrid welding of magnesium alloy, Opt. Lasers Eng. 100 (2018) 208–215.
[21] F. Chaussé, P. Paillard, E. Bertrand, G. Rückert, Hybrid laser-metal active gas welding of S460ML steel plates: weldability and consequences of dropping on weld quality, J. Laser Appl. 28 (2) (2018), 022416.
[22] X. Zhang, G.Y. Mi, C.M. Wang, Microstructure and performance of hybrid laser-arc welded high-strength low alloy steel and austenitic stainless steel dissimilar joint, Opt. Laser Technol. 122 (2020) 105878.

[23] A. Fellman, V. Kujanpaa, The effect of shielding gas composition on welding performance and weld properties in hybrid CO2 laser-gas metal arc welding of carbon manganese steel, J. Laser Appl. 18 (2006) 12–20.
[24] T. Kim, Y. Suga, T. Koike, Welding of thin steel plates by hybrid welding process combined TIG arc with YAG laser, JSME Int. J. Ser. A 46A (2003) 202–207.
[25] Y.P. Kim, N. Alam, H.S. Bang, Observation of hybrid (cw Nd:YAG laser + MIG) welding phenomenon in AA5083 butt joints with different gap conditions, Sci. Technol. Weld. Joining 11 (3) (2006) 295–307.
[26] K. Murakami, Y. Mitooka, M. Hino, H. Iogawa, H. Ono, S. Katayama, J. Japn. Inst. Met. 70 (2006) 134–137.
[27] G.L. Qin, Z. Lei, S.Y. Lin, Effects of Nd:YAG laser + pulsed MAG arc hybrid welding parameters on its weld shape, Sci. Technol. Weld. Joining 12 (2) (2007) 79–86.
[28] P.T. Swanson, C.J. Page, E. Read, H.Z. Wu, Plasma augmented laser welding of 6 mm steel plate, Sci. Technol. Weld. Joining 12 (2) (2007) 153–160.
[29] F. Möller, H. Kügler, S. Kötschau, S.F. Goecke, Gap bridging ability in laser GMA hybrid welding of thin 22MnB5 sheets, Phys. Procedia 56 (2014) 620–629.
[30] J. Lamas, J. Frostevarg, A.F.H. Kaplan, Gap bridging for two modes of laser arc hybrid welding, J. Mater. Process. Technol. 224 (2015) 73–79.
[31] O. Seffer, R. Lahdo, A. Springer, S. Kaierle, Laser-GMA hybrid welding of API 5L X70 with 23 mm plate thickness using 16 kW disk laser and two GMA welding power sources, J. Laser Appl. 26 (4) (2014), 042005.
[32] Ö. Üstündağ, A. Fritzsche, V. Avilov, A. Gumenyuk, M. Rethmeier, Hybrid laser-arc welding of thick-walled ferromagnetic steels with electromagnetic weld pool support, Welding in the World Le Soudage Dans Le Monde 62 (4) (2018) 767–774.
[33] W.M. Steen, Arc-augmented laser processing of materials, J. Appl. Phys. 51 (1980) 5636–5641.
[34] M. Eboo, W.M. Steen, J. Clarke, Arc-augmented laser welding, in: Advances in Welding Processes, Proc. of the 4th Int. Conf., Harrogate, UK, 1978, pp. 257–265.
[35] W.M. Steen, M. Eboo, Arc augmented laser welding, Metal Constr. 11 (1979) 332–335.
[36] C. Bagger, F.O. Olson, Review of laser hybrid welding, J. Laser Appl. 17 (1) (2005) 2–14.
[37] H. Cui, I. Decker, J. Ruge, Wechselwirkungen zwischen WIGSchweißlichtbogen und fokussiertem Laserstrahl, in: Proc. of the Conference Laser'89, Springer, Berlin, 1989, pp. 577–581.
[38] B.R. Finke, F. Stern, I. Decker, Auswirkungen eines unterstützenden Laserstrahls auf den WIG-Schweißprozess, DVS-Ber. 135 (1991) 149–152.
[39] H. Cui, I. Decker, H. Pursch, J. Ruge, J. Wendelstorf, H. Wohlfahrt, Laserinduziertes Fokussieren des WIG-Lichtbogens, DVS-Ber. 146 (1992) 139–143.
[40] U. Dilthey, F. Lüder, A. Wieschemann, Expanded capabilities in the welding of aluminium alloys with the laser-MIG hybrid process, Aluminium 75 (1999) 64–75.
[41] T. Graf, H. Staufer, Laser hybrid process at Volkswagen, International Institute of Welding, 1999. IIW-Doc. XII-1730-02.
[42] T. Jokinen, T. Vihervä, H. Riikonen, V. Kujanpaä, Welding of ship structural steel A36 using a Nd:YAG laser and gas metal arc welding, J. Laser Appl. 12 (2000) 185–188.
[43] Höfemann, M., Fersini, M., Szinyur, J., Haferkamp, H., and Cordini, P., 'Laser-GMA-hybrid welding of zinc-coated steel for hydroforming applications', Proc. of the 1st Int. WLT-Conf. on Lasers in Manufacturing, Munich, Germany, 466–472, 2001.

[44] U. Dilthey, H. Keller, Prospects in laser GMA hybrid welding of steel, in: Proc. of the 1st Int. WLT-Conf. on Lasers in Manufacturing, Munich, Germany, 2001, pp. 453–465.

[45] C. Waltz, T. Seefeld, G. Sepold, Laser-GMA welding and its influence on bead geometry and process stability, in: Proc. of the 1st Int. WLT-Conf. on Lasers in Manufacturing, Munich, Germany, 2001, pp. 444–452.

[46] K. Shibata, H. Sakamoto, T. Iwase, Laser-MIG hybrid welding of aluminium alloys, in: Proc. of the 1st Int. WLT-Conf. on Lasers in Manufacturing, Munich, Germany, 2001, pp. 436–443.

[47] T. Jokinen, P. Jernström, M. Karhu, I. Vanttaja, V. Kujanpää, Optimisation of parameters in hybrid welding of aluminium alloy, in: Proc. Of the 1st Int. Symp. on High Power Laser Macroprocessing, 4831, Proc. SPIE, 2003, pp. 307–312.

[48] T. Ishide, S. Tsubota, M. Watanabe, Latest MIG, TIG arc-YAG laser hybrid welding systems for various welding products, in: Proc. of the 1st Int. Symp. on High Power Laser Macroprocessing, 2003, 4831, Proc. SPIE, 2003, pp. 347–352.

[49] C.V. Hyatt, K.H. Magee, J.F. Porter, V.E. Merchant, J.R. Matthews, Laser-assisted gas metal arc welding of 25-mm-thick HY-80 plate, in: Welding Journal Symposium, Miami, FL, U.S.A, 2001, pp. 163s–172s.

[50] A. Fellman, P. Jernstrøm, V. Kujanpää, The effect of shielding gas composition in hybrid welding of carbon steel, in: Proceedings 9th NOLAMP, Conference on Laser Materials Processing in the Nordic Countries, Trondheim, Norway, 2003, pp. 103–112.

[51] J. Zhou, H.L. Tsai, Modeling of transport phenomena in hybrid laser-MIG keyhole welding, Int. J. Heat Mass Transf. 51 (2008) 4353–4366.

[52] C.J. Page, T. Devermann, J. Biffin, N. Blundell, Plasma augmented laser welding and its applications, Sci. Technol. Weld. Joining 7 (1) (2002) 1–10.

[53] S.M. Joo, Y.P. Kim, C.S. Ro, H.S. Bang, J.U. Park, Basic experiments by hybrid welding to steels for shipbuilding, Adv. Nondestr. Eval. 270–273 (2004) 2383–2388.

[54] T. Graf, H. Staufer, Laser-hybrid welding drives VW improvements, Weld. J. 82 (2003) 42–48.

[55] M. Ono, Y. Shinbo, A. Yoshitake, M. Ohmura, Welding properties of thin steel sheets by laser-arc hybrid welding: laser-focused arc welding, in: Proc. 1st Int. Symp. on High-Power Laser Macroprocessing, Osaka, Japan, 2002, pp. 369–374.

[56] G. Song, L.M. Liu, M.S. Chi, J.F. Wang, Investigations of laser-TIG hybrid welding of magnesium alloys, Mater. Sci. Forum 488–489 (2005) 371–375.

[57] K.C. Chiang, H.L. Tsai, Shrinkage-induced fluid flow and domain change in two-dimensional alloy solidification, Int. J. Heat Mass Transfer 35 (1992) 1763–1769.

[58] J. Zhou, H.L. Tsai, P.C. Wang, Transport phenomena and keyhole dynamics during pulsed laser welding, ASME J. Heat Transfer 128 (7) (2006) 680–690.

[59] Y. Wang, Modeling of Three-Dimensional Gas Metal Arc Welding Process, Ph.D. Thesis, University of Missouri-Rolla, 1999.

[60] J.J. Lowke, P. Kovitya, H.P. Schmidt, Theory of free-burning arc columns including the influence of the cathode, J. Phys. D: Appl. Phys. 25 (1992) 1600–1606.

[61] J. Haidar, A theoretical model for gas metal arc welding and gas tungsten arc welding, J. Appl. Phys. 84 (1998) 3518–3529.

[62] J. Haidar, J.J. Lowke, Predictions of metal droplet formation in arc welding, J. Appl. Phys. D: Appl. Phys. 29 (1996) 2951–2960.

[63] J. Haidar, An analysis of the formation of metal droplets in arc welding, J. Phys. D: Appl. Phys. 31 (1998) 1233–1244.

[64] J. Haidar, Prediction of metal droplet formation in gas metal arc welding. II, J. Appl. Phys. 84 (1998) 3530–3540.

[65] J. Haidar, An analysis of heat transfer and fume production in gas metal arc welding. III, J. Appl. Phys. 85 (1998) 3448–3459.

[66] J.J. Lowke, P. Kovitya, H.P. Schmidt, Theory of free-burning arc columns including the influence of the cathode, J. Phys. D: Appl. Phys. 25 (1995) 1600–1606.

[67] P. Zhu, J.J. Lowke, R. Morrow, A unified theory of free burning arcs, cathode sheaths and cathodes, J. Phys. D: Appl. Phys. 25 (1992) 1221–1230.

[68] J.U. Brackbill, D.B. Kothe, C. Zemach, A continuum method for modeling surface tension, J. Comput. Phys. 100 (1992) 335–354.

[69] H.G. Fan, R. Kovacevic, Droplet formation, detachment, and impingement on the molten pool in gas metal arc welding, Metall. Trans. 30B (1999) 791–801.

[70] J. Hu, H.L. Tsai, Heat and mass transfer in gas metal arc welding. Part I: the arc, Int. J. Heat Mass Transf. 50 (2007) 833–846.

[71] S.I. Anisimov, Vaporization of metal absorbing laser radiation, Sov. Phys. JETP 27 (1) (1968) 182–183.

[72] C.J. Knight, Theoretical modeling of rapid surface vaporization with back pressure, AIAA J. 17 (5) (1979) 519–523.

[73] Kanouff, M.P., Kassinos, A., Noble, D.R., Schunk, P.R., 2000. Laser Spot Weld Modeling Using an ALE Finite Element Method. Sandia National Laboratories Report.

[74] A. Kaplan, A model of deep penetration laser welding based in calculation of the keyhole profile, J. Phys. D: Appl. Phys. 27 (1994) 1805–1814.

[75] W. Duley, Laser Welding, John Wiley & Sons, 1999.

[76] R. Ducharme, K. Williams, P. Kapadia, J. Dowden, B. Steen, M. Glowacki, The laser welding of thin metal sheets: an integrated keyhole and weld pool model with supporting experiments, J. Phys. D: Appl. Phys. 27 (1994) 1619–1627.

[77] J.R. Ho, C.P. Grigoropoulos, J.A.C. Humphrey, Gas dynamics and radiation heat transfer in the vapor plume produced by pulsed laser irradiation of aluminum, J. Appl. Phys. 79 (1996) 7205–7215.

[78] J. Zhou, W.H. Zhang, H.L. Tsai, S.P. Marin, P.C. Wang, R. Menassa, Modeling the transport phenomena during hybrid laser-MIG welding process, in: Proceedings IMECE'03, ASME International Mechanical Engineering Congress & Exposition, Washington, DC, 2003, pp. 1–8. IMECE2003-41763.

[79] J. Zhou, H.L. Tsai, Investigation of mixing and diffusion processes in hybrid spot laser—MIG keyhole welding, J. Phys. D: Appl. Phys. 42 (2009) 1–15.

[80] D.Q. Zhang, X. Jin, L.X. Gao, H.G. Joo, K.Y. Lee, Effect of laser-arc hybrid welding on fracture and corrosion behavior of AA6061-T6 alloy, Mater. Sci. Eng. A 528 (2011) 2748–2754.

[81] Y.F. Meng, G. Ming, X.Y. Zeng, Quantitative analysis of synergic effects during laser-arc hybrid welding of AZ31 magnesium alloy, Opt. Lasers Eng. 111 (2018) 183–192.

[82] K.N. Braszczyńska-Malik, M. Mróz, Gas-tungsten arc welding of AZ91 magnesium alloy, J. Alloys Compd. 509 (41) (2011) 9951–9958.

[83] G. Song, Z. Diao, X.Z. Lv, L.M. Liu, TIG and laser–TIG hybrid filler wire welding of casting and wrought dissimilar magnesium alloy, J. Manuf. Proc. 34 (Pt. A) (2018) 204–214.

[84] M. Gao, H.G. Tang, X.F. Chen, X.Y. Zeng, High power fiber laser arc hybrid welding of AZ31B magnesium alloy, Mater. Des. 42 (2012) 46–54.

[85] L.M. Liu, G. Song, M.S. Chi, Laser-tungsten inert gas hybrid welding of dissimilar AZ based magnesium alloys, Mater. Sci. Technol. 21 (9) (2005) 1078–1082.

[86] M. Gao, X.Y. Zeng, B. Tan, J.C. Feng, Study of laser MIG hybrid welded AZ31 magnesium alloy, Sci. Technol. Weld. Join. 14 (4) (2009) 581–592.

[87] X.D. Qi, G. Song, Interfacial structure of the joints between magnesium alloy and mild steel with nickel as interlayer by hybrid laser-TIG welding, Mater. Des. 31 (2010) 605–609.
[88] L.M. Liu, X.D. Qi, Effects of copper addition on microstructure and strength of the hybrid laser-TIG welded joints between magnesium alloy and mild steel, J. Mater. Sci. 44 (2009) 5905–5911.
[89] L. Liu, X. Qi, Strengthening effect of nickel and copper interlayers on hybrid laser-TIG welded joints between magnesium alloy and mild steel, Mater. Des. 31 (8) (2010) 3960–3963.
[90] C. Li, D.Y. He, F. Guo, X.Y. Li, J.M. Jiang, Effect of fiber laser-MIG hybrid process parameters on weld bead shape and tensile properties of commercially pure titanium, Mater. Manuf. Process. 25 (2010) 1309–1316.
[91] Q.M. Liu, Z.H. Zhang, S.F. Liu, Application and development of titanium alloy in aerospace and military hardware, J. Iron Steel Res. 29 (2015) 4–10.
[92] H. Sui, X. Zhang, D. Zhang, Feasibility study of high-speed ultrasonic vibration cutting titanium alloy, J. Mech. Eng. 247 (2017) 120–127.
[93] Z. Li, H. Guo, L. Li, Study on surface quality and tool life in ultrasonic vibration countersinking of titanium alloys (Ti6Al4V), Int. J. Adv. Manuf. Technol. 103 (2019) 1119–1137.
[94] K. Gangwar, M. Ramulu, Friction stir welding of titanium alloys: a review, Mater. Des. 141 (2017) 230–255.
[95] J.P. Shi, G. Song, J.Y. Chi, Effect of active gas on weld appearance and performance in laser-TIG hybrid welded titanium alloy, Int. J. Lightweight Mater. Manuf. 1 (1) (2018) 47–53.
[96] M. Junaid, M.N. Baig, M. Shamir, F.N. Khan, K. Rehman, J. Haider, A comparative study of pulsed laser and pulsed TIG welding of Ti-5Al-2.5Sn titanium alloy sheet, J. Mater. Process. Technol. 242 (2017) 24–38.
[97] M. Balasubramanian, V. Jayabalan, V. Balasubramanian, Effect of microstructure on impact toughness of pulsed current GTA welded α–β titanium alloy, Mater. Lett. 62 (6–7) (2008) 1102–1106.
[98] J.W. Mao, W.J. Lv, L.Q. Wang, D. Zhang, J.N. Qin, Microstructure and mechanical properties of GTA weldments of titanium matrix composites prepared with or without current pulsing, Trans. Nonferrous Metals Soc. China 24 (5) (2014) 1393–1399.
[99] R.F. Li, Z.G. Li, Y.Y. Zhu, L. Rong, A comparative study of laser beam welding and laser–MIG hybrid welding of Ti–Al–Zr–Fe titanium alloy, Mater. Sci. Eng. A 528 (3) (2014) 1138–1142.
[100] P.M. Mashinini, I. Dinaharan, J. David Raja Selvam, D.G. Hattingh, Microstructure evolution and mechanical characterization of friction stir welded titanium alloy Ti–6Al–4V using lanthanated tungsten tool, Mater. Charact. 139 (2018) 328–336.
[101] V. Esfahani Yeganeh, P.J. Li, Effect of beam offset on microstructure and mechanical properties of dissimilar electron beam welded high temperature titanium alloys, Mater. Des. 124 (2017) 78–86.
[102] C. Li, K. Muneharua, S. Takao, H. Kouji, Fiber laser-GMA hybrid welding of commercially pure titanium, Mater. Des. 30 (1) (2009) 109–114.
[103] L.M. Liu, X.F. Hao, X. Du, Microstructure characteristics and mechanical properties of laser-TIG hybrid welding joint of TA15 titanium alloy, Mater. Res. Innov. 12 (3) (2008) 114–118.
[104] C.M. Allen, G. Verhaeghe, P.A. Hilton, Laser and hybrid laser-MIG welding of 6.35 and 12.7 mm thick aluminium aerospace alloy', Mater. Sci. Forum 519 (2006) 1139–1144.

[105] G. Casalino, M. Mortello, P. Leo, Study on arc and laser powers in the hybrid welding of AA5754 Al-alloy, Mater. Des. 61 (2014) 191–198.
[106] M.M. Atabaki, J. Ma, W. Liu, Hybrid laser/arc welding of advanced high strength steel to aluminum alloy by using structural transition insert, Mater. Des. 75 (2015) 120–135.
[107] C. Zhang, M. Gao, X. Zeng, Influences of synergy effect between laser and arc on laser-arc hybrid welding of aluminum alloys, Opt. Laser Technol. 120 (2019) 105766.
[108] W.F. Xu, X.K. Wu, J. Ma, Abnormal fracture of 7085 high strength aluminum alloy thick plate joint via friction stir welding, J. Mater. Res. Technol. 8 (2019) 6029–6040.
[109] J. Wang, G. Peng, L. Li, 30 kW-level laser welding characteristics of 5A06 aluminum alloy thick plate under subatmospheric pressure, Opt. Laser Technol. 119 (2019) 105668.
[110] A.T. Kermanidis, A. Tzamtzis, An experimental approach for estimating the effect of heat affected zone (HAZ) microstructural gradient on fatigue crack growth rate in aluminum alloy FSW, Mater. Sci. Eng. A 691 (2017) 110–120.
[111] M.M. Atabaki, M. Nikodinovski, P. Chenier, Experimental and numerical investigations of hybrid laser arc welding of aluminum alloys in the thick T-joint configuration, Opt. Laser Technol. 59 (2014) 68–92.
[112] J. Yan, X.Y. Zeng, M. Gao, J. Lai, T.X. Lin, Effect of welding wires on microstructure and mechanical properties of 2A12 aluminum alloy in CO2 laser-MIG hybrid welding, Appl. Surf. Sci. 225 (2009) 7307–7313.
[113] D.Q. Zhang, X. Jin, L.X. Gao, Effect of laser–arc hybrid welding on fracture and corrosion behaviour of AA6061-T6 alloy, Mater. Sci. Eng. A 528 (6) (2011) 2748–2754.
[114] D.Q. Zhang, J. Li, H.G. Joo, Corrosion properties of Nd: YAG laser–GMA hybrid welded AA6061 Al alloy and its microstructure, Corros. Sci. 51 (6) (2009) 1399–1404.
[115] D. Cai, S. Han, S. Zheng, Microstructure and corrosion resistance of Al5083 alloy hybrid plasma-MIG weld, J. Mater. Process. Technol. 255 (2018) 530–535.

Electron beam welding—Techniques and trends☆

6

M.St. Weglowski[a], S. Błacha[a], and A. Phillips[b]
[a]Lukasiewicz - Institute of Welding, Gliwice, Poland, [b]Cambridge Vacuum Engineering, Denny Industrial Centre Pembroke Avenue Waterbeach, Cambridge, United Kingdom

6.1 Introduction

Constantly growing industry is looking for new technological solutions that will ensure reduction of production costs, introducing a new product on the market or improving an existing product. One such approach is to use electron beam welding.

Although the technology itself has been known for over 60 years, in the last decade there has been growing interest on the part of individuals in science and industry. Some earlier dates are included commencing with the first recorded demonstration in 1879 by Sir W. Crookes that an electron beam could melt metal. Sir Crookes fused the platinum anode in a cathode ray tube but it was nearly 20 years later when, in 1897, J.J. Thompson showed that cathode rays were in fact a stream of electrons. The first serious attempt to use the electron beam as a tool for melting appears to be that recorded by Pirani in 1907 [1, 2]. The first electron beam drilling experiment of an unintended hole of Ø 7 mm in a platinum aperture, as well as the first electron beam welding actually bead on plate by K.H. Steigerwald were carried out in 1949 [1]. A concentrated beam of electrons was first used for welding in a vacuum chamber in the 1950s in Germany and France [1]. For convenient reference, some of the key dates in the history of electron beam welding are given in Table 6.1.

The first reference and photograph of a weld section with a depth to width ratio of the order of 20:1 appears in an anonymous article in Metalworking Production of 20 November 1959, i.e. 2 years after Stohr's disclosure of the process. As mentioned earlier (Table 6.1), the first commercial equipment in the UK was produced toward the end of 1958. All early units available were essentially low-voltage equipment following the original system disclosed by Stohr. Equally, most of the early units were installed at establishments concerned with nuclear energy applications although they were occasionally used to examine the electron beam weldability of more common materials to ascertain the potential of the technique [2]. Since the 1960s, there has been a rapid development of technology of electron beam welding in many countries, among others: USA, ZSRR (Russia, Ukraine), Germany, France, Japan, Great Britain and Poland. In the following paper features, applications and devices will be presented.

☆ The paper was presented at the 9th Symposium on Vacuum based Science and Technology, held on November 17–19, 2015 in Kolobrzeg, Poland.

Welding and Joining of Aerospace Materials. https://doi.org/10.1016/B978-0-12-819140-8.00006-7
Copyright © 2012 Woodhead Publishing Ltd. All rights reserved.

Table 6.1 Beginnings of electron beam welding development [1–4].

Year	Event
1879	Sir W. Crookes: fused a platinum anode in cathode ray
1897	J.J. Thompson: showed that cathode rays were electron beams
1907	M. Pirani: patent of electron beam melting
1938	M. Ardenne: use of electron beams as a work tool
1951	British Patent 727,460: Fine wire spot welding by electron beam; K.H. Steigerwald constructs electron beam welding machines, primarily for drilling
1954	J.A. Stohr: begins work on electron beam welding
1956	Heraeus: produce a nonvacuum electron gun
1957	J.A. Stohr: J. A. Stohr's presentation on the results of practical electron beam welding of zirconium at the Paris international "Fuel Elements Conference"
1958	W.L. Wyman report work in USA; Zeiss equipment supplied to Bettis; N.A. Olshansky reports work in USSR; first commercial equipment available in UK
1959	First report of welds with very high depth to width ratio using Steigerwald equipment; first report of works in Japan with JEOL machine
1968	First electron beam welding machine in Poland

As an introduction, a number of special characteristics of electron beam welding are summarized and discussed in greater detail in the following sections. In comparison with other methods of joining, electron beam welding is characterized by Ref. [5]:

- an extremely high power density of about 10^7 W cm^{-2} at the focus of the beam;
- energy transfer occurs by conduction of heat across the surface of the workpiece itself;
- as no edge preparation is necessary, regardless of the thickness of the workpiece, filler metal is not generally required;
- the high welding speed results in narrow welds and heat affected zones with little distortion of the workpiece;
- inertia free oscillation of the electron beam makes it possible in many cases to join materials otherwise considered unsuitable for welding;
- the variable working distance allows workpieces of widely differing shapes to be welded;
- as welding is carried out under vacuum, no consumables (gases, fluxes) are required to protect the weld pool from oxidation;
- short evacuating times can be achieved by adopting the working chamber to suit the number and size of workpieces;
- computer monitoring and control of the electrical and mechanical welding parameters is possible;
- the welding parameters, and thus the quality of the welds produced, are highly reproducible and consistent;
- at accelerating voltages above 60 kV lead shielding is directly bonded to the welding machine in order to prevent X-ray emission;
- beam powers of far less than 1 kW up to 300 kW are available for welding materials thicknesses of less than 0.5 mm up to 300 mm;
- machines are available for welding variable one-off components as well as for use in mass production operations as in the automotive industry;

- simple longitudinal weld seams can be welded as well as complicated three dimensional components requiring use of programed welding parameters and workpiece manipulation;
- structural steels, alloy steels, nonferrous metals and even gas sensitive special metals can be successfully welded.

Unfortunately there are no statistics available to show exactly how many electron beam welding machines are in use worldwide. World distribution of electron beam machines (Fig. 6.1) and their age, covering the last 50 years, demonstrates clearly the shift of electron beam use from the old industrialized countries to the emerging markets in Asia. There is little information on the former COMECON area. It can be estimated that the total number of electron beam welding machines is about 3500 in the whole world.

An electron beam as a heat source has been utilized for many years. However, its use in various technological processes, particularly in welding, is still dynamically developing both in terms of research and industrial applications. The electron beam is also a useful energy source in surface modification, cladding, surfacing, rapid prototyping, engraving as well as texturing [7].

In spite of quickly developing competitive laser welding technologies, electron beam welding remains indispensable in many applications due to the possibility of obtaining greater penetration depth and the metallurgical purity of a weld, as well as because of its greater welding rate.

The discussion about the advantages and disadvantages of both the beam welding processes (i.e. electron and laser beam welding) is frequently held in an emotional

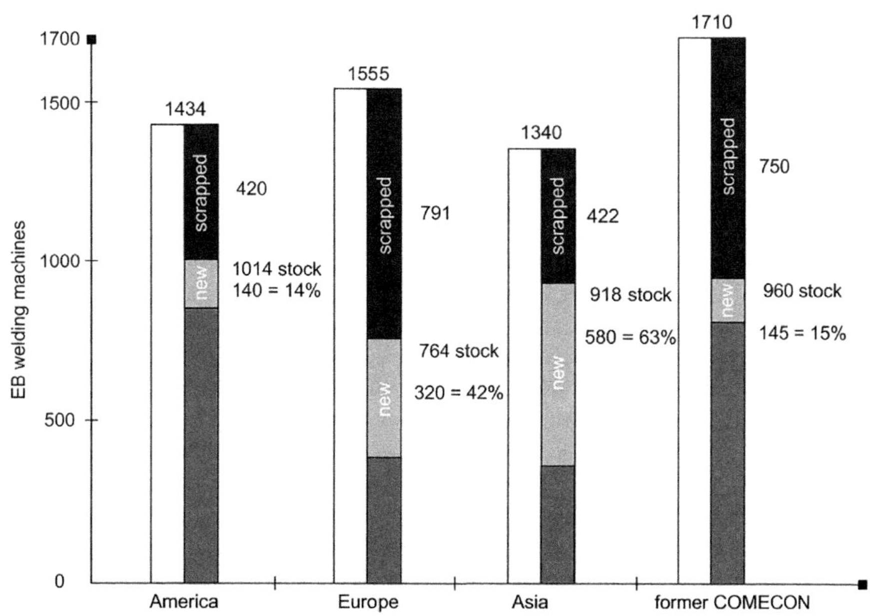

Fig. 6.1 The market for electron beam equipment shifts from America to Asia [6].

form. Supporters of both technologies find various advantages and disadvantages of the respective processes. This detailed discussion should not be held in this paper.

However the main drawbacks of electron beam welding in vacuum are as follows [2]:

- the vacuum limitation, the dimensions of the vacuum chamber impose an upper limit on the size of component to be handled, since the practice is to place the whole component in the vacuum chamber;
- magnetic materials, an electron beam is readily deflected by small magnetic fields. Thus welding of materials with retained magnetism presents special problems since the alignment of the beam is likely to be distributed by the component's own magnetic field;
- profile 3D welding, the majority of EBW machines are provided with manipulators (x-y table, rotary tilt etc...) capable of producing linear, circular as well as 2D profile welds. Such welds represent by far the largest proportion of all welds produced in industry, but there will be occasions when the design demands a 3D profile welds;
- accessibility, areas to be reached by the electron beam must be "seen" by the gun; in other words the path of the electrons from the gun to the area to be welded is a straight line;
- criticality of machining and assembly, the heat produced by a well-focused beam is extremely localized; this is the main asset of the process. It is, however, not without its penalties. Abutment faces have to be machined to close tolerance since a gap above 0.1 mm cannot be tolerated.
- cleaning of the welded elements, vacuum technologies are quick to point out the importance of cleanliness in association with vacuum plant. The introduction of impurities into the vacuum system can do much to impair the pumping speed and general efficiency of the pumping unit. The other issue is the quality of the welded joints.
- economic consideration, many factors contribute to the cost of producing an electron beam welds such as: the machine hour rate, cost factors, productive time, welding time, etc.

Simultaneously, in many works details analysis can be found [8–18].

6.2 Electron beam welding

Electron Beam is capable of welding work pieces, with keyhole technique, with thickness from 0.01 to 250 mm of steel and to 500 mm of aluminum [19]. The electron beam welding process is a versatile technology. For example, different grades of steel can be joined with each other by means of electron beam welding as well as hardly weldable by other welding methods metals: refractory metals (tungsten, molybdenum, niobium) and chemically active metals (titanium, zirconium, beryllium). Electron beam welding is also able to join dissimilar metals. This process can be also applied to so-called black-and white joints. Similar statements apply to the various nonferrous metals. However, the situation can be more complicated when totally different metals (e.g. chemical composition, thermal conductivity, solidification properties, coefficient of thermal expansion etc.) have to be joined to each other [19]. Fig. 6.2 presents possible electron beam weldable material combinations [20].

From the practical point of view the good example is electron beam welding of steel and aluminum which cannot be jointed in the fusion welding process because intermetallic phases are formed depending on the mixing ratio. In individual cases,

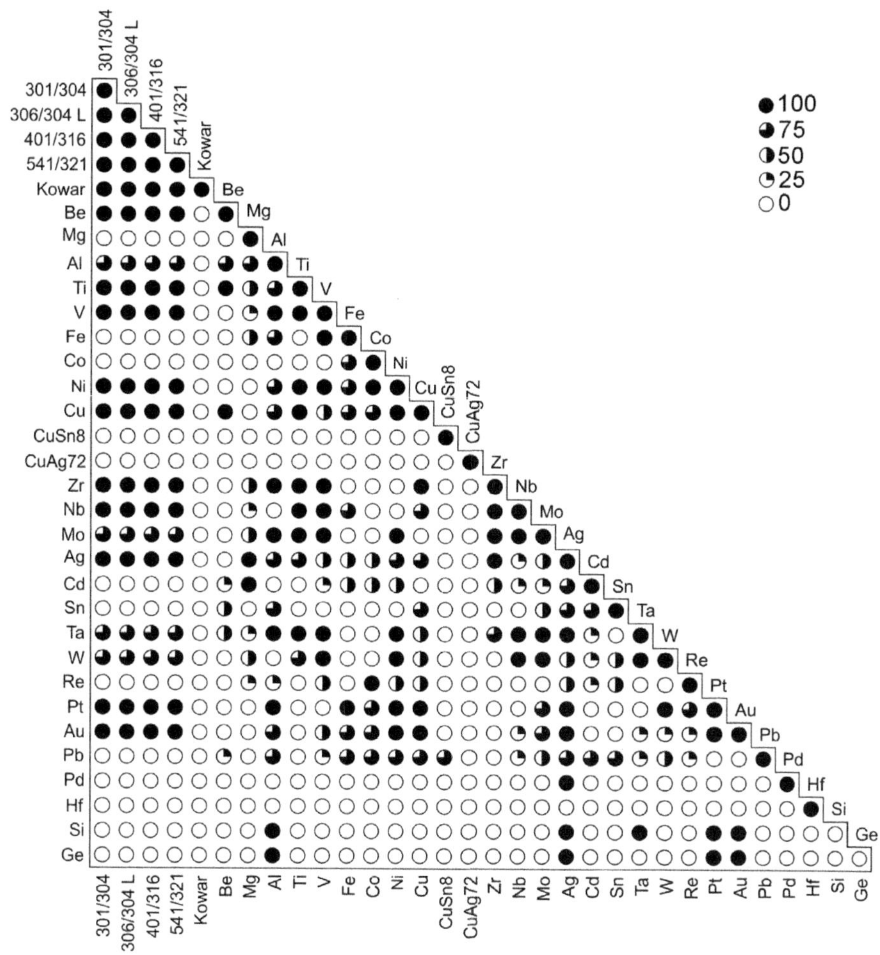

Fig. 6.2 Weldability of structural materials as regards electron beam welding [20], 100—EB weldable, solid solubility exists in all alloy combinations, very desirable combination or already approved successfully; 75—solid solubility exists, strongly different melting points; EB welding with special approved beam management method; 50—insufficient data for proper evaluation, EB welding could be possible; 25—no data available, low probability of weldability by EB; 0—intermetallic compounds formed, undesirable combination, not EB weldable.

success may nevertheless be achieved because, for example, these phases are so finely distributed by the high molten pool dynamics typical of the EB that, as small islands, they are not predominant in a ductile structure [7].

The EB welded joints covered: continuous, pulse as well as spot welds. Classical fillet welds with convex reinforcement, are feasible only by welding with filler material in the form of a wire. Electron beam welding is basically a welding process without filler material. The thermal shrinkage of the weld metal protected the welded joint against the incompletely filled groove [19]. Therefore, it often arises that convex reinforcement and penetration bead of the weld can be observed. The formation of the convex reinforcement is also associated with a change in the density of the weld relative to the base metal. The technique of welding joints in one pass, however, leads to disadvantages of excessive and uneven height of the reinforcement, damaging the material underneath the welded object. The gas bubbles and incomplete fusion in the area of the root of weld can be also observed. These problems by the use of multipass welding techniques with or without the filler material can be solved. In the multipass welding technique without filler material the edges are prepared with a narrow groove and small bevel angle or parallel walls. The thermal shrinkage phenomenon to perform high-quality welded joints is adopted [21].

In the EB welding processes, in order to obtain a high quality welded joint of materials characterized by the limited or difficult weldability of a metal, the filler material is advisable or necessary. The thin foil of suitable material between edges can be adopted. This filler metal should be metallurgical compatible with the parent materials. Also the difference between electrochemical potential of filler metal as well as parent material should be limited. Basically, for practical application the experimental analysis should be carry out [19].

If from a technological point of view, the filler material to fill a groove of thick welded joints or low accuracy preparing parts is required, the solid wire is fed using similar system as in GTA welding methods. The welded joint is beveled at a slight angle (7.5°) and the welding parameters as well as the filler material to obtain the highest possible properties of the welded joint are selected.

With the EB welding in the flat position of welded joints of all weldable metals thicker than 30–40 mm, metal backing to prevent leakage of molten weld metal from the cylindrical keyhole weld is recommended. Therefore, for metal characterized by low viscosity the proper welding parameters are strongly recommended [19].

The upper limit on the thickness range of a welded joint that can be welded in the flat position is also a result of impacts of vapor and gases to escape only through the upper opening of the weld on the electron beam metal. The electron beam also comes out from the part of welded joint. EB welding in the flat position of the welded joints of a thickness greater than 125 mm, irrespective of the power of the electron beam cannot be guaranteed. This phenomenon is particularly for metals with a high pressure metal vapor, for example: magnesium alloys, aluminum alloys, copper alloy, steel with high content of the interstitial elements, in particular oxygen and nitrogen can be observed.

To avoid flaws in the initial and final area of the longitudinal weld joints, the run-on and run-off plates are recommended. In the case of peripheral welded joints, start and finish of the EB welding should be carried out by programmable power electron beam, current or beam focusing.

The different, heat source-dependent manner of heating workpieces is decisive for the shape and dimensions of welds. Fig. 6.3 presents the comparison of a butt weld made using various welding technologies. The high density heat input and narrow weld cause electron beam welding to be accompanied by small workpiece deformations. The level of deformation is reduced in comparison to TIG, MIG and plasma welding, thus for many practical application post-weld finishing can be eliminated. The amount of heat necessary for melting weld metal and heat losses in the welding

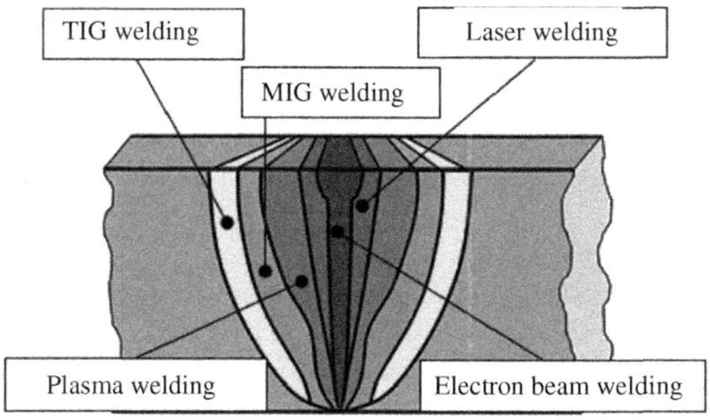

Fig. 6.3 Geometry of a butt joint made using various welding technologies.

process define the total amount of heat which must be supplied by the heat source in order to make a weld. As regards welding process-accompanying heat losses, the electron beam has a significant advantage over other heat sources. While comparing laser beam and electron beam welding processes, it is possible to assume that for solid-state lasers approximately 4% (Nd:YAG) and 25–50% (fiber and disc lasers) of the energy supplied are used in the welding process. The efficiency of electron-beam based devices is between 60% and 70% [22]. However the efficiency of laser beam welding strongly depends on physical properties and chemical composition of the parent materials, as well as reflectivity of the surface [19].

A comparison of welding using an electron beam with other processes is presented in Table 6.2. The calculations were performed for the steel welded joint of a thickness of 150 mm.

Electron beam technologies are particularly useful in the following industries [6, 19, 24, 25]:

Table 6.2 Comparison of arc welding process with EB welding, steel welded joint with 150 mm in thickness [23].

	Welding process			
Parameter	**Electron beam welding**	**Narrow gap GMA**	**Narrow gap SAW**	**SAW**
Current	0.27 A	260 A	650 A	510 A
Voltage groove area	150 kV 800 mm^2	30 V 2100 mm^2	30 V 4900 mm^2	28 V 5900 mm^2
Number of passes	1	35	81	143
Filler metal melting	0	23 kg	54 kg	66 kg
efficiency	7.7 kg/h	5 kg/h	13 kg/h	9 kg/h
Welding time	27 min	4 h 35 min	4 h 11 min	7 h 27 min

- automotive; welding elements of gear transmissions, engine housings, detectors, radiators, crankshafts, piston rods, valve heads, filters, catalysts, turbo-compressors, wheel rims, airbags and many others;
- aviation; welding of titanium tanks for propylene used in satellites and rockets, aluminum containers, thrust jets, fuel injections, fuselage elements made of titanium, armrests, stators, turbine blades and housings, discs in drum rotors of axial-flow compressors, perforated bottoms in rockets;
- power engineering and electrical power engineering; welding of turbine elements, blades, high-current flexible connectors, containers for nuclear waste;
- mechanical engineering; welding of transport hook elements in marine engines, furnaces for destroying used ammunition, wheels and gear transmissions, hydraulic cylinders, sensors of temperature and deformations, microscope frames made of aluminum alloys, commutators, bandsawing machines, drilling tools, catalysts for continuous steel casting, high-pressure valves;
- medical; welding of toothed gear elements in artificial limbs, quick prototyping of endoprosthesis elements, implant surface processing;
- railway; welding of carriage elements, e.g. cross-beams made of aluminum alloys, tow hooks, bearing sleeves, points elements, universal couplings.

The range of application of electron beam welding technology by the industrial sector is shown in Fig. 6.4.

It should be mentioned that modern devices enable the conduction of electron beam welding with one beam or with many beams almost at the same time (a beam of a high frequency, higher than the human eye can spot, is activated and deflected, which in consequence creates the "presence" of many beams), e.g. in welding toothed wheels or particulates filters [22]. Multiple-beam welding is used, among others, in welding with preheating (e.g. when cold crack avoidance is important) (Fig. 6.5A), or in

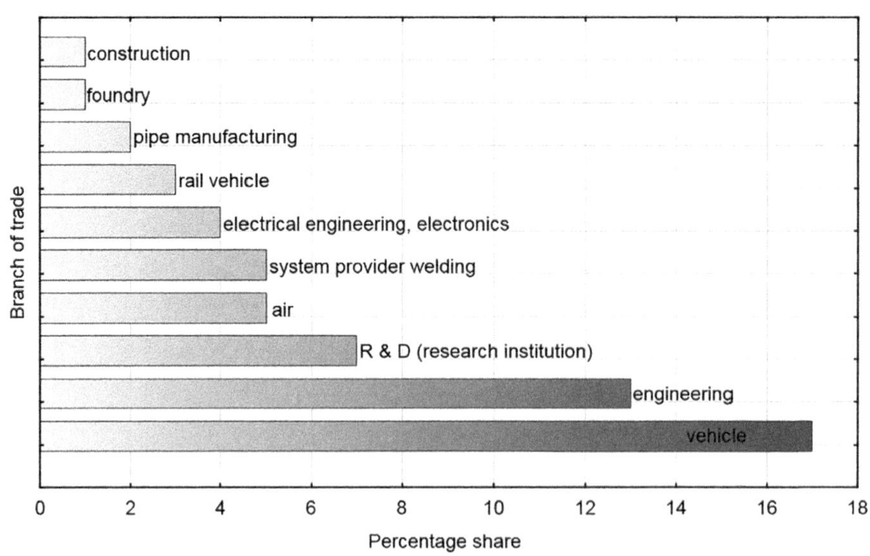

Fig. 6.4 Chart of application areas of the electron beam welding [26].

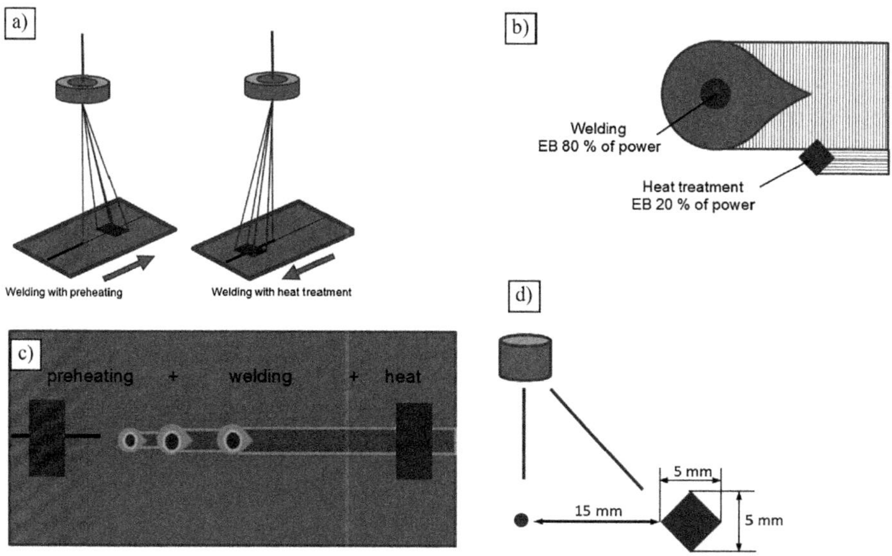

Fig. 6.5 Electron beam welding process (A) with preheating or postweld heat treatment, (B) with preheating and postweld heat treatment, (C) with heat treatment conducted outside the weld axis, (D) exemplary geometrical dimensions of the additional beam [27, 28]. EB, electron beam.

postweld heat treatment processes (in order to eliminate brittle quenched microstructures such as martensite) (Fig. 6.5B) or for welding "accompanied by" preheating or postweld heat treatment (Fig. 6.5C). Multiple electron beam welding, i.e. utilizing several programed-shape beams is also used in order to reduce welding stresses while welding (Fig. 6.6).

Electron beam welding technology can be also adopted for the high or ultra-high strength steel. A few authors presented interesting results on beam welding of high or ultra-high strength steel. Zhang et al. [30] studied the EB welded 300 M under as-weld and postweld heat treatment (PWHT) conditions and they have shown that the PWHT can significantly improve the mechanical properties of welded joints. Elliot [31] investigated the microstructural variations in carbon manganese steels produced upon changes in the EBW process conditions. The investigation has shown that the choice of the appropriate EBW conditions can significantly affect the resultant weld metal

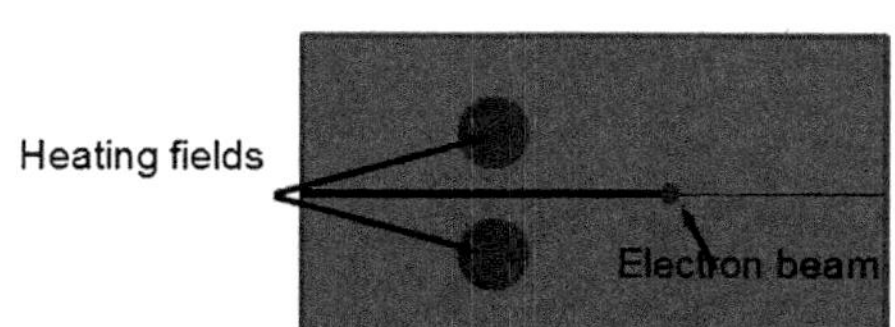

Fig. 6.6 Examples of additional electron beam uses in reducing welding stresses; arrangement of beams while making (A) linear welds, (B) all-round welds [29].

microstructure and properties, especially toughness. Maurer et al. [32] studied the influence of the soft zone on the tensile strength of welded joints. A microalloyed, high strength steel was joined by EBW at different energy input levels. The authors proved that by changing the EBW parameters the microstructure and properties of the weld could be altered in a broad range. Also Weglowski et al. [33–35] studied the electron beam welding of high strength steel grades S690QL, and S960QL. The tensile tests of welded joints revealed that the tensile strengths approach Rm ¼ 850 MPa (S690QL) and 1074 MPa (S960QL) and are at the same level as in the base material (850 MPa and 1065 MPa).

The welded joints achieved 180° bend angle and impact energy higher than 40 J at −40 °C.

6.3 Characteristics of the electron beam welding process

Electron beam welding occurs by joining metallic materials by melting the area of a welded joint in high vacuum in the range from 10^{-3} to 10^{-6} mbar, in a medium or reduced vacuum from 75 to 1000 mbar, with or without vacuum (at atmospheric pressure). All the processes and installations related to electron beam technology are characterized by the utilization of powerful electron beam generators—electron beam guns, with beam power up to max 100 kW. An electron beam emitted from a tungsten cathode, the most common directly heated cathode, is accelerated by a high voltage in the range of 10–200 kV to speeds over 100 km/s. In the electron beam generator, the electrons emitted by the cathode are bundled into a beam by accelerating and focusing electrostatic fields. This task is performed by a beam generator which is composed of the cathode, the control electrode and the anode diaphragm located at the earth potential. The beam emerges through the anode diaphragm and spreads in a conical shape until it enters the magnetic field of the focusing lens. After the focusing by the system of focusing lens, the electrons flow through the system of deflecting coils with which the electrons are deflected electromagnetically in a quasi-inertia-free form and can thus be guided over the workpiece [8, 19], (Fig. 6.7). Bombardment of the area of the weld joint with strongly focused electron beam having a diameter in the range of 0.1–0.8 mm causes the conversion of the kinetic energy of the electrons into the

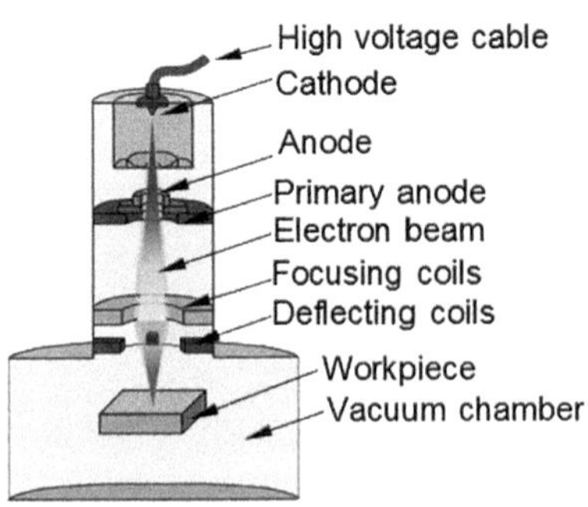

Fig. 6.7 Schematic of electron beam machine for welding and surface modification [36].

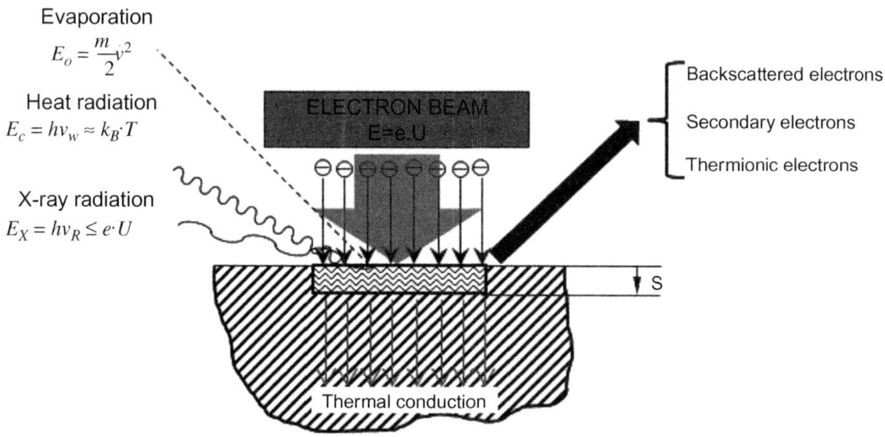

Fig. 6.8 Schematic of interaction between electron beam and surface of the welded material.

heat, which is absorbed and melts the welded joint area structure at the energy density up to 10^{12} W/m^2 [19]. The interaction of the electron beam with the surface of the material is shown in Fig. 6.8.

The electron beam strikes the workpiece surface, interacts with the electrons of the material and causes emissions of [37]:

- back scattered electron (reflected),
- secondary electrons,
- Auger electrons,
- X-rays,
- ray fluorescence,
- thermal radiation.

Each of the types of radiation is emitted from different depths of the material and varies its volume, always with increasing energy of the primary electron beam (i.e., the increase of the accelerating voltage) increases the depth and volume of the broadcast area. At the same time the voltage increases the diameter of the area that is greater than the diameter of the beam. Additionally, the depth of penetration of electrons is affected by the atomic number of the material of the welded components. From the lowest depth the Auger electrons are emitted—they are issued by a few atoms, i.e. about 1 nm, thick subsurface layer. These secondary electrons have a characteristic energy depending on the type of scattering atoms. The signal from this radiation is used in electron microscopes to measure the emission of heat radiation from the surface of the material.

After reaching of the electron beam power density at the level of 10^{10} W/m^2 in the site of beam concentration, the amount of supplied energy is so high that the heat abstraction is not possible by the thermal conductivity of the material, convection or by heat radiation. The welded metal in the point of the electron beam impact not only melts but begins to evaporate. A narrow channel is then formed, filled

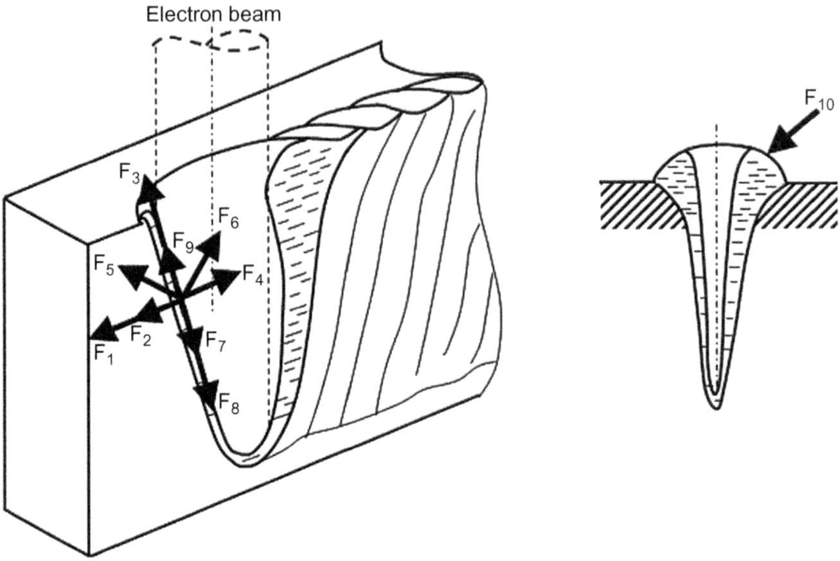

Fig. 6.9 Forces which maintain the cavity in equilibrium: F1—force due to evaporated metal atoms, F2—force due to vapor pressure in cavity, F3—friction between stream of vapor and liquid film, F4—surface tension, F5—electromagnetic force, F6—thermocapillary force due to difference between temperature of front and back walls of cavity, F7—force due to electron pressure, F8—force due to hydrostatic pressure, F9—thermocapillary force due to temperature gradient of front wall, F10—surface tension [35].

and surrounded by a vapor of molten metal envelope—being a capillary through which the electron beam penetrates into the deeper layers of the material [38].

The balance of forces retaining channel in equilibrium is shown in Fig. 6.9. The capillary (gasodynamic channel) is formed mainly due to impacts of the metallic vapor pressures and recoil forces. A smaller share in the impact has direct evaporation of metal (approx. 2% of the volume capillary) and the pressure of the electron beam, study the chemical composition of a thin layer. From a greater depth (5–50 nm) radiation is emitted by secondary electrons, these are the electrons knocked out from the atomic orbits by primary electrons. From greater depth reflected electrons are emitted, these primary electrons are undergoing back-reflection and the intensity of the emission depends mainly on the atomic number of the atoms of the material.

Emission of photons, that is fluorescence in the visible light occurs when secondary electrons recombine with holes arising in some materials as a result of scattering. Another effect of electron interaction with the matter is the emission of X-rays—braking (continuous) and distinctive radiation. A continuous spectrum is produced by braking the incident electrons in the electrostatic field of atomic nuclei. In contrast, the characteristic X-ray emission arises as a result of energy emission through an electron jumping from a higher orbit to the free space on the lower orbit, taking place of the electron being knocked out by the electron of the primary beam. The wavelength and energy of the radiation depend on the atomic number Z of the element and are not dependent on the physical and chemical state of the material. Based on the analysis of wavelength or intensity of energy of the radiation, the chemical composition of the material (in electron microscopes) quantitatively and qualitatively can be determined [37]. Moreover, the primary electrons excite oscillations of the crystal lattice, the result according to measurements [39], an electron beam with an accelerating voltage of 30 kV and current of 150 mA has an impact on nonalloy steel with a force of 1G. Electromagnetic force with the most commonly used ranges of beam current is of negligible size [40]. The properly selected welding speed allows achieving the capillary of the "dynamic stability" during movement in welded material. On the front wall of the capillaries the material is continuously melting, on the back part of it—the solidification process is observed. The envelope of liquid metal surrounding the capillary has an uneven thickness. It is very thin from the side of the melting zone and wider in the area of the weld solidification, and from the interaction surface of the beam between the point of incidence of the beam and the zone of solidification sometimes takes the shape of expanding scales. In a capillary liquid envelope transport of the material occurs. Moving metal from the front wall to the rear is due to intensive evaporation of metal on the front side, supporting the metal transfer to the zone of solidification and thermocapillary force [40]. Thermocapillary force occurs because of differences in surface tension forces caused by the temperature gradient between the zone of the melting and solidification. Transport of liquid metal in the area of the passage takes place around the bundle (in the plane perpendicular to the electron beam) in a direction opposite to the welding direction and in the direction of the electron beam and the counterclockwise [38]. The thermocapillary convection due to the temperature gradient is known as the Marangoni effect (also the Gibbs-Marangoni effect) [41, 42].

An electron beam acting on the welded materials loses, during movement toward the bottom of the capillary, its energy by collisions with the ionized metal vapor, molten metal splashes from the envelope of the capillary tube, and especially by interaction with the surface of the liquid metal envelope of the capillary—mainly on its bottom and side walls. This causes a periodic change of vapor pressure in the channel and the channel can be closed under the force of surface tension by a thin, (having a thickness of several micrometers), film of liquid metal. The previous studies carried out by the technique of X-ray (X-ray capillary) showed that the capillary can be closed also by the liquid metal pushed out of the bottom of capillary. Closing takes place then in the place where the kinetic energy of the displaced fluid flow decreases to zero. Further, the continuous interaction of the electron beam in turn causes the melting of metal layers closing the capillary. This process can be done periodically with a frequency of 150 Hz. Thus, the beam impact point on the welded materials changes its position. This point under certain conditions can move at speeds up to 6.5 m/s. These oscillations occur both at the point of stationary impact of a beam on material, and during welding with full and incomplete penetration [38]. The influence of welding parameters on deep penetration are presented in previous paper [42].

6.4 Machines for electron beam welding and other processes

Electron beam welding processes are usually carried out in a vacuum, yet there are also nonvacuum welding machines available. Modern electron beam welding machines are controlled by PLCs equipped with working tables or numerically controlled welding positioners enabling the automation of welding processes and provided with various control and safety systems aimed to maximize the operator's protection against radiation and carry out technological processes in the vacuum [43]. The division of electron beam welding machines is shown in Fig. 6.10. Types of electron beam welding machines due to working chamber design are shown in Fig. 6.11. The example of a universal and specialist electron beam welding machine for welding are shown in Figs. 6.12 and 6.13, respectively. Due to technological,

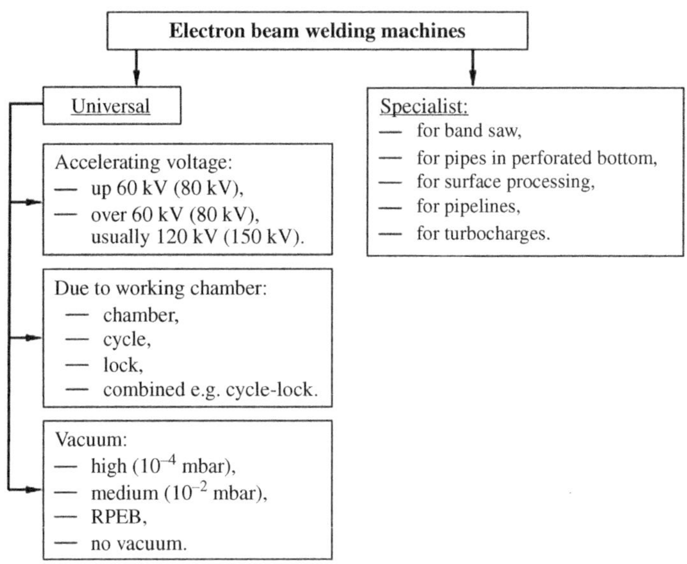

Fig. 6.10 Division of electron beam welding machines [43].

Fig. 6.11 Types of electron beam welding machines due to working chamber design (A) universal welding machine with chamber, (B) cycle machine, (C) continuous operation welding machine (lock).

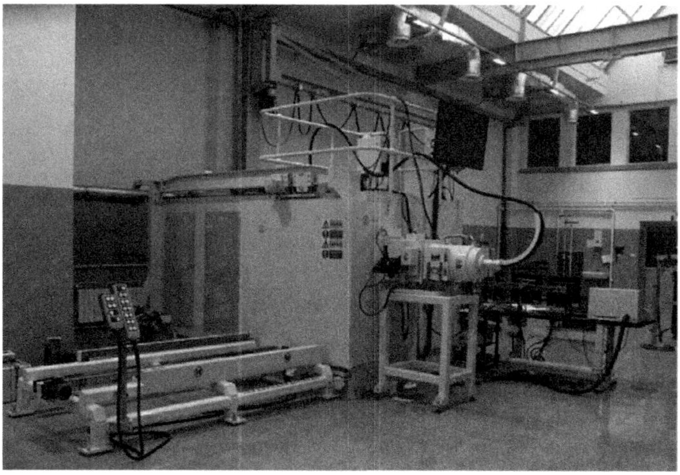

Fig. 6.12 A universal electron beam welding machine model XW150:30/756 for welding and surface modification installed at the Institute of Welding in the year of 2014, (30 kW, 150 kV), produced by Cambridge Vacuum Engineering (CVE).

Fig. 6.13 Example of a specialist electron beam welding machine for welding subsea pipelines.

production-related and metallurgical requirements electron beam welding solutions include the following machines [19]:

- universal high pressure welding machines, with a gun fixed inside or outside the working chamber. After appropriate tooling modification such devices enable welding a vast range of products and elements;
- special high pressure welding machines, intended for welding specific elements such as, for instance band saws, toothed wheels, turbo-compressor rotors;
- reduced pressure electron beam welding machines (RPEB) with local chambers fixed on a structure being welded. The electron gun is located outside a local chamber of small volume, covering only a section of a flat or girth joint being welded, e.g. of storage tanks. The vacuum is only maintained in the small chamber. Such a solution minimizes the time needed to generate the vacuum;
- nonvacuum electron beam welding machines (NVEBW). The electron beam is generated in high vacuum and at high accelerating voltage of 150–220 kV. The beam is moved toward the workpiece by the system of vacuum passes, i.e. the system of nozzles gradually reducing vacuum to atmospheric pressure. Welding at atmospheric pressure almost entirely eliminates problems related to the size of a structure being welded. The NVEBW machines are provided with high-efficiency pumps and special electron beam discharge orifices to ensure the highest vacuum decrease gradient between the electron gun and atmosphere.

Electron beam welding machines—irrespective of their size, intended use and manufacturer—are composed of [7]:

- an evacuable working chamber,
- located inside it, a movement system for the workpiece (if necessary, for the generator as well),
- the working chamber pumping unit,
- the EB generator,
- the EB generator pumping unit,
- the high voltage supply, grid and filament supply,
- the machine controller (frequently with an interface to higher-level systems),
- a cooling system for the power components,
- a compressed air system for certain tooling and valve movements.

As regards increasing the welding process efficiency, ensuring the best quality of welded joints and facilitating operators' work, electron beam welding equipment manufacturers offer many additional systems, including the following [44]:

- automatic beam alignment system,
- electron-optical monitoring system,
- automatic seam tracking,
- control of the process,
- fast deflection generator,
- wire as well as powder feeder.

To achieve narrow weld and deep penetration, it is essential to focus the beam on the workpiece very accurately. However, the focus position of the electron beam is dependent on various parameters as well as the accuracy of the gun assembly. The electron beam focus is typically achieved by controlling the accelerating voltage, beam

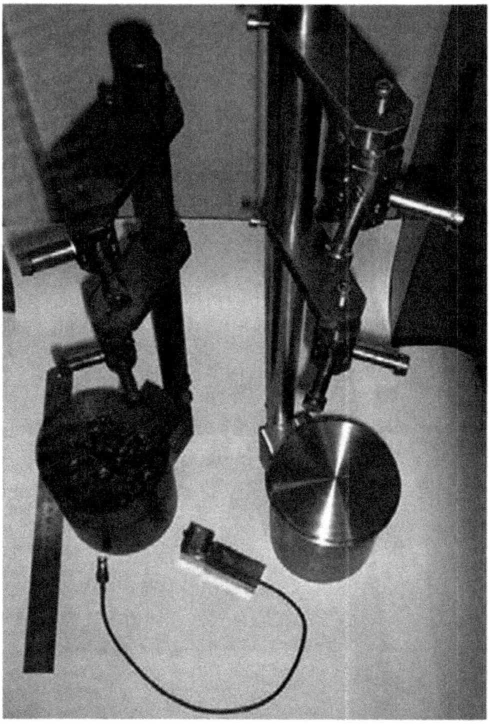

Fig. 6.14 Electron beam quality control device Beam Probe by CVE and TWI.

current, focus coil current, vacuum levels in the gun and in the chamber, and the working distance. These parameters finally control the beam power density. Therefore, these parameters are tightly controlled in the welding process. It should be underlined that the erosion of cathode over time, changes in the alignment of various magnetic lenses, etc. may also cause variations in beam characteristic. The surface focus of the beam at low beam current levels is usually checked by machine operators prior to carrying out welding. Elmer [45] pointed out that this could substantially vary from operator to operator and from machine to machine. This observation was also reported by Giedt and Tallerico [46] where they found that there may be $\pm 20\%$ to $\pm 40\%$ variations in the weld depth just due to manual focus adjustment by different operators. Hence, the independent measuring beam characteristics like beam current, beam current density distribution, beam-width, beam brightness system become more popular in industry (Fig. 6.14) [47].

6.5 Standardization

In welding processes that are not under direct manual control, such as electron beam welding, requirements for various machine parameters, operators as well as welding procedure qualification are established. Standards for electron beam welding covers a very wide range of issues. ISO 14744 standards series [48–53] on acceptance inspection of electron beam welding machines are based on the concept that the production of continuously high-quality welds is ensured if, among other things, the settings, within defined limits, are reproducible during the operating period. These standards specify details of the machine parameters (accelerating voltage, beam current, lens current and welding speed characteristics) together with deviation permitted in short-term or long-term operation. It also includes requirements regarding the run-out accuracy of the devices positioning the workpiece and regarding the stability of the spot position of the electron beam. Details on quality requirements, production welding facilities as well as the weldability of some materials in EN 1011-7 [54] standard are given. According to this standard the quality requirements should be given in the design specification prior to the beginning of the welding work. They should be based on EN ISO 3834 [55–59] and ISO 13919 [60, 61] unless relevant application standards are available e.g. automotive or aerospace industry. The ISO 13919 standard provides guidance on levels of imperfections in electron beam welded joints in steels, aluminum and its alloys. The three quality levels are arbitrarily identified as D, C and B and are intended to cover the majority of practical applications. The levels refer to production quality and not to the fitness-for-purpose of the product manufactured. The choice of quality level for any application should take into account design considerations, subsequent processing, e.g. surfacing, mode of stressing (e.g. static, dynamic), service conditions (e.g. temperature, environment), and consequences of failure. Economic factors are also important and should include not only the cost of welding but also that of inspection, test and repair.

Requirements of welding procedure specification are given in EN ISO 15609 standard [62]. This standard is part of a series of standards, and details of this series are given in EN ISO 15607:2003, annex A [63]. The welding procedure specification (WPS) shall provide all information required to make a weld. Welding procedure specifications may cover a certain range of thicknesses of the joined parts and may also cover a range of parent metals and even filler metals. Some manufacturers may additionally prefer to prepare work instructions for each specific job as part of the detailed production planning.

Qualification of electron beam welding technology can be carried out by using the following qualification methodology: based on previous welding experience [64], by adoption of a standard welding procedure [65], based on preproduction welding test [66] or based on welding procedure test [67].

The requirements for operators of EBW machines are also described at international standard. The most important is ISO 14732 standard [68] which provides the basis for the mutual recognition by examining bodies for approval related to the competence of welding operators in the various fields of application. Tests should be carried out in accordance with this International Standard unless more severe tests are specified by the relevant application standard when these shall be applied. Only welding operators responsible for the setting up and/or adjustment during welding have to be approved. Personnel exclusively performing the programming of the welding unit or operation of the welding unit do not need any particular approval.

It is noteworthy that the national standards as well as recommendation are also very useful for development of the electron beam welding technology. One of the most popular is the American Welding Society (AWS) code C7.1 M/C7.1 [69]. The second one is Deutscher Verband für Schweiben—German Welding Society (DVS) Technical codes [70]. The summary of standards and other regulation in the field of electron beam welding processes are provided in Table 6.3.

Table 6.3 Standards and other regulation in the field of electron beam welding processes.

No	Standard reference	Title	Scope of regulation
1	ISO 14744	Welding. Acceptance inspection of electron beam welding machines	Specifies requirements for acceptance inspection of electron beam welding machines
2	EN 1011-7	Welding. Recommendations for welding of metallic materials—Part 7: Electron beam welding	Details on quality requirements, production welding facilities as well as the weldability of some materials
3	EN ISO 3834		
4	ISO 13919	Quality requirements for fusion welding of metallic materials	Defines comprehensive quality requirements for fusion welding of metallic materials both in workshops and at field installation sites
		Welding. Electron and laser-beam welded joints. Guidance on quality	Gives guidance on quality levels of imperfections in electron and laser beam welded
5	ISO 15609-3		
6	ISO 15607		
7	ISO 15611		
8	ISO levels for imperfections	Specification and qualification of welding procedures for metallic materials. Welding procedure specification. Part 3: Electron beam welding	
		Specification and qualification of welding procedures for metallic materials. General rules	
		Specification and qualification of welding procedures for metallic materials. Qualification based on previous welding experience	
		Specification and qualification of welding procedures for metallic joints in steel. Three levels are given which refer to production	

Continued

Table 6.3 Continued

No	Standard reference	Title	Scope of regulation
		quality. Specifies requirements for the content of welding procedure specifications for electron beam welding Defines general rules for the specification and qualification of welding procedures for metallic materials Gives the necessary information to explain the requirements referenced in ISO 15607 concerning the qualification of welding procedures based on previous welding experience Gives the information necessary to explain the requirements referenced in ISO 15612 materials. Qualification by adoption of a standard welding procedure 15,607 about qualification by adoption of a standard welding procedure, and establishes the conditions, limits and ranges of qualification necessary for the use of a standard welding procedure	
9	ISO 15613		
10	ISO 15614-11		
11	ISO 14732	Specification and qualification of welding procedures for metallic materials. Qualification based on preproduction welding test Specification and qualification of welding procedures for metallic materials. Welding procedure test. Part 11: Electron and laser beam welding Welding personnel. Qualification testing of welding operators and weld setters for mechanized and	

Table 6.3 Continued

No	Standard reference	Title	Scope of regulation
		automatic welding of metallic materials Specifies how a preliminary welding procedure specification is qualified based on preproduction welding tests Defines the conditions for the execution of welding procedure qualification tests and the limits of validity of a qualified welding procedure for all practical welding operations Specifies requirements for qualification of welding operators and also weld setters for mechanized and automatic welding	
12	AWS	An American National Standard. Recommended practice for electron Provides process definitions and requirements, safe practices, and inspection criteria C7.1/MC7.1 beam welding and allied processes for electron beam welding. Also covers common applications of electron beam welding. Includes numerous power curves for various alloys	
13	DVS	Technical codes of electron beam welding	Documents cover recommendations for the cleaning of the joining zone, test procedures for the quality assurance and wedge specimen for the verification of electron beam welds as well as fundamental principles for the designing of components and utilization of nonvacuum electron beam welding

6.6 Other application of electron beam welding

6.6.1 Electron beam cladding and surface modification

Modification of the surface with an electron beam is mainly conducted by using a deflection system. This allows modification of larger surfaces using a single set up electron gun or, lead multibeam process. Fig. 6.15 shows the selected deflection techniques of electron beam for surface modification.

The cladding process using an electron beam is a unique technology allowing insertion additive material in the form of a powder or a wire directly into a pool of liquid metal. Additionally, refining is carried out as the melting process typically is carried out in a vacuum. Modern devices can precisely adjust parameters such as beam power

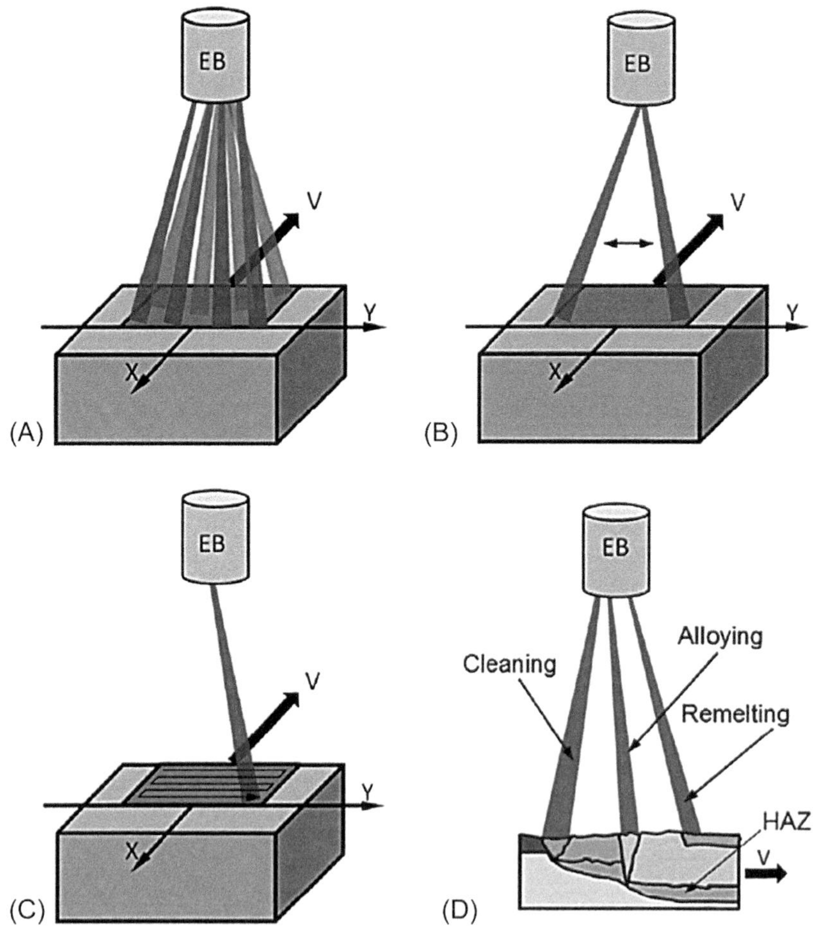

Fig. 6.15 Deflection techniques of electron beam for surface modification (A) multitrack technique, (B) scan technique, (C) meander technique, (D) multiprocess technique [71].

or the beam diameter. This is particularly important in order to ensure minimal penetration or the minimum volume of molten metal at the electron beam energy density up to 10^{12} W/m^2.

Deposition processes may be conducted in a single pass or multiple times, which enables optimization of the chemical composition and properties of the resulting layer.

Gnyusov and Tarasov [72] show the results of the deposition process using a powder in which electromagnetic powder feeder provided efficiency of 40 g/min. Studies included the production of layers deposited on a substrate of alloy steel by cladding powders with a particle size in the range from 50 to 350 mm with the following chemical composition:

- Fe-20%Mn-4%V-4%Mo þ 15%WC,
- Fe-20%Ni-4%V-4%Mo þ 15%WC.

The cladding layers were subjected to an annealing process (T ¼ 500, 600 and 700 °C for 1, 2, 3, 5, and 10 h cooling in air) and heat-treated immediately in a vacuum chamber (self-aged coatings). Fig. 6.16 shows the impact of specific technology solutions on wear resistance.

In Morimoto and coauthors article [73] the results of the deposition process with powder: Cr_3C_2-50%, Ni-Cr-50% (Ni-78.8% Cr-19.34% Fe-0.34%, C <0.01% Mn <0.99%, Si-0.47%) and grains in the range of 5.5 ÷ 88 mm, steel grade S235JR are presented. A powder of Cr-17% Be4% Si-5% C-0.9% Fe <5% Ni-residue, a granularity of 53 ÷ 63 mm as a buffer layer deposit was used. Studies have shown that the process of deposition of the powder allowed obtaining a hardness of 791 HV.

One of the main advantages of electron beam cladding process using powder is the possibility of precise dosing of the additive material and adjusting the thickness of the layers. Galchenko and coauthors article [74] the results of the electron beam cladding with powders FeB þ FeTi and steel grade S235 as substrate are presented. The results

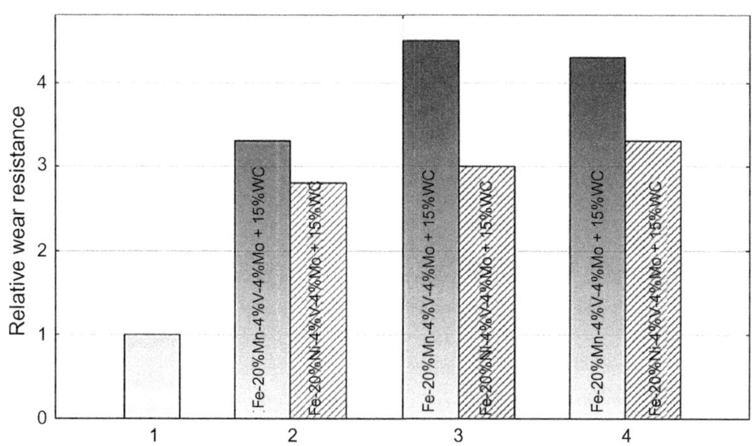

Fig. 6.16 Relative wear resistance ε of electron beam clad coatings determined by loose. Particle abrasive test. 1 day reference 0.45 wt% C steel; 2 day electron beam coatings; 3 day coatings aged at 600 °C for 1 h; and 4 day self-aged coatings [72].

have shown that the hardness of cladding layers is 10 times higher than that of substrate material.

Electron beam deposition process is also possible using a wire as a filler metal. Morimoto et al. [73] presents the results of surfacing austenitic steel grade X6CrNiMoTi17-12-2 and duplex steel grade X2CrNiMoN22-5-3 with two wires:

Fe-79.2%, C-04%, Cr-17.9%, Mo-1.2%, Si <1%, Mn-0.7%,
Co-59.2%, C-1.8%, Cr-31%, W-8%.

The authors obtained a significant increase in hardness (up to 550HV0, 3) and increase in corrosion resistance.

It is noteworthy that the deposition process using electron beam in vacuum is particularly important for reactive materials such as titanium, niobium, molybdenum. In Tomie et al. article [75] the results of cladding with titanium strip having a thickness in the range of 0.1–0.3 mm and the austenitic stainless steel grade 304 (thickness 3.0 mm) as substrate are presented. The authors revealed that a precise control of energy density is fundamental issue. The proper quality of deposit at energy density in the range of 0.05–0.12 kW/mm^2 was obtained.

There is also the possibility of producing cladding layers by melting the powder layer previously disposed on the surface of the substrate (Fig. 6.17). The results of melting of the powder layer WC-Co þ Ni with a thickness of 3 mm on the substrate steel grade S275JR in the Abe et al. paper [76] are presented. The remelting process allows obtaining the deposit with a hardness above 1400 HV.

Electron beam cladding at atmospheric pressure is also possible. Krivezhenko et al. [77] present the results of the powder deposition process. A powder mixture with the composition 40% B4C þ10% Fe was used to form a coating. It was established by the X-ray diffraction analysis that the major phases in the hardened layer were a-iron, iron boride Fe_2B and borocementite $Fe_3B0.6C0.4$. The wear resistance of cladding layers increases two times if compared to the base material.

6.6.2 Hardening

Transformation hardening technology is an electron beam modification of surface layers and consists of short-term (from approx. 1 mS to approx. 1 s) heating up at a rate of 10^3–3×10^3 K/s, of the surface layer to a temperature above the transition

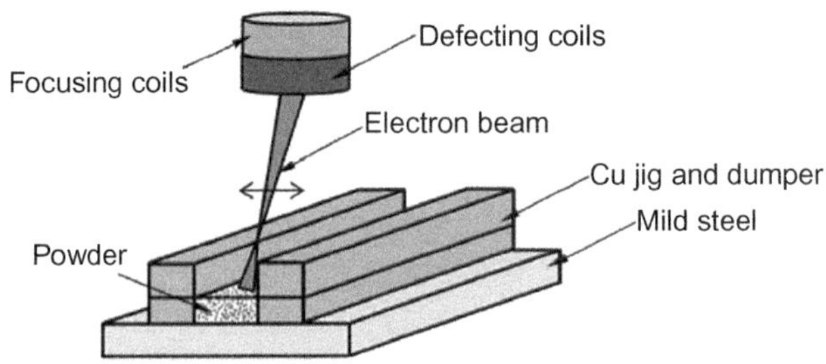

Fig. 6.17 Schematic of the production of cladding layers based on remelting of powder layer [76].

temperature martensitic, but lower than the melting point. Usually power density is about a few kW/m^2.

As a result of the intensive cooling at a rate of 10^4 to more than 10^5 K/s, a very fine crystalline microstructure having a hardness of a few degrees Rockwell higher than the hardness obtained by conventional methods (carried out by inductive heating, flame or plasma) can be observed. The traditional processes allow achieving a maximum rate up to 1000 K/s. The surface roughness of the electron-beam hardening is the same as before hardening, because there is no melting of the workpiece. The most commonly used single layer process significantly improves the tribological properties of structural and tool steels. The thickness of the hardened layer is from several microns to several millimeters. Electron beam allows hardening of the following materials: unalloyed structural steels and alloys, bearings, tools, cast iron gray and white. In addition to higher hardness than conventional hardening an electron beam allows a precise heating of the selected locations of the workpiece even with very small dimensions. Maintaining the tight tolerances and the thickness of the hardened layer with less stress hardening, allowing the hardening shaped parts on standby without changing the dimensions of [78, 79].

On the other hand, hardening with melting causes deterioration of surface roughness with respect to the output surface, especially in intensive remelting. However, it allows better performance characteristics than after transformation hardening, particularly tribology. The main reason is the increase in the hardness into double figure percentage and the favorable distribution of stresses.

Efficiency is estimated to be about 150 cm^2/min [79, 80].

6.6.3 Alloying

The alloying process involves remelting surface layers with alloying elements totally or partially soluble in the substrate. It is performed at higher power densities and longer heating times than while hardening. The alloying process deteriorates the surface roughness in comparison to the raw surface, while significantly improving tribological properties as well as corrosion resistance. Alloying can be achieved by re-melting of layers applied in any way (e.g.: electrolytic or thermal spraying) onto a substrate coating or injection into a molten pool [79]. Alloying and melting processes can be carried out in a single pass or multiple times to allow optimization of the chemical composition and properties of the resulting layers. Multiple passes cause the temperature field on the modified material to also enable the production of graded materials with a specific gradient property. Particularly promising from the point of view of carrying out the modification process in a vacuum, is to modify the surface of reactive materials such as titanium and its alloys. In the case of modification of titanium alloys attention should be paid to the porosity of the modified areas, which leads to deterioration of the quality and reduce the functional properties of the newly created areas.

Selection of optimum parameters of the process of electron beam melting of titanium and its alloys allows for the modification without porosity of the surface. Studies to date have shown that even during welding of titanium alloys in a vacuum, porosity may occur. The tendency to porosity is related to the solubility of hydrogen in the titanium which is a function of temperature. It should be stressed that the effect of the porosity is the amount of heat introduced into the modified area and is directly related to the rate of melting and alloying. Parameters of the process affect the nucleation, growth and degas the liquid metal. Limited porosity at low melting speeds is due to a sufficiently long time required for degassing, while at high speeds there is limited time to nucleation of bubbles of gas and degassing the modified region [80, 81].

While remelting of titanium alloy (Ti-45 Al-2Nb-2Mn-1 B) using an electron beam at a rate of 16 mm/s, the lamellar structure of substrate transforms into a dendritic structure composed of initial a2 (Ti_3Al) dendrites and an interdendritic phase of the g (TiAl) [82]. The alloying process using an electron beam allows for a significant increase in the wear resistance. Utu and coauthors [83] have demonstrated that introduction of the top layer of pure titanium powder $Al_8Si_{20}BN$ and melting the layer using electron beam results in a substantial increase in the wear resistance. It should be emphasized that the corrosion resistance (H_2SO_4) of modified layer is not deteriorated as compared to pure titanium.

It should also be noted that it is possible to operate the modification of titanium alloys by using an electron beam at atmospheric pressure at energy 1.4 MeV.

Golkovskia et al. [84], while alloying pure titanium with powder TieTa obtained a layer with a thickness of 2.0–2.5 mm with a content of tantalum in the range of 3.9–22.4%. The microstructure was composed of a phase (a^0)þb. The tensile strength was 735 MPa and resistance to corrosion in the environment of boiling nitric acid was significantly higher than that of pure Ti (weight loss 190 times smaller) at content of 22.4% Ta. Technological conditions, heating and cooling speed at which new material has been produced, are very different from equilibrium. For this reason, the microstructure has high dispersing components, inhomogeneity of chemical composition and the formation of b phase at a lower Ti content of tantalum.

Oh and Lee [85]. Present the results of the process of alloying titanium alloy Tie 6 Al—4 V with powders TiC, SiC and TiC þ SiC. The study shows that it is possible to produce composite material with a hardness above 700 HV0.5 (base material hardness was 320 HV0.5) on the surface of the titanium alloy. In contrast, alloying with TiN powder [86] allow for obtaining hardness of 937 HV0.5.

It is also possible to increase the hardness of pure titanium by introducing a powder of TiC. The surface layer has the greatest wear resistance approx. 60% and is characterized by a hardness 500 HV0.1 (titanium 180 HV0.1) [87].

The alloying processes for steel are also conducted. In Neagu paper [88] presents the results of the steel grade 10NiCr180 NiCr alloyed with powder. The process of alloying allows obtaining a layer with a hardness of 1300 HV (base material 220 HV).

One of the promising applications of electron beam surface engineering is surface texturing. The process involves the formation on characteristic microprotrusions on the surface (Fig. 6.18). It is used in medicine to improve biocompatibility and for joining e.g. metals with plastics [89–91].

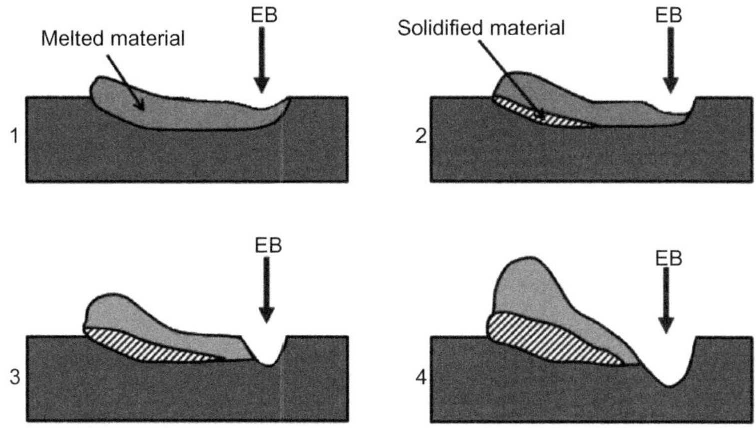

Fig. 6.18 Process scheme of the electron beam use for surface texturing [89].

6.6.4 Electron beam rapid prototyping

In the initial phases of product development, the most important components of the production costs relate to the functions and structures and the materials and technology. The most important stages in the development of a new product usually end up with obtaining of the prototype, which is still in the development phase, and it still requires estimation of its functional properties, and sometimes even the market to evaluate its characteristics. Prototyping using traditional methods is long and costly. Traditional methods typically require a lot of manual labor and a highly skilled workforce, which significantly raises the final cost. Hand-made models also disrupt the cycle of electronic information flow between the design level, and the level of actual production [92]. Rapid prototyping technologies are also useful in the short run production which is not cost effective when using the forging or casting processes. Metal rapid prototyping processes most likely will not replace existing processes, but can offer benefits in certain super alloy applications in terms of cost savings or providing otherwise unavailable processing capabilities, e.g. dual alloy deposition and functionally graded materials. Such capabilities enable innovative design e.g. for future aerospace [93].

The principle of the process of rapid prototyping with a focused electron beam is very simple (Fig. 6.19).

In electron beam rapid prototyping, the metallic feed wires are heated and melted by the electron beam and changed as periodical molten droplets or continuous liquid bridge in a vacuum chamber. Layer by layer deposits, using CNC to control the deposition path, are produced when the droplets of liquid bridge are transferred into the high temperature weld pool in the substrates or previous layer. After the deposits are cooled and solidified, 3D metallic components can be produced. The other option is to use powdered metal which is melted by the electron beam under high vacuum

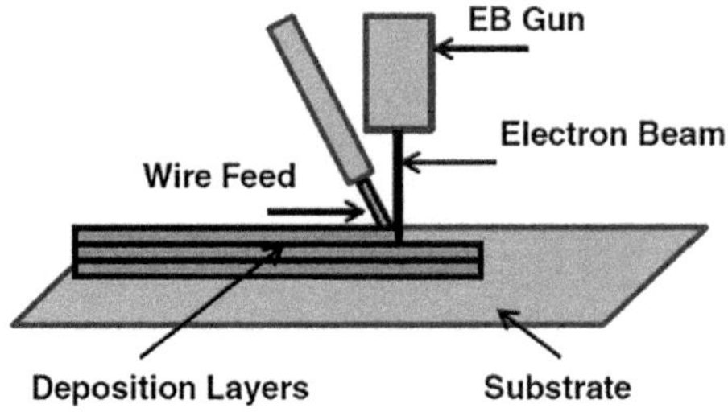

Fig. 6.19 Schematic of rapid prototyping process with wire and laser or electron beam as a heat source [94].

atmosphere. When one layer is finished, the powder bed moves down and a new layer of material is added which is again selectively melted. Like this, the form of the model is built up in 3D layer-by-layer. So, EBM is one of the 3D printing techniques which can be used for high stress aerospace applications and medical grade implants. Different titanium alloys as well as cobalt chrome are used primarily in EBM services. Electron beam melting equipment is different than other laser or heat based sintering processes. In generally, EBM which produces objects do not require any additional thermal heat treatment to achieve the full mechanical properties of the part. EBM prototype services produce parts with lower solid densities and no internal stresses due to the high temperature of electron beam melting. Therefore it allows companies to use EBM prototyping services to produce both prototypes and low cost, high quality parts [95].

Recently, General Electric company informed about plan of EB 3D print of the turbines in their upcoming GE9X engine, which will be the driving force behind Boeing's new 777 × passenger aircraft, set to begin production in 2017 [96].

The electron beam rapid prototyping process is governed by numerous parameters that control the resulting microstructure and shape of the deposit. Three of the most significant of these parameters, which are easily controlled, are the translation speed, wire feed rate and beam power (accelerating voltage and beam current). Running several thermal cycles leads to the melting of the substrate material, which is then followed by heat treatment, both of which can in effect significantly affect the microstructural changes in the structure and consequently the mechanical properties. Previous work has indicated that the highest cooling rate occurs in the liquid ahead of the solid/liquid interface such that at the melting temperature and just below, the cooling rate approaches 103 $K\ s^{-1}$ for 316 stainless steel substrate [97].

Thermal cycles also cause the material produced by rapid prototyping technology to be characterized by a nonuniform microstructure at the cross-section. The combination effect of rapid solidification, directional cooling, and phase transformations induced by repeated thermal cycles, has a profound influence on the microstructures of the deposited materials. Rapid solidification reduces elemental partitioning and extends solid solubility limit and can result in the formation of metastable phases. Directional heat extraction may result in preferred directional grain growth. Repeated thermal cycles have a possible complex set of effects, including microstructural banding i.e., microstructural differences between depositions layers [93]. It is noteworthy that rapid prototyping alloys have a complex thermal history involving directional heat extraction, and repeated melting and rapid solidification. Typically, rapidly prototyped alloys also experience repeated solid state phase transformation. These factors introduce complexities, which are not commonly found in conventional processes.

Microstructural homogenization of the deposited alloys, thereby producing a uniform mechanical properties in the cross section can be increased by the use of suitable process conditions of rapid prototyping. Rapid prototyping processes by using powder usually lead to nonuniform density of the material. Furthermore, powder fills a given volume to a lower extent than ideal [94].

The early investigation revealed [98] that in contrast to more conventional Ni-base super alloys in cast and wrought forms, and with a variety of thermomechanical processing schedules, the EBM fabricated and processed Inconel 625 demonstrated unusual microstructures and microstructural (columnar precipitate) architecture. As fabricated and hot isostatically pressed cylinders exhibited an initial EBM developed g" (fcc) Ni_3Nb precipitate platelet columnar architecture within columnar [200] textured g (fcc) Ni-Cr grains aligned in the cylinder axis, parallel to the EBM build direction. Upon annealing, these precipitate columns dissolve and the columnar g grains recrystallized forming generally equiaxed grains, containing $NbCr_2$ Laves-phase precipitates.

These microstructural features were correlated with unique mechanical properties, especially elongations at room temperature, which exceeded worked wrought alloy 625 product by more than 283% and annealed wrought alloy 625 products by 57% for EBM fabricated and annealed components. The unusual microstructure/microstructural architecture features in the EBM-fabricated illustrated the potential for EBM product with controlled microstructural architectures and their engineering applications [98].

The Tie6Ale4V ELI specimens fabricated using the electron beam melting (EBM) system has a sound microstructure and mechanical properties. Grindability is comparable to those of wrought and cast Tie6Ale4V ELI specimens. The specimen had much finer a-b lamellar structures than the cast specimen [99].

Antonysamy [100] has shown that under the standard process conditions in the bulk section, all the additive manufacturing platforms showed primary columnar b grains with a $\langle 001 \rangle$b jj Nz fiber texture for electron beam selective melting (EBSM) processes (Fig. 6.20).

However, it is noteworthy that application of a solid filler metal becomes not only a simpler alternative to rapid prototyping with powder, but also has the potential for a higher quality deposit, i.e. build-ups, as well as: (1) lower porosity and (2) fewer inclusions from the natural oxide present at the surface of powder materials for aerospace grades, (3) high energy efficiency, (4) high speed deposition rate, (5) clean forming environment. However, controlling the interaction between the electron beams and a filler wire in the fusion zone requires particular regulation of the wire feeding rate, angle and height in relation to the beam parameters in order to obtain a sound deposit [97, 101].

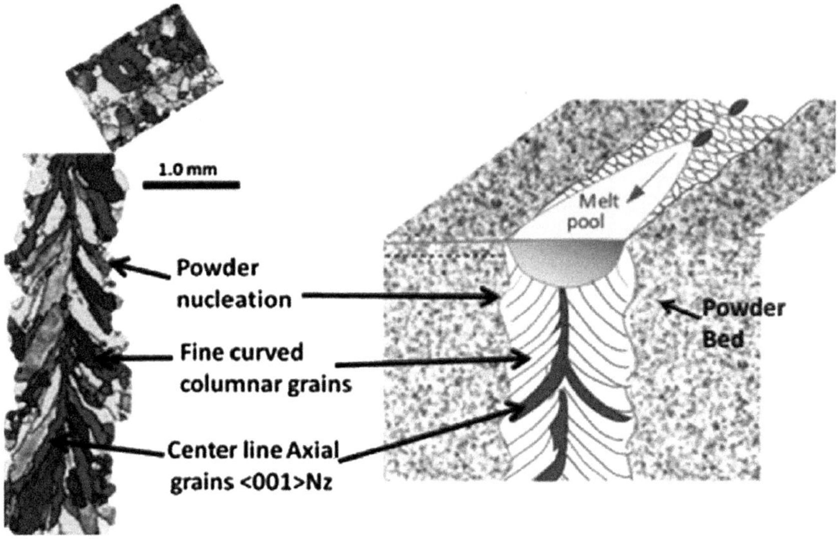

Fig. 6.20 Schematic diagram showing the b-grain structures generated by a single beam contour pass in the 1 mm wall produced by the EBSM process [100].

The preliminary experiments revealed that for the aerospace industry Inconel, aluminum or titanium alloys can be used for rapid prototyping with wire and electron beam as a heat source [97]. In considering the conditions necessary for manufacturing rapid prototyping for the aerospace industry, both the source and deposition process must be critically selected with the objective of achieving defect free components. To this end, the basic knowledge about the process is essential. The overall columnar microstructure of the build-up, which consisted of bands of dendrites, was apparent across the entire length of the deposit manufactured with the solid wire feed. The predominantly columnar grains in microstructure of the deposit were observed to have grown epitaxially toward the heat gradients, which is inevitably related to the high cooling rate experienced.

The earlier investigations have also demonstrated that controlling the heat input through careful selection of the traveling speed, wire feed rate, and beam power can influence the microstructure that is developed in the deposited material. Finer grained, equiaxed microstructures are obtained at the lower heat input conditions, which typically corresponds to narrower deposits and lower deposition rates. Larger grains, including epitaxial growth from the baseplate in the titanium alloy (Tie6Ale4V) and pervasive dendritic microstructures within the aluminum alloy (2219) grains and in the interpass regions, develop during builds in which the heat inputs tend to be higher to achieve higher deposition rates. This demonstrates that there is a trade-off between high deposition rates and fine-grained microstructure [102]. Mechanical properties obtained in aluminum and titanium alloys have demonstrated values comparable to those of wrought product forms.

The main benefits of rapid prototyping using electron beam and wire are as follows [103]:

- lower material costs: Wastes less material than traditional subtractive manufacturing (i.e., machining) processes;
- high strength parts: Uses real engineering alloys, such as Tie6Ale4V and Al 2219, in standard wire forms, rather than specialized metallic powders required with direct metal laser sintering processes;
- compound material properties: Allows deposition of multiple alloys to create parts where material chemistry can vary according to functional part requirements (e.g., changing cross-sectional strength on an airplane wing spar);
- lower power, higher safety: Requires less power than machines derived from traditional e-beam welders, which lowers operating power costs and radiation, thereby increasing operator safety;
- versatile part envelope: Produces a wide range of part sizes, from a few inches to tens of feet;
- mobile and rugged: Can be relocated and has been successfully demonstrated on an aircraft without requiring timeconsuming alignment procedures.

6.6.5 Repairing technology

Electron beam welding is characterized by the high density heat input, low distortion and ability to weld dissimilar metals. Thus, can be adopted to repair welding. In most cases the parts that have worn out or are broken are critical and may paralyze the production of the whole factory. Simultaneously, fabrication of the new part could take many weeks of even months. Therefore repairing technology can provide and economic and satisfactory solution [24]. Even large parts can be refitted by EB welding such as: rotor of a gas turbine, quill bearing boxes, compressor blades, engines chamber. The second option is deposition welding which offers flexible solutions for the repairing technology. The use of electron beam technology is particularly suitable for this purpose. The exactly reproducible adjustability of the parameters makes possible a dosed power input. The layer thickness and fusion penetration can be correspondingly adapted to the requirements and distortions can be minimized. One special advantage is the minimizable degree of dilution between the base material and the filler, as a result of the layer thickness or number of layers to be applied can be reduced in comparison to the other methods. In contrast to conventionally processed systems, electron beam processed components feature a weld deposit overlay that is close to final contours. This reduces the cost of finishing [5, 104]. The deposition welding for roller bearing seats shafts, turbine blades, die blocks, splines can be adopted.

6.7 Trends in electron beam welding

A number of advances have been made in the electron beam welding process during the past few years [105–108]. The main achievements in productivity and machine design can be observed, as well as application of EB to other technologies.

The productivity of the welding process is directly linked with design of a machine. The main aim is to reduce pump down time as well as improve the control system of the EB machine. The main requirements for high speed vacuum systems of EB-machines are the following [106]:

- kinematic flexibility,
- mechanical precision (minimal number of mechanical interfaces),
- reliability (minimal number of vacuum seals and bearings in vacuum),
- short pump down time,
- minimized pump time drift of contaminated chambers,
- dry (hydrocarbon-free) vacuum,
- good final vacuum,
- low maintenance,
- energy efficiency,
- low noise emission,
- low investment and running cost.

It should be noted that a high productivity rate is especially important in automotive industry. Cambridge Vacuum Engineering, a company from Great Britain determined the new concept of fast cycle EB welding machine for turbocharger components. They assumed that the new machine had to achieve efficiency about 180 parts per hour (1 part produced in less than 20 s). To this aim, the chamber has to be kept continuously under vacuum, a small foot print and small vacuum pumps should be provided. Besides less maintenance issues associated with filament life and vacuum gauges, the bias and HV systems have to be more reliable [107].

If the universal electron beam welding machine is considered, the major developments cover the control system and tooling, as well as:

- real time seam tracking (Fig. 6.21), the collector system adjusts the beam position (deflection) back to the joint in real time as the part is being welded;
- back scattered imaging (Fig. 6.22), in most vacuum chambers the illumination is quite poor, the field of view is also quite small. The backscattered image technique is a very useful tool where normal viewing is difficult due to the field of view and with poor illumination;
- joint finding and joint mapping (Fig. 6.23). A typical joint trace, allows the joint to be captured and the deflection value set into the machine, as a result the beam is accurately placed on the joint;
- full Quality Assurance (QA) package including pass and fail system. The idea is to monitor key functions of the machine during welding and report any problems in terms of a brief report and more comprehensive data accessible off line. This consists of a data capture card and in this case a separate PC independent of the machine system;
- quick change tooling. For a universal EB machine the rotary tilt and *z* axis tables can be set up accurate to microns and parts of a degree, meeting the ISO standards. Wire feed systems allow EB welding, cladding as well as rapid prototyping, the additional degree of freedom is provided. Multiple viewing and gun positions including horizontal and vertical position are provided (Fig. 6.24).

However, the development of single EB welding machines is directly linked with application and demands of end users as well as final investment cost. Some new solutions related to parts of the electron beam welding gun are also offered. The authors

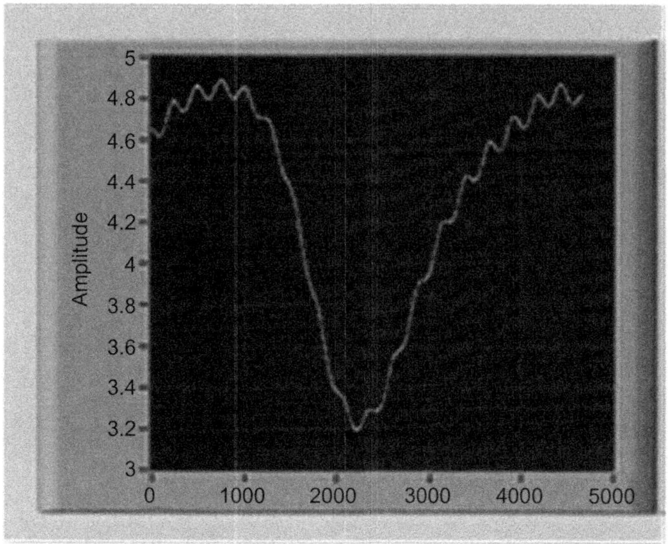

Fig. 6.21 Real time seam tracking. Typical good joint trace using the back scattered collector plate.

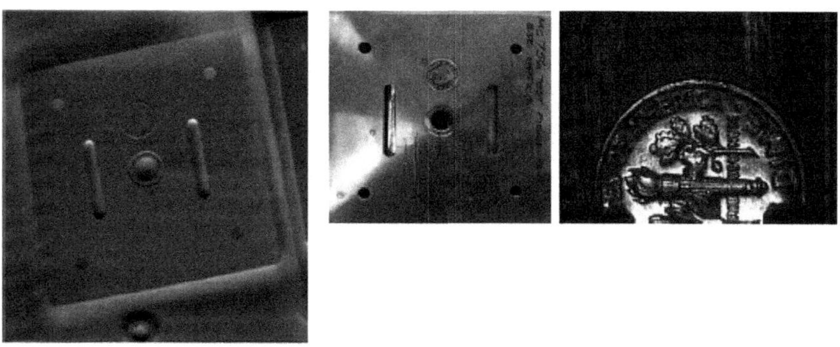

Fig. 6.22 Back scattered imaging. Special test piece made to highlight various features of an area 100 mm × 100 mm. The center picture is a block made to show the imaging can work over an area 100 mm × 100 mm. This picture is taken on a bench outside the chamber in a well illuminated area. The picture on the right is a CCTV image inside the chamber with chamber illumination, in most chambers the illumination is quite poor, the field of view is quite small. The LHS view is a backscattered image of the middle part.

[108] present results of development of the electron gun with replaceable feed-through insulators. A 24 kW EB welding gun with designed replaceable insulators is fabricated in BARC. Emphasis during the design of the gun had been laid to reduce the use of imported components to zero.

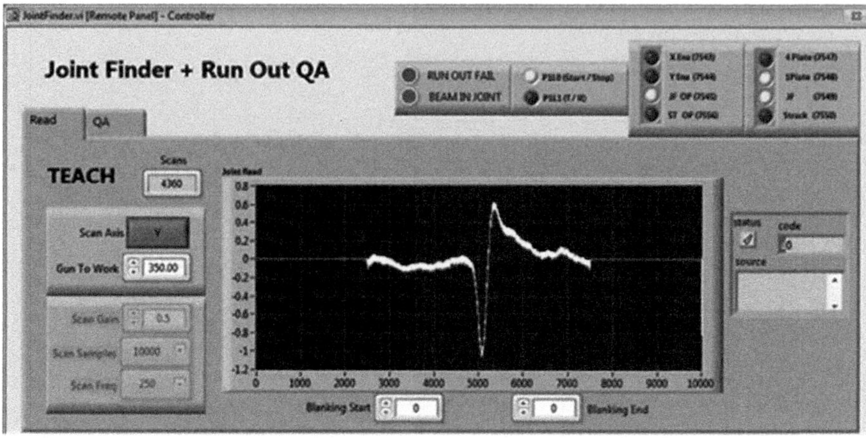

Fig. 6.23 Joint finding and joint mapping.

Fig. 6.24 Quick tooling change, possibility to set up electron gun in vertical and horizontal position, EB machine at Instytut Spawalnictwa (Institute of Welding) in Poland.

New and future developments in range of EB machines covered the following items [107]:

- dynamic sliding seal on EB machine,
- reduced pressure welding,
- beam shaping and analysis including auto calibration of focus, beam current, beam alignment and filament,
- EB machine integrated with pin hole probe analyzer,
- improved electron optics for the heavy duty gun,
- auto peaking and setting of the filament current,
- improve the Back Scattered Electron (BSE) imaging to do advanced joint finding as well as mapping,
- introduce an electrical drawing package to help standardize circuits and engineering information,
- design and build a development machine for developing "additive manufacturing process" using wire feed system. This machine may also be used for surface modification (macro level) for the semi-conductor industry.

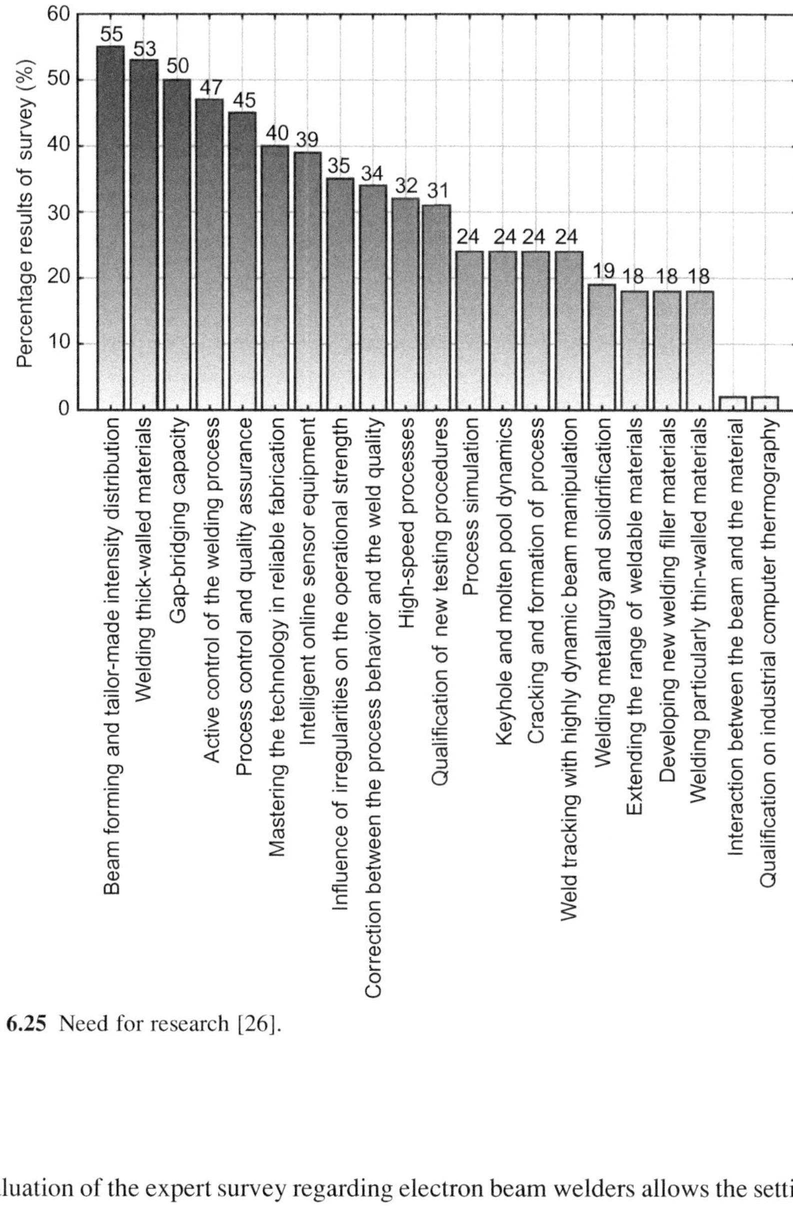

Fig. 6.25 Need for research [26].

Evaluation of the expert survey regarding electron beam welders allows the setting up of the requirements of end-users in industry. It should be noted that these requirements can be different and depend on country, range of production and many others factors. Fig. 6.25 presents the needs of research in the electron beam welding technology from the German industry point of view. It is worth noting that over 50% of companies needed improved beam forming, welding thick-wall materials as well as gap bridged capacity. Only 2% of them are interested in qualification of industrial computer thermography.

The research subjects which could be extracted from the opinions of the EBW experts are listed below [8]:

- optimizing the threading-out region in the case of all-round welds (avoiding spiking) in conjunction with online control/joining of cast Ni materials which are difficult to fusion-weld;
- EB-welding (thick-walled) duplex steels;
- fundamentals of magnetism and demagnetization; beam behavior in the case of high-current welds (>200 mA);
- process-dependent, thermophysical and thermomechanical properties;
- NVEBW technology with a plasma cathode as well as NVEB welding and cutting of all materials.

With regard to the question about improvements in EBW as a result of research efforts, freedom from pores was specified with 55% and freedom from cracks with 50% of answers. The results of survey are presented in Fig. 6.26.

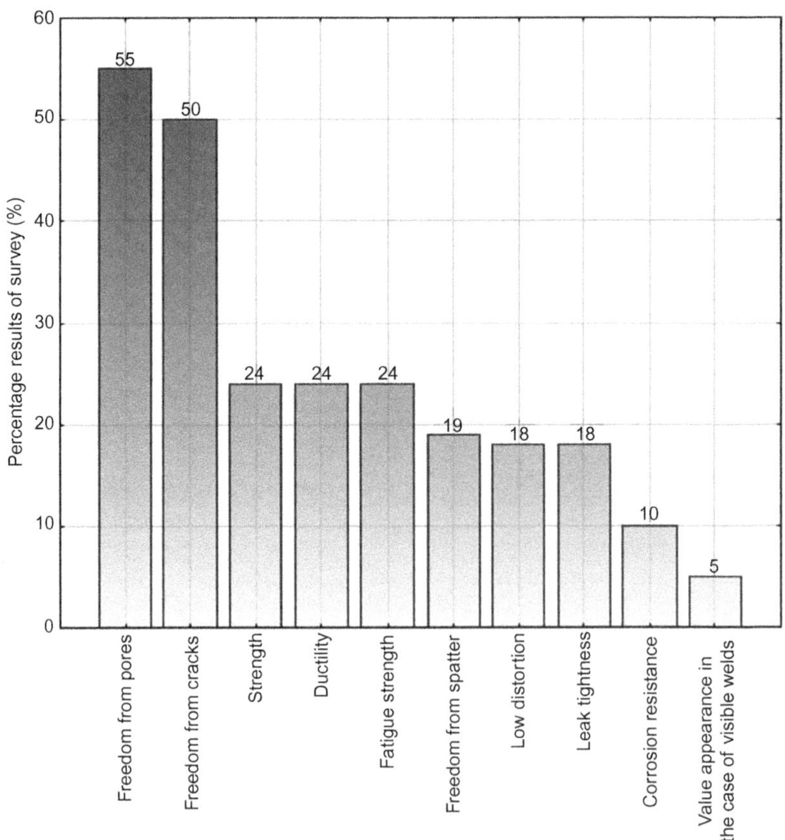

Fig. 6.26 Which characteristics of an EB weld should be improved by R&D efforts? [26].

6.8 Summary

The state of the art of electron beam welding and other technologies including surface materials modification, rapid prototyping, cladding etc., based on own experience, coworkers and coauthors results, as well literature review has been reported. The research data available in the literature in the past 50 years has demonstrated numerous applications of electron beam in the surface materials processing over the world. Especially, the availability of universal electron beam welding machines created the revolution in the area of electron beam welding and materials processing. These machines provide illimitable possibilities of their use in the surface processing. The examples presented in this paper are selective. It could be concluded that overall state of research on electron beam application in the welding and surface engineering is well developed and still growing industrial application is observed.

References

[1] D. Dobeneck, An International History of Electron Beam Welding, Pro-beam AG & Co. KGaA, 2007, pp. 1–e20.

[2] A.H. Meleka, Electron Beam Welding, McGRAW eHILL, London, 1971, pp. 11–e12.

[3] British Patent 727460, Fine Wire Spot Welding by Electron Beam, 1951.

[4] J. Dworak, M.S. Weglowski, Electron Beam Welding, Instytut Spawalnictwa, Gliwice, 2014, pp. 119–e132.

[5] H. Schultz, Electron Beam Welding, Abington Publishing, Cambridge, 1993.

[6] D. Dobeneck, Electron Beam Welding a Key Technology to Construct Vehicles for Road, Rail, Sea, Air, Pro-beam AG & Co. KGaA, 2007, pp. 1–e60.

[7] K.R. Schulze, Electron Beam Technologies, vol. 1e, Compact Knowledge, DVS Media GmbH, Düsseldorf, 2012, pp. 1–e76.

[8] S. Bo€hm, The Electron Beam as a Tool for Joining Technology, DVS Media GmbH, Düsseldorf, 2014, pp. 1–e100.

[9] P. Anderl, Comparison of laser with electron beam welding, Chem. Ing. Tech. 61 (1989) 767–e774.

[10] P. Anderl, D. Boehme, Laser and Electron Beam Welding. A Comparison, 123, DVS Ber, 1989, pp. 143–e148.

[11] Efficiency in Electron Beam and Laser Welding/Paper-nr. DVS 3209, DVS-Merkblatter, 1995, p. 3209.

[12] G. Cam, V. Ventzke, J.F. Dos Santos, et al., Characterization of laser and electron beam welded Al alloys, Prakt. Metallogr. 37 (2000) 59–e89.

[13] Consideration of Costs for the Electron Beam and Laser Beam Welding Pro- cesses, DVS-Merkblatter, DVS 3211, 1998.

[14] F. Hauser, S. Keitel, E. Marquardt, Comparison of the techniques of laser beam welding, non-vacuum electron beam welding, and plasma welding for joining aluminium alloys, Schweien Schneid. 186 (1997) 179–e181.

[15] M. Mueller, Comparison between laser and electron beam welding, in: Praxis Des Laserstrahl-Schweissens von Metallen, DVS-VDI (ADB), Fachtagung, 1987, pp. 1–e18. Paper-Nr. 4.

[16] J. Overath, Investigations on laser and electron beam welding of die-cast aluminium components, Schweien Schneid. 52 (2000) 434.

[17] U. Reisgen, S. Olschok, S. Jakobs, A comparison of electron beam welding with laser beam welding in vacuum, in: International Electron Beam Welding Conference, Dusseldorf, 2012, pp. 119–e127.
[18] U. Reisgen, S. Olschok, S. Longerich, Laser beam welding in vacuum—a comparison with electron beam welding, Weld. Cut. 9 (2010) 224–e230.
[19] A. Klimpel, Welding Handbook, vol. 1, Silesian University of Technology, Gliwice, 2013, pp. 601–e654.
[20] Promotional materials of Focus GmbH, Electron Beam Welding with the Focus MEBW-60, 2013.
[21] J.L. Murphy, T.M. Mustaleski, L.C. Watson, Multipass, autogenous electron beam, welding, Weld. J. 67 (1988) 187–e195.
[22] D.V. Dobeneck, Electron Beam Welding, Pro-beam AG & Co. KGaA, 2007.
[23] A. Saxena, Electron beam welding, in: Training Materials for Students, Sant Longowal Institute of Engineering & Technology, 2014.
[24] D.V. Dobeneck, Examples of 30 Years Jop Shop Experience, Pro-beam AG & Co. KGaA, 2007, pp. 1–e60.
[25] G. Ripper, Electron beam welding, in: Actual Applications in the Aerospace Industry. International Electron Beam Welding Conference, Aachen, 2012.
[26] S. Bo€hm, The electron beam as a welding tool: German study on the state of the art and future requirements, in: Proceedings—67th IIW International Conference, Seoul, 2014.
[27] P. Fu, Z. Mao, W. Yu, S. Gong, Z. Tang, Microstructure and properties of simultaneous EBW and heat treatment with multi-pools for near titanium alloy, in: International Electron Beam Welding Conference, Aachen, 2012.
[28] K. Rüthrich, R. Zenker, Characteristics and prospects of process integrated thermal field heat treatment for electron beam welding of cast iron, in: International Electron Beam Welding Conference, Aachen, 2012.
[29] U. Reisgen, S. Olschok, A. Backhause, Reduction of weld residual stresses with the electron beam, in: International Electron Beam Welding Conference, Aachen, 2012.
[30] G. Zhang, X. Yang, X. He, J. Li, Enhancement of mechanical properties and failure mechanism of electron beam welded 300 M ultrahigh strength steel joints, Mater. Des. 45 (2013) 56–e65.
[31] S. Elliot, Electron beam welding of C/Mn steels—toughness and fatigue properties, Weld J. 63 (1984) 8–e16.
[32] W. Maurer, W. Ernst, R. Rauch, S. Kapl, A. Pohl, T. Krüssel, R. Vallant, N. Enzinger, Electron beam welding of a TMCP steel with 700 MPa yield strength, Weld. World 56 (2012) 85–e94.
[33] M.S. Weglowski, S. Blacha, S. Dymek, M. Kopyscianki, Electron beam welding of high strength quenched and tempered steel, Mater. Sci. Forum 879 (2016) 2078–2083.
[34] M.S. Weglowski, S. Blacha, S. Dymek, M. Kopyscianki, Microstructural characterization and mechanical properties of electron beam welded joint of high strength steel grade S690QL, Arch. Metall. Mater. 61 (2016) 297–e304.
[35] M.S. Weglowski, M. Zeman, A. Grocholewski, Effect of welding thermal cycles on microstructure and mechanical properties of simulated heat affected zone for a we ldox 1300 ultra-high strength alloy steel, Arch. Metall. Mater. 61 (2016) 127–e132.
[36] Technical Scheets, Electron Beam Welding, EWF Mater, 2007.
[37] A. Barbacki, Electron Microscopie, Pozna University of Technology, Poznan, 2007.
[38] J. Dworak, Electron beam welding of heat treatment steel with oscillating beam, Biul. Inst. Spaw. 36 (1992) 54–e57.

[39] Y. Arata, K. Terai, S. Matsuda, Study on characteristics of weld defects and its prevention in electron beam welding (report 3), Trans. Jpn. Weld. Res. Inst. 3 (1973) 81–e83.
[40] G. Nikolajev, Special Welding Techiques, Maszinostrojenie, Moskva, 1975.
[41] R.W. Meyer, W. Scheffels, K.H. Steigerwald, A study of the formation and the energy balance of the capillary in electron beam deep penetration welding, in: Proceedings of the Electron and Laser Beam Symposium, 1965, pp. 531–e539.
[42] M.S. Weglowski, S. Blacha, J. Dworak, Electron beam welding and its characteristics, Inst. Weld. Bull. 58 (2014) 5–e14.
[43] M.S. Weglowski, S. Blacha, J. Dworak, Electron beam welding—equipment and accessories, Inst. Weld. Bull. 58 (2014) 22–e30.
[44] D. Dobeneck, Electron Beam Welding the Fundamentals of a Fascinating Technology, Pro-beam AG & Co. KGaA, 2007.
[45] J. Elmer, Characterization of Defocused Electron Beams and Welds in Stain- less Steel and Refractory Metals Using the Enhanced Modified Faraday Cup Diagnostic, Lawrence Livermore National Laboratory, 2009, pp. 1–e9.
[46] W. Giedt, L. Tallerico, Prediction of electron beam depth of penetration, Weld. J. 67 (1998) 299–e305.
[47] A. Kaur, C. Ribton, W. Balachandaran, Electron beam characterisation methods and devices for welding equipment, J. Mater. Process. Technol. 221 (2015) 225–e232.
[48] ISO 14744e1, Welding. Acceptance Inspection of Electron Beam Welding Machines. Part 1: Principles and Acceptance Conditions, 2008.
[49] ISO 14744e2, Welding. Acceptance Inspection of Electron Beam Welding Machines. Part 2: Measurement of Accelerating Voltage Characteristics, 2000.
[50] ISO 14744e3, Welding. Acceptance Inspection of Electron Beam Welding Machines. Part 3: Measurement of Beam Current Characteristics, 2000.
[51] ISO 14744e4, Welding. Acceptance Inspection of Electron Beam Welding Machines. Part 4: Measurement of Welding Speed, 2000.
[52] ISO 14744e5, Welding. Acceptance Inspection of Electron Beam Welding Machines. Part 5: Measurement of Run-Out Accuracy, 2000.
[53] ISO 14744e6, Welding. Acceptance Inspection of Electron Beam Welding Machines. Part 6: Measurement of Stability of Spot Position, 2000.
[54] EN 1011e7, Welding. Recommendations for Welding of Metallic Materials—Part 7: Electron Beam Welding, 2004.
[55] EN ISO 3834e1, Quality Requirements for Fusion Welding of Metallic Materials—Part 1: Criteria for the Selection of the Appropriate Level of Quality Requirements, 2005.
[56] EN ISO 3834e2, Quality Requirements for Fusion Welding of Metallic Materials—Part 2: Comprehensive Quality Requirements, 2005.
[57] EN ISO 3834e3, Quality Requirements for Fusion Welding of Metallic Materials—Part 3: Standard quality requirements, 2005.
[58] EN ISO 3834e4, Quality Requirements for Fusion Welding of Metallic Materials—Part 4: Elementary Quality Requirements, 2005.
[59] EN ISO 3834-5, Quality Requirements for Fusion Welding of Metallic Materials—Part 5: Documents with which it is Necessary to Conform to Claim Conformity to the Quality Requirements of ISO 3834-2, ISO 3834-3 or ISO 3834e4, 2015.
[60] ISO 13919e1, Welding. Electron and Laser-Beam Welded Joints. Guidance on Quality Levels for Imperfections. Part 1: Steel, 1996.
[61] ISO 13919e2, Welding. Electron and Laser Beam Welded Joints. Guidance on Quality Levels for Imperfections. Part 2: Aluminium and its Weldable Alloys, 2001.

[62] ISO 15609e3, Specification and Qualification of Welding Procedures for Metallic Materials. Welding Procedure Specification. Part 3: Electron Beam Welding, 2004.
[63] ISO 15607, Specification and Qualification of Welding Procedures for Metallic Materials. General Rules, 2003.
[64] ISO 15611, Specification and Qualification of Welding Procedures for Metallic Materials. Qualification Based on Previous Welding Experience, 2003.
[65] ISO 15612, Specification and Qualification of Welding Procedures for Metallic Materials. Qualification by Adoption of a Standard Welding Procedure, 2004.
[66] ISO 15613, Specification and Qualification of Welding Procedures for Metallic Materials. Qualification Based on Pre-Production Welding Test, 2004.
[67] ISO 15614e11, Specification and Qualification of Welding Procedures for Metallic Materials. Welding Procedure Test. Part 11: Electron and Laser Beam Welding, 2002.
[68] ISO 14732, Welding Personnel. Qualification Testing of Welding Operators and Weld Setters for Mechanized and Automatic Welding of Metallic Materials, 2013.
[69] AWS C7.1/MC7.1, An American National Standard. Recommended Practice for Electron Beam Welding and Allied Processes, fourth ed., 2013.
[70] DVS Technical Codes of Electron Beam Welding, English Edition, vol. 8, DVS Media, 2013.
[71] R. Franke, I. Haase, et al., Friction and wear behaviour of electron beam surface treated aluminium alloys AlSi10Mg(Cu) and AlSi35, Wear 269 (2010) 921–e929.
[72] S.F. Gnyusov, S. Tarasov, Structural phase states and heat aging of composite electron-beam clad coatings, Surf. Coat. Technol. 232 (2013) 775–e783.
[73] J. Morimoto, N. Abe, et al., Formation of a Cr3C2/Ni-Cr alloy layer by an electron beam cladding method and evaluation of the layer properties, Vacuum 62 (2001) 203–e210.
[74] N.K. Galchenko, K.A. Kolesnikova, et al., Structure and properties of boride coatings synthesized from thermo-reactive powders during electron-beam surfacing, Adv. Mat. Res. 880 (2014) 265–e271.
[75] M. Tomie, N. Abe, et al., Electron beam cladding of titanium on stainless steel plate, Trans. JWRI 19 (1) (1990) 51–e55.
[76] N. Abe, J. Morimoto, et al., Formation of WC-Co layers by an electron beam cladding method and evaluation of the layer properties, Vacuum 59 (2000) 373–e380.
[77] D.S. Krivezhenko, L.S. Laptev, et al., Electron-beam cladding of boron carbide on low-alloyed steel at the air atmosphere, Appl. Mech. Mater. 698 (2015) 369–e373.
[78] R. Zenker, Modern thermal electron beam processes—research results and industrial application, La Metall. Ital. 4 (2009) 1–8.
[79] P. Adamiec, J. Dzibinski, Fabrication and Properties of Surface Layers of Parts for Transportation Machines, Silesian University of Technology, Gliwice, 2005.
[80] T. Mohandas, D. Banerjee, et al., Fusion zone microstructure and porosity in electron beam welds of an aþb titanium alloy, Metall. Mater. Trans. A 30A (1990) 789–e798.
[81] J. Huang, N. Warnken, Hydrogen transport and rationalization of porosity formation during welding of titanium alloys, Metall. Mater. Trans. A 43A (2012) 582–e591.
[82] A.R. Rastkar, B. Shokri, Surface transformation of Tie45Ale2Nbe2Mne1B titanium aluminide by electron beam melting, Surf. Coat. Technol. 204 (2010) 1817–e1822.
[83] D. Utu, G. Marginean, et al., Improvement of the wear resistance of titanium alloyed with boron nitride by electron beam irradiation, Surf. Coat. Technol. 201 (2007) 6387–e6391.

[84] M.G. Golkovskia, I.A. Bataev, et al., Atmospheric electron-beam surface alloying of titanium with tantalum, Mater. Sci. Eng. A 578 (2013) 310–e317.

[85] J.C. Oh, S. Lee, Correlation of microstructure with hardness and fracture properties of (TiC,SiC)/Tie6Ale4V surface composites fabricated by high- energy electron-beam irradiation, Surf. Coat. Technol. 179 (2004) 340–e348.

[86] E. Yun, K. Lee, et al., Correlation of microstructure with high-temperature hardness of (TiC,TiN)/Tie6Ale4V surface composites fabricated by high-energy electron-beam irradiation, Surf. Coat. Technol. 191 (2005) 83–e89.

[87] O.G. Lenivtseva, D.V. Lazurenko, et al., The structure and wear resistance of the surface layers obtained by the atmospheric electron beam cladding of TiC on titanium substrates, Appl. Mech. Mater. 682 (2014) 14–e20.

[88] D. Neagu, Alloying with electron beam of some surfaces on pieces made by non-strengthen materials, Nonconv. Technol. Rev. 10 (2010) 31–e35.

[89] W. Xiong, B. Blackman, et al., The effect of composite orientation on the mechanical properties of hybrid joints strengthened by surfi-sculpt, Compos. Struct. 134 (2015) 587–e592.

[90] B.G.I. Dance, A.L. Buxton, An introduction to Surfi-Sculpt® technology—new opportunities, new challenges, in: 7th International Conference on Beam Technology, Halle, Germany, April, 2007, pp. 17–e19.

[91] X. Wang, J. Ahn, et al., Effect of forming parameters on electron beam Surfi- Sculpt protrusion for Tie6Ale4V, Mater. Des. 76 (2015) 202–e206.

[92] E. Chlebus, Computer Techniques CAx in Engineering Production, WNT, Warsaw, 2000.

[93] M.E. Kinsella, Additive Manufacturing of Superalloys for Aerospace Application, Air Force Research Laboratory Report, 2008.

[94] W.E. Frazier, Metal additive manufacturing: a review, J. Mater. Eng. Perform. 23 (2014) 1917–e1928.

[95] http://www.3dprinterpro.com/electron-beam-melting-service/.

[96] https://3dprint.com/12262/ge-ebm-3d-printing/.

[97] P. Wanjara, M. Brochu, et al., Electron beam free forming of stainless steel using solid wire feed, Mater. Des. 28 (2015) 2278–e2286.

[98] L.E. Murr, E. Martinez, et al., Microstructural architecture, microstructures, and mechanical properties for nickel base superalloy fabricated by electron beam melting, Metall. Mater. Trans. A 42A (2011) 3491–e3508.

[99] M. Koike, P. Greerc, et al., Evaluation of titanium alloys fabricated using rapid prototyping technologiesdElectron beam melting and laser beam melting, Materials 4 (2011) 1776–e1792.

[100] A.A. Antonysamy, Microstructure, Texture and Mechanical Property Evolution during Additive Manufacturing of Ti6Al4V Alloy for Aerospace Applications, Thesis submitted to the, University of Manchester, 2012.

[101] Q. Tang, S. Pang, B. Chen, H. Sou, J. Zhou, A three dimensional transient model for heat transfer and fluid flow pf weld pool during electron beam freeform fabrication of Ti-6Al-4V alloy, Int. J. Heat Mass Transf. 78 (2014) 203–e215.

[102] K.M. Taminger, R.A. Hafley, Electron Beam Freeform Fabrication for Cost Effective Near Net Shape Manufacturing, Document NATO AVT-139, (2006).

[103] National Aeronautics, Space Administration, NASA's EBF3: The Future of Art-to-Part Manufacturing, National Aeronautics and Space Administration, 2011.

[104] A. Richter, T. Krüssel, C. Just, Deposition welding with the electron beam as repair technology, in: International Electron Beam Welding Conference, Aachen, 2012.

[105] C.L.M. Conttrell, Electron beam welding—a critical review, Mater. Des. 6 (1985) 285–e291.
[106] M. Wahl, Efficient high speed vacuum systems for electron beam machines, in: Proceedings of an International Electron Beam Welding Conference, Aachen, 2012.
[107] A. Phillips, CVE Materials, Cambridge Vacuum Engineering (CVE), 2014.
[108] T.K. Saha, M. Mascarenhas, E. Kandaswamy, Development of a high power electron beam welding gun with replaceable high voltage feed-through insulators, in: The International Symposia on Discharge and Electrical Insulation in Vacuum (ISDEIV), Mumbai, 2014, pp. 601–e604.

Heat-affected zone cracking in nickel-based superalloys and the role of minor elements

7

O.A. Ojo, N.L. Richards, and K.R. Vishwakarma
Department of Mechanical and Manufacturing Engineering, University of Manitoba, Winnipeg, MB, Canada

7.1 Introduction

The formation of heat affected zone cracking in fusion-welded materials is a major concern in the design and manufacture of nickel-based superalloy welded assemblies. It is a general weldability problem that affects a large number of advanced highly alloyed cast and wrought nickel base superalloys, particularly, those strengthened by ordered $L1_2$ intermetallic Ni_3(Al, Ti or Ta) γ' precipitates. While the problem of fusion zone cracking is also encountered in many of these alloys, it does not pose as great a challenge as HAZ liquation cracking because it can be essentially managed effectively by proper selection of filler materials and appropriate welding procedures. HAZ liquation cracking is, however, more insidious since the factors and phenomena contributing to its occurrence are often related to the composition of the material and its microstructure, both of which have been optimized to achieve desirable high temperature base metal properties. The HAZ cracking in the superalloys is generally intergranular and it usually associated with the formation of liquid film on HAZ grain boundaries during welding (Figs. 7.1 and 7.2). The inability of this film to accommodate thermally and/or mechanically induced stresses experience during cooling results in grain boundary microfissuring through decohesion along one of the solid-liquid interfaces on the grain boundary and, thus, it is sometimes referred to as liquation cracking, hot cracking or hot tearing. Liquid film stage is the common element in various manifestations of hot tear near the complete solidification point of metals. The cooling cycle of HAZ intergranular liquid is somewhat similar to the final stages of solidification of castings and fusion zone in welds, hence, to a first approximation, the criteria that govern weld solidification cracking can be adopted to explain liquation cracking in the HAZ of weldments, and these are considered next.

7.2 Characteristics of crack-inducing intergranular liquid

A number of studies have been performed on hot cracking in metallic materials [1–18] and several theories have been proposed to explain its occurrence, including, the Shrinkage-Brittleness theory [1], the Strain theory [2] and the Generalized theory

Welding and Joining of Aerospace Materials. https://doi.org/10.1016/B978-0-12-819140-8.00007-9
Copyright © 2021 Elsevier Ltd. All rights reserved.

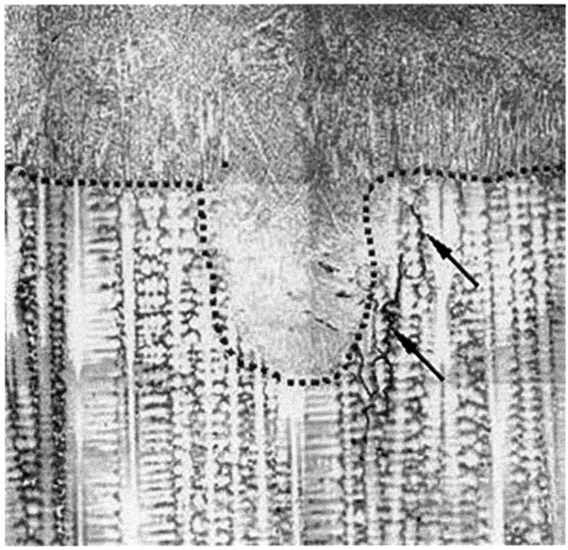

Fig. 7.1 Optical micrograph of an electron beam weld region in directionally solidified TMS-75 superalloy showing HAZ cracks (red dotted points show delineation of FZ from HAZ).

[3, 4]. All the theories are based on the premise of liquid phase enhanced micro-fissuring under thermo-mechanical stresses. The Shrinkage-Brittleness theory is based on the view that cracking occurs in the brittle temperature range (BTR) of an alloy. It is considered that shrinkage strains develop during on-cooling semi-solid stage between the liquidus and solidus temperatures, and if a critical accommodation strain is exceeded, a tear would result and persists if there is insufficient liquid remaining to "heal" it. The extent of BTR is dependent on the composition of the alloy since it lies within the solidification range. Alloys with a wide solidification temperature range tend to be more prone to solidification cracking than alloys that solidify over a narrow temperature range.

The proponents of the strain theory of hot tearing [2] proposed that hot cracking is caused by localized strains set up by thermal gradients, which tend to tear apart masses of solidifying material separated by "essentially continuous liquid films." It was postulated that cracking only takes place when a film stage is reached and localized strains are exceedingly high. In passing through the liquidus—solidus temperature range, irrespective of whether the process entails heating or cooling, an alloy develops a condition of essentially continuous liquid film, which reduces both material hot strength and ductility. The time/temperature period during which the "film" exists is important since it determines total cumulative strain that acts on the film, which increases with cooling temperature. The ability of a liquid phase to effectively wet and spread out along grain boundary and forming a continuous or semi-continuous film has been generally recognized to directly control the propensity of a material to liquation cracking during welding. In a complementary way, Borland [7] in his work on fundamentals of solidification cracking in welds, recognized that while

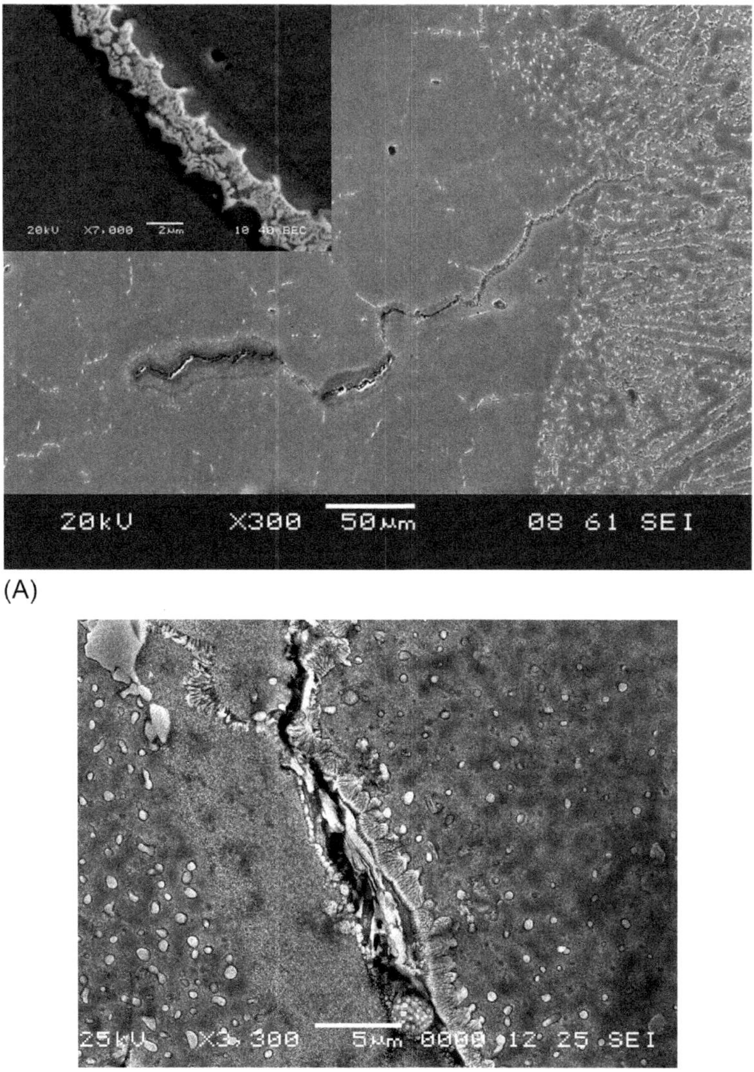

Fig. 7.2 (A) Scanning electron micrograph show a grain boundary liquation crack in a HAZ in IN 718 superalloy (inset shows eutectic-type re-solidified product). (B) Scanning electron micrograph of a HAZ liquation crack in a TIG welded IN 738 superalloy.

continuous interfacial liquid phase provides easy extension path for an existing crack, the presence of the liquid does not necessarily account for an easy formation of the initial crack. He considered the energies required for formation of a crack by decohesion/separation of solid-liquid components and that required to cause liquid penetration of grain boundary under the influence of stress. It was mentioned that when the liquid phase is constrained internally and is not exposed to the surface, the various possibilities are:

(i) Separation at the solid-liquid interface requiring $\gamma_{LV} + \gamma_{SV} + \gamma_{SL}$
(ii) Separation within the liquid requiring $2\gamma_{LV}$
(iii) Penetration of liquid along the grain boundaries requiring $2\gamma_{SL} - \gamma_b$

where γ_{LV} is the liquid-vapor interfacial energy, γ_{SV} is the solid-vapor interfacial energy, γ_{SL} is the solid-liquid interfacial energy, and γ_b is the solid-solid interfacial energy at the grain boundary. In order for the liquid to separate without first "necking" down (since it is constrained by the solid walls), i.e., case (ii) above, the fracture stress would be exceedingly high and only a little less than the theoretical strength of solid. The remaining possibilities then are (i) and (iii) above, i.e., that fracture initiates at the solid-liquid interface or that grain boundary penetration of the liquid occurs. He showed by assuming $\gamma_{SV} = 3\gamma_b$ and taking $\gamma_{LV} = 0$, that grain boundary liquid penetration is more likely for systems that exhibit low solid-liquid interfacial energy γ_{SL}.

For a given grain boundary energy, the lower the solid-liquid interface energy, the better the wetting behavior and the easier it is for such a liquid to wet and spread and penetrate grain boundary [10]. Absorption of surface-active alloying elements into intergranular liquid often reduces solid-liquid interfacial energy, which significantly aids wetting properties. Alternatively, for a given solid-liquid interfacial energy, the higher the grain boundary energy, the higher the tendency for it to be wetted and penetrated by liquid phase. It has been found that grain boundaries in cast nickel-based superalloys are more of a "random" nature with higher order of "Σ" values, than special boundaries [19]. Random high angle grain boundaries are inherently of higher energy than the special boundaries, including twin boundaries, which may enhance grain boundary wetting and penetration by liquid film in the HAZ of cast superalloy materials during welding. Guo et al. [20] and Kokawa et al. [21] have confirmed that liquid penetration at grain boundary was greatest at high angle boundaries and was relatively insignificant at low angle (twin) boundaries.

The amount/thickness of HAZ intergranular liquid is another important factor that controls the occurrence of weld HAZ liquation cracking. The thickness of grain boundary liquid film can affect an alloy's resistance to cracking through its influence on (i) the re-solidification behavior; (ii) the critical level of stress required to cause microfissuring at a given temperature during weld cooling; and (iii) stress relaxation by grain boundary liquid penetration and liquid healing of incipient cracks.

7.2.1 Re-solidification behavior

The rate at which grain boundary liquid is eliminated prior to the occurrence of sufficient welding stresses during cooling influences resistance to liquation cracking by affecting the range of re-solidification temperature. There are three main mechanisms through which intergranular liquid film could be relieved of its excess solute concentration, and, thus, solidify during welding thermal cycle. These are: normal solidification involving significant solute microsegregation, solidification controlled by solid-state back-diffusion of solute into the adjacent grain matrix, and by rapid solidification through liquid film migration (LFM). The most common re-solidification mode of HAZ intergranular liquid involves significant solute microsegregation ahead of solid/liquid interface, which is sometimes accompanied with the formation of eutectic microconstituents. Re-solidification of thick intergranular liquid by this mode may results in extension of solidification temperature range and, thus, BTR with concomitant buildup of high welding stresses across liquated grain boundaries. Exclusive re-solidification of intergranular liquid via solute back-diffusion of excess solute atoms in the adjacent grain matrix may also occur during welding but it is largely limited due to insufficient available diffusion time to eliminate the liquid phase, particularly, for thick grain boundary liquid. It becomes even more precarious in cast nickel-based superalloy weldments due to (a) microsegregation induced solute rich interdendric zones in pre-weld material, which could reduce solute concentration gradient expected to drive such diffusional process and (b) limited grain boundary surface area available for diffusion flux due to the comparatively large grain size of cast alloys. Rapid re-solidification of metastable HAZ grain boundary liquid film can occur by liquid film migration process [22]. This mode of solidification is controlled by a high diffusion rate in the liquid and it is an alternative to lattice back-diffusion controlled and normal dendritic solidification types, and as such has been reported to be beneficial to HAZ liquation cracking resistance [23]. There are two major driving forces that have been reported to be responsible for LFM. They are (i) diffusional coherency strain energy, which requires sufficient size difference between the diffusing solute in a metastable liquid and the matrix atoms in order to develop substantial coherency strains resulting from lattice mismatch [22–24] and (ii) asymmetry of surface tension at the two solid-liquid interfaces. The latter requires the occurrence of appreciable curvature at grain interfaces in order to set up a substantial concentration gradient within the liquid needed for solute diffusive flux during LFM [23–25]. Grain boundary curvature has been found to be a major driving force behind occurrence of LFM in the HAZ of IN 903 weldment, which effectively improved resistance to HAZ liquation cracking [23]. Despite the large grain size of cast superalloys, some of these alloys tend to exhibit serrated grain boundaries with appreciable local boundary curvature [26] that may aid the occurrence of LFM. Considerable LFM was observed and related to serrated grain boundary morphology in the HAZ of tungsten inert gas welded cast IN 738 superalloy [27] (Fig. 7.3). The effect of rapid re-solidification by LFM in precluding HAZ liquation cracking may, however, be hampered by the formation of thick intergranular liquid film. Thick intergranular liquid would not only increase the time required for complete elimination of the liquid phase but it could also decrease the driving force for the process as migration distance increases, as was observed during the LFM study in an Al-Cu system [28]. Hence, in situations where rapid re-solidification via liquid film migration (LFM) is enhanced by substantial local grain boundary curvature and/or diffusional coherency strain energy, thick intergranular liquid may render it ineffective in preventing cracking as was observed in IN 738 [27].

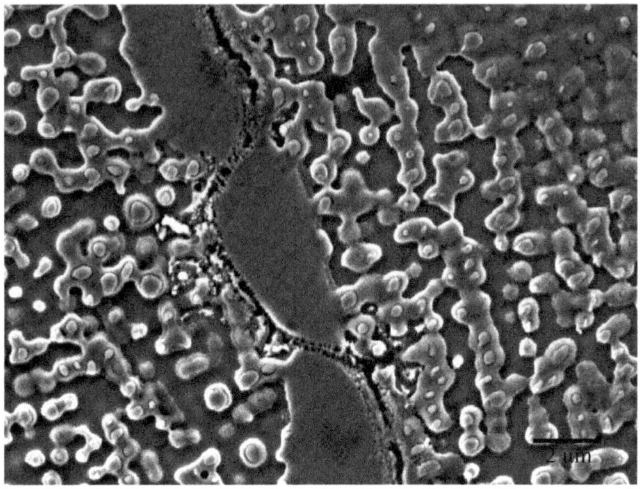

Fig. 7.3 HAZ grain boundary liquid film migration in TIG welded IN 738 superalloy.

7.2.2 Critical stress/strain level

Solidification shrinkage and thermal contraction imposed stresses/strains on weld HAZ are crucial in crack formation and propagation. According to Miller and Chadwick [8], the critical tensile stress required to overcome surface tension γ_{SL} on a grain boundary containing liquid film of thickness h, is given by:

$$\sigma = 2\gamma_{SL}/h \tag{7.1}$$

which indicates that an increase in the intergranular liquid film thickness would lower the stress/strain required to cause decohesion at such a solid-liquid interface. Nevertheless, it does not necessarily mean that, in a given alloy system, the actual occurrence of HAZ liquation cracking will monotonically increase with increase in liquid film thickness. For a given alloy, hot cracking susceptibility would most likely increase with increase in intergranular liquid thickness up to an amount, where the phenomenon of liquid healing begins with an attendant decrease in the extent of cracking. Vero [5] introduced the concept of "healing" whereby incipient cracks are considered to be filled with liquid and their harmful effects are, thus, overcome. He postulated that if a solidifying metal mass contains more than a critical volume of liquid, any crack formed by the contraction of the primary grains would be healed by the inflow of the liquid. Liquid healing occurs when a large enough amount of intergranular liquid is present to effectively backfill and heal incipient cracks as they form. Many braze and solder alloys implicitly use the phenomenon of liquid healing and thus are virtually immune from the problem of hot cracking during solidification. The amount of liquid necessary to fully affect liquid healing can be alloy dependent. Clyne and Davies [18] suggest that at least 10 vol.% is required, which is supported by the work of Arata et al. [9]. Likewise, Cross et al. [29] in their study of hot cracking in aluminum alloys reported an increase in total crack length with increase in the amount of terminal eutectic liquid up to a peak value beyond which further increase in the amount of the liquid resulted in a decrease in the observed total crack length. It has been also suggested by some authors that it is not just strain or stress but the strain rate that is a critical factor for causing hot cracking. The rationale for this is that the strain rate during solidification is limited by the minimum strain rate at which a material will crack. Prokhorov [17] was the first to suggest a criterion based on this concept and more recently, a strain rate-based criterion is proposed by Rappaz et al. [15].

Recent advances in alternate joining methods like friction stir welding and processing provide more insight in to how the type and amount of stresses present during welding can influence the extent of microfissuring in liquation prone nickel base superalloys. Dissolution kinetics of γ' during welding can be influenced by presence of external compressive stress as observed in Linear friction welding of IN 738. Inter and intragranular liquation observed in the Linear friction welded IN 738 was attributed to non-equilibrium reaction between gamma prime particles and the surrounding matrix, nevertheless there was no microfissuring associated with it. The lack of cracking was attributed to the presence of compressive stress during the forging stage of LFW, which overcomes the tensile stresses developed during cooling of the weld. It was also suggested that the external stress/strain can enhance atomic diffusion leading to rapid thermal and strain assisted dissolution of gamma prime particles. The external strain also increased the liquid film migration velocity leading to rapid elimination of liquid formed in the joint [30]. In another research on Friction stir processing (FSP) of IN738 alloy, it was found that due to severe plastic deformation and the heat generated in the process, the microstructure of as-cast IN 738 changes drastically. The grain size is refined and the secondary precipitates like carbides and gamma prime undergo size reduction. It was also calculated that on first approximation, the reduction in size of gamma prime particles from 0.45 to 0.05 μm in friction stir processed material, increases the rate of dissolution of the particles in FSP by 80 times. Hence γ' particles will undergo solid-state dissolution before reaching the $\gamma - \gamma'$ eutectic temperature, hence preventing their constitutional liquation [31]. Similar reduction in HAZ microfissuring due to strain assisted elimination of liquation products was found in Astroloy during friction welding by Oluwasegun et al. [32].

7.2.3 Stress relaxation by intergranular liquid

Borland [3, 7] was among the first to suggest that a thick intergranular liquid film can resists liquation cracking due to its ability to accommodate strains. He reported that in a good wetting system, stresses can be relaxed by liquid penetration along the boundary regions. This concept has been recently confirmed by other investigators who have shown that in a wetting system, externally applied shear stress can be significantly relaxed by grain boundary liquid penetration. In such systems, the work done by the applied force per unit time ($\dot{E}_s$) is used to expand the solid-liquid interface and reduce the solid-solid interface along the grain boundary, which can be expressed as [33]:

$$\dot{E}_s = -\int_{\sum sf} u\frac{\gamma_{ss} - 2\gamma_{sf}}{w}\mathrm{d}s \tag{7.2}$$

where u is the velocity of the advancing liquid, γ_{ss} and γ_{sf} are the energies of the solid-solid and solid-liquid interfaces, respectively, and w is the width of the liquid film. According to Eq. (7.2), a decrease in the intergranular liquid thickness ("thin" liquid films), would hamper stress relaxation by grain boundary liquid penetration since a higher level of stress is required to replace the solid-solid grain boundary interface with a solid-liquid interface in such situation. This could lead to a buildup of stresses across such a liquated grain boundary that could eventually result in cracking by solid-liquid interface decohesion. This is in agreement with the "Strain theory" of hot cracking developed by Pellini [2] and the "Generalized theory" of hot cracking in welds and castings developed by Borland [3, 4]. Based on these theories, cracking occurs during the thin film stage of solidification that allows high stresses to build up locally across adjoining grains. It is considered that thick intergranular liquid would resist cracking through better accommodation of welding or casting strains (stresses). Therefore, whether a HAZ crack will open up or not depends on re-solidification temperature range (BTR), the magnitude of welding stresses/strains, amount of intergranular liquid and ease of flow of liquid, as controlled by the wetting property. If a large amount of liquid is present and has good wettability, welding stresses could be appreciably relaxed by the penetration of intergranular liquid and at the same time heal any incipient fissure. Owczarski et al. [34] in their detailed study of HAZ cracking in nickel base superalloys noted that cracking occurred predominantly in HAZ region that contained small amount of liquid compared to the more extensively liquated region (i.e., area closer to the fusion zone boundary). The extensively liquated region with the widest BTR, however, has been recognized to experience the most damage to ductility among the HAZ crack susceptible regions, as determined by Gleeble hot ductility test [35]. This indicates that in spite of a wide BTR, formation of thick intergranular liquid could limit HAZ cracking. A similar observation was reported in IN 718 superalloy where highly liquated HAZ grain boundaries appeared uncracked after welding while cracking was predominant in HAZ regions that exhibited limited amount of liquation [36]. An occurrence of thick intergranular grain boundary liquid was also observed to produce reduced HAZ cracking in welded IN 738 superalloy [37].

Another factor that has been generally recognized to influence HAZ cracking in nickel-based superalloys is the grain size. It has been generally reported that the resistance of nickel-based superalloys to HAZ cracking could be improved by reducing their grain size as a bigger grain size, in the range of 20–200 μm [38, 39]. It is believed to reduce susceptibility to cracking due to: a longer interface sliding length, larger strain at grain boundary triple points and a larger stress concentration at the grain boundaries, a more extensive LFM, in materials with fine grains. Recent studies have, however, showed that an increase in grain size does not always imply concomitant increase in susceptibility to HAZ cracking [40, 41]. It was observed that above certain grain size in cast superalloys, increase in grain size could actually improve resistance to HAZ cracking. This is related to reduction in the number of crack susceptible liquated grain boundaries that intercept fusion zone boundary in large grain cast and directionally solidified (DS) superalloys. Such an increase in grain size could even override the hardening effect of higher volume fraction of γ' precipitates on increased HAZ cracking in superalloys [41] (Fig. 7.4).

Besides grain size, grain boundary misorientation is known to effect the liquation of grain boundaries. In a recent study by Chen et al. [42], laser additive manufacturing of Inconel 718 powder was done on a polycrystalline Inconel 718 as a means of repair welding. It was observed that liquation cracking was observed where the grain boundary misorientation angle was about 30°. It was found that liquation on large angle grain boundaries can remain stable at lower temperatures and as a result localized thermal stresses can effectively separate the solid/liquid interface [42]. Besides the grain size and grain orientation, it has been found that the serrated grain boundaries also influence microfissuring. Serrated grain boundaries were created in hot rolled Alloy 263 by a proprietary heat treatment of 1150 °C for 30 min and slow cooled at a rate of 10 °C/min to 800 °C and finally subsequently aged for 8 h. In this condition, 85% of grain boundaries were serrated and devoid of any γ' or $M_{23}C_6$ precipitate, but discontinuous planar MC carbides were observed. Serrated grain boundary is preferable than flat grain boundary from the viewpoint of interfacial free energy, which is responsible for the driving force of grain boundary serration. Serrated grain boundaries have lower equilibrium segregation of boron because of their lower grain boundary interfacial energy. In addition, serrated grain boundaries have lower wettability as compared to random flat grain boundaries. Non-equilibrium segregation of boron is also less as serrated grain boundaries are special grain boundaries unlike random grain

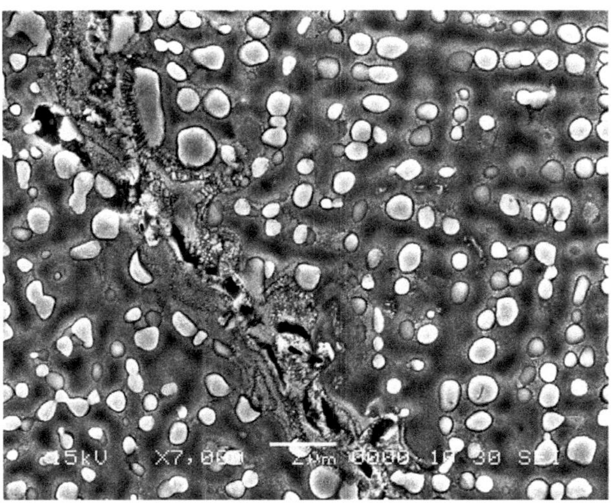

Fig. 7.4 Re-solidified products formed from uncracked thick intergranular liquid in a HAZ in IN 738 superalloy.

boundaries that are excellent vacancy sinks. The results were also supported by Gleeble hot ductility measurements where serrated samples showed lower susceptibility to liquation cracking [43].

7.3 Formation of HAZ grain boundary liquid

Since the presence of intergranular liquid film constitutes a fundamental factor that causes HAZ cracking, it is in order to consider different ways by which it is generated during welding. HAZ intergranular liquation is known to occur either by non-equilibrium phase transformation below the bulk solidus temperature of an alloy or by supersolidus melting above the equilibrium solidus temperature of the alloy. Sub-solidus HAZ liquation is generally considered more detrimental in relation to cracking resistance in that it does not only extend the effective melting temperature range of an alloy but could also influence the nature of supersolidus melting by establishing non-equilibrium film at a lower temperature that could alter the reaction kinetics during subsequent heating [34]. In addition, the metastable liquid produced by sub-solidus liquation reacts with adjacent solid grain through back-diffusion of solute across the solid-liquid interface, which results in low non-equilibrium solid-liquid interfacial energy [44] that aids intergranular wetting and liquid penetration. A more recent theoretical model [45] developed for describing penetration of liquid phase along grain boundary has shown that grain boundary penetration requires an under-saturated solid grains, which is essentially the prevalent situation in the sub-solidus portion of weld HAZ. There are two main mechanisms that are generally used to describe the occurrence of sub-solidus liquation in the HAZs; grain boundary penetration mechanism and grain boundary segregation mechanism. Grain boundary penetration mechanism involves a phenomenon known as constitutional liquation of second phase particles in the HAZ during welding operation and subsequent penetration of the grain boundary regions by the resulting liquid film [45]. This liquation results from dissolution of constituent particle at elevated temperatures and subsequent eutectic-type reaction at the particle-matrix interface. Grain boundary segregation mechanism involves the segregation of surface-active elements, which in most cases are also melting point depressants, to the grain boundaries leading to a localized composition with a lower melting point. This mechanism could be important in many single-phase materials that are comparatively free of intermetallic and constituent particles. In practice, both grain boundary penetration and segregation mechanisms could be active in the HAZs of a material.

7.4 Constitutional liquation of second phase particles in nickel-based superalloys

The theory of constitutional liquation was first proposed and published by Pepe and Savage with supporting experimentation on 18Ni Maraging steel [46]. The theory has gained wide acceptance since its introduction and it has been used to explain the occurrence of non-equilibrium melting in a variety of alloy systems. It describes the formation of liquid pockets and the subsequent intergranular liquid film by a eutectic-type reaction between a second phase particle and the surrounding matrix. Under conditions of slow heating (i.e., equilibrium heating) particles can be dissolved by diffusional processes before they have an opportunity to react with matrix. However, most welding processes are characterized by very rapid heating rates (non-equilibrium), which could preclude complete dissolution of the particle prior to reaching a temperature at which eutectic reaction occurs between the dissolving particle and the surrounding matrix. The amount of liquid that forms along the interface depends on the heating rate, initial particle size and dissolution kinetics of the constituent particle at elevated temperatures. A less readily dissolvable alloy carbides or intermetallic compounds makes constitutional liquation almost unavoidable, except for welding conditions producing extremely slow heating rates. The occurrence of constitutional liquation in multi-component nickel-based superalloys is generally reported to involve MC-type carbides [47–51], M_3B_2 borides [51, 52] and M_2SC sulfo-carbides [51], which are essentially solidification reaction products formed during ingot casting. Besides these particles, it has been also found that the main strengthening phase of most precipitation hardened nickel-based superalloys, γ' precipitates, formed by solid-state precipitation reaction, can also constitutionally liquate in the HAZ during welding and contribute to intergranular liquation cracking [51].

In a recent study of gas tungsten arc welded joints of a newly developed cast alloy K4750 (with Al + Ti content of 4.4%), grain boundary liquation and constitutional liquation of MC carbides was observed in HAZ region, adjacent to fusion zone, but it did not result in cracking. This was also attributed to low volume fraction and finer absence γ' and absence of η phase and borides due to rapid cooling stage of the solution heat treatment prior to welding. The other trace elements like sulfur, zirconium and boron were also kept to a minimum to avoid grain boundary segregation induced cracking [53].

7.4.1 Constitutional liquation of γ' precipitates

It has been long recognized that the crucial weldability problem, HAZ cracking, in precipitation hardened nickel-based superalloys becomes increasingly worse with an increase in concentrations of γ' forming elements [54–56]. The role of γ' precipitate particles in promoting susceptibility to HAZ microfissuring has been generally reported to be through their rapid re-precipitation behavior during cooling from the welding temperatures, which induces large shrinkage stresses along with a significant intragranular strengthening. The latter resists stress relaxation thereby causing welding stresses to be concentrated on liquated grain boundaries and, thus, increases the driving force for cracking [54–56]. Hence, attempts in reducing influence of γ' particles on HAZ cracking in these alloys have been mainly based on their solid-state re-precipitation behavior during weld cooling. In contrast, however, very limited information was available about the actual on-heating dissolution behavior of γ' precipitate particles during welding cycle, which can be expected to not only influence their mode of re-precipitation during cooling but also contribute to grain boundary embrittlement phenomenon. It has been implicitly assumed that γ' precipitate particles in nickel-based superalloys undergo complete solid-state dissolution in the HAZ regions which experience peak temperatures above γ' equilibrium solvus temperature during welding cycles. Nevertheless, a very important possibility, that on-heating to the welding temperatures these particles could persist to temperatures at which they could react with the austenitic γ matrix producing a liquid phase by a eutectic-type reaction, along with its concomitant consequences, has essentially not received due consideration.

It is generally accepted that $\gamma - \gamma'$ eutectic reaction occurs during solidification of most γ' precipitation hardened superalloys. Ni-Al-Cr and Ni-Al-Ti are the two ternary systems mostly used to represent this type of superalloys. The configuration for low Ti concentrations in Ni-Al-Ti is similar to that of Ni-Al-Cr system [57]. Therefore from a more theoretical point of view, the Ni-Al-Ti system is a good reference for the discussion of $\gamma - \gamma'$ equilibria and the solidification path for this class of superalloys. Willemin and Durand-Charre [58] have studied liquid-solid equilibria in the nickel rich corner of the Ni-Al-Ti system and have proposed a projection of the liquid surface for this system. According to the Ni-Al-Ti ternary system data (Fig. 7.5), the eutectic reaction between γ and γ' phases are found to be monovariant in nature. This implies that $\gamma - \gamma'$ eutectic reaction in the nickel superalloys based on this system occurs over a range of temperature. Hence, there exists a temperature range in γ' precipitation hardened nickel-based alloys, within which $\gamma - \gamma'$ eutectic reaction occurs and persistence of γ' particles to this temperature range during continuous heating could result in their constitutional liquation. Ojo et al. [51] have observed that, besides the generally reported coarsening and complete solid-state dissolution of γ' particles, these particles liquated in the HAZ of nickel-based superalloy IN 738LC weldment. A review of micrographs in published literature shows that constitutional liquation of γ' precipitates most likely occurs in other γ' strengthened nickel-based superalloys, but was not recognized. Considering that γ' precipitate particles are essential and principal strengthening phase of most precipitation hardened nickel-based superalloys, and new generation of nickel-based superalloys continue to contain even higher volume

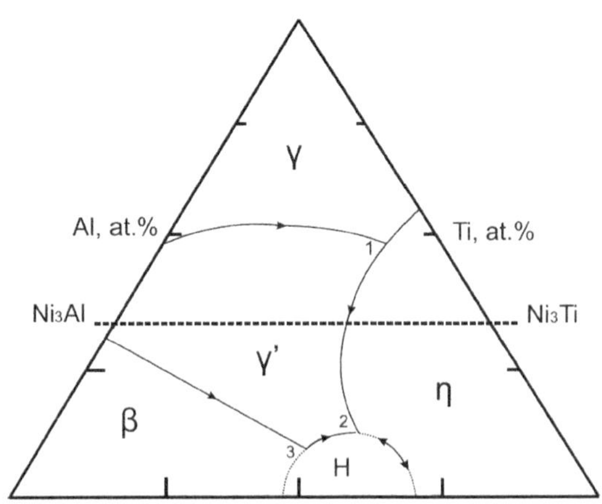

Fig. 7.5 Projection of liquidus surface of the Ni rich corner of Ni-Ti-Al ternary system [58].

fraction of these particles, it is deemed appropriate to understand better the role of these precipitates in weld HAZ microfissuring.

In the theory of constitutional liquation by Pepe and Savage [46] it is implicitly assumed that thermodynamic equilibrium exists at the precipitate-matrix interface during the precipitate dissolution, as well as at the precipitate-liquid and liquid-matrix interface during liquation. This implies that the precipitate dissolution and liquation reactions are diffusion controlled. It has, however, been suggested that in a situation where the solid-state dissolution is fully interface controlled (such as of coherent precipitates) the enrichment of solutes at the particle-matrix interface required for constitutional liquation might be significantly restricted [59]. Considering that γ' particles are coherent with γ matrix in most nickel base superalloys, it is reasonable to consider if the dissolution behavior of these particles would limit their liquation by reducing the solute concentration at the particle-matrix interface to a level which is below the equilibrium value, when these particles are heated to a temperature where they are thermodynamically capable of liquating.

It is commonly assumed that in superalloys the γ' phase precipitates are fully ordered up to their solutionizing temperature and that dissolution is accompanied by migration of the interface between the ordered γ' phase and γ matrix. In such a situation, the above deviation from equilibrium solute concentration at the particle-matrix interface during dissolution can be expected. However, a recent high temperature X-ray micro-diffraction study of dissolution behavior of γ' phase in AM1 single crystal nickel base superalloy has shown a significant variation from the above assumption [60]. It was reported that partial structural disordering of γ' phase occurs at γ/γ' interface at temperatures above 800 °C. As the temperature increases, the layer of disordered γ' increases in thickness from the γ/γ' interface inwards within the γ' particle, reducing the volume fraction of the ordered phase without changing the composition enough to transform the γ' to γ phase. It was indicated that dissolution of the γ' particle occurred by partial disordering reaction before transformation to γ phase, i.e.,

$$\text{ordered}\,\gamma' \rightarrow \text{disordered}\,\gamma' \rightarrow \gamma$$

A similar observation has been also reported in other single crystal nickel base superalloys by high temperature X-ray micro-diffraction investigation [61]. Therefore, the concept of solid-state dissolution of γ' in which the ordered γ' phase is surrounded by solute rich γ phase formed by disordering of γ', show that, contrary to an interface controlled dissolution prediction, solute concentration at the γ/γ' interface can be commensurate with near thermodynamic equilibrium values. This implies that liquation can be expected once a γ' particle survives to a temperature where a $\gamma - \gamma'$ eutectic-type reaction is possible, as dictated by local thermodynamic equilibrium. Besides IN 738 superalloy, γ' liquation has been also subsequently observed in the HAZs of welded IN 718 [62], DS Rene 80 [41] (Fig. 7.6), DS TMS-75 superalloy and a Ni_3Al-based alloy IC 6 [63]. Constitutional liquation of γ' particles in nickel-based superalloys can aid susceptibility to HAZ cracking not only due to the good wetting behavior of liquid produced by the non-equilibrium phase reaction as discussed earlier, but also by extending the BTR. Using the liquidus surface projection

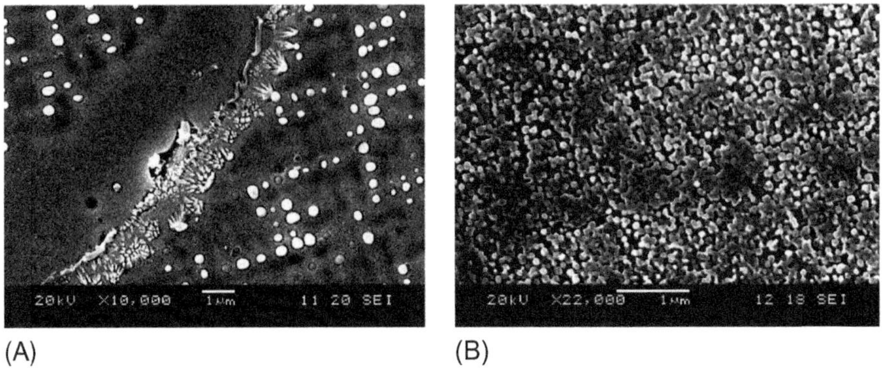

Fig. 7.6 Constitutionally liquated γ' particles in (A) DS Rene 80 superalloy and (B) a proprietary single crystal superalloy.

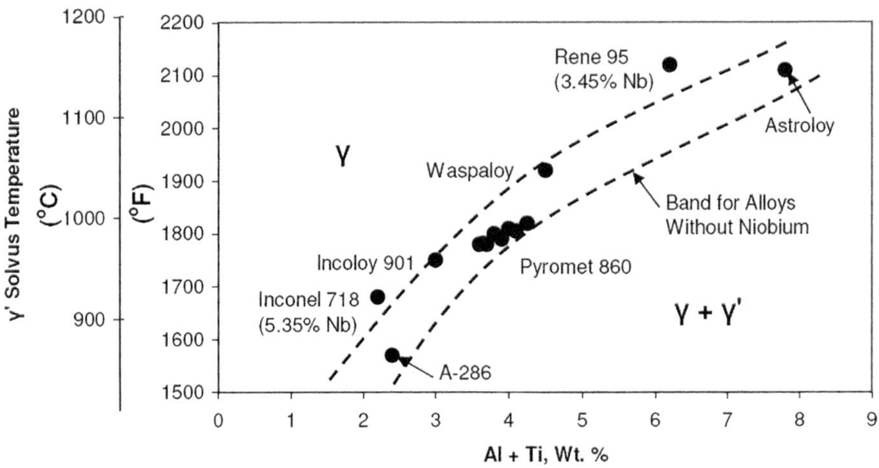

Fig. 7.7 Variation of γ' solvus temperature of seven superalloys with Al + Ti content as determined by DTA [64].

of Ni-Ti-C ternary phase diagram, Ojo et al [51] showed that constitutional liquation of γ' particle occurs at temperatures below the liquation temperatures of MC-type carbides, which underscores its importance in weld HAZ cracking of carbon bearing superalloys. Furthermore, the high volume fraction of γ' particles capable of liquating and contributing to intergranular liquid volume in precipitation hardened superalloys could limit the effectiveness of LFM in preventing HAZ cracking as previously discussed.

Increase in concentration of γ' forming elements in nickel-based superalloys generally reduces resistance to HAZ cracking. DTA analysis [64] has shown that increase in γ' forming elements produces not only an increase in volume fraction of γ' phase but also a significant increase in the γ' solvus temperature T_{SV} (Fig. 7.7), which causes a resultant decrease in the temperature range between T_{SV} and $\gamma - \gamma'$ eutectic temperature. New generations of single crystal and directionally solidified nickel-based superalloys containing higher volume fraction of γ' phase are known to exhibit significantly higher values of γ' T_{SV} [57], which drives it closer to the $\gamma - \gamma'$ eutectic temperatures. Another factor that has been recognized to affect γ' T_{SV} is the Co content, a reduction in Co concentration producing a substantial increase in T_{SV} [65]. It is interesting to note that a decrease in the temperature difference between T_{SV} and $\gamma - \gamma'$ eutectic reaction temperature, either due to an increase in γ' forming elements or decrease in Co concentration, would result in a decrease in the minimum heating rate required for the γ' particles of a given size to persist to a temperature where it could react with γ matrix to produce liquation via a eutectic-type reaction. That is, it would reduce the value of minimum heating rate required to cause constitutional liquation of γ' particles. Likewise, it will also reduce the maximum γ' particle size that can be tolerated without liquating during continuous heating at a particular heating rate. Therefore, increase in γ' forming elements in nickel-based superalloy would not only increase the volume fraction of γ' particles capable of liquating but could also increase susceptibility to constitutional liquation by reducing the minimum heating rate and minimum γ' particle size required for such occurrence.

Based on the concept of the effect of rapid on-cooling γ' re-precipitation, it has been suggested that in order to prevent HAZ microfissuring, γ' particles in nickel-based superalloys should be overaged prior to welding [39, 55]. However, based on the present discussion, for a specific heating rate, increase in γ' particle size would increase their tendency to survive during continuous heating to a temperature where they could constitutionally liquate and contribute to grain boundary liquation. It should be noted that liquation of γ' phase in nickel-based superalloys is not only potentially deleterious in relation to HAZ microfissuring during welding, but also during subsequent post weld heat treatment (PWHT), which is known to commonly affect these alloys. Liquated HAZ grain boundary regions have been recognized to be most susceptible to PWHT cracking and also to the formation of deleterious TCP phases during service [66]. Liquation of γ' precipitates has not been generally considered, perhaps unintentionally, in discussions on metallurgical processes taking place during heating of nickel-based superalloys to elevated temperatures.

7.5 Role of minor elements on HAZ intergranular liquation cracking

The effects of small percentages of minor elements on the weldability of nickel-based superalloys, added intentionally, or present in tramp amounts, have intrigued researchers since the 1950s. The present review builds on previously published research [67], and also on more recent results obtained over the last ten years or so. Elements considered in affecting the weldability of nickel-based superalloys include the minor elements boron, carbon, phosphors and sulfur. Magnesium, zirconium and the rare earths are also briefly considered for their influence on weldability, but data on these elements is limited and generally from the 1960s, often on air melted alloys. Generally, in the review we will consider element additions of less than about 0.1 wt.%, and very often in parts per million effects as in the case of boron additions

With the development of the nickel-based superalloys from the 1930s onwards, the effect of these elements on weldability soon attracted researchers to evaluate their effects on mechanical properties. Pease [68] showed that the elements, boron, phosphorus and sulfur adversely affected weldability. Carbon was considered to have a variable effect, while magnesium was considered beneficial, though zirconium was considered detrimental. Variable effects of sulfur on weldability are frequently seen, with some authors showing no effect and others a detrimental effect. Furthermore, boron has been considered by some researchers to be unimportant, while others have shown a very strong influence from the element. Similarly, carbon has been considered to have no effect on weldability, though constitutional liquation of carbides was shown by Pepe and Savage [69] to be responsible for HAZ microfissuring. Up until the 1990s the effect of phosphorus was not clear though with the low levels normally found in superalloys the element was not generally considered to be a problem. However, alloy 718Plus [70] contains deliberate additions of phosphorus, and exerts an influence on weldability.

Problems in the understanding of the effects of these elements include:

(a) The percentage addition made relative to the actual amount of element recovered, as well as optimum additions to the alloys in affecting the weldability [71].
(b) The conflicting results obtained by various investigations, resulting in sometimes no effect observed, or a positive effect observed as a single addition, and also multiple effects seen from combinations of additions.
(c) Heats from the 1950–1960 period may have been air melted.
(d) A further complication is the effect of heat treatment on grain growth and segregation of the minor elements to grain boundaries in affecting weldability.

Initially we will consider the effects of the individual elements, followed by any combinational or synergistic effects on weldability. In addition, potential metallurgical reasons for the elemental behavior will also be discussed.

7.5.1 Effect of boron

In the early years of superalloy development boron was found to be beneficial to the creep properties even with additions as low as 0.005 wt.% showing a beneficial effect. Thus over the years the addition of boron has been considered mandatory for high temperature properties in excess of half the melting point of an alloy. Pease [68] considered that boron was harmful to weldability, as did Owczarski et al. [34], with the latter authors considering the element also to be responsible for HAZ microfissuring. Vincent [72] also suggested a link between boron and weldability via metallographic analysis, during investigation of the metallography of HAZ cracking in wrought alloy 718. Kelly [73, 74] also in a statistically designed experiments, considered B to be detrimental to weldability. Further research by Kelly [74] also concluded that minor additions of boron, carbon and sulfur on the weldability of cast alloy 718, again resulted in only boron over the range 0.001–0.01 wt.% being detrimental. Thompson et al. [75] however did not observed boron to be a problem in his investigations of alloy 718, instead favoring sulfur as the problem element, due to its tendency to segregate at the grain boundaries during heat treatment. The weldability of cast 718 [76] was found to be directly related to the boron segregation on grain boundaries, being a function of the relative amount of equilibrium and non-equilibrium from heat treatment. Chen et al. [36] have also examined the weldability of wrought plate 718 at two boron levels (11 and 43 ppm). Test pieces were electron beam welded at 44 kV, 30 ma, 77 cm/min using a sharp focus. Table 7.1 shows the Av. TCL values obtained for the two B containing alloys in the two cooling conditions, viz., air cooled, and quenched from the solution heat treatment temperatures. The table shows that in the higher boron alloy, the air cooled material yielded a higher Av. TCL values than the water quenched test pieces, the lower boron alloy showed the same trend as well. Comparison of the low to high boron alloy also showed the expected trend with the high B alloy exhibiting larger Av. TCL values. Chen et al. [77] have confirmed the non-equilibrium segregation effect of boron in samples homogenized at 1200 °C for 2 h, followed by air and water quenching. Air cooling was again found to result in more non-equilibrium boron segregation than in the water quenched material. Analysis using SIMS showed that the HAZ cracking was due to the segregation of B on grain boundaries (Fig. 7.8).

Table 7.1 Effect of B concentration on Av. TCL values of alloy 718 based alloys.

Heat treatment	AC or WQ	TCL (μm)
High B	AC	3134 ± 975[a]
High B	WQ	1770 ± 433
Low B	AC	2348 ± 129
Low B	WQ	1515 ± 84

[a] 95% Confidence limit.

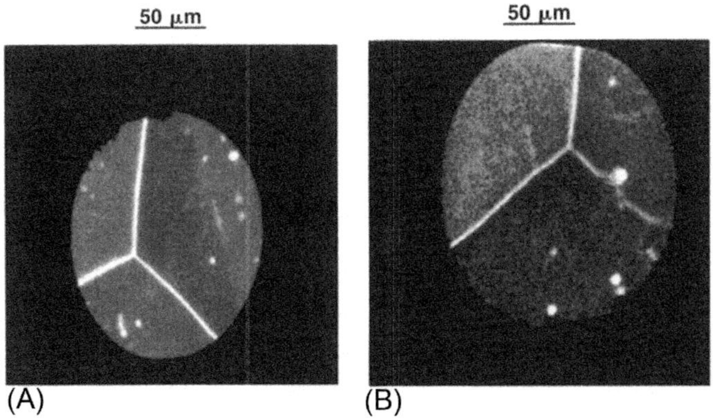

Fig. 7.8 SIMS image of boron in (A) air cooled and (B) water quenched alloy IN 718 solution treated at 1200 °C/2 h [77].

Table 7.2 Result of crack length measurement [78].

Number	C (wt.%)	B (wt.%)	P wt. T	Change in Av TCL microns WQ	Change in Av TCL microns AC
1	0.008	<0.001	<0.001	Base	Base
2	0.031	<0.001	<0.001	0	+43%
3	0.005	0.013	<0.001	+338%	+178%
4	0.031	0.012	<0.001	+158%	+87%
5	0.009	0.010	0.022	+1675%	+1256%
6	0.030	<0.001	0.022	+71%	+165%
7	0.033	0.011	0.022	+142%	+291%

Benhaddad et al. [78] also showed a boron effect, Table 7.2, with a 0.013 wt.% boron addition to the base alloy increasing the Av. TCL by 338% AC and 178% WQ. Thus the trend of increasing boron concentration on microfissuring is to increase the tendency of cracking, with boron having a highly significant effect. Thus from the earliest investigations on the effect of B on weldability, to the present day, it can be concluded that boron is detrimental to weldability. The SIMS data has been quantitatively corroborated by the research of Karlsson and Norden [79] who measured grain boundary boron levels using an atom probe of up to 8 atomic percent in a 316 type stainless steel.

Mo et al. [80] studied ductility dip cracking in butt welded carbon steel weldments by gas tungsten arc welding using two versions of filler alloy NiCrFe-7 with (46 ppm B) and without boron. They reported that boron extensively segregates to the grain boundary in the weld metal; aggravate grain boundary embrittlement and even promotes formation of continuous $M_{23}C_6$ carbide chains and M_2B boride phases. $M_{23}C_6$ phase being incoherent serves as crack initiation site and lowers ductility dip cracking resistance and tensile strength. For weldment with upto 46 ppm boron, the intergranular cracks were not only longer and wider but also about four times more when compared to the weldment without any boron [80].

7.5.2 *Effect of carbon*

Carbides form the main precipitate product of carbon, either as primary MC-type carbides, where M is a metal component, or as secondary carbides. These latter carbides such as $M_{23}C_6$ form usually during heat treatment, having a strong effect on the mechanical properties of the alloys. Rundell [81] in air melted alloy RA333, Table 7.3 showed a reduction in circular patch values from the base alloy containing 0.06 wt.% carbon of 291°, to 167° for a 0.20 wt.% carbon alloy. Kelly (9) using the spot-varestraint test, did not observe any effect of the presence of carbon on the total crack length (TCL) in cast alloy 718. Using the Gleeble thermal simulator however, Owczarski et al. [34] concluded that MC carbides were a factor in causing liquation cracking in Waspalloy and Undimet 700. Similarly, Thompson et al. [82], Fig. 7.9A and B showed that an increase in carbon level from 0.02 to 0.06 raised

Table 7.3 Circular patch crack sensitivity in degrees [81].

Element	C (wt.%)	S (wt.%)	P (wt.%)	Ce (wt.%)	Mg (wt.%)	Zr (wt.%)	Cracking resistance (degrees)
Base	0.06	0.0022	0.012				291
Base + carbon	0.20						167
Base + sulfur		0.042					279
Base + phosphorus			0.035				171
Base + cerium				0.01			207
Base + magnesium					0.07		193
Base + zirconium						0.01	242

the TCL value by 23%. In an investigation by Benhaddad et al. [78], Table 7.2, the effect of carbon on the electron beam (EB) weldabilty of a series of IN718 based alloys again concluded that carbon had a limited effect on weldabilty. The Average TCL values however were still low compared to the base alloy and thus the carbon addition is considered to have a minor effect on weldability, which is in agreement with the conclusions of Kelly [74]. Thus it would appear, therefore, that carbon has an effect on microfissuring, but within the normal limits for carbon concentration (<0.1 wt.%), its influence is only minor. Note however that continuous grain boundary carbide films due to processing are detrimental to mechanical properties, whereas optimum carbide distributions, however, assist in restricting grain boundary sliding during high temperature creep.

United Kingdom experience with welding of Nimonic alloys [83] also showed no effect of carbon on weldability over the normal range of carbon additions (up to 0.15/ 0.20 wt.%). Reduced carbon values to improve weldability has not been proposed however usually as within the normal carbon ranges mentioned above, the effect of carbon at this present concentration on weldability is limited. High temperature properties such as creep would likely be reduced by carbon levels outside the common lower values used today.

7.5.3 Effect of sulfur and phosphorous

Both sulfur and phosphorus are normally considered to be tramp elements and in the literature. Table 7.3 shows the individual effects of these elements relative to the base alloy in a circular patch test with sulfur having minimum effect and phosphorus a large effect. In an investigation of the weldability of alloy 600, Savage et al. [84] found that varying the sulfur from 0.004 to 0.016 wt.% using the sub-scale varestraint test, increased the TCL values from 7645 to 22,149 μm; in addition a variation in the phosphorus level from 0.001 to 0.01 increased the TCL from 7645 to 13,310 μm. A combination effect of sulfur and phosphorus also occurred with an increase in the TCL to 24,185 μm. In alloy 21-6-9, Brooks [85] reported a combined effect of both elements. With phosphorus plus sulfur values of 0.1 wt.%, cracking occurred at zero augmented strain, whereas at 0.06 wt.% sulfur plus phosphorus, a strain of 0.5% was needed to

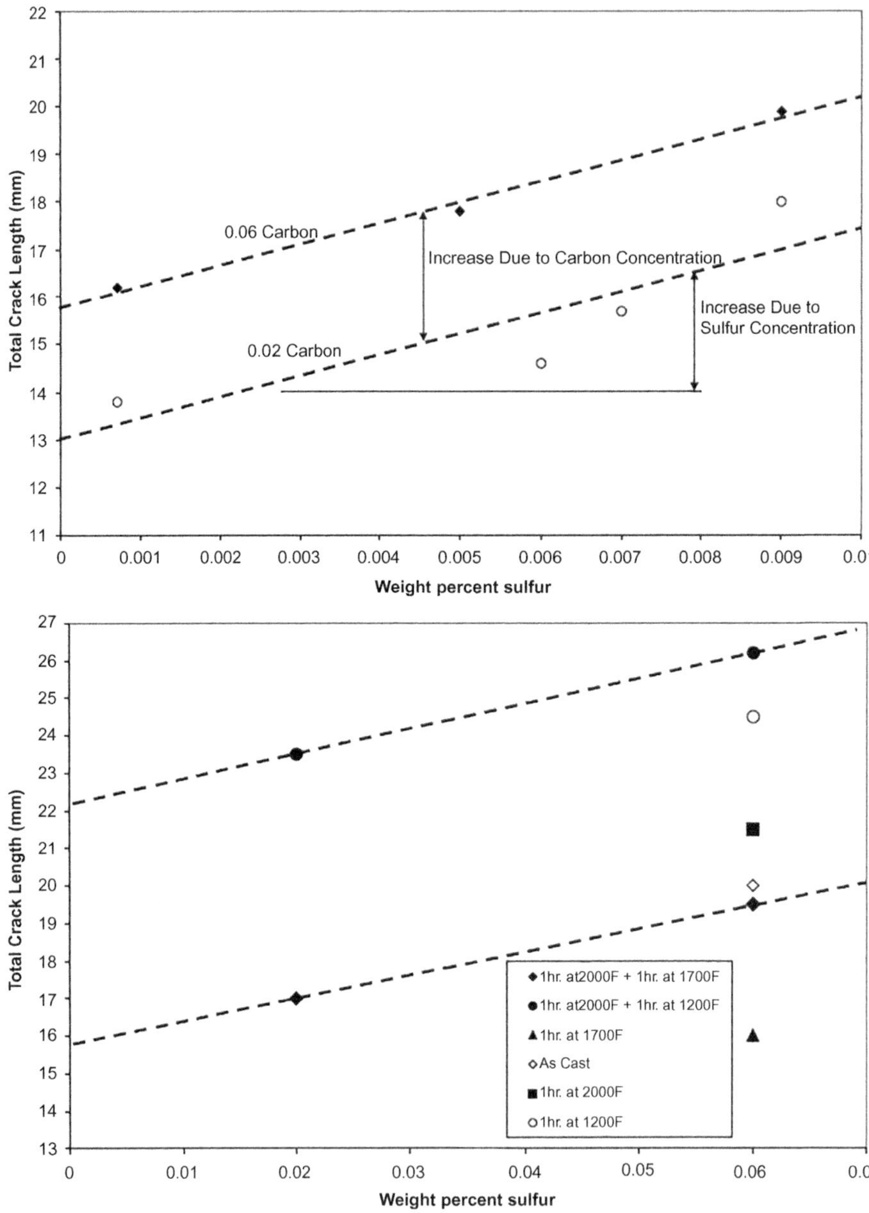

Fig. 7.9 (A) Intergranular hot cracking in alloy 718 vs sulfur at two carbon values [82] and (B) intergranular hot cracking in alloy 718 vs carbon and heat treatment [82].

initiate cracking. Thompson et al. [86] using the spot-varestraint testing on cast alloy 718, observed that a 10-fold increase in sulfur raised the TCL by 17%. In an alloy of 718 type, Guo et al. [87] varied the sulfur level from 7 to 110 ppm. Gleeble testing showed that the ductile recovery temperature (DRT) values were reduced,

Table 7.4 Av. TCL and DRT values vs sulfur level [87].

Ppm sulfur	Cooling rate	Av. TCL (μm)	DRT (°C)	Peak temp.-DRT
7	WQ	165	1165	55
19	WQ	190	1140	70
37	WQ	195	1110	100
110	WQ	325	1080	130

Table 7.4 (table 5, from review) (weldability reduced), with an increase in sulfur concentration. Changing the sulfur concentration by 0.01 wt.% also doubled the Av. TCL values from electron beam welds, with the Av. TCL/sulfur relationship followed an approximately straight line relationship.

No benefit has been observed from the presence of sulfur in a superalloy and only practical considerations dictate the lowest levels of the element that can be achieved in modern processing. The free energy of formation of sulfides ΔG vary from −22 kcal/mol for Ti_2S to −143 for ZrS_2. Sulfur is also known to occur as a carbo-sulfide, as reported by Wallace and co-workers [88], having the general composition M_2SC, with M being either Ti, Zr, or Nb, depending on the alloy. The influence of S, however, shown by Thompson et al. [86] on the weldability of alloy 718 via the TCL measurements, by Guo et al. [87] via Gleeble tests and TCL measurements in EB welds, are evidence of an effect of sulfur, although it seems to be somewhat modest. To combat the effects of sulfur, one can either add magnesium or rare earths to have sulfur present as a refractory rare earth sulfide.

The observations regarding sulfur made in the last paragraph would normally be applicable to phosphorus in that the element has no beneficial effect on the properties of superalloys. Research by Cao and Kennedy [89], however has shown a beneficial effect of P on creep properties, with additions of P up to about 0.025 wt.% to alloy 718, improving the creep rupture properties of the alloy by a factor of 2–2.5 times. Vishwarkama and Chaturvedi [90] however in an investigation of the weldabilty of low (0.006) and high (0.013) phosphorus additions in the presence of 0.022 and 0.028 wt.% carbon and 0.003 and 0.005 wt.%, boron showed increased cracking compared with a standard wrought 718 composition, Fig 7.10 (fig. 17 superalloys 2008).

Thus it can be concluded that no beneficial effect has been observed for S in superalloys and its effect on microfissuring is detrimental rather than beneficial. Thus one should keep the element as low as possible from a weldability point of view as well as from a long time creep type property point of view. In the case of phosphorus, however, elevated levels of the element improve the creep properties of alloy 718Plus, but adversely affect weldability.

7.5.4 Magnesium, zirconium, and the rare earths

Morrison et al. [91] in alloy 718, showed the beneficial effect of Mg additions on weldability at concentrations greater than 20 ppm. Rundell [81] on the other hand, Table 7.3 showed a reduction in circular patch values with a magnesium addition,

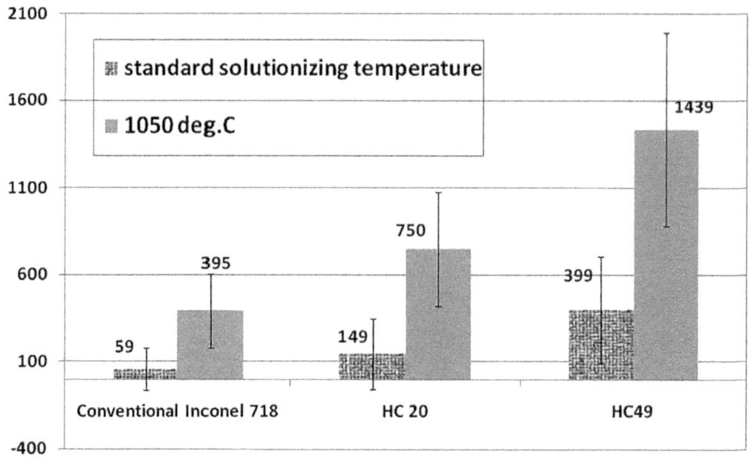

Fig. 7.10 Average TCL vs alloy compositions with two heat treatments [90].
From K.R. Vishwarkarma, M.C. Chaturvedi, Superalloys, Ed., R.C. Reed, et al., Pennsylvania, USA, 2008, pp. 241–250. Used with permission of The Minerals, Metals & Materials Society.

and a similar effect from cerium. Morrison et al. [91] also examined the effect of rare earth additions of up to 0.4% on the on-cooling ductility of Gleeble tested alloy 718. Rare earth additions of 0.1 wt.% rare earths lowered the DRT value of the alloy from 1127 °C to about 1032 °C. Furthermore, a 0.05% magnesium addition also reduced the DRT to about 1093 °C compared to the base alloy. In an investigation of 0.01% cerium addition to alloy RA333, Rundell [81] showed a beneficial effect compared to zero cerium addition. The cracking resistance as measured by a Circular Patch test, was reduced from 291° in the control sample (no deliberate additions) to 207°, for the 0.01% cerium addition.

Yeniscavitch and Fox [92] showed a beneficial effect on the DRT of Hastelloy X from adding up to 0.029 wt.% zirconium, but a negative effect with a 0.019 wt.% magnesium. Normally one would think however that magnesium would be beneficial in tying up sulfur.

In experimentation on the castability of DS alloy 792 by Zhang [93] zirconium additions to the alloy concluded that without boron, zirconium additions did not reduce castability. In the presence of boron, small additions of both elements did not greatly affect castability, whereas higher levels did affect the castability of the alloy. Keeping the boron addition below 150 ppm was recommended. Further testing using the Gleeble on the base alloy without boron and zirconium was compared with additions of 200 ppm boron and 200 ppm zirconium, Fig 7.11. The base alloy had improved ductility relative to the minor element casting and recovered its ductility at a higher temperature. As a means of tying up sulfur also, magnesium additions are frequently used in alloy 718 during VIM [94]. Some of this magnesium is lost during processing, but up to 30 ppm can remain in the wrought product. Consequently, sulfur should not present itself as a problem in practice, unless via a high temperature

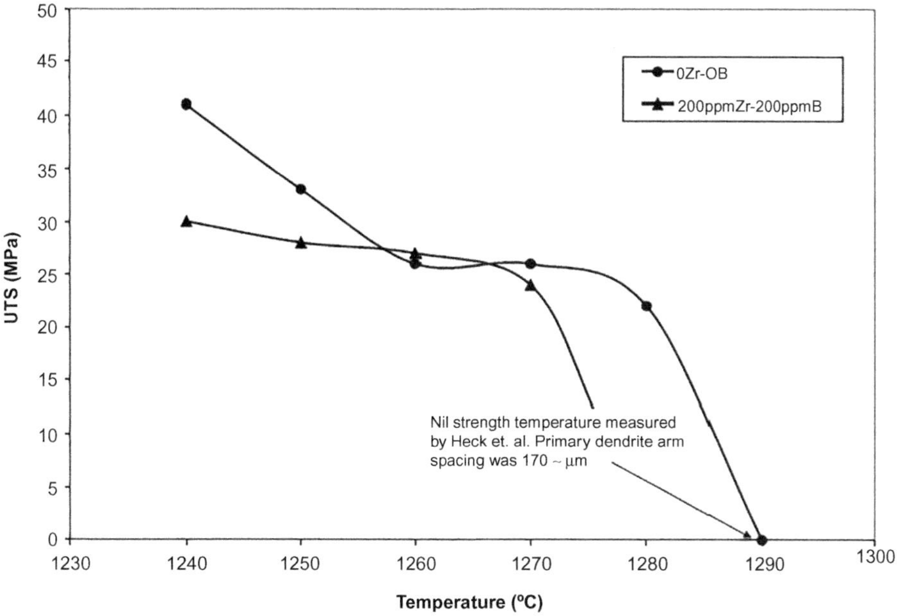

Fig. 7.11 UTS from Gleeble tested for B and Zr additions compared to base alloy 792 without additions [93].
From J. Zhang, Superalloys, Ed., K.A. Green, et al., T.M.S., Pennsylvania, USA, 2004, pp. 727–733. Used with permission of The Minerals, Metals & Materials Society.

heat treatment or during welding, the element is released from combination in a compound into an elemental form, which is able to segregate to free boundaries, e.g., grain boundaries or carbide/matrix interfaces.

Note however that while rare earth additions, as reported in a WRC publication in 1967, [95], showed a reduction in the DRT values, the importance of the actual rare earth additions made as suggested by Doherty et al. [71], needs to be given cognizance. Further research as related to superalloys would likely prove fruitful on today's manufactured alloys.

7.5.5 Combinational effects of several elements

From Yeniscavitch and Fox's [92] data, two element combinations of carbon, boron, sulfur, phosphorus, magnesium and zirconium, resulted in less detrimental effects on the Gleeble zero ductility range than single additions. This can probably be explained by synergistic effects [78]. The interaction and effect of several elements has been investigated by some researchers, the work of Savage and co-workers [84, 96] mentioned earlier being relevant; also Brooks [85] investigated the effect of sulfur and phosphorus in alloy 21-6-9.

A study however by Benhaddad et al. [78] researched the systematic addition of phosphorus to a base 718 alloy with varying minor element levels. The boron, carbon and phosphorus levels of the alloys are shown in Table 7.2 and were produced by Allvac of Monroe, N.C. The Av. TCL after EBW (44 kV, 30 ma, 77 cm/min) compared to the base alloy is shown in Table 7.2. Analysis of the data by comparing the change in Av. TCL relative to the base alloy, 988, which has low S, P, B and C in it, shows the following.

An increase in carbon over the base alloy showed only a modest increase in the Av. TCL with a very low boron content. Increasing the boron content shows a large effect however, while a phosphorus addition shows a huge increase in cracking. Interestingly the higher carbon addition mitigates against the detrimental effect of phosphorus, both in low and high boron alloys. Gleeble testing for hot ductility showed that the addition of phosphorus to the low boron alloy, reduced the ductile recovery temperature (DRT) of alloy 718 by 60 °C compared to the air cooled base alloy. An 0.03 wt.% carbon addition caused the ductility recovery temperature, DRT, to decrease from 1180 to 1100 °C in the WQ material and a minimal effect from 1160 to 1150 °C in the AC condition. In the low carbon versions, phosphorus additions to a boron containing alloy reduced the DRT value from 1090 °C (AC and WQ), to 1040 and 1050 °C, respectively, for AC and WQ specimens.

Alam et al. [97] studied the combined effect of boron, carbon and phosphorus by comparing segregation effects in two Inconel 718 variations—WE91 with 0.006 wt.% C (220 ppm P and 110 ppm B) and G727 with 0.33 wt.% C (220 ppm P and 120 ppm B). Atom probe tomography and NanoSIMS analysis were used to study the segregation behavior at the grain boundaries. It was found that higher C alloy G727, drastically reduced the segregation of boron while not affecting the phosphorus segregation. G727 showed improved welding performance which was attributed to insufficient segregation of boron atoms to produce liquation of the grain boundaries. Lower segregation of boron due to presence of carbon was suggested to be due to changes in the interaction coefficients in the matrix between boron, phosphorus and carbon opposed to site competition [97].

In an investigation by Nishimoto et al. [98] on micro-cracking in multipass welds of alloy 690, filler variations of the minor elements boron, phosphorus and sulfur were made as shown in Table 7.5. The TCL values were reduced in the order FF1 > FF2 > FF3 though there was not a major difference between the fillers due probably to the low percentage of additions used. The brittle temperature range also decreased in the same fashion as the TCL values.

Table 7.5 Chemical analysis of filler alloy minor elements [98].

Alloy	Filler	Ni	B as ppm	C as wt.%	P as wt.%	S as wt.%
690 base	–	Balance	<10	0.020	0.009	0.002
Base + filler	FF1	Balance	<1	0.020	0.005	0.0016
Base + filler	FF3	Balance	12	0.008	0.0009	0.0037
Base + filler	FF5	Balance	24	0.015	0.002	0.0001

Thus considering only one addition in its effect on weldability may not give a true picture when more than one addition is made, or varying levels of additions in terms of lows and highs occur.

7.5.6 Heat treatment effect

The first reference to the effect of heat treatment on microfissuring in alloy 718 was shown by Valdez and Steinman [95]. These authors showed that an increase in solution treatment temperature from 954 to 1066 °C increased cracking during Varestraint testing by a factor of two and a half times. Thompson et al. [86] reported that Gleeble test specimens solution treated at 1066 °C had a ductility of 70% reduction in area, whereas those heat treated at1204 °C had zero ductility. Similarly Morrison et al. [91] in autogenous fusion gas tungsten arc welds on wrought alloy 718, showed that cracking increased by an order of magnitude on changing the heat treatment temperature from 1066 to 2132 °C. Thompson et al. [86] on cast 718 using the spot-varestraint test, showed that the TCL, after solution treatment at 1093 °C, increased by about 20% compared to the as-cast material. A lower solution treatment at 927 °C reduced the TCL by 21% compared to the as-cast alloy and by 29% compared to the 982 °C heat treatment. Similarly Kelly [74] using a bead-on-plate GTA test, showed that the TCL of cast 718 solution treated at 1093 °C was reduced by 24% compared to a 1163 °C treatment. Varying the heat treatment temperature from 950 to 1163 °C showed that the TCL of cast 718 varied in a U-fashion when the pre-weld heat treatment temperature varied from 950 to 1163 °C. It was not possible to correlate their behavior with the re-solution tendency of various phases and their volume fraction after solution treatment at different temperatures in the 950–1163 °C range. The effect was explained in terms of equilibrium and non-equilibrium segregation of boron and confirmed by SIMS analysis. It was observed that of all the elements investigated (boron, carbon, phosphorus and sulfur, plus niobium) only boron segregated to grain boundaries.

Khakzadshahandashti et al. [99] heat treated gamma prime strengthened superalloy ZhS6U in two different pre-weld heat treatment conditions of 1140 °C/2 h/furnace cooling +1050 °C/16 h/water quenching and 1140 °C/2 h/furnace cooling + 1050 °C/16 h/furnace cooling. The second pre-weld heat treatment involving the final step followed by furnace cooling made the gamma prime particles coarser (from 0.74 to 0.86 μm) and lowered the hardness of the base metal compared to the first pre-weld heat treatment by more than 100 VPN. The samples with first pre-weld heat treated condition suffered from HAZ liquation cracking associated with liquation of secondary phases, mainly MC-type carbides. The welded coupons with second pre-weld heat treatment were free of any liquation cracking in the HAZ and it was attributed to the better ability of the base metal to release welding tensile stresses generated during cooling [99].

On the study of laser weldability of IN 738, Egbewande et al. [100] modified the standard heat treatment of 1120 °C for 2 h followed by furnace cooling with a prolonged hold of 24 h at 1120 °C followed by furnace cooling. It was found by decreasing the equilibrium segregation of boron and elimination of intragranular borides, this new heat treatment significantly reduced HAZ microfissuring. In an attempt to make a more industrially viable pre-weld heat treatment, the hold time was reduced from 24 to 16 h followed by furnace cooling in hybrid fiber laser and gas metal arc welding of IN 738 LC. This pre-weld heat treatment further improved the resistance to HAZ microfissuring by complete dissolution of coarse and fine borides at the hold temperature and by minimum non-equilibrium segregation of boron on very slow cooling in the furnace. [101]. Similar results were obtained by adopting a pre-weld heat treatment (1204 °C for 2 h followed by furnace cooling and subsequently solutionizing at 1120 °C for 2 h followed by furnace cooling) that resulted in complete elimination of M_5B_3 boride particles and concomitantly induced moderate base alloy hardness in laser welded Rene 80 DS superalloy, resulted in nearly a crack free microstructure after laser beam welding [102].

The method of cooling from the solutionizing temperature needs careful consideration as it affects non-equilibrium segregation of boron. While water quenching and furnace cooling have shown to have a positive effect, air cooling from 1120 °C in laser weldability investigation of laser welded Haynes 282 showed increased susceptibility to microfissuring. A balance was seen in solutionizing between 1080 and 1100 °C temperature followed by water quenching for this alloy [103].

Heat treatment can also influence microfissuring via grain size changes on homogenizing at higher temperatures. Thompson et al. [75] showed that heat treatment influenced the grain size with the TCL value increasing by a factor of about three over the range 30–200 μm. Guo et al. [104] also showed similar results in that over the same range of grain sizes the values of the Av. TCL increased by a factor of 2, Fig 7.12.

Also evident from their work in Fig. 7.12 is the effect of boron level (11 and 43 ppm) on the Av. TCL and the similar behavior at both B levels with respect to grain size. The reduction of grain size from 140 to 40 μm resulted in drastic reduction in liquation and elimination of cracking even when the pre-weld microstructure had M_5B_3 borides present in laser welded Haynes 282 [103]. Thus heat treatment has a major effect on HAZ microfissure occurrence due to its multiple effects on the properties of superalloys, in terms of segregation, grain growth, precipitation behavior, etc. Therefore, it is necessary to consider the potential effects of heat treatment in any study of HAZ cracking.

7.5.7 Segregation behavior of minor elements

Both Mclean type equilibrium and non-equilibrium segregation to grain boundaries have been observed in nickel-based superalloys [105–107]. In addition, temperature effects from heat treatment can also affect segregation behavior via grain growth and via the distribution ratio of equilibrium to non-equilibrium percentages.

McLean [105] developed the following expression for equilibrium segregation:

$$C_b = \frac{C_m e^{E_b/kt}}{1 + C_m{}^{E_b/kt}}$$

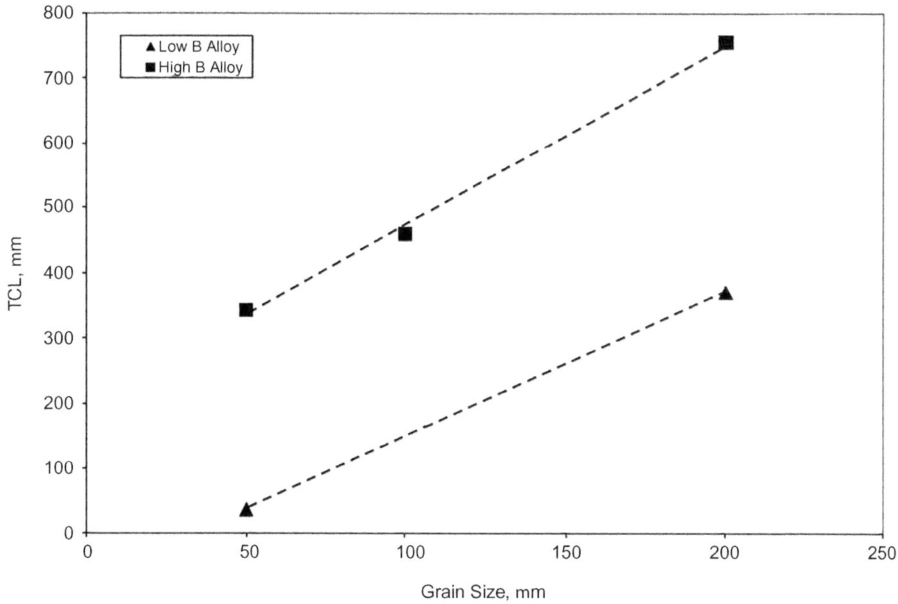

Fig. 7.12 Av. TCL vs grain size for two boron levels in alloy 718 [103].

where C_m is the solute bulk concentration, C_b is the solute grain boundary concentration, E_b is the binding energy, k is the Boltzmann's constant, and T is the temperature in Kelvin. An increase in temperature results in less segregation and is most commonly observed in the well-known case of temper embrittlement due to the segregation of P, Sn, Sb, etc., in low alloy steels after slow cooling in the 400–600 °C, temperature range.

Westbrook and Aust [106] developed the concept of non-equilibrium segregation, where instead of segregation occurring over the distance of a few atoms at the grain boundary, segregation can extend up to several microns into the grains. In this case, the strain energy of the solute atom in the matrix is reduced by interaction with a vacancy to form a vacancy-solute complex, resulting in a reduction in the free energy of the system, especially where the binding energy of the complex is appreciable, i.e., $\gg$kT. In the case of a B-vacancy complex the binding energy is well above kT at about 0.5 eV. As shown by Huang et al. [108] the two segregation mechanisms are complimentary, since equilibrium segregation decreases with increasing temperature and non-equilibrium segregation increases with increasing temperature due to an increase in vacancy concentration. Consideration needs to be given to the effects of conventional (McLean type) segregation and non-equilibrium segregation on HAZ microfissuring behavior. The non-equilibrium effect can be particularly noticeable during welding due to the nature of the high temperatures achieved during the weld heating/cooling period.

References

[1] W.I. Pumphrey, P.H. Jennings, J. Inst. Metals 75 (1948) 235.
[2] W.S. Pellinni, Foundry 125 (1952). pp. 124, 192, 194, 916, 199.
[3] J.C. Borland, Br. Weld. J. 7 (1960) 508.
[4] J.C. Borland, Welding and Metal Fabrication, vol. 47, Media Scores Ltd., 1979, p. 19.
[5] J. Vero, Metal Ind. (Lond.) 48 (1936) 431–491.
[6] B.I. Medovar, Automatic Welding, vol. 7, International Association (Welding), 1954, p. 12.
[7] J.C. Borland, Welding and Metal Fabrication, vol. 47, Media Scores Ltd., 1979, p. 99.
[8] W.A. Miller, G.A. Chadwick, Acta Met. 15 (1967) 607.
[9] Y. Arata, F. Matsuda, S. Katayama, Trans. JWRI 6 (1977) 623.
[10] C.S. Smith, Trans. AIME 175 (1948) 15.
[11] T.W. Clyne, G.J. Davies, Br. Foundrymen 68 (1975) 238.
[12] G.K. Sigworth, AFS Trans. 104 (1996) 1053.
[13] B. Rogberg, Scan. J. Met. 12 (1983) 51.
[14] S.A. Metz, M.C. Flemings, AFS Trans. 78 (1970) 453.
[15] M. Rappaz, J.M. Drezet, M. Gremaud, Metall. Mater. Trans. 30A (1999) 449.
[16] I. Farup, J.M. Drezet, M. Rappaz, Acta Mater. 49 (2001) 1261.
[17] N.N. Prokhorov, Russian Casting Production, 2, British Cast Iron Research Association, Birmingham, 1962, p. 1972.
[18] T.W. Clyne, G.J. Davies, TMA Proceedings of International Conference on Solidification, Sheffield. Metals Society (1979) 275.
[19] E.M. Lehockey, G. Palumbo, P. Lin, Metall. Mater. Trans. 29A (1998) 3069.
[20] H. Guo, M.C. Chaturvedi, N.L. Richards, G.S. McMahon, Scripta Mater. 40 (1999) 383.
[21] H. Kokawa, C.H. Lee, T.H. North, Metall. Trans. 22A (1991) 1627.
[22] B. Radhakrishnan, R.G. Thompson, Scripta Metall. Mater. 24 (1990) 537.
[23] R. Nakkalil, N.L. Richards, M.C. Chaturvedi, Acta Metall. Mater. 41 (1993) 3381.
[24] C.A. Handwerker, J.W. Cahn, D.N. Yoon, J.E. Blendell, Diffusion in Solids—Recent Developments, T.M.S.-A.I.M.E., 1986, p. 275.
[25] J.S. Kirkaldy, Acta Mater. 46 (1998) 5127.
[26] A.K. Koul, R. Thamburaj, Metall. Trans. 16A (1985) 17.
[27] O.A. Ojo, N.L. Richards, M.C. Chaturvedi, Scripta Mater. 51 (2004) 141.
[28] S.W. Baker, G.R. Purdy, Acta Mater. 46 (1998) 511.
[29] Cross, C. E., Tack, W. T. and Loechel, L. W., Proceedings, Materials Weldability Symposium, Detroit, MI; Oct.; 1990, ASM International, Metal Park.
[30] O.T. Ola, O.A. Ojo, P. Wanjara, M.C. Chaturvedi, Metall. Trans. 42A (2011) 3761.
[31] S.M. Mousavizade, F.M. Chaini, M.J. Torkamany, J. Sabbaghzadeh, A. Abdollah-zadeh, Scripta Mater. 60 (2009) 244.
[32] K.M. Oluwasegun, J.O. Olawale, O.O. Ige, M.D. Shittu, A.A. Adeleke, B.O. Malomo, J. Mater. Eng. Perform. 23 (2014) 2834.
[33] S. Hier-Majumder, P.H. Leo, D.L. Kohlstedt, Acta Mater. 52 (2004) 3425.
[34] W.A. Owczarski, D.S. Duvall, C.P. Sullivan, Weld. J. 45 (1966) 145-s–155-s.
[35] W. Lin, J.C. Lippold, W.A.I.I.I. Baeslack, Weld. J. (1993) 135s.
[36] W. Chen, M.C. Chaturvedi, N.L. Richards, Metall. Mater. Trans. 32A (2001) 931–939.
[37] O.A. Ojo, M.C. Chaturvedi, Metall. Mater. Trans. 38A (2007) 356.
[38] R.G. Thompson, J. Metals 7 (1988) 44–48.
[39] R. Thamburaj, W. Wallace, J.A. Goldak, Int. Met. Rev. 28 (1) (1983) 1–22.

[40] R.K. Sidhu, O.A. Ojo, M.C. Chaturvedi, Metall. Mater. Trans. 38A (2007) 858.
[41] R.K. Sidhu, O.A. Ojo, M.C. Chaturvedi, Metall. Mater. Trans. 40A (2009) 150.
[42] Y. Chen, F. Lu, K. Zhang, P. Nie, S. Hosseini, K. Feng, J. Alloys Compd. 670 (2016) 312.
[43] H.Y. Hong, I.S. Kim, B.G. Choi, Y.S. Yoo, C.Y. Jo, Met. Trans. A 43A (2012) 173.
[44] I.A. Aksay, C.E. Hoge, J.A. Pask, J. Phys. Chem. 78 (1974) 1178.
[45] D. Chatain, E. Rabkin, J. Derenne, J. Bernardin, Acta Mater. 49 (2001) 1123.
[46] J.J. Pepe, W.F. Savage, Weld. J. 46 (1967) 411-s.
[47] B. Radhakrishnan, R.G. Thompson, Met. Trans. A 24A (1993) 1409.
[48] B. Weiss, G.E. Grotke, R. Stickler, Weld. J. 49 (1970) 471s.
[49] J.A. Brooks, Weld. J. 53 (1974) 517s.
[50] R. Nakkalil, N.L. Richards, M.C. Chaturvedi, Metall. Trans. 24A (1993) 1169.
[51] O.A. Ojo, N.L. Richards, M.C. Chaturvedi, Metall. Mater. Trans. 37A (2006) 421.
[52] W.A. Owczarski, D.S. Duvall, C.P. Sullivan, Weld. J. 46 (1967) 423s.
[53] J. Xie, Y. Ma, W. Xing, M. Ou, L. Zhang, K. Liu, J. Mater. Sci. 54 (2019) 3558.
[54] M. Prager, C.S. Shira, Welding Research Council Bulletin No. 128, 1968.
[55] G. Cam, M. Kocak, Int. Mater. Rev. 43 (1998) 1.
[56] M.B. Henderson, D. Arrell, M. Larsson, G. Marchant, Sci. Technol. Weld. Join. 9 (2004) 13.
[57] M. Durand-Charre, The Microstructure of Superalloys, Gordon and Breach Science Publication, 1997.
[58] P. Willemin, M. Durand-Charre, J. Mater. Sci. 25 (1990) 168.
[59] B. Radhakrishnan, Interface Sci. 1 (1993) 175.
[60] A. Royer, P. Bastie, M. Veron, Scripta Mater. 40 (1999) 955.
[61] N.N. Stepanova, V.A. Sazonova, D.P. Rodionov, Scripta Mater. 40 (1999) 581.
[62] O.A. Idowu, O.A. Ojo, M.C. Chaturvedi, Weld. J. 88 (2009).
[63] O.A. Ojo, R.G. Ding, M.C. Chaturvedi, Scripta Mater. 54 (2006) 2131.
[64] D.L. Sponseller, Superalloys Conf., USA (1996) 259.
[65] M.V. Nathal, L.J. Ebert, Met. Trans. A 16A (1985) 1849.
[66] W.A. Owczarski, Welding Research Council, New York, 6, 1969.
[67] N.L. Richards, M.C. Chaturvedi, Int. Mater. Rev. 45 (3) (2000) 109–129.
[68] G.R. Pease, Weld. J. 36 (1957 July) 330-s–334-s.
[69] J.J. Pepe, W.F. Savage, Weld J. 49 (12) (1970) 545-s–553-s.
[70] K.R. Vishwarkarma, N.L. Richards, M.C. Chaturvedi, Mater. Sci. Eng. A 480 (2008) 517–528.
[71] J.E. Doherty, A.F. Giamei, B.H. Kear, Can. Met. Quart. 13 (1) (1974) 229–236.
[72] R. Vincent, Acta. Met. 33 (7) (1985) 1205–1216.
[73] T.J. Kelly, Advances in welding science & technology, in: S.A. David (Ed.), Proceedings of the International Conference on Trends in Welding Research, Gatlinburg, ASM, May 1986, pp. 623–627.
[74] T.J. Kelly, Weld. J. 68 (Feb. 1989) 44-s–51-s.
[75] R.G. Thompson, J.R. Dobbs, D.E. Mayo, Weld. J. 65 (11) (Nov. 1986) 299-s–304-s.
[76] M.C. Chaturvedi, N.L. Richards, A. Saranchuk, Superalloys (1997) 743–751. 718, 625, 706; Various Derivatives, Ed. E. Loria, T.M.S.
[77] W. Chen, M.C. Chaturvedi, N.L. Richards, G. McMahon, Met. Trans. 29A (1998) 1947–1954.
[78] S. Benhaddad, N.L. Richards, M.C. Chaturvedi, Met. Mater. Trans. A 2002 (2005-2017) 33A.
[79] Karlsson, H. Norden, Acta Metall. 36 (1) (1988) 35–48.
[80] W. Mo, X. Hu, S. Lu, D. Li, Y. Li, J. Mater. Sci. Technol. 31 (2015) 1258–1267.

[81] G. Rundell, Effects of Minor Elements on the Weldability of Nickel Alloys, W.R.C., 1967, pp. 36–46.
[82] R.G. Thompson, B. Radhakrishnan, D.E. Mayo, J. Phys. Colloq. 49 (C5 #10) (1988) 471–479.
[83] W. Betteridge, J. Heslop, The Nimonic Alloys, second ed., E. Arnold, 1974.
[84] W.F. Savage, E.F. Nippes, G.M. Goodwin, Weld. J. 56 (Aug. 1977) 245-s–253-s.
[85] J.A. Brooks, Weld. J. 54 (June 1975) 189-s–195-s.
[86] R.G. Thompson, J.R. Dobbs, D.E. Mayo, Weld. J. Res. Suppl. (Nov. 1986) 299-2–304s.
[87] H. Guo, M.C. Chaturvedi, N.L. Richards, Sci. Technol. Weld. Join. 5 (6) (2000) 378–384.
[88] W. Wallace, R.T. Holt, T. Terada, Metallography 6 (1973) 511–518.
[89] W.-D. Cao, R.L. Kennedy, Superalloys (1994) 463–477. 718, 625, 706; Various Derivatives, Ed. E. A. Loria, The Minerals, Metals & Materials Society.
[90] K.R. Vishwarkama, M.C. Chaturvedi, Superalloys (2008) 241–250. Ed., R.C. Reed et al.
[91] T.J. Morrison, C.S. Shira, L.A. Wisenberg, Effects of Minor Elements on the Weldability of Nickel Alloys, W.R.C., 1967, pp. 47–67.
[92] W. Yeniscavitch, C.W. Fox, Effects of Minor Elements on the Weldability of Nickel Alloys, W.R.C., 1967, pp. 24–35.
[93] J. Zhang, in: K.A. Green, et al. (Eds.), Superalloys, 2004, TMS, 2004, pp. 727–733.
[94] G. Maurer (Special Metals Corporation, New Haven, N.Y., USA), private communication to N. L. Richards, Oct. 1997.
[95] P.J. Valdez, J.B. Steinman, Effects of Minor Elements on the Weldability of High Nickel Alloys, W.R.C., 1967, pp. 93–120.
[96] W.F. Savage, B. Krantz, Weld. J. 45 (1966, Jan.) 13-s–25-s.
[97] P.J.F. Alam, M. Chaturvedi, L.T. Stephenson, M.R. Kilburn, J.M. Cairney, Metall. Trans. 43A (2012) 2183.
[98] K. Nishimoto, K. Saida, H. Okauchi, Sci. Technol. Weld. 11 (4) (2006) 455–461.
[99] A. Khakzadshahandashti, M.R. Rahimipour, K. Shirvani, M. Razavi, Int. J. Miner. Metall. Mater. 26 (2019) 251.
[100] A.T. Egbewande, H.R. Zhang, R.K. Sidhu, O.A. Ojo, Metall. Trans. 40A (2009) 2694.
[101] O.T. Ola, O.A. Ojo, M.C. Chaturvedi, Philos. Mag. 94 (29) (2014) 3295.
[102] L.O. Osoba, R.K. Sidhu, O.A. Ojo, Mater. Sci. Technol. 27 (2011) 897.
[103] L.O. Osoba, R.G. Ding, O.A. Ojo, Metall. Mater. Trans. A 43A (2012) 4281.
[104] H. Guo, M.C. Chaturvedi, N.L. Richards, Sci. Technol. Weld. Join. 4 (1999) 257–263.
[105] D. McLean, Grain Boundaries in Metals, Clarendon Press, Oxford, 1957.
[106] J.H. Westbrook, K. Aust, Acta. Met. 11 (10) (1963) 1151–1163.
[107] R.G. Faulkner, J. Mater. Sci. 16 (1981) 373–383.
[108] X. Huang, M.C. Chaturvedi, N.L. Richards, J. Jackman, Acta Mater. 45 (8) (1997) 3095–3107.

Improvements in bonding metals for aerospace and other applications

8

A. Kwakernaak, J. Hofstede, J. Poulis, and R. Benedictus
Delft University of Technology, Delft, The Netherlands

8.1 Introduction: Key problems in metal bonding

Adhesive bonding using natural materials was applied as a joining technology in ancient times. The first known application of adhesive is the use of bitumen (a natural substance that contains hydrocarbons found on the surface of the earth in tar or asphalt pits) about 36,000 years ago [1]. Various adhesive materials of animal or vegetable origin were used in ancient cultures.

With the development of synthetic polymeric materials, higher loaded joints in more demanding applications became possible. The first adhesive-bonded joints between metal parts emerged in two different ways. In one, they were developed from vibration damping structures, which used rubber layers vulcanized between the metal parts. In the other, Norman A. de Bruyne developed phenolic adhesives suitable for metal bonding based on the development of synthetic adhesives for bonding wood [2].

The development of epoxy resins is another milestone in the history of metal bonding, with the launch of epoxy adhesives into the market in 1946. Owing to the major advantages with regard to fatigue and damage tolerance, and the inherent potential weight saving of a bonded metallic structure over a mechanically fastened one, the technology was rapidly adapted by the aircraft industry. Unfortunately, the operational experience of structures with the first generation of epoxy adhesives, in combination with etched surfaces, showed limited durability. Only when chromic-acid anodizing (CAA) was applied as a surface treatment, as in the European aerospace industry, was sufficient long-term durability obtained [3, 4]. Also, when adhesive bonding is applied in other structural engineering applications, the rapid deterioration of the mechanical properties of the bonded joint upon exposure to environmental influences is often directly related to unstable interfacial durability. Evidently the durability of metal-bonded joints under humid or corrosive environments depends on the surface treatment of the metallic substrate before bonding.

Over the years many surface treatments have been developed for various types of metallic materials with the objective of providing a durable adhesive-bonded joint. Not all of these methods are environmentally friendly and some are highly toxic, so there is a need for continued development to replace these methods with safe

Welding and Joining of Aerospace Materials. https://doi.org/10.1016/B978-0-12-819140-8.00008-0
Copyright © 2012 Woodhead Publishing Ltd. All rights reserved.

alternatives. Manufacturing processes have been developed depending on the type of adhesive chemistry and the foreseen application.

In general, thermosetting adhesives are used for structural metal-bonding applications. The autoclave process has been developed to provide the elevated temperature and pressure required for curing one-component adhesives. Owing to the high capital cost and the long cure cycles related to the use of autoclaves, there is a need for development of out-of-autoclave technology, that is low-temperature and low-pressure processes. In the early days of structural bonding, simple analytical joint-strength prediction techniques had already been developed. Although these analytical techniques have been further developed over the years, they are generally limited to simple joint configurations. For complex joint geometries nowadays finite element models (FEMs) are used to predict bonded joint behavior. However, the reliability of the joint-strength prediction is not very high and a useful method of predicting long-term behavior and durability is lacking.

8.2 Developments in the range of adhesives for metal

Structural bonding of metals became possible [2] with the development, in 1942, of the modified phenol-formaldehyde adhesive Redux 775. The phenolformaldehyde is toughened with a thermoplastic polyvinyl formal powder that is chemically bound to the phenolic network, providing a high-strength adhesive material with excellent environmental resistance. This adhesive system was applied by coating the areas to be bonded with liquid resin and sprinkling the thermoplastic powder on top of it. Later on the liquid/powder adhesive system also became available as a more user-friendly film adhesive, ensuring a more constant quality of the applied adhesive layer.

8.2.1 *Modified phenolic adhesives*

The modified phenolic adhesive Redux 775 was very successful in structural applications in the aircraft industry (see Fig. 8.1) and is still in use today.

Other adhesive suppliers followed with the development of phenolic adhesive systems. Similar to the polyvinyl formal, polyvinyl butyral is also used as a toughening agent [5]. Blending a phenolic resin with nitrile rubber produces nitrile-phenolic adhesive films. The ratio of nitrile rubber to the phenolic resin can be varied resulting in adhesives with different properties. Formulations with a relatively high rubber content have high flexibility and are used in vibration damping and acoustic fatigue applications. The flexible nitrile-phenolics with high peel strength are also used for seal-bonding applications in aircraft integral fuel tanks. The nitrile-phenolic adhesives with lower rubber content have lower peel strength but improved high-temperature characteristics. Types with operational temperatures up to 260 °C are used to bond missile components.

Another adhesive type used for these high-temperature applications is the epoxy-phenolic adhesives. These adhesives have good shear strength, but are more brittle, which is demonstrated in the relatively low peel strength. Although the phenolic adhesive shows very good durability, it does require high cure temperature and high

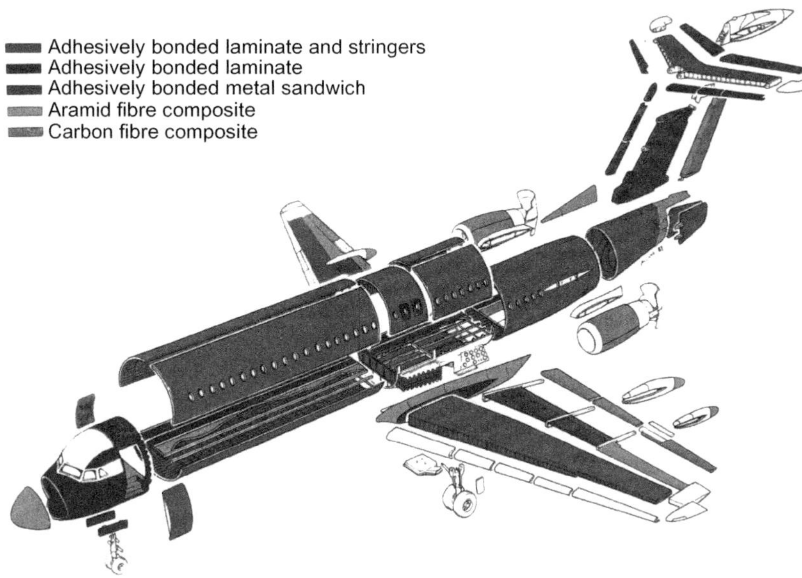

Fig. 8.1 Extent of adhesive bonding in the Fokker 100 aircraft. All metal laminate and stringers are bonded with Redux 775. Sandwich structures are bonded with toughened epoxy adhesives.

pressure to prevent the formation of porous bond lines caused by water from the polycondensation cure reaction.

8.2.2 Epoxy adhesives

Since their introduction in the 1950s, epoxy adhesives have gained an increasingly strong foothold in the field of structural metal bonding. A wide range of epoxy resins and curing agents is available for formulating adhesives with specific properties for a specific application. Epoxies can be formulated as liquid or paste two-component systems that cure at room temperature, but also as premixed one-component systems that require heat to cure and form the adhesive bond. In general, epoxy bonds are rigid and of high strength and they fill gaps well with little shrinkage. To enhance their mechanical properties, epoxies are often modified to meet a wide variety of bonding needs. The major advantage of epoxy adhesives is that they are suitable for bonding metals and provide good adhesion to many plastics. In general, they have very high resistance to physical and chemical influences and show high long-term stability with a limited tendency to undergo creep. Epoxy adhesives can withstand continuous temperatures from −55 to 100 °C or, depending on the type, up to a maximum of 200 °C.

8.2.2.1 Modified epoxy adhesives

Unmodified epoxies have good strength but low toughness. Introducing more flexibility into epoxy systems allows the adhesive to deform more under stress and distribute loads over a larger area. Furthermore, it increases the capability to compensate for differences in thermal expansion or elastic moduli of the substrates and will improve both the peel and the impact strength. Flexibility can be provided through the resin or hardener constituents by incorporating large groups in the molecular chain, which increase the distance between cross-links. Another method of increasing flexibility is by blending the primary epoxy resin with other, more elastic polymers (e.g., nylon or nitrile rubber). The disadvantage of increasing the flexibility is the negative influences on other properties, such as lower tensile strength, lower temperature resistance or less chemical and moisture resistance [5–7].

The development of toughened epoxy systems has overcome this problem and resulted in epoxy adhesives with high impact and peel strength, while maintaining chemical, moisture and temperature resistance. Toughened epoxy adhesives generally have two distinct phases: the larger phase is the base resin and the other phase consists of small (in the order of 1 μm in diameter), distributed elastomeric entities. The addition of the second-phase modifiers significantly improves fracture toughness by providing crack pinning and stress redistribution mechanisms within the material. A variety of toughening agents have been used to modify epoxy adhesives to improve peel and fracture toughness without significantly affecting other properties of the epoxy base resin. Epoxies are often modified by the addition of reactive liquid elastomers (e.g., carboxyl-terminated butadiene-acrylonitrile, CTBN) or functionally terminated thermoplastics (e.g., polyether sulfone, PES) [6–8]. Recent investigations have shown new ways of improving the toughness of epoxy adhesives by incorporation of additional functionalized nanoparticles in the epoxy matrix or by creating a flexible polymer structure that interpenetrates the epoxy network at nanoscale level [9–11].

8.2.2.2 Out-of-autoclave curing methods

In the early application of one-component structural adhesives, curing was performed using hot presses. This had the disadvantage that bonded components were limited in size and that special tooling was needed for curved parts and stringer bonding. By using an autoclave, a more flexible and less tool-intensive manufacturing process was born. The high capital cost of large autoclaves and the long cure cycles have driven the development of out-of-autoclave technology.

Owing to their polyaddition cure mechanism, the epoxy adhesives need less pressure during cure than the phenolic type of adhesives. This makes it possible to use only vacuum to create sufficient pressure for simple components during the cure cycle. The assembly to be bonded is put together with the adhesive in a vacuum bag and the cure cycle can be performed in an oven at the required temperature. Care should be taken that the vacuum pressure is not too high because a small residue of solvents in the adhesive could lead to the formation of small voids in the adhesive bondline by the combination of vacuum and cure temperature. Generally a vacuum pressure of around 60 kPa with an adhesive free of solvents will give good results.

An alternative to this process is the use of the Quickstep technology [12], which has been developed for out-of-autoclave curing of composite parts, but that can also be used for curing adhesive-bonded parts. The Quickstep process involves using fluid-filled heated floating-mold technology for curing. Flexible membranes separate the product and mold from a circulating liquid that transfers heat and provides pressure. Application of this technology is limited to the size of the tool needed and the complexity of the component to be bonded. Another alternative is the use of low-viscosity adhesives and a liquid-adhesive injection process, such as that used in composites technology (RTM: resin transfer molding or VARTM: vacuum-assisted resin transfer molding). This is sometimes used for smaller parts with complex shapes.

8.2.2.3 Automotive bonding

Special adhesives have been developed for use in automotive structures, which are capable of bonding to oily steel sheets without cleaning. In the automotive body, more and more adhesive beads are applied in order to increase structural stiffness and rigidity to reduce noise-vibration-harshness (NVH) and increase final car body performance. Closure panels are generally adhesively bonded with hem flange adhesives. These automotive structures with uncured or partly cured adhesives are conveyed to the paint shop where surface oils are removed by degreasing, the steel is surface treated for corrosion protection and paint adhesion, followed by electrocoating by cataforesis. The adhesives are cured during the paint-bake cycle.

8.2.2.4 Assembly bonding

Adhesive bonding is often applied in assembling components in combination with mechanical fastening. The advantage of this combination of joining methods is to lower the manufacturing costs by reducing the number of process steps. In fact, the adhesive can support the assembly when drilling holes and fastener installation are conducted after cure of the adhesive. This makes "out-of-jig" drilling possible, as well as the elimination of process steps (such as disassembly for deburring operations). Owing to the added strength of the bondline in the assembly bonded joint, it is possible to reduce the number of fasteners, giving additional weight and cost savings. The added stiffness of the bondline also greatly reduces the stress concentrations near the rivet holes, thereby improving the life of fatigue-critical joints considerably. One of the earliest applications of assembly bonding was in the aluminum alloy fuselage panel joints of the Fokker F28 [13], and later also in the Fokker 100. The longitudinal splices between adhesive-bonded fuselage panels were bonded with a RT-curing epoxy paste adhesive in assembly, cured and subsequently drilled and riveted. The effect of the assembly bonding of this critical joint is a dramatic improvement in fatigue life [14] (see Fig. 8.2) and a significant reduction of weight and manufacturing costs. Nowadays bond-assisted assembly is mostly applied to flaps, ailerons, rudders and the like for reasons of cost saving. This can effectively be applied both in metallic and in composite structures, with cost savings of 10–20%.

The reduction of mechanical fasteners from the rivet-bonded joint is limited owing to the low-temperature requirements of aircraft applications. The toughness of the two-component room-temperature curing epoxy adhesives is less than that of the one-component film adhesives that cure at higher temperatures. Fig. 8.3 shows a comparison of the shear stress-strain characteristics of an epoxy film adhesive and

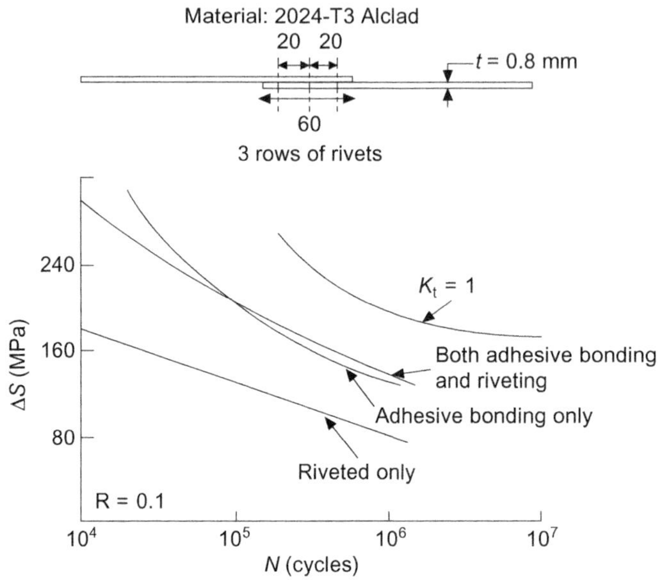

Fig. 8.2 S – N curves of lap joints.

a two-component epoxy paste adhesive. The curves tested at −55 °C, RT and 80 °C are shown both after manufacturing and after ageing for 30 days at 70 °C and 95% RH. It is clear that the toughness of the film adhesive is considerably higher than the paste adhesive. This is even more evident in tests at low temperatures. Typically, at −55 °C two-component epoxy paste adhesives do not have the peel strength that epoxy film adhesives have, so the use of fasteners to take up any peel loads remains essential in aircraft applications. Obviously, there is a need for further development in low-temperature curing epoxy adhesives with high toughness, to obtain similar properties to the high-toughness epoxy films.

8.2.3 Polyurethane adhesives

Two-component polyurethanes are also used in industrial assembly. Curing in these adhesive systems is initiated by mixing together the resin (polyglycols or PUR (polyurethane) prepolymer with terminal OH groups) and the hardener (modified isocyanate). At room temperature curing can take from a few hours to several days. Heating can accelerate this process and also increases the strength of the bond. After curing, the adhesive ranges from tough and hard to rubber-like and flexible depending on the raw materials used. The strength of these adhesives is about one-third that of good epoxy adhesives. They have better low-temperature strength than other adhesives; some types can be used for cryogenic applications. They have good chemical resistance, although generally not as good as epoxies or acrylics, and their strength will drop considerably with moisture absorption. Their properties at elevated temperature drop off rapidly above 60 °C. Two-component polyurethanes are used for

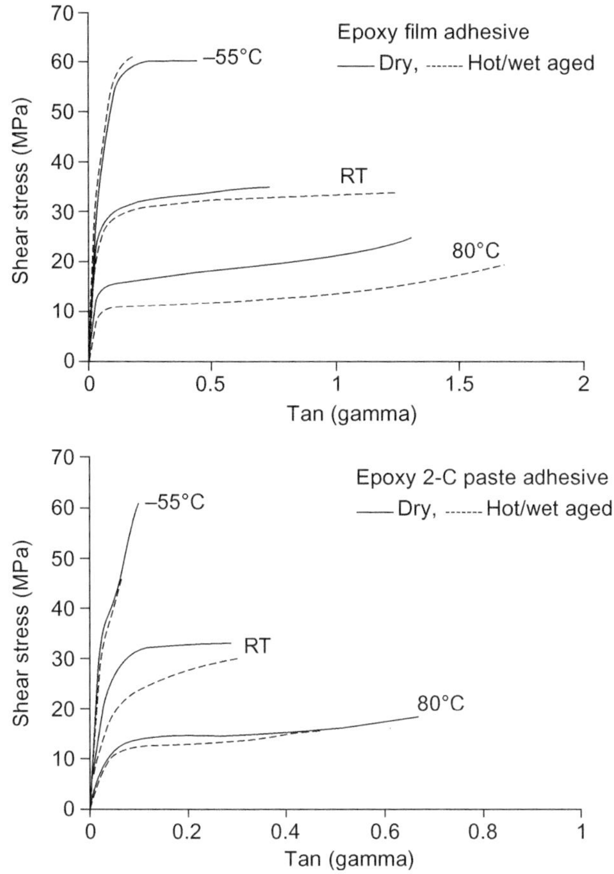

Fig. 8.3 Shear stress-strain curves of epoxy film and 2-C room-temperature curing epoxy adhesives.

large-surface adhesive bonds in land vehicle structures (semi-trailers and train sandwich structures), building elements (sandwich panels), ship building and container structures.

8.2.4 Methyl methacrylate adhesives

Another group of two-component adhesives, whose strength is between the two-component epoxies and polyurethanes, are the methyl methacrylate adhesives. Two-component methyl methacrylate adhesives or reactive acrylic adhesives consist of two major ingredients, the monomer and the rubber toughener. Reactive acrylics are based mainly on monofunctional monomers, for example, methyl methacrylate or cyclohexyl methacrylate, giving these adhesives their typical penetrating odor. Low-odor versions have been introduced by using higher molecular-weight monomers. The reactive acrylic adhesives have good toughness with high impact and peel strengths because of the rubber component, generally a chlorosulfonated polyethylene rubber. These adhesives are used for a wide range of applications in the marine, automotive, recreational vehicle, transportation and manufacturing industries. Applications include often dissimilar substrate bonding, for example, engineering plastics, SMCs, and fiberglass to metal bonding, particularly when fast curing with limited surface preparation is required.

8.2.5 Adhesives with high flexibility

Next to the rigid one- and two-component adhesives, very flexible moisture-curing one-component adhesives (polyurethane, MS polymer, silicone) are also used in assembly. These flexible adhesives provide joints with a more uniform stress distribution and less of a difference between average and maximum stress. These adhesives distribute peel and shear stresses over a larger area, thereby improving joint efficiency. However, as adhesives with high flexibility and elongation typically have lower cohesive strength than more rigid adhesives, the advantage of flexibility and high elongation is usually compromised. In order to transfer the same load, a much larger overlap is needed, as shown in Fig. 8.4.

8.2.6 Improvements in temperature resistance of adhesives

All polymers are degraded to some extent by exposure to high temperature. Physical properties, such as stiffness and strength, are lower at high temperatures (softening), but they also degrade during thermal aging.

Adhesives that are resistant to high temperature usually have rigid polymeric structures, high softening temperatures and stable chemical groups. These same factors make the adhesive very difficult to process and they usually show low peel strength. Any form of added toughener generally increases the peel strength but lowers the temperature resistance. Addition of fillers, such as aluminum powder, silica or ceramic material, can increase the temperature resistance, but at the expense of peel strength.

8.2.6.1 Glass transition temperature

An important parameter is the glass transition temperature (T_g), the temperature at which the resin begins to soften and its mechanical properties degrade [15]. For thermoset resins the upper service temperature for structural adhesives is typically limited by the resin modulus that falls off rapidly above the glass temperature, as depicted in Fig. 8.5. Therefore structural adhesives must have a T_g higher than the maximum operating temperature to avoid a cohesively weak bond and creep problems [15].

In Table 8.1 the typical glass transition temperatures of various thermoset adhesives are compared. The two-component toughened epoxies are limited to about 80 °C. However, specific formulations of two-component epoxies exist that show higher T_g typically at the expense of lower peel strength. The widely used toughened (nitrile) epoxy film adhesives have a T_g up to 121 °C and therefore are not feasible for application at higher temperatures. A higher T_g is found for unmodified epoxies and

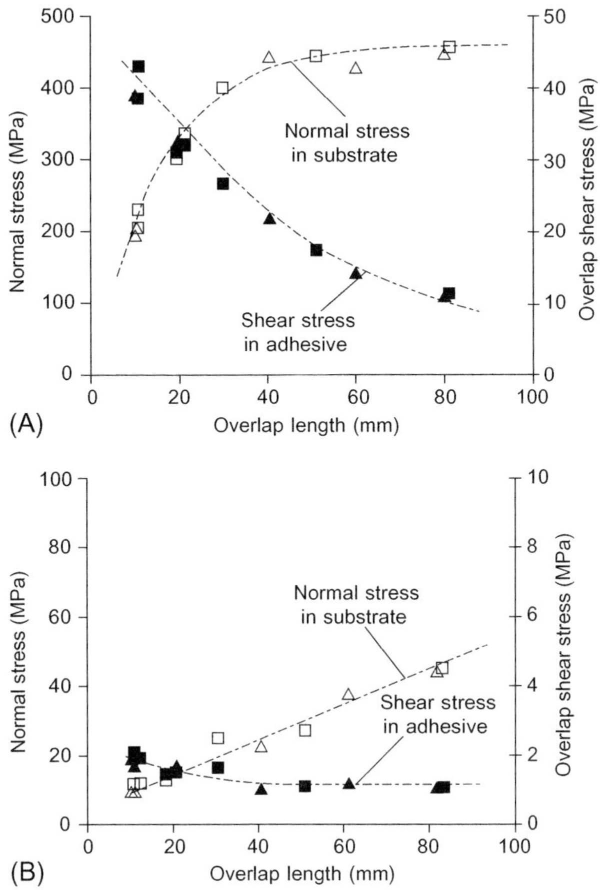

Fig. 8.4 Comparison between (A) epoxy (rigid) and (B) MS polymer (flexible) adhesive overlap joints

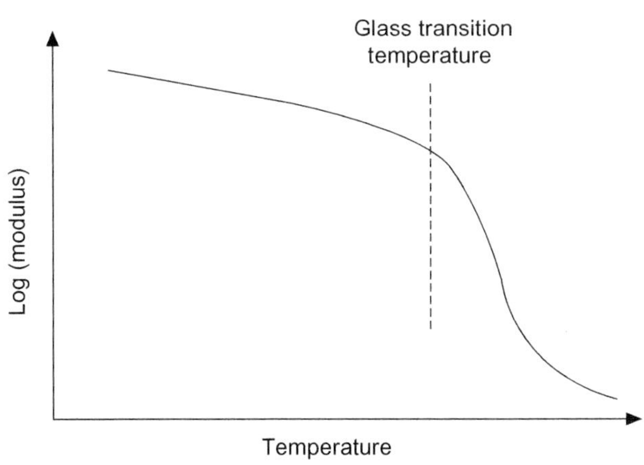

Fig. 8.5 Decrease in the adhesive modulus with temperature, with indication of the glass transition temperature.

Table 8.1 Range of typical glass transition temperatures for various structural adhesives.

Adhesive type	Glass transition temperature (°C)
Polyurethane	<80
Acrylate	<80
Toughened epoxy 2-C (RT cure)	60–80
Nitrile epoxy 1-C (120 °C cure)	90–120
Modified epoxies 1-C (180 °C cure)	100–150
Unmodified epoxies	100–150
Epoxy phenolic/nitrile phenolic	150–200
BMI	200–300
Cyanate-ester	250–350
Polyimide	280–330

nitrile-phenolic, epoxy-phenolic and other high-temperature adhesives. Polyaromics (among others, polyimides) show the highest thermal resistance of the organic adhesives. Only the ceramic (inorganic) adhesives perform better (up to 1000 °C), but are very brittle. Under wet conditions the ingress of moisture softens the adhesive, which lowers T_g (by about 20–30 °C for toughened epoxy adhesives).

8.2.6.2 Overlap shear strength at elevated temperature

Table 8.2 shows that at a service temperature of around 125 °C, the standard structural adhesives, such as polyurethanes and rubber-modified epoxies, have already lost most of their overlap shear strength. Stable mechanical properties between 125 and 175 °C are seen for the so-called high-temperature structural adhesives, such as nitrile-phenolic, epoxy-phenolic, heat-resistant one-part epoxy, bismaleimide (BMI) and polyimide. Despite the significant differences in overlap shear strength at RT, it is found that most commercial high-temperature adhesives have an overlap shear

Table 8.2 Strength of various adhesive types at high temperature.

	Overlap shear strength (MPa) at			
Adhesive type	**RT**	**125 °C**	**175 °C**	**200 °C**
Standard epoxy film 120 °C cure	42	10	–	–
Elevated-temperature epoxy 180 °C cure	35	28	–	–
High-temperature epoxy 180 °C cure	28	27	25	15
Nitrile-phenolic 180 °C cure	25	21	16	8
Epoxy-phenolic 180 °C cure	25	22	19	13
Bismaleimide 180 °C cure	20	17	17	17
Polyimide 180 °C cure	25	20	20	20

Note: Strength values depend on cure cycle, high-temperature properties will need a postcure.

strength of around 20 MPa at 125 °C. Bismaleimide and polyimide adhesives with a free standing postcure will have a T_g above 280 °C and overlap shear strength values above 10 MPa at 280 °C. The values in Table 8.2 are typical values obtained from static, short-term tests, whereas the applied loads may be continuous. The viscoelastic behavior of the adhesive may result in creep failure under long-term sustained load, especially at temperatures near or above the T_g.

8.3 Developments in surface treatment techniques for metal

An adhesive bond consists of a layer of adhesive that adheres to the contact areas of the surfaces of the parts that are joined. Therefore, the strength of the joint depends on the strength of the adhesive material (cohesion) and on the level of adhesion strength between the adhesive and the bonded surfaces (adhesion).

The adhesion strength is more complex and depends on the adherend surface-adhesive interaction. The most common surface forces that originate at the adhesive-adherend interface are Van der Waals forces. In addition, covalent bonding, acid-base interactions and hydrogen bonds, generally considered a type of acid-base interaction, may also contribute to intrinsic adhesion forces.

Surface treatments are often required to provide maximum adhesion strength, not only to remove contaminants, but also to increase the difference in surface energy between adhesive and substrate, so good wetting and adsorption of the adhesive is obtained. The surface treatment should form surface layers with sufficient mechanical strength to transfer the loads through the bonded joint during the service life of that joint (durability).

8.3.1 Surface treatment of aluminum alloys

Aluminum alloys are generally considered to be "difficult to bond." This is true in the sense that without proper surface treatment of the aluminum surface before the bonding takes place, the strength and especially the retention of strength during the lifetime of the joint will be poor.

The aerospace industry recognized this in the early stages of the application of adhesives in metal-bonded structures, and this resulted in surface treatments that are very well adapted to the specific demands in this industry. By anodizing, long-term durability of bonded structures is obtained, even under the extreme environmental and chemical exposures that the products are exposed to. In Table 8.3 a short overview is given of the most well-known and widely used surface treatments for metals.

A degreasing step is the basic step for all treatments and should be performed in all cases. This can be done either by wiping with a cloth or by immersing the material in a tank with an alkaline degreasing agent. The process may increase bond strength, but only degreasing is generally not enough to obtain good strength. The natural oxide layer that is still present on the surface mainly causes this. This layer has irregular properties and the mechanical strength can be relatively low. Good strength between

Table 8.3 Short overview of surface treatment methods for aluminum alloys.

Category	Surface treatment
Chemical and electrochemical	Degreasing Etching/pickling (CSA/FPL)[a] Anodizing (PAA/CAA/PSA)[a] Conversion coatings (chromate, titanate, zirconate) Grinding, scouring, brushing Grit-blasting with corundum (aluminum oxide)
Mechanical	Application of silanes, sol-gels, primers
Application of adhesion promoters	

[a] CAA, chromic acid anodizing; CSA, chromic and sulfuric acid; FPL, forest products laboratory; PAA, phosphoric acid anodizing; PSA, phosphoric-sulfuric acid anodizing.

the adhesive and the oxide can be obtained, but the bonded joint will often fail owing to failure of the oxide layer itself. There are some cases, when certain types of adhesives are used and when the adhesive bond will not be exposed to harsh environments, where only degreasing will suffice.

8.3.1.1 Etching processes for aluminum alloys

In most cases, degreasing is not sufficient for good adhesion and more treatment is necessary. In the aerospace industry, an etching treatment is conducted that will remove the natural oxide layer of the aluminum, leaving only a very thin-but-closed oxide. When the adhesive bonding takes place soon after the etching treatment, good initial bond strength can be obtained. The process developed in Europe based on a mixture of chromic and sulfuric acid (CSA) was introduced in the 1940s [16] and used in combination with the early REDUX bonding. This CSA etch is used at 60–65 °C and shows good initial adhesion and durability in combination with phenolic adhesives. A similar process mainly used in the USA became known as FPL etch, named after the Forest Products Laboratory that developed the process initially [17]. Based on research by Bethune [18], the process is improved to become the "optimised FPL etch," which is closer to the composition of the CSA etch. It is essential in these etching procedures to use process conditions to obtain good adhesion.

Fig. 8.6 shows surface configurations, and the adhesion quality measured by peel strength for various process conditions in sulfuric acid-sodium dichromate solutions. It is clear that the microscopic-etch pitsurface morphology showed optimal adhesion. In both CSA and FPL, a good microstructure is obtained with a sufficient level of chromate ions and some solution ageing is present with Al and Cu ions.

Attempts have been made to replace the oxidizing power of the chromate ions by other oxidizing components. The application of ferric sulfate in sulfuric acid solution was successful [19, 20]. This process, known as the P2 etch, shows a similar microscopic surface morphology as the CSA and optimized FPL etch.

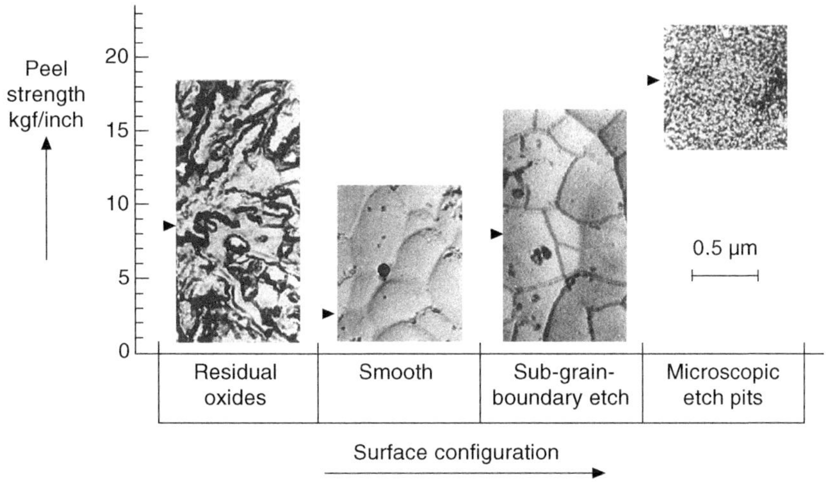

Fig. 8.6 Microscopic surface configuration after various etch process conditions in sulfuric acid-sodium dichromate solutions.

Using a low anodic voltage in combination with sulfuric acid at 50 °C [21] or phosphoric acid at room temperature will result in similar microstructures and good initial adhesion. However, the durability of the etched surfaces is limited in combination with epoxy adhesives. The etched surface treatments showed poor long-term durability with epoxy adhesives in aircraft operational use, especially on clad alloys. Alternative etching procedures based on acid or alkaline solutions can be used but will generally result in lower adhesion.

8.3.1.2 Anodizing processes for aluminum alloys

To improve durability the etching treatment has to be followed by an anodizing treatment. This is an electrochemical method for creating a surface structure suitable for adhesion. The created oxide layer consists of many pores (see Fig. 8.7) in which the adhesive is able to penetrate, before it cures completely. This results in so-called mechanical interlocking or hooking of the adhesive in the substrate. In addition, the total bonding area is enlarged by the porous structure.

Several different anodizing processes are available. The CAA process is used extensively by the aerospace industry in Europe [22, 23] and has a proven track record of long-term durability in combination with both clad and bare aluminum alloys. In the United States, another CAA process was used for adhesive bonding with partial sealing in a chromate-containing rinse after anodizing [20]. After the durability problems with FPL-etched and bonded components in the USA, the phosphoric acid anodizing (PAA) was adopted [24]. Where the CAA coating is thicker, much less vulnerable and gives better corrosion protection than the PAA layer, it has one major disadvantage: in the process, hexavalent chromium is used, which is an environmentally unfriendly chemical.

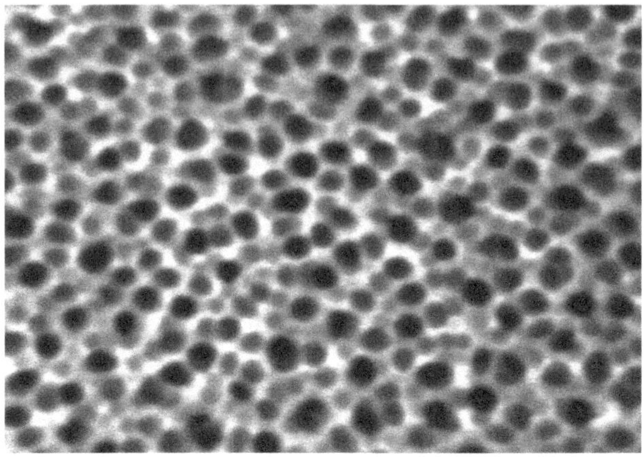

Fig. 8.7 Electron micrograph of CAA oxide (average pore diameter 30 nm).

The sulfuric-acid anodizing (SAA) process widely used for corrosion protection of aluminum alloys is less suited for structural adhesive bonding with rigid adhesives. The pores are narrower (10 nm), so adhesives cannot fully penetrate resulting in relatively low-strength interfaces. However, in industrial applications combined with flexible adhesives, thin sulfuric-acid anodic layers are applied successfully.

The above mentioned CAA and PAA anodizing treatments for aluminum generally result in good initial strength of the joint, as well as excellent durability (especially when an extra primer is applied to the surface, which acts as a corrosion inhibitor). The anodizing process of aluminum is however a relatively complex and expensive process. This has often led to the conclusion that adhesive bonding of aluminum is not economically viable in applications outside the aerospace industry. However, there are developments in surface treatment specifically aimed at reducing the cost, without sacrificing bond strength and durability too much. In this context, conversion coatings have to be mentioned.

8.3.1.3 Conversion coatings for aluminum alloys

Chromate conversion coatings are traditionally used in the corrosion protection and paint-pretreating industries, with excellent results for corrosion protection. Because of the different loading situation on the adhesive compared with coatings, it is not possible to use these systems for adhesive bonding, especially when peel stresses are acting on the adhesive. The performance is relatively weak and low joint strengths are obtained when these traditional conversion coatings are used.

Much better results are found when more recent conversion coatings, based on titanate and/or zirconate, are used [25, 26]. The peel performance is enhanced and the durability of such systems is good.

8.3.1.4 Chromium-free anodizing treatments for aluminum

As an alternative to the CAA and PAA treatments, a new process is currently under development, the phosphoric-sulfuric acid anodizing (PSA) treatment [27, 28]. This results in an oxide layer with the adhesion performance of the CAA treatment without the environmental penalty of the chromium. Current research aims to find the proper process parameters to obtain the optimal oxide layer for a wide range of aluminum alloys. Anodic layers with good adhesion and durability have been obtained in a PSA solution with 125 g/L H_3PO_4 and 75 g/L H_2SO_4 at 20–22 °C and 18–20 V for 20 min (see Fig. 8.8). Other reports [29] with other PSA process conditions also showed good adhesion and durability.

8.3.2 Surface treatment of steel and stainless steel

Steels are used in many automotive, shipbuilding and other industrial applications. In many applications steel is adhesive bonded instead of welded because of improved corrosion resistance, joining of dissimilar materials, increased joint stiffness and fatigue resistance, less heat distortion and, often, more cost effectiveness.

In the case where the adhesive-bonded joint experiences no environmental or chemical exposure, a surface treatment for degreasing and cleaning thoroughly may be sufficient to provide a medium-strength bonded joint.

8.3.2.1 Surface treatment of carbon steel

Brockmann [30] suggests that, in contrast to aluminum and titanium alloys, where the surfaces are usually treated by chemical methods, etching procedures for different types of steel are not recommended. Good bonding results are usually obtained by using abrasive or mechanical roughening techniques, such as grinding or grit-blasting.

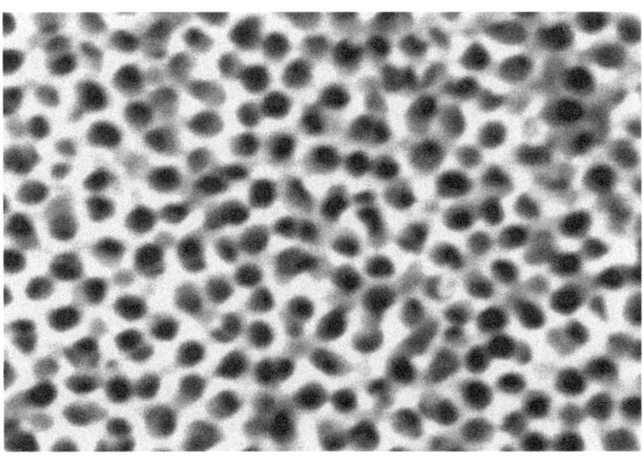

Fig. 8.8 Electron micrograph of PSA oxide (average pore diameter 30 nm).

The best results are obtained with 99.5% pure alumina grit (Al_2O_3, corundum) with a particle size between 150 and 250 μm. It has to be performed on a clean dry surface in order to prevent contamination of the grit-blasting medium with organic material. The grit-blasting should be carried out with equipment provided with oil- and water-separators. After the grit-blasting process, any dust on the surface has to be removed by blowing off the surface with dry and oil-free compressed air.

Abrasive treatments of thin sheet metal may result in warping. In which case an acid-etch solution might be preferred. A successful method is an etching technique in a nitric-phosphoric acid solution at room temperature for 5–7 min that produces a micro-rough surface morphology and good adhesion and durability on carbon steels. The solution in deionized water contains 25 g/L H_3PO_4 and 2 g/L HNO_3 with a small addition of a suitable surfactant [20].

As with corrosion-resistant conversion coating on steel, generally phosphate layers are used as a pretreatment for paint adhesion. These conversion coatings will result in good adhesion but low strength in bonded joints. The cause is the low strength of the phosphate layer, which will fracture when the bonded joint is loaded.

8.3.2.2 Surface treatment of stainless steel

Stainless or corrosion-resistant steels (CRES) are steel alloys containing over 11% chromium. They are applied in various types of instruments and appliances, industrial equipment, as an automotive and aerospace structural alloy and construction materials in large buildings, for their decorative properties and chemical and corrosion resistance.

Abrasive treatments used as a surface treatment for adhesive bonding with carbon steels do not have good results with stainless steels. Grit-blasting with alumina improves adhesion, but it is also detrimental to the passive layer that protects the stainless steel against corrosion. It can sometimes be used for applications that are not exposed to moisture or a corrosive environment.

A number of chemical and electrochemical processes improve adhesion on stainless steels. Various strong-acid etchants are sometimes used to improve adhesion, resulting in carbon smut layers on the surface of the stainless steel. By brushing off the black deposit or by desmutting in a passivation solution, high-strength bonds can be obtained. However, the peel strength of passivated layers is sometimes low.

After degreasing, an oxalic (100 g/L)-sulfuric (100 g/L) acid mixture at 90 °C can be used to etch the surface [5, 30], followed by smut removal by brushing off the deposit. Another process often used to obtain good adhesion is a sulfuric-acid etch followed by desmutting and passivating in a sulfuric acid-sodium dichromate mixture [5, 20, 31, 32]. The etch process is best performed at 80 °C for 10 min in a solution of 250 g/L H_2SO_4, and the smut removal in a solution of 300 g/L sulfuric acid and 30 g/L sodium dichromate at 65 °C for 5 min.

Good adhesion and durability results are also reported with a highly concentrated mixture of sulfuric acid and sodium dichromate at 80 °C for 60 min [31, 32]. The toxicity of these chromate-containing etching mixtures is however a major drawback.

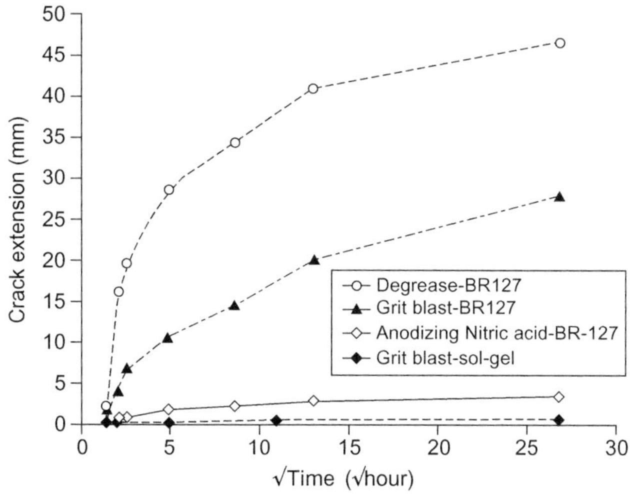

Fig. 8.9 Crack extensions of wedge test specimens of stainless steel bonded with epoxy film adhesive after various surface treatments.

A surface treatment method that does not have this disadvantage is the nitric-acid anodizing process [31, 32]. After degreasing, the anodizing is performed at a current density of 0.5 A/dm^2 in a 45–50% volume nitric-acid solution at 50 °C for 60 min. The adhesion and durability of bonded joints on surfaces formed in this process are excellent. The surface of the stainless steel has a microporous morphology and is chromium enriched. Fig. 8.9 shows the results of wedge test specimens exposed in a salt-spray cabinet (ASTM B117) for up to 30 days. Four surface treatments are compared: alkaline cleaning; alkaline cleaning followed by grit-blasting with alumina and cleaning; alkaline cleaning and anodizing in nitric acid; cleaning, grit-blasting and cleaning followed by AC 130 sol-gel treatment (see Section 8.3.4). After surface treatment the specimens are primed with a corrosion-inhibiting bond primer, except the sol-gel treated specimens. All specimens are bonded within 2 h after surface treatment with an epoxy adhesive film and autoclave cured at 120 °C. The anodized and the sol-gel-treated specimens showed excellent durability with only cohesive and very limited crack growth. Although grit-blasting resulted in slower crack growth compared with only degreased specimens, both treatments resulted in interfacial delamination.

8.3.3 Surface treatment of titanium

Titanium and its alloys have been used in the aircraft industry because of their low density and good high-temperature properties and nowadays in carbon composite structures as local reinforcement for improved bearing strength. Titanium is sometimes used in industrial and medical applications for its very good corrosion and chemical resistance. A range of surface treatments of titanium has been developed over the years [5, 20, 33]. Treatments that give the titanium surface macro- and micro-roughness show good results in initial adhesion and durability.

8.3.3.1 Etching and conversion treatments

Early etching treatments based on nitric-hydrofluoric acid etching provided adequate initial adhesion but very poor durability. Phosphate-fluoride came in use as an etching and oxide conversion layer, but under hot wet exposure the durability results were poor. The oxide layer is made more stable by the modified phosphate-fluoride process. Further improvements were made with commercial alkaline etchants, Turco 5578, DAPCOtreat and acid treatment Pasajell 107, which provided more macro-roughness to the surface of the titanium.

Titanium treated for 20 min in an alkaline peroxide etch, an oxidizing mixture of sodium hydroxide (2%) and hydrogen peroxide (2.2%) at 50–70 °C, results in micro-roughness and good bond durability. The problem with this process is the instability of the hydrogen peroxide at the elevated temperature of the solution.

8.3.3.2 Anodizing surface treatments

A number of anodizing treatments have been developed that outperform all other treatments. Generally an etching process is used to remove old oxide scales before anodizing.

Boeing developed a CAA process containing hydrofluoric acid [34]. Anodizing is performed in a solution of 50 g/L chromic acid for 25 min at 5 V at room temperature. The hydrofluoric acid is added to obtain sufficient current density (0.2 A/dm^2) in the process. This process results in surfaces with a fine microstructure and good adhesion, and bonded joints show excellent durability.

Fokker uses a chromic acid solution (40 g/L) such as that used for anodizing aluminum alloys without adding fluorides. Anodizing is performed at 50 °C for 40 min at 15 V and bonded joints also show excellent durability.

Anodizing in alkaline peroxide solutions has also been used successfully as a surface treatment for durable bonded joints. Further research [35] indicates that anodizing in a solution of sodium hydroxide (200 g/L) without hydrogen peroxide showed excellent durability. The anodizing in this process is performed at 10 V at room temperature.

8.3.4 Development of sol-gel surface treatment for aluminum, steel and titanium

The increased continuation of the life of both military and civil aircraft has resulted in a need for improved repair techniques. Conventionally, the structure is repaired by removing the cracked area and by riveting a patch on to the structure. However, the rivets act as stress concentrations and will limit the life of the repair. A more efficient method is the adhesively bonded patch repair (see Section 8.4.9).

In a bonded repair solution, surface treatment on the aged aircraft structure is crucial in order to obtain good adhesion and durability of the repair [36]. The conventional approaches for preparing metal surfaces for bonding (etching and anodizing) are difficult to implement on existing structures. Mechanical abrasion of the metallic structure only is not sufficient for a durable bonded repair. In situ surface-treatment processes have been developed to obtain sufficient chemical modification of the surface. By using chromic acid or phosphoric acid in gel form local-area anodizing in a so-called brush anodizing process, thin anodic layers can be formed on aluminum alloys providing good adhesion and durability.

Good results are also obtained by a PAA containment system (PACS) using a double vacuum bag to contain the processing liquids [37]. These acid processes have a number of drawbacks such as leakage, spillage and corrosion initiation in crevices and between dissimilar metals such as fasteners. AMRL Melbourne developed the process often referred to as the gritblast silane (GBS) process [38, 39]. After cleaning, the area to be prepared is grit blasted with alumina. After removal of the blasting debris the surface is treated with an aqueous solution of an organo-functional silane-based coupling agent. The coupling agent found to be most suitable for epoxy adhesives is the epoxy-terminated silane, γ-GPS [38]. In combination with a corrosion-inhibiting bond primer and an epoxy film adhesive, durable bonds to aluminum alloys, steel and titanium alloys are obtained.

A more recent process has been developed at Boeing based on sol-gel chemistry [40]. After cleaning, abrasion (either grit-blast or abrasive paper) and removal of abrasive residues, the sol, which is prepared from a mixture of glycidoxyl functional silane and zirconium alkoxide, is applied. The thin film formed on the surface provides chemical bonds to the metal side, and has active sites that bond during cure with an epoxy primer. Good durability results are obtained in combination with a water-based corrosion-inhibiting bond primer [41, 42]. The sol-gel material is commercially available as AC-130 from AC TECH, Inc. Fig. 8.9 shows the excellent results of a wedge test durability experiment on the sol-gel treatment on stainless steel, even without a primer. The durability result is excellent on aluminum and titanium alloys and compares well with good anodizing treatment.

8.3.5 Developments in bonding primers

Adhesive primers generally function to conserve the surface of a material that has to be bonded in a later stage, thus providing more flexibility in the manufacturing process. Generally, adhesive primers can be considered to be a strongly (with organic solvent) diluted adhesive, often combined with a coupling agent such as a silane. They have the main function to wet the freshly prepared surface easily and, after drying or curing, to stabilize the surface until the adhesive is applied (which may take as long as a year). For structural bonding most primers are epoxy-based, available as a liquid and are sprayed onto the surface as a layer to a thickness of about 4–10 mm. One-component primers and two-component systems are both available. One-component primers have to be cured at elevated temperature, either precured or cocured with an elevated-temperature curing adhesive. Two-component primers have to be mixed before application.

A corrosion-inhibiting compound, usually a chromate, is sometimes added to the adhesive primer formulation to protect the adherend against corrosion. To minimize production of volatile organic compounds, similar water-based versions have been developed. A steady change from solvent-based to water-based primers has taken place, driven by ecological concern. In addition, the small amount of a hexavalent chromium salt in corrosion-inhibiting primers makes them environmentally unfriendly. Recently primers have been developed that are chromate free and contain other corrosion inhibitors.

8.4 Developments in joint design

In contrast to other joining methods, such as riveting and bolting, adhesive bonding has no adverse effect on the material characteristics of the surfaces to be joined, for example, there are no holes that damage the joined parts and create stress concentrations. As can be seen in Fig. 8.10, the load transfer is uniformly distributed along the bonded joint. By preference, the load is transferred by shear stresses in the adhesive layer, whereas tensile, peel or cleavage loads should be avoided or minimized as much as possible. The strength of the joint not only depends on the shear strength of the adhesive itself, but also on the shear and peel stress distribution along the length of the overlap.

8.4.1 Shear stress distribution of bonded overlap joints

The metal adherends are characterized by their stiffness, modulus of elasticity and can even show plastic deformation at high stress levels. Because of this finite stiffness the adherends will elongate under a load and this elongation (strain) is proportional to the applied load (Fig. 8.11A). As the adhesive transfers load along the overlap, the axial strain in one adherend is gradually reduced and increases in the other adherend. When both adherends have the same axial stiffness, in the middle of the overlap both adherends have the same strain, whereas at the ends a large difference exists. This causes the adhesive layer to deform additionally and the shear stress distribution shown in Fig. 8.11B is obtained. This mechanism is generally referred to

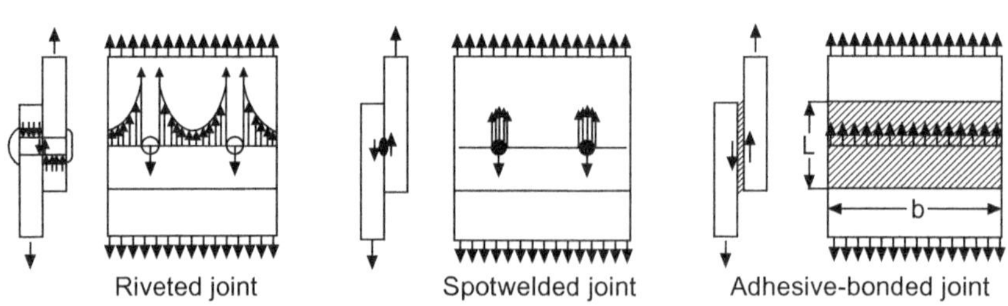

Fig. 8.10 Comparison of load transfer in various types of joints.

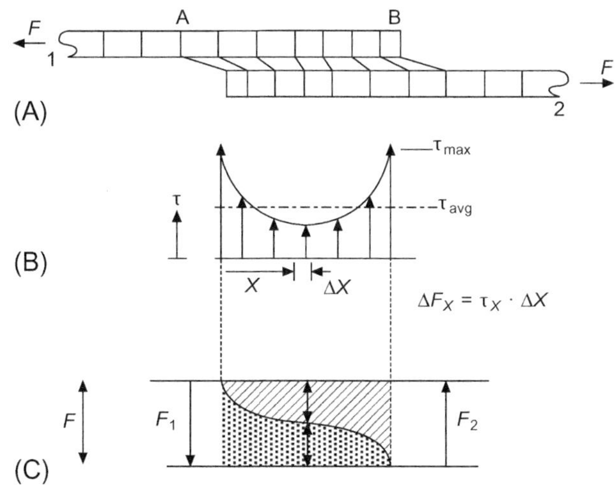

Fig. 8.11 Elastic deformations of the adherends resulting in peak shear stresses in the adhesive. (A) Lap joint; (B) shear stress; (C) load transfer.

as shear lag. The highly stressed ends of the overlap will transfer most of the load between the two adherends, whereas the lower stressed middle part contributes far less (see Fig. 8.11C). This explains why the nonuniform shear stress distribution has an unfavorable effect on the efficiency of the bonded joint. Joint failure is determined by the stress concentrations at the ends of the overlap, which depend on joint geometry, adherend material and type of adhesive.

8.4.2 Effect of joint geometry, material and type of adhesive for lap joints

Following the principle of shear lag, an increase in adherend stiffness, either by more thickness or a higher Young's modulus, results in more uniform distribution of shear stresses along the overlap length, accompanied by lower stress concentrations at the ends. Thus the joint efficiency is increased. At high load levels the yield stress of the adherend will be reached, thereby (locally) reducing its stiffness. The strains in the adhesive will increase more than proportionally with the increased load, which has a disadvantageous effect on the stress distribution and thus on the joint efficiency. Joint strength does not increase proportionally with the overlap length, as the middle part of the bonded joint becomes less and less effective in transferring load (see Fig. 8.12).

Joint efficiency decreases, as the average shear stress decreases, whereas the stress concentrations at the end of the overlap remain the same. More flexible adhesives provide joints with a more uniform stress distribution, that is higher joint efficiency, but stiffer adhesives generally provide greater strength by virtue of their higher degree of cross-linking. Asymmetrical lap joints in which the adherends are not of equal stiffness and show a non-symmetrical stress distribution, have the highest stresses at the

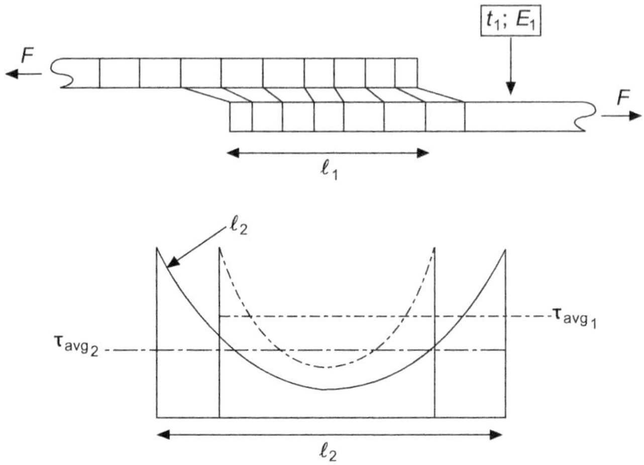

Fig. 8.12 Effect of the overlap length on the shear stress distribution.

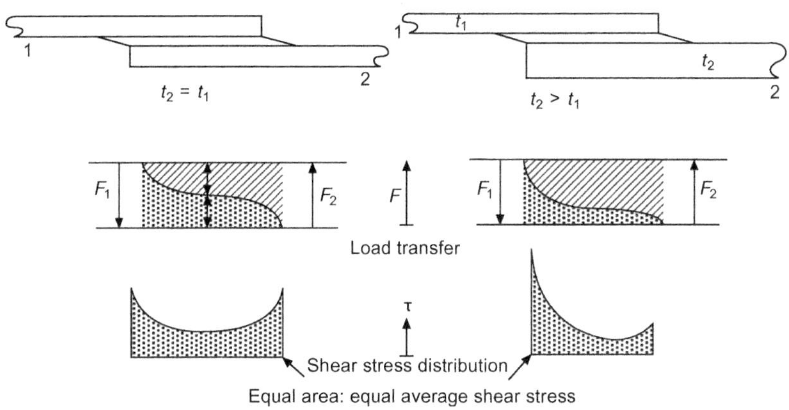

Fig. 8.13 Effect of unequal axial stiffness adherends on the shear stress distribution.

side of the less-stiff adherend, owing to the higher axial strain (see Fig. 8.13). Obviously the joint efficiency is lower than for the equal-stiffness lap joint.

8.4.3 Eccentricity in overlap joints

Eccentricity of the loading forces on an overlap joint results in bending deformation of the joint, thereby introducing additional normal stresses in the bondline. These stresses have a tendency to peel the two adherends apart (see Fig. 8.14). With increasing adherend thickness the eccentricity increases, that is slightly higher peel stresses are introduced, which partly counteract the positive effect that the larger adherend thickness has on the shear stress distribution. More importantly, the induced bending

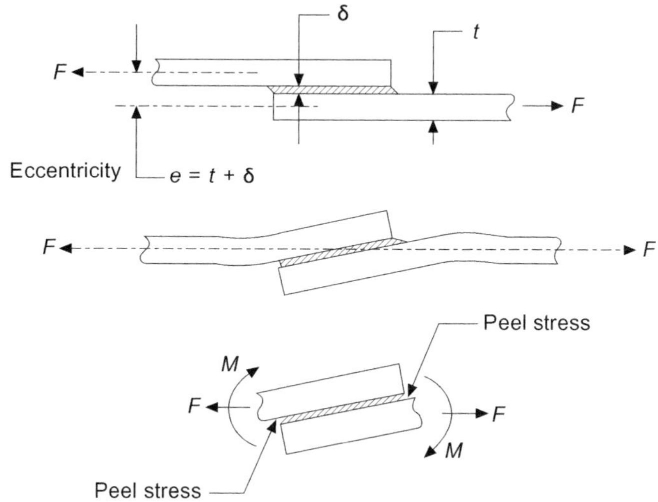

Fig. 8.14 Peel stresses at the edge of the bond layer caused by eccentricity in the joint.

deformation also causes a stress concentration in the adherend just in front of the thickness change. This stress concentration, which increases with the thickness of the bonded doubler and the stiffness of the adhesive, can be detrimental to the fatigue strength of the structure.

8.4.4 Joint optimization

Joint design can be improved by adapting the geometry such that lower shear and peel stress peaks are obtained. Decreasing the eccentricity in the joint leads to lower bending moments at the end of the lap joint, and thus lower peel stress. Replacing the single lap joint by a double lap joint or double strap joint can reduce this eccentricity. Designing adherends for constant elongation in the joint area will diminish the shear stress peaks and result in constant load transfer over the length of the lap joint. In order to obtain a constant deformation in the adherend over the length of the overlap, the axial stiffness of the adherend should decrease linearly. This can be effected by beveling the adherend over the length of the joint from the undisturbed thickness to zero. The ideal lap joint is the result of a combination of tapering and bringing the adherends in line; the scarf joint (see Fig. 8.15).

The lack of eccentricity results in peel stresses, and the shear stress is constant owing to the linear decrease in thickness of the adherend. However the bondline remains loaded under a small tension stress owing to the angle between the bondline and the adherends. This angle should be relatively small in order to limit the level of the tension stress in the adhesive. The application of tapered adherends is limited owing to manufacturing and cost constraints. Sometimes tapering is approximated by stepped adherend edges or stepped laminates.

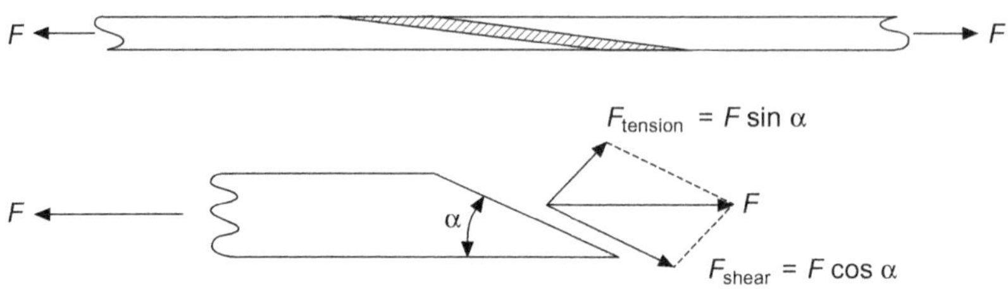

Fig. 8.15 The scarf joint.

In Fig. 8.16 an example is given of a bonded-edge reinforcement for load introduction. An optimized step lamination is used to reduce the nominal stress from 200 MPa in the basic sheet down to 50 MPa at the edge. With a step thickness of the first doubler of 20% of the basic sheet, the secondary bending and the stress concentration are kept acceptably small. A similar effect can be obtained by doubling the doubler thickness at each next step.

The resulting higher stress concentration is compensated by the reduced nominal stress caused by the increase in total thickness of the package. The minimum step length required to redistribute the stress evenly over the laminate thickness after the step has been determined by photo-elastic research [43]. For a stiff adhesive, the step length has to be at least 10 times the added thickness step. In practice this step length is generally much longer for reasons of quality control, acceptable defects and reparability.

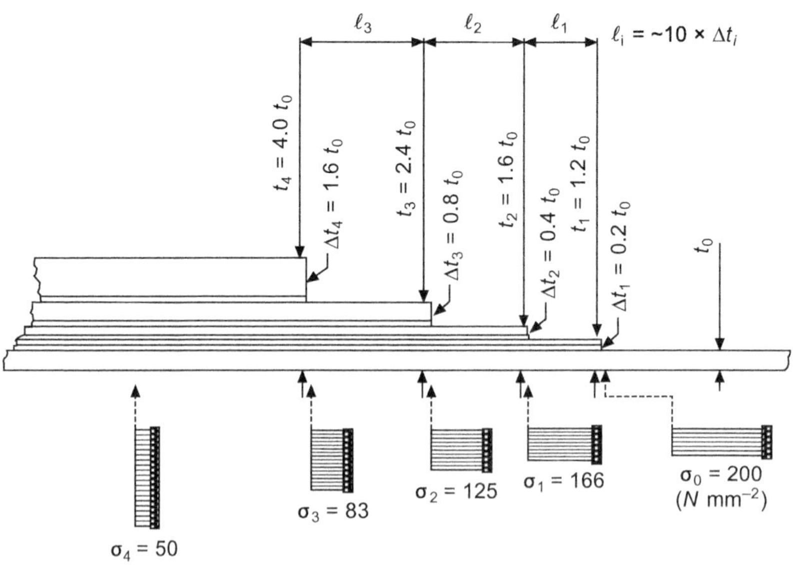

Fig. 8.16 Adhesive-bonded stepped lamination with equal stress concentration at each step.

8.4.5 Adhesive-bonded laminates

Adhesive bonding of thin layers of a material to build up a laminate has been used with various types of wood (plywood) for thousands of years. By laminating wood, the best properties can be obtained using the highest quality materials available; it reduces prices and improves the stability of structures. By combining laminated-wood fiber orientations, properties could be tailored similar to those of advanced composites and hybrid laminates today. Also, by bonding laminations in metal, better properties are obtained than for monolithic materials. Structural reinforcements can be placed where needed, optimizing both weight and component cost. This adhesively bonded build-up has a big advantage over integrally machined structures with regard to fatigue properties. A fatigue crack initiated in the skin will be retarded in its growth when it reaches a doubler or stringer flange. Furthermore, it will take time to initiate cracks in the doubler or the stringer flange. In integrally machined-stiffened skins a crack will be retarded to some extent by thickness steps or stiffeners, but the panels will have a considerably larger weight when designed for the same inspection interval or fatigue life.

In metallic structures a bonded laminated skin benefits from the improved properties of a thin sheet compared with thick monolithic material. An aluminum laminate has better fatigue properties than a solid material of the same thickness. First, a metal laminate shows longer fatigue life because the thinner sheet material shows a lower crack growth rate. Second, a fatigue crack in an outer layer will not start a crack immediately in the next layer. There is crack arrest in the bond layer for part-through cracks (see Fig. 8.17).

In thick monolithic material the plastic deformation at the crack tip is restricted because the surrounding material limits contraction in the thickness direction, that

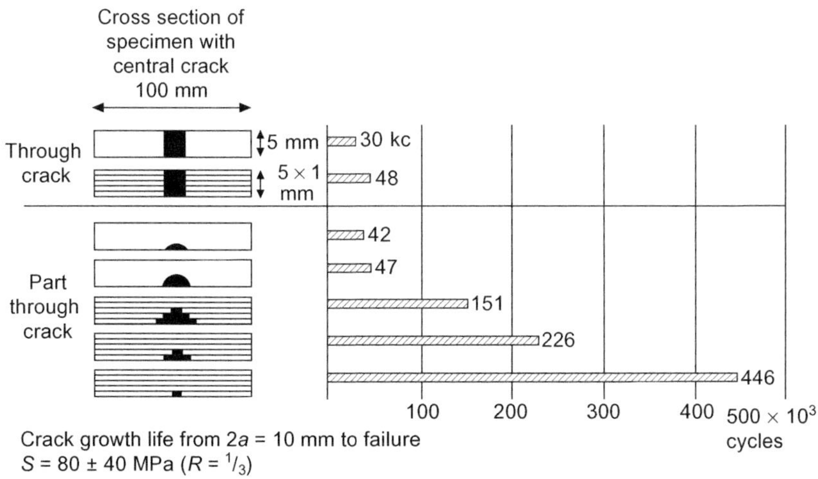

Fig. 8.17 Fatigue-crack-growth life improvement of laminated specimens with full and part-through central cracks in comparison to solid specimens [44].

is a plain strain condition, whereas in thin sheets this contraction is not limited, that is a plane stress condition. The plain strain condition in thick material results in secondary stresses at the crack tip, which results in a higher crack growth rate at the same fatigue load, as well as lower fracture toughness if compared with thin sheets. In a laminate the relatively low modulus adhesive material allows unrestricted contraction of the individual sheets, and the laminate can benefit from the better properties of thinner sheets.

Fig. 8.17 shows the difference in fatigue-crack growth between solid 2024 T3-clad aluminum and laminated aluminum with through cracks. The improvement is far more significant for part-through cracks, i.e., only one, two or three layers with a through crack and the remaining layers intact [44]. This shows the advantage for laminated material, in the situation in which the fatigue crack has to be initiated in every layer. In a 'damage-tolerant' design, the behavior in the cracked condition is also important. The higher fracture toughness of thin sheets results in higher residual strength for laminated material [45].

Fig. 8.18 shows favorable results for laminates in AA 7075-T6 in compact-tension specimen tests, compared with monolithic more damage tolerant AA 2024 T3 material. This advantageous effect is independent of the type of adhesive. Only in the case of a very low stiffness bond, such as use of an interfaying sealant, the separate lamellae are allowed to buckle, resulting in lower toughness values.

The fracture of a cracked bonded laminate clearly shows the ductile behavior, whereas the solid specimen shows a flat brittle fracture surface (see Fig. 8.19). The wing skins of Fokker and SAAB aircraft have this laminated structure, with the advantage of improved fatigue and damage tolerance in the highly loaded area of the wing box. Fig. 8.20 shows an example of a lower-wing-skin cross-section near the root of the outer wing. Although laminates provide the stated improvement in fatigue-crack

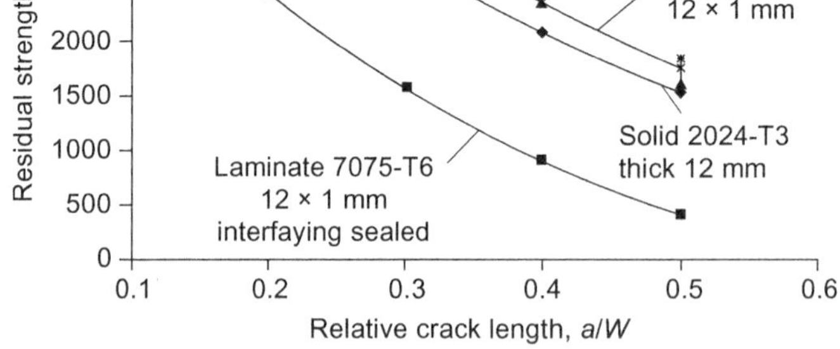

Fig. 8.18 Residual strength of laminated AA 7075-T6 material with three adhesives (▲, FM123-5; ×EC2216; × FM 1000) compared with solid AA 2024-T3 specimens, where a is the crack length and W is the width of the test specimen.

Fig. 8.19 Fracture surface of bonded aluminum-alloy laminate compared with solid aluminum-alloy specimen.

Fig. 8.20 Detail of cross-section of Fokker F28 outer wing skin.

initiation and high resistance to fracture, the fatigue-crack-growth behavior of laminates in through cracks is only slightly improved compared with the monolithic plate. The fatigue and fracture toughness properties are further improved by incorporating high-strength fibers into the adhesive layers.

8.4.6 Fiber metal laminates (FMLs)

Research at Delft University of Technology further optimized bonded laminates by incorporating high-strength fibers for fatigue and damage-tolerance properties. ARALL, Aramid Reinforced Aluminum Laminates, was the first material developed targeted primarily for wing applications, having all fibers in the spanwise direction.

A potential weight saving of 30% can be achieved for the lower wing skin panel compared with a bonded wing design [46]. An application of ARALL in the C-17 cargo-door skin showed a weight saving of 26% [47]. The development of ARALL was followed by the development of GLARE (GLAss-fiber REinforced aluminum laminate), which further improved the properties with the use of high-strength S2 glass fibers. For fuselage applications a dedicated variant was developed, called Glare 3, with biaxial fiber layers. The excellent fatigue properties of a fiber metal laminate (FML) are owing to the fact that the high-strength fibers "bridge" the crack. Loads from the cracking metal layer are transmitted via the adhesive to the fiber, unloading the metal layer and slowing down crack growth (see Fig. 8.21). Fatigue loading causes adhesive delamination to occur around the crack, preventing the fibers from breaking. Generally FMLs consist of thin metal layers (0.2–0.5 mm) that allow more fiber layers in order to reduce the shear stress in the adhesive between the fibers and the metal. This controls the delamination and subsequently the crack growth characteristics.

Fig. 8.22 shows the fatigue-crack-growth characteristics of two types of GLARE® laminates compared with AA 2024-T3. All the fibers of Glare 2 are in one direction perpendicular to the crack, and Glare 3 has cross-ply fiber layers. In addition to good fatigue and residual strength properties, very good impact properties are also found compared with solid aluminum, owing to the fibers supporting the metal.

The stress-strain curve (see Fig. 8.23) shows considerable capability to transform energy into plastic deformation (area under the stress-strain curve). With the development of GLARE® and improvements in manufacturing technology [48, 49], the first large-scale application of FML material in a civil aircraft, the Airbus A380, was realized. Manufacture of these larger panels was made possible by the introduction of

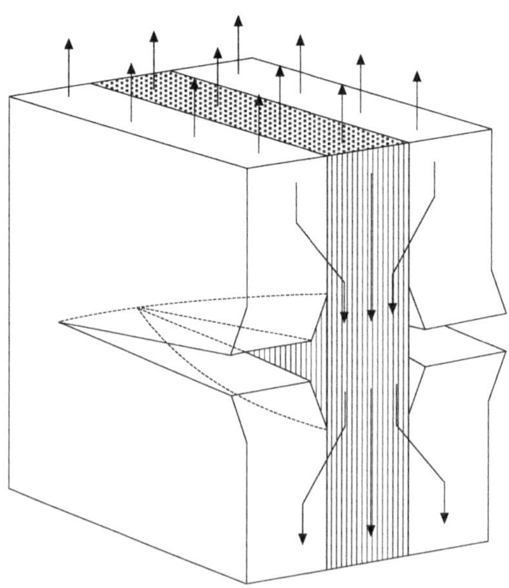

Fig. 8.21 Bridging fatigue cracks with high-strength fibers.

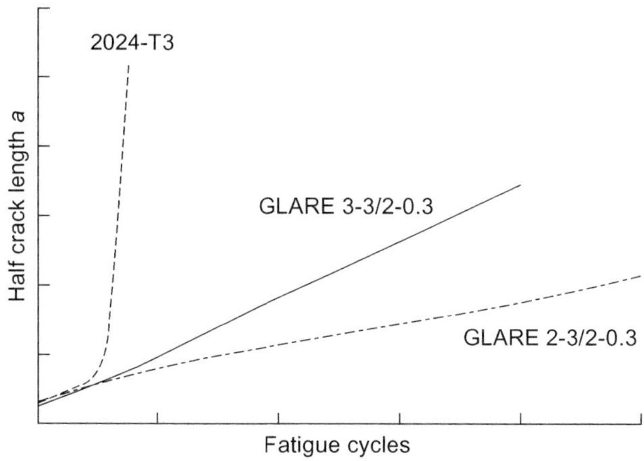

Fig. 8.22 Fatigue-crack growth of GLARE® compared with AA 2024-T3.

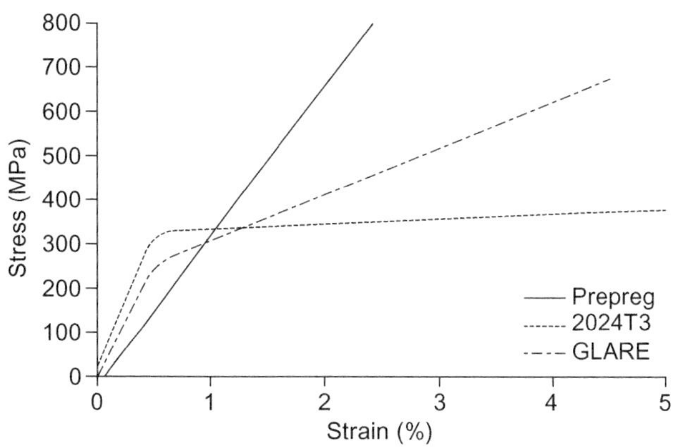

Fig. 8.23 Stress-strain curve of GLARE®.

splices in the panel to overcome the limitations of thin-sheet width dimensions. The high-strength glass-fiber-reinforced adhesive layer is continuous in the splice area, whereas the thin metal sheets are joined by overlaps.

Fig. 8.24 shows a fuselage panel with the bonded-fiber metal laminate GLARE®, reinforcing doublers around the door cut-out and adhesive-bonded stringers. Next to these large fuselage panels, the leading edges of the vertical and horizontal tail plane are also made of the FML material, because of its good impact properties [50, 51] and protection against bird strikes.

Fig. 8.25A shows one of the GLARE® leading edges of the vertical tail plane with cut-outs for the location of antennas. Fig. 8.25B shows the result of a bird-strike test on a typical leading-edge laminate.

Fig. 8.24 Large fuselage panel with spliced GLARE® skin with bonded doublers and stringers.

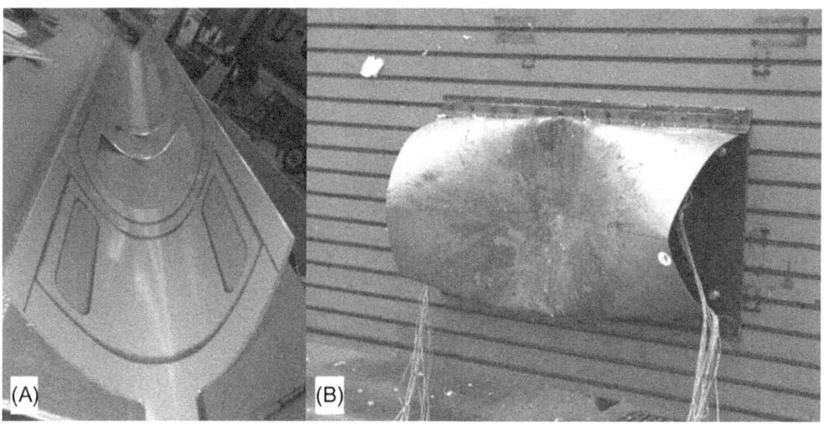

Fig. 8.25 VTP leading edge. (A) One of the LE sections; (B) bird-strike test result.

8.4.7 Weight and cost reduction

Weight and cost reductions can be obtained in advanced designs, owing to the improved properties of adhesive-bonded laminates and fiber metal laminates, compared with conventionally riveted monolithic structures. Cost can be reduced for designs in GLARE® by integrating local reinforcements, doublers, splices and thickness steps into a one-shot cured large panel. Automation of lay-up of thin aluminum sheets and adhesive prepreg will further reduce costs. No preforming of thin sheets is needed as the lay-up will follow the contour of the mold and take on the (compound) curvature of the mold after consolidation. This integrated manufacturing approach leads to a decrease in manufacturing costs. The higher price of the constituent materials is balanced by this manufacturing efficiency and by the weight saving. Weight savings between 12% and 26% are mentioned [47, 48, 52] owing to the higher allowable design stress, integrated splices and efficient designs.

8.4.8 Sandwich structures

A sandwich is built up from two face sheets with high mechanical properties, which are bonded to a low density core, generally of aluminum honeycomb or foam. This type of structure has a high bending stiffness in the longitudinal and, in contrast to a stringer-sheet-panel structure, also in the transverse direction. The relatively thin sandwich skins are perfectly supported by the core so that local buckling of these skins under compressive loads is prevented. These favorable stiffness properties make the sandwich structure especially effective for compression panels. The application of sandwich panels in aircraft is nowadays mainly restricted to secondary applications, owing to problems for highly loaded sandwiches caused by the complexity of panel coupling, attachments to sandwich panels and durability problems of some of the bonded metallic sandwiches.

Sandwich structures are often used in control surface applications. The use of thin, lightweight skin sheets bonded to low-weight core materials, enables a structure to be obtained that has a high bending and torsion stiffness that maintains accurately the aerodynamic shape under load, thanks to the perfect stabilization of the skin against buckling. Other aircraft parts, which are usually built in the sandwich structure, are the trailing-edge panels of the wing box and cabin interior parts and floor panels. Most of these parts nowadays are no longer metal structures, but composite panels with a Nomex® honeycomb core. In space applications, such as satellite and solar-array structures, sandwiches provide the stiffness required, originally in thin-skin metallic sandwiches, but nowadays in aluminum honeycomb cores with carbon-fiber composite skins.

In other industries many sandwich applications can be found as walls and panels in semitrailers, trains, building and construction, ship building and container structures. Sandwiches are sometimes also applied as energy-absorbing structures.

8.4.9 Bonded repairs

Aging aircraft structures require safe, damage-tolerant and cost-effective repair techniques. Fatigue and corrosion problems become an important topic in maintenance owing to the intensive use and long life of aircraft. Compared with conventional riveted repairs, bonded-patch repairs have the advantage of providing more uniform and efficient load transfer [53]. They do not have high stress concentrations at the mechanical fasteners, which nucleate new fatigue cracks leading to even larger repairs. Further, the much higher joint stiffness allows bonded patches, unlike riveted ones, to restrain crack opening and stop the fatigue crack from growing further. The bridging of the crack reduces the stress intensity at the crack tip, similar to the mechanism seen for FMLs. The patch is designed such that the repaired stress intensity factor (K) at the crack tip is below the threshold value, and stresses and strains in the bondline, skin and patch are not critical.

Design procedures developed by the RAAF (Royal Australian Air Force) and USAF mostly use the analytical model developed by Rose and coworkers [54], which is a two-dimensional continuum analysis based on the theory of elasticity. It considers an infinitely wide-center cracked isotropic plate, with a one-sided bonded orthotropic plate remotely loaded by a biaxial system. First the repair is modeled as an equivalent inclusion to calculate the stress redistribution in the plate caused by the bonded doubler. Then the crack is introduced and the stress intensity factor, *K*, at the crack tip is calculated.

The Rose model has been further extended to include bending caused by eccentricity of a one-sided patch and to include thermal stresses induced by curing the adhesive [54, 55]. Owing to the shift in neutral axis, single-sided patches inevitably induce bending stresses, which can be as high as 50% of the stresses at the critical locations [53]. The calculation of secondary bending requires (geometrically) non-linear analysis, as linear analysis largely overestimates bending stresses.

The neutral line method (NLM) provides a geometrically nonlinear closed-form solution. Secondary bending stresses can be reduced by gradually changing the neutral axis, that is by tapering or stepping the edge of the patch. The difference in the coefficient of thermal expansion (CTE) between patch material and parent material plays a crucial role in patch effectiveness, considering that temperature differences between cure and operating temperatures can become as high as 180 °C. Thermal residual stresses for low CTE patches are compressive in nature, both in the patch and in the skin near the patch edge, but are tensile in nature in the skin at the crack tip. The latter lowers the crack-tip stress intensity factor reduction and the patch becomes less effective at lower operating temperatures [53]. As only the repair area is locally heated, the surrounding structure restrains its expansion, which has been modeled by Fredell [55] using fully clamped edges of the patch area, and by Wang/Rose (see page 137 in Ref. 54) with a distribution of springs.

A complex analytical model is available in specially developed software packages, such as CalcuRep or CRAS, which calculate the critical design parameters as follows:

- the repaired stress intensity factor (*K*) at the crack tip
- the maximum stress in the patch (at the crack)
- the maximum skin stress (at the patch tip)
- the maximum shear strain in the adhesive
- the load transfer length in the bondline
- thermal residual stresses from curing
- bending stresses caused by eccentricity

The patch material and geometry can quickly be optimized using an iterative design procedure, based on conservative engineering guidelines and past experience. This allows non-specialists, such as maintenance engineers, to design safe and damage-tolerant bonded repairs quickly. The crack patching bonding technique was first applied to the fatigue-critical D6AC-steel wing pivot fitting of the F-111 bomber, and later to many other aircraft, such as the C-141, C-5A, F-16, F-18, Mirage and Lockheed Tristar [54]. Repair materials mainly studied and used by the RAAF and USAF are carbon- and boron-reinforced epoxy, as their high stiffness makes thin patches possible. The application of bonded GLARE patches has been extensively investigated in a joint cooperation between the USAF and Delft University of Technology, as this repair material has a much lower CTE mismatch from the parent (metal) structure [53, 56].

8.4.10 Bonded window frames

Passenger windows require adequate reinforcement around the edge of the cut-out, which is in a highly loaded area of the fuselage. In a typical conventional design a window pan (T-shaped forged frame) is riveted to the skin using about one hundred fasteners per frame. It is obvious that drilling so many holes in such a heavily loaded area weakens the skin significantly with the danger of fatigue-crack initiation at the rivet holes. With dozens of windows one can imagine the costs and effort required in drilling, deburring and rivet installation, as well as inspection of every rivet hole during maintenance. Bonded window frames do not require mechanical fasteners, therefore drastically reducing the amount of potential fatigue-crack initiation locations in the skin and window frame. Combining the bonding of components such as skins, doublers, window frames and stringers offers a cost-effective way of manufacturing large fuselage shells. The higher joint efficiency makes the window pan a more effective edge reinforcement, allowing a reduced frame cross-section and thinner skins. Bonded window reinforcements were first applied in the SAAB 2000 aircraft and are now in production for the Airbus A380.

8.5 Developments in modeling and testing the effectiveness of adhesive-bonded metal joints

The shear lag theory, first used by Volkersen [57], and later extended by Goland and Reissner [58] to include peel stresses, showed that the stress distribution in a bonded overlap joint is highly non-linear. As shown in Fig. 8.26, high shear and peel stresses are found at the overlap ends and low stresses in the middle, which means that almost all load is carried by the first few millimeters from the edges [59, 60]. The fact that high strain gradients are present in relatively small areas makes it most difficult to model accurately, or determine experimentally, the stresses and strains inside the bondline and in the adherends.

8.5.1 Analytical solutions

The governing differential equations for the stresses in bonded joints can be derived based on simplifying assumptions concerning the behavior of the adherends and the adhesive. Many different solutions are available that deviate with respect to the assumptions made to simplify the problem and the boundary conditions applied. Analytical closed-form solutions are possible for simple, linear geometries and linear

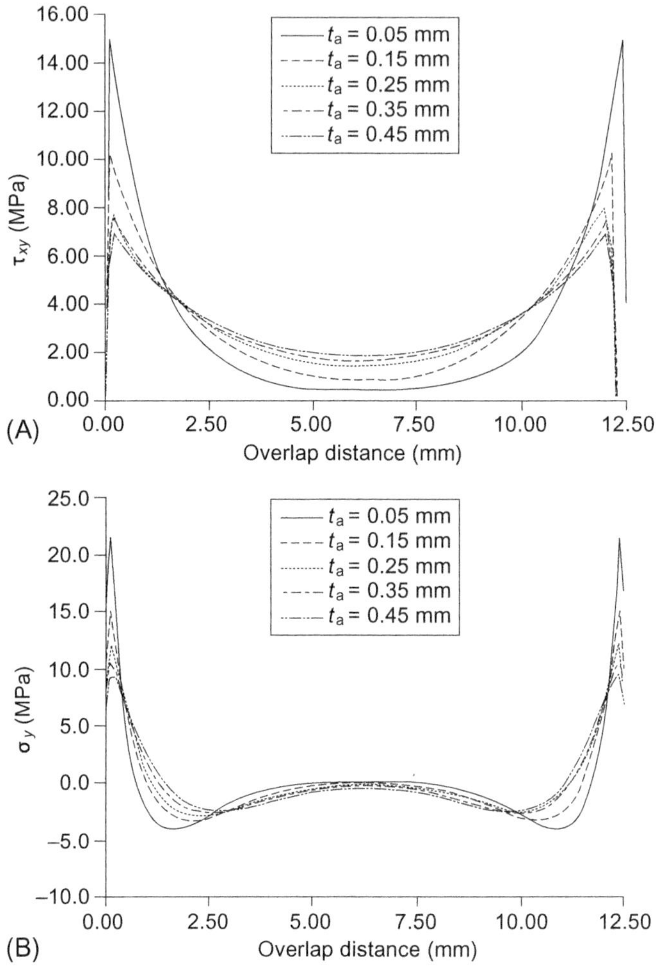

Fig. 8.26 Typical shear (A) and peel (B) stresses in a bonded lap joint.

material behavior (Gleich [61], van Ingen [62], Adams and Wake [63], Kinloch [15]). The shear lag theory published by Volkersen in 1938 [57] assumes that the adherends only carry tensile stresses and the adhesive-only shear stresses. Adhesive stresses are assumed to be constant through the thickness. Peel stresses are not taken into account, nor is eccentricity. The use of the Volkersen method therefore is very limited, yet it does give important insight into the basic understanding that the shear stress distribution in a bonded overlap joint is highly non-linear. Peak stresses arise near the ends of an overlap, whereas low shear stresses are found in the mid-section. This is accompanied by a variation of the normal stress in the adherend.

Goland and Reissner in 1944 [58] included peel stresses in their solution as well as the effect of the load eccentricity. Eccentricity results in additional shear and bending loads, which depend on the actual loading condition caused by the bending rotation of the joint. Hart-Smith [64] used a similar approach and included adhesive plasticity in shear. The adhesive is assumed to behave as a perfect elastic-plastic material and to be plastic only in small zones at the overlap edges. In these plastic zones the shear stress has a constant value, but the shear strain varies.

Ojalvo and Eidinoff [65] developed a theory in 1978 that allows for linear variation of the shear stress over the bondline thickness. The analysis for asymmetrical joints was modeled by Williams [66] in 1975 and Bigwood and Crocombe in 1989 [67]. Transverse shear effects in the adherend have been included by Delale et al. [68] and Yuceoglu and Updike [69]. The latter was developed for symmetrical configurations, the former for dissimilar adherends. The zero stress condition in the adhesive at the edge of the overlap has been taken into account by Allman [70], Du Chen [71] and Renton and Vinson [72]. More recently, analysis methods for three-dimensional (3D) problems under more general loading conditions were developed, for instance, by extending the Goland and Reissner model [73]. Mortensen developed a unified approach for which good agreement is found, in comparison with Erdogan's analysis and with the FEM (Zhang) [74].

8.5.2 Numerical tools

For complex geometries or non-linear analysis a closed-form analytical solution will be difficult or impossible to find. The governing differential equations can then be solved numerically by the finite difference method (FDM) [61]. Herewith it should be noted that it is only the solving method that is different. The governing equations are still derived based on simplifying assumptions, so similar limitations to the closed-form solutions apply [62]. Implementation in, for instance, Matlab or equivalent software programs makes FDM a powerful tool for linear elastic analysis and, as such, an excellent engineering design tool [75].

The FEM approach divides the joint into smaller building blocks, called finite elements, each with their own (simplified) governing differential equations. Together they describe the behavior of the joint. This opens up the possibility of analyzing complex joint geometries, such as spew fillet, adherend tapering and 3D, and of including geometrical and material non-linear behavior. However, the method is time-consuming and not easily applicable to routine design work [62]. The thin bondline and the high strain gradients at the edges require a large number of small elements to obtain sufficiently accurate results. Different approaches to modeling the bonded joints have been reported (Jones [76], Carver and Wooley [77], Adams [63], NASA [78], Barut [73], Goncalves [79]):

- two-dimensional FEM: using plate/shell elements for substrate and adhesive
- shear spring method: 3D plate/shell element for the adherend + one-dimensional spring, bar or beam elements for the adhesive
- three-layer method: the adherends and adhesive are modeled as 3D plate/shell layers rigidly connected
- full 3D model: adherends and adhesives are modeled as 3D solids

To limit the number of elements the following approaches are used (Jones [76], Zhu and Kedward [80], Engelstadt [81], Cornec [82]):

- a coarse global model with a fine local model for detailed stress analysis
- FE codes that use a higher-degree polynomial (p-based)
- specific interface elements representing the bondline (e.g., cohesive zone model, CZM)

8.5.3 Failure load prediction

The prediction of joint failure depends highly on the accuracy of the calculated stresses and strains in combination with a suitable failure criterion. Strength predictions based on linear elastic analysis are inadequate, as this is not the case at the moment of failure. Both material and geometrical non-linear behavior should be taken into account. The accuracy of the calculated stresses and strains in the bondline is sensitive to variations in material properties and small changes in local geometry (see Fig. 8.27). On the one hand it is very difficult to determine accurately the stress-strain properties for the adhesive materials. On the other hand the stress field at the edge of the joint is strongly influenced by local geometrical parameters. Two distinct approaches to strength predictions can be identified: maximum-value criteria and fracture mechanics [83].

The use of maximum-value criteria was advocated in nonlinear stress analysis, and good correlation with experimental results was reported for specific cases. However, this approach could not be generalized to other joint-strength predictions, although it was shown that for adhesive joints the fracture energy is not independent of the joint geometry and as such cannot be treated as an adhesive property [83]. Many failure criteria have been proposed by various authors and good agreement with test results is reported, yet only for specific cases (Gleich [61], Odi and Friend [83]). So far, none of the following criteria were found to be universally applicable:

- maximum stress/strain
- elastic-plastic curve with maximum strain (Hart-Smith) [60, 84]
- modified Von Mises (Zhu and Kedward) [80]
- yield criteria, e.g., Tresca (Wang and Chalkley [85], Ignjatovic [86])

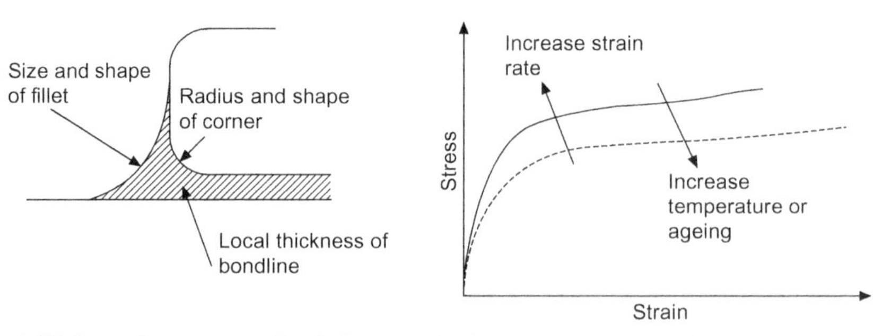

Fig. 8.27 Several parameters that influence calculated stresses and strains.

- linear elastic fracture mechanics (LEFM)
- stress singularity (Gleich [61], Zhu and Kedward [80])
- strain energy density (Hart-Smith) [84]
- strain invariant failure (SIFT) (Engelstadt) [81]
- CZM (Cornec) [82]

A good failure criterion should predict both failure mode as well as failure load [61]. To date, however, the accurate prediction of bond strength has been limited by the lack of a suitable, universally applicable failure criterion and insufficient accuracy in calculating bondline stresses and strains.

8.5.4 Fracture mechanics approach

The principle of fracture mechanics predicts failure propagation based on the assumption of a preexisting crack (or delamination) in the bonded joint at the most crucial location, usually the highly loaded edges. The strain energy release rate, G, or the stress intensity factor, K, is checked against a critical value, that is Gc or Kc respectively, in order to check for further crack growth or failure (Broughton [87], Kinloch [15], Groth [88]). Applied to bonded-joint durability, that is fatigue and environmental effects [89], the virtual crack-closure technique (VCCT) predicts the delamination growth on the basis of the work required to close a virtual crack [90]. VCCT is available in a commercially available FE code [91]. An interface-fracture finite element for predicting the delamination propagation, damage tolerance and residual strength is reported by Engelstadt [81]. The use of a traction-separation law for the decohesion is the fundamental idea of CZM. In this way, the unrealistic continuum mechanics stress singularity at the crack tip is avoided [82].

8.5.5 Improved analytical methods for fatigue-crack-growth prediction in FML

During crack growth in a FML, such as Glare®, the fibers transfer part of the load around the crack tip in the aluminum layers, thereby reducing crack growth. This is accompanied by delamination at the interface between the aluminum and glass-fiber/adhesive layers, in which size and shape play an important role in fiber-bridging effectiveness. An accurate analytical prediction model was developed by Alderliesten [92], which accounts for delamination growth and fiber bridging. Similar to models developed for monolithic aluminum, the stress intensity at the crack tip in the metal layers is taken as the factor determining the extension of that crack under cyclic loading. The stress intensity factor consists of a crack-opening contribution caused by far field stresses in the aluminum layers, and a crack-closing contribution of the intact fibers in the wake of the crack. The stresses in these fiber layers determine the delamination growth. The stress intensity factor is described by LEFM, including the contribution of the fiber layers and the crack growth associated delamination behavior in the prepreg layers in the wake of the propagating crack.

The bridging stress along the crack length is calculated on the basis of the crack-opening relations for the individual mechanisms. It is then used to calculate the delamination extension, using a correlation between the delamination growth rate and the energy release rate. Once the stress intensity factor at the crack tip in the aluminum layers is known, the fatigue-crack-growth rate can be calculated using an empirical Paris relation. A good correlation between predicted and experimental crack growth rates, crack-opening contours and delamination shapes has been obtained using a wide range of test data.

8.5.6 Testing adhesive-bonded joints

Most test methods and specimens used for adhesive bonding are coupon tests related to the quality evaluation of either the adhesive material or the surface pretreatment (see Table 8.4).

Others are used for the determination of adhesive mechanical properties, that is shear stress-strain curve or fracture energy (Minford [93], Kinloch [15], deVries and Anderson [94], ASTM [95] or ISO/EN standards). These coupon tests, however, are less useful for design purpose or tool validation [60]. Most case studies on stress analysis of bonded joints use the single or double overlap test specimen. This choice, however, is driven by the fact that results are widely available and the simplicity of the test itself [62]. A major difficulty is that the test results depend highly on factors such as substrate and bondline thickness, surface treatment and overlap length, and that the failure mode is a combination of peel and shear. With the absence of a suitable failure criterion [61], the one-test result, that is failure load, cannot be linked directly to the analysis results.

Any comparison reported between test and analysis is based on indirect results, such as deflection of the specimen or strains measured on the substrate's outer surface. Until recently local strains inside the bondline simply could not be measured accurately, although digital optical methods supported by image correlation software can now be used for exactly that purpose. As simple coupon tests are unsuitable, the validation of calculation tools requires specially developed specimens. Typical

Table 8.4 Overview of some of the most used test methods for adhesive bonding.

Description	Tested item	Specification	
Single overlap shear test	Quality control	ASTM D1002	EN 2243–1
Floating roller peel test	Surface treatment	ASTM D3167	EN 2243–2
Climbing drum peel test	Surface treatment	ASTM D1781	EN 2243–3
Wedge test	Surface treatment	ASTM D3762	
Thick adherend test	Shear stress-strain curve	ASTM D5656	EN 2243–6
Double cantilever beam	Fracture energy	ASTM D3807	ASTM D3433

examples are doubler run-out, cracked lap-shear specimen (CLS) or stringer run-out specimens [81]. Their principle is based on crack initiation and growth in relation to fracture energy, both static and dynamic. By changing geometrical parameters the load on the joint can be varied. Scaling up is severely limited [60], as all dimensions of a bonded joint can be scaled up, yet bondline thickness cannot. Further, the strength of the substrate is proportional to its thickness, whereas bond strength is proportional to the square root of the substrate thickness.

8.5.7 Determination of bondline strains by fiber-optic sensors

Fiber-optic sensors are currently used as transducers for various physical phenomena. The embedding of fiber-optic sensors in structural materials presents the possibility of structural health monitoring. In adhesive-bonded joints they can be used to monitor changes in strains in the bondline and the onset of delamination [96]. Various types of fiber-optic sensors are available on the market. Fiber Bragg gratings (FBG) are mostly applied in monitoring composite structures and adhesive bondlines. Most applications are used in the area of bonded composite repair patches, where the change in the stress field is monitored to detect cracks propagating beneath the patch [97, 98].

The measured wavelength shift in a Bragg grating sensor is proportional to the linear combination of the principle strains and the temperature. Furthermore the effect of the fiber coating, adhesion quality between coating and adhesive material and effect of the fiber diameter hinder the accuracy of the strain measurement. Commonly the fibers have a diameter between 50 and 100 μm, whereas the adhesive bondline thickness of a structural joint is between 0.1 and 0.2 mm. This makes it likely that fibers embedded in the adhesive layer influence the strain distribution in the bonded joint and influence the failure behavior.

8.5.8 Development of optical digital video microscopy to measure bondline strains

Strain fields in bonded joints are characterized by high gradients and local concentrations inside thin bondlines that for a long time prohibited accurate measurement. The development of high-resolution optical measurement systems, together with sophisticated image correlation software, makes detailed material behavior visible and can provide detailed local information about the strain field without the need for physical contact with the specimen [99]. Thick adherend tests are typically used for the determination of adhesive mechanical properties under shear loading. In the test, the relative displacement of the substrates with respect to each other needs to be measured to obtain the shear angle. For some time, the standard method is based on the work by Krieger [100], in which a specially developed type of mechanical extensometer is attached to both sides of the specimen. Pins are positioned on the substrates as close as possible to the bondline. The accuracy of the measurements is lowered owing to the following difficulties with the mechanical extensometers: [101]

- rotation of the bondline caused by secondary bending
- pins are located some distance away from the interface, so the shear deformation of the aluminum adherend needs to be filtered
- slippage of the pins results in inaccurate readings

By using a noncontact optical method these drawbacks are eliminated, as the bondline deformation is measured directly. Slippage does not occur and adherend deformation and rotation do not influence the measurements. Further it is possible to measure local strain fields, whereas an extensometer only measures the average strain over a large gauge length. High accuracy is possible, although it is dependent on the quality of the images, the pattern of the surface and the light used. Digital image correlation (DIC) is the analysis of a large number of images taken from a test specimen during the test, in which one specific part of the first image is correlated to each consecutive image to establish the displacement of one or more points of the image. By using higher-degree polynomial interpolation, sub-pixel accuracy of down to 0.01 pixels can be obtained. Following the relative displacement of multiple points, the local strain field (or any other deformation related property) is calculated. For the thick adherend test the shear angle is obtained from the relative displacement of both substrates, which is determined for multiple points on the adhesive-substrate interface. Each image is linked to the load data recorded by the tensile machine and subsequently the shear stress-strain curve is plotted. Similar optical methods are used for Mode I crack-propagation tests using double cantilever beam (DCB) specimens, or for the determination of the strain field around a crack tip [99]. By zooming out, the strain field around bonded doublers can be determined to verify design calculations, where out-of-plane deformation (3D) can be determined by using two cameras simultaneously.

8.6 Future trends

In the development of adhesive-bonded structures, long-term durability and reliability of bonded joints is a continued area of attention. The requirements for strength and durability of the total system, the combination of adherend material, surface treatment, primer and adhesive are of continued concern. The need for environmentally safe materials and processes will result in further development of low volatile organic components (VOC), chromate-free adhesive materials and surface treatments. The search for more toughness in high-strength structural adhesive systems is a challenge, especially in room-temperature or moderate-temperature curing adhesives. The developments in nanostructured materials may further increase adhesive toughness. These improved adhesives should enable the curing of large structures without the use of an autoclave whereas the structural joint characteristics at both low as well as high temperatures are maintained. Developments in nanomaterials may also lead to improved high-temperature adhesive materials and even to adhesives with improved fire properties.

Structural development in bonded laminates and FMLs will result in new applications. Recently a new FML called CentrAl (centrally reinforced aluminum) was

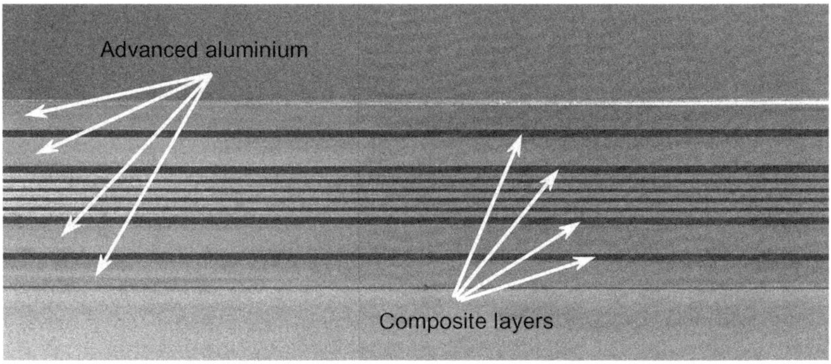

Fig. 8.28 The CentrAl concept.

introduced for application in aircraft wings [102–104]. The CentrAl concept comprises a central layer of FML (GLARE), sandwiched between one or more thick layers of new-generation damage-tolerant aluminum alloys (see Fig. 8.28). Fiber-reinforced adhesive layers called bondpreg™ also bond the outer layers in this concept. This creates a robust structural material, which is not only exceptionally strong but also insensitive to fatigue. Because the hybrid material is practically immune to fatigue, wing panels can be designed that do not need frequent inspection and repair of cracks during the life of the aircraft, in other words they have a "care-free" economic life. The new CentrAl structures are stronger than carbon-fiber-reinforced plastic (CFRP) structures [102]. CentrAl allows higher stress levels, and by using it in lower wing structures the weight can be reduced by 20% compared with CFRP structures. The application of CentrAl will result in considerably lower manufacturing and maintenance costs.

Rapid developments in sophisticated computer hardware and software will result in more accurate strength prediction models and will boost the development of digital optical methods for strain measurements, which can then be used for model validation. This will result in more reliable design methods for adhesively bonded metal joints.

8.7 Sources of further information and advice

Many references have been given in this chapter, but the most important sources of further information are summarized below. A general review of the multidisciplined subject of adhesion and adhesives is given by Kinloch in *Adhesion and Adhesives—Science and Technology* [15], which covers the first principles of surface chemistry, physics and adhesive chemistry up to the engineering design of joints and the service life considerations. The terminology most used in the field of adhesive bonding is described in the *Handbook of Adhesion* [105], which provides a basic understanding via short, self-contained articles on scientific, engineering and industrial aspects of adhesion.

A true landmark reference, which reviews more than 4500 articles, is a comprehensive discussion on every important aspect of aluminum bonding by Minford in the

Handbook of Aluminium Bonding Technology and Data [93]. The wide variety of different adhesive types, their properties and applications, as well as the many surface treatments available for various substrates is described by Landrock in the *Adhesives Technology Handbook* [5]. *Structural Adhesive Joints in Engineering* by Adams [63] provides basic engineering design knowledge, with the focus on understanding the way stresses are transferred from one member to another.

The most important way to ensure long-term durability of structural adhesive joints is discussed in *Durability of Structural Adhesives* by Kinloch [106]. Included is the kinetics and mechanism of environmental attack, as well as the durability of aluminum, steel and titanium-bonded substrates. The mechanical testing of adhesives and bonded joints, as well as stress analysis and failure mechanisms, is described in "*Adhesively Bonded Joints: testing, analysis and design*." [107] Thrall and Shannon collected in *Adhesive Bonding of Aluminum Alloys* the lessons learned during the PABST program about adhesives, surface treatments, mechanical properties, environmental durability, structural analysis and tooling design for adhesively bonded primary aircraft structures [108].

An overview of all essential aspects of bonded-patch repair, including materials and processes, design of repairs, certification issues and many example cases, is given in *Advances in the Bonded Composite Repair of Metallic Aircraft Structures* compiled by Baker, Rose and Jones [54]. To learn more about the development of FMLs and their static, fatigue and impact properties, as well as design, production and maintenance of FML-based aircraft structures, the reader is referred to the book edited by Gunnink and Vlot called *Fiber Metal Laminates: an introduction* [109].

References

[1] E. Boëda, J. Connan, D. Dessort, S. Muhesen, N. Mercier, M. Valladas, N. Tisnerat, Bitumen as hafting material on Middle Palaeolithic artefacts, Nature 380 (1996) 336–338.

[2] J.A. Bishopp, The history of Redux® and the Redux bonding process, Int. J. Adhes. Adhes. 17 (1997) 287–301.

[3] R.J. Schliekelmann, Operational experience with adhesive bonded structures, in: AGARD-LS-102, Bonded Joints and Preparation for Bonding, AGARD-NATO, Paris, 1979, pp. 1-1–1-30.

[4] A. Kwakernaak, More than 40 years experience with primary adhesive bonded structures, in: J. Hognat, R. Pinzelli, E. Gillard (Eds.), 50 Years of Advanced Materials or 'Back to the Future', SAMPE Europe, Switzerland, 1994, pp. 67–78.

[5] A.H. Landrock, Adhesives Technology Handbook, Noyes Publications, Park Ridge, NJ, 1985.

[6] E.W. Garnish, Advances in epoxy adhesive technology, in: W.C. Wake (Ed.), Developments in Adhesives-1, Applied Science Publishers, London, 1977.

[7] A.J. Kinloch, Toughening epoxy adhesives to meet today's challenges, MRS Bull. (2003) 445–448.

[8] J.A. Bishopp, The chemistry and properties of a new generation of toughened epoxy matrices, Int. J. Adhes. Adhes. 12 (1992) 178–184.

[9] A.J. Kinloch, J.H. Lee, A.C. Taylor, S. Sprenger, C. Eger, D. Egan, Toughening structural adhesives using nano- and micro-phase inclusions, in: M.K. Chaudhury (Ed.), 27th

Annual Meeting of the Adhesion Society, Adhesion Society, Wilmington, NC, 2004, pp. 96–98.

[10] R.D. Brooker, B.R.K. Blackman, A.J. Kinloch, A.C. Taylor, Nano-reinforcement of epoxy/thermoplastic blends, in: G. Anderson (Ed.), 31st Annual Meeting of The Adhesion Society, Adhesion Society, Austin, TX, 2008, pp. 250–252.

[11] R. Barsotti, R. Inoubli, S. Schmidt, N. Macy, S. Magnet, T. Fine, C. Navarro, M. Wells, Block copolymers for epoxy toughening, in: SAMPE Fall Technical Conference Proceedings: Multifunctional Materials: Working Smart Together, Memphis, TN, September 8–11, Society for the Advancement of Material and Process Engineering, Covina, CA, USA, CD-ROM, 2008, p. 8.

[12] N. Noble, D. Brosius, J. Schlimbach, An advanced out-of-autoclave curing technology for prepregs and resin infusion, in: SAMPE Asia Conference 2008 Proceedings, Bangkok, Thailand, 11–13 February, 2008, Society for the Advancement of Material and Process Engineering, Covina, CA, USA, CD-ROM, 2008, p. 10.

[13] R.J. Schliekelmann, Adhesive bonding in the Fokker-VFW F-28 'Fellowship', in: Fokker report K-67, NTIS A06–99–028392, 1973.

[14] A. Hartman, Fatigue Tests on Single Lap Joints in Clad 2024-T3 Aluminium Alloy Manufactured by a Combination of Riveting and Adhesive Bonding, NLR, 1966. Report M.2170.

[15] A.J. Kinloch, Adhesion and Adhesives: Science and Technology, Chapman & Hall, London, 1987.

[16] Anon, Process for Cleaning Aluminium Alloy Plating Prior to Painting, Process Specification DTD 915, UK Ministry of Aircraft Production, 1941.

[17] H.W. Eickner, N.E. Schowalter, A Study of Methods for Preparing Clad 24S-T3 Aluminum Alloy Sheet Surfaces for Adhesive Bonding, Forest Products Laboratory Report No. 1813, 1950.

[18] A.W. Bethune, Durability of bonded aluminum structure, SAMPE J. 11 (3) (1975) 4–10.

[19] W.J. Russel, E.A. Garnis, A chromate-free low toxicity method of preparing aluminium surfaces for adhesive bonding, SAMPE J. 17 (3) (1981) 19–23.

[20] R.F. Wegman, Surface Preparation Techniques for Adhesive Bonding, Noyes Publications, Park Ridge, NJ, 1989.

[21] P.F.A. Bijlmer, Pickling Aluminium, 1977 (US patent 4042475).

[22] Anon, Chromic acid anodising of aluminium and aluminium alloys, in: Process Specification Def Stan 03–24, Issue 5, UK Ministry of Defence, Defence Standard, 2008.

[23] P.F.A. Bijlmer, Adhesive bonding on anodised aluminium, Met. Finish. 70 (4) (1972) 30–34.

[24] J.C. McMillan, R.A. Davis, J.T. Quinlivan, Phosphoric acid anodizing of aluminium for structural bonding, SAMPE Q. 7 (3) (1976) 13–18.

[25] A. Kwakernaak, A. van den Berg, Adhesive bonded joints in aluminium structures, Acta Tech. Belg. Metall. 42 (1–2) (2002) 103–108.

[26] A. van den Berg, A. Kwakernaak, Conversion coating as a pretreatment for adhesive bonding, in: Proceedings 10th International Symposium Swiss Bonding, 27–29 May, 2002, Rapperswil, Swibotech, Switzerland, 2002.

[27] E. Kock, V. Muss, C. Matz, F. De Wit, Verfahren zur anodischen Oxidation, 1993 (Patent EP0607579 A1, 16 December 1993).

[28] A. Kwakernaak, The importance of anodic oxide morphology in relation to adhesion and durability of bonded joints, in: Workshop in Bremen: Anodisation in the Aircraft Manufacturing Industry, 15 April 2004, IFAM, Bremen, Germany, 2004.

[29] G. Critchlow, Alternatives to chromic acid anodizing for structural bonding applications, in: Symposium, Disruptive Technologies for Light Metals, IOM3, London, UK, 2006.

[30] W. Brockmann, Steel adherends, in: A.J. Kinloch (Ed.), Durability of Structural Adhesives, Elsevier Applied Science, London, 1987, p. 306. Chapter 7.

[31] R.P. Haak, T. Smith, Surface treatment of AM355 stainless steel for adhesive bonding, Int. J. Adhes. Adhes. 3 (1983) 15–23.

[32] F. Bouquet, J.M. Cuntz, C. Coddet, Influence of surface treatment on the durability of stainless steel sheets bonded with epoxy, J. Adhes. Sci. Technol. 6 (2) (1992) 233–242.

[33] A. Mahoon, Titanium adherends, in: Durability of Structural Adhesives, Elsevier Applied Science, London, 1987.

[34] M.C. Locke, K.M. Harriman, D.B. Arnold, Optimization of chromic acid—fluoride anodizing for titanium prebond surface treatment, in: Proceedings 25th National SAMPE Symposium, San Diego, CA, USA, 6–8 May 1980, Society for the Advancement of Material and Process Engineering, Covina, CA, USA, 1980, pp. 1–12.

[35] A.C. Kennedy, R. Kohler, P. Poole, A sodium hydroxide anodize surface pretreatment for the adhesive bonding of titanium alloys, Int. J. Adhes. Adhes. 3 (2) (1983) 133–139.

[36] D. Arnott, A. Rider, J. Mazza, Surface treatment and repair bonding, in: A.A. Baker, L.R. F. Rose, R. Jones (Eds.), Advances in the Bonded Composite Repairs of Metallic Aircraft Structure, Elsevier Science, Oxford, 2002.

[37] L. Bergan, On-aircraft phosphoric acid anodising, Int. J. Adhes. Adhes. 19 (1999) 199–204.

[38] A.A. Baker, R.J. Chester, Minimum surface treatments for adhesively bonded repairs, Int. J. Adhes. Adhes. 121 (2) (1992) 73–78.

[39] R.J. Kuhbander, J.J. Mazza, Understanding the Australian silane surface treatment, in: Proceedings of 38th International SAMPE Symposium and Exhibition, Anaheim, CA, Society for the Advancement of Material and Process Engineering, Covina, CA, USA, 1993.

[40] Blohowiak, K.Y., Osborne, J.H. and Krienke, K.A. (1998–2000). Sol for Bonding Epoxies to Aluminum or Titanium Alloys, US Patent 6,037,060 (2000); Sol-gel Coated Metal, US Patent 5,958,578 (1999); Sol-gel Coated Metal, US Patent 5,939,197 (1999); Surface Pretreatment for Sol Coating of Metals, US Patent 5,869,141 (1999); Surface Pretreatment of Metals to Activate the Surface for Sol-gel Coating, US Patent 5,869,140 (1999); Sol Coating of Metals, US Patent 5,849,110 (1998); Sol for Coating Metals, US Patent 5,814,137 (1998).

[41] D.B. McCray, J.J. Mazza, Optimization of sol-gel surface preparations for repair bonding of aluminum alloys, in: Proceedings of 45th International SAMPE Symposium and Exhibition, Long Beach CA, May, Society for the Advancement of Material and Process Engineering, Covina, CA, 2000, pp. 53–54.

[42] K.Y. Blohowiak, J. Grob, W.B. Grace, N. Cejka, D. Berg, Improvements in sol-gel surface preparation methods for metal bonding applications, in: Proceedings of 39th International SAMPE Technical Conference, Cincinnati, OH, Society for the Advancement of Material and Process Engineering, Covina, CA, USA, CD-ROM, 2007, p. 15.

[43] C.D.H. Harteveld, Optimal Adhesive Bonded Panel Edge Configurations Determined by Photoelastic Investigation (in Dutch), 1971. Fokker Report R-1429.

[44] J. Schijve, Fatigue Properties of Adhesive Bonded Laminated Sheet Material of Aluminium Alloys, Aerospace Engineering, TU Delft, 1978. Report LR-276.

[45] P.F.A. Bijlmer, Fracture toughness of multiply layer adhesive aluminium alloy sheets, in: Proceedings 11th ICAS Conference, Lisbon, 1978, 1978, pp. 544–554.

[46] L.H. van Veggel, A.A. Jongebreur, J.W. Gunnink, Damage tolerance aspects of an experimental ARALL F27 lower wing skin panel, in: Proceedings 14th ICAF Conference, Ottawa, 1987.

[47] T.J. Matway, Producibility and cost effectiveness of ARALL laminates, C17 aft cargo door skin: a case study, in: Proceedings 12th European SAMPE Conference, Maastricht, 1991, pp. 373–379.

[48] J.W. Gunnink, L.B. Vogelesang, Fibre metal laminates and the very large civil transport, in: Proceedings 15th International SAMPE Europe Conference, Toulouse, 1994, pp. 93–102.

[49] P.A. Hooijmeijer, Cost reduction of Glare components, in: Proceedings 35th International SAMPE Technical Conference, Dayton, OH. Society for the Advancement of Material and Process Engineering, Covina, CA, USA, CD-ROM, 2003, p. 10.

[50] P.A. Hooijmeijer, Impact on Glare fibre metal laminates, in: Proceedings 26th International SAMPE Europe Conference, Paris, 2005, pp. 482–487.

[51] K. Poston, A.C. Mattousch, T.J. Matway, Impact properties and related applications of fiber metal laminates, in: Proceedings 15th International SAMPE Europe Conference, Toulouse, 1994, pp. 103–113.

[52] H.-J. Schmidt, B. Schmidt-Brandecker, G. Tober, Design of modern aircraft structure and the role of NDI, in: Proceedings 7th European Conference on Non-Destructive Testing, Copenhagen, www.NDT.net, June 1999, 4(6), 1998.

[53] H.J.M. Woerden, W.J. Mortier, C.B. Guijt, S. Verhoeven, Bonded repair patches, in: A. Vlot, J.W. Gunnink (Eds.), Fibre Metal Laminates: An Introduction, Kluwer Academic, Dordrecht, Netherland, 2001, pp. 451–475.

[54] A.A. Baker, R. Jones, L.R.F. Rose (Eds.), Advances in the Bonded Composite Repair of Metallic Aircraft Structures, Elsevier Science, Oxford, UK, 2002.

[55] R.S. Fredell, Damage Tolerant Repair Techniques for Pressurized Aircraft Fuselages, PhD Thesis, Delft University of Technology, 1994.

[56] A. Vlot, S. Verhoeven, G. Ipenburg, D.R.C. Simpersad, H.J.M. Woerden, Stress concentrations around bonded repairs, Fatigue Fract. Eng. Mater. Struct. 23 (3) (2000) 263–276.

[57] O. Volkersen, Die Nietkraft verteiling in zugbeanspruchten Nietverbindungen mit konstanten laschen Querschnitten, Luftfahrtforschung 15 (1938) 41–47.

[58] M. Goland, E. Reissner, The stresses in cemented joints, J. Appl. Mech. 11 (1944) A17–A27.

[59] A. Vlot, S. Verhoeven, P.J.M. Nijssen, Bonded repairs for aircraft fuselages, in: Series 07 Aerospace Materials, TU Delft, 1998.

[60] L.J. Hart-Smith, The design of adhesively bonded joints, in: Symposium on Innovations in Bonded Structures, Adhesion Institute TU Delft, April 2006, 2006.

[61] D. Gleich, Stress Analysis of Structural Bonded Joints, 2002. PhD Thesis, TU Delft.

[62] J.W. van Ingen, A. Vlot, Stress Analysis of Adhesively Bonded Single Lap Joints, 1993. Report LR-740, TU Delft.

[63] R.D. Adams, W.C. Wake, Structural Adhesive Joints in Engineering, Elsevier Applied Science, London, UK, 1984.

[64] L.J. Hart-Smith, Adhesive-Bonded Single Lap Joints, 1973. NASA Report CR 112236.

[65] I.U. Ojalvo, H.L. Eidinoff, Bond thickness effects upon stresses in single lap adhesive joints, AIAA J. 16 (3) (1978) 204–211.

[66] J.H. Williams, Stresses in adhesives between dissimilar adherends, J. Adhes. Dent. 7 (1975) 97–107.

[67] D.A. Bigwood, A.D. Crocombe, Elastic analysis and engineering design formulae for bonded joints, Int. J. Adhes. Adhes. 9 (4) (1989) 229–242.

[68] F. Delale, F. Erdogan, M.N. Aydinoglu, Stresses in adhesively bonded joints: a closed form solution, J. Compos. Mater. 15 (1981) 249–271.

[69] U. Yuceoglu, D.P. Updike, Bending and shear deformation effects in lap joints, J. Eng. Mech. Div. ASCE 107 (1) (1981) 55–76.

[70] D.J. Allman, A theory for elastic stresses in adhesive bonded lap joints, Q. J. Mech. Appl. Math. 30 (1977) 415–436.

[71] D. Chen, S. Cheng, An analysis of adhesive-bonded single-lap joints, J. Appl. Mech. 50 (1983) 109–115.

[72] W.J. Renton, J.R. Vinson, Analysis of adhesively bonded joints between panels of composite materials, J. Appl. Mech. 44 (1977) 101–106.

[73] A. Barut, J. Hanauska, E. Modeni, D.R. Anbur, Analysis method for bonded patch repair of a skin with a cut-out, Compos. Struct. 55 (2002) 277–294.

[74] J. Zhang, B.A. Bednarcyk, C. Collier, P. Yarrington, Y. Bansal, M.-J. Pindera, 3D stress analysis of adhesively bonded composite joints, in: 49th AIAA/ASME/ASCE/AHS/ASC Structures, Structural Dynamics & Materials Conference, Published by AIAA on a CD-ROM, 10(6–7), 2005–21, 2005. 35 pp.

[75] Z.C. Roza, M.J.L. van Tooren, Finite Difference Methods for Stress Analysis of Adhesive Bonded Joints. The Design of a MATLAB Adhesive Toolbox, Aerospace Materials Series 07, 1998. No. 10, TU Delft.

[76] R. Jones, Numerical analysis and design, in: A. Baker, F. Rose, R. Jones (Eds.), Advances in the Bonded Composite Repair of Metallic Aircraft Structures, vol. 1, Elsevier Science, Oxford, UK, 2002.

[77] D.R. Carver, G.R. Wooley, Stress concentration factors for bonded lap joints, J. Aircr. 8 (1971) 817–820.

[78] http://femci.gsfc.nasa.gov/.

[79] J.P.M. Goncalves, M.F.S.F. De Moura, P.M.S.T. De Castro, A three dimensional finite element model for stress analysis of adhesive joints, Int. J. Adhes. Adhes. 22 (2002) 357–365.

[80] Y. Zhu, K. Kedward, Methods of Analysis and Failure Prediction for Adhesively Bonded Joints of Uniform and Variable Bondline Thickness, DOT/FAA/AR-05/12, US Department of Transportation, Federal Aviation Administration, 2005.

[81] S.P. Engelstadt, O.T. Berry, G.D. Renieri, L.R. Deobald, G.E. Mabson, B. Dopker, High fidelity composite bonded joint analysis validation study—Part 1 analysis, in: 49th AIAA/ASME/ASCE/AHS/ASC Structures, Structural Dynamics & Materials Conference, Published by AIAA on a CD-ROM, 10(6–7), 2005–2166, 2005. 16 pp.

[82] A. Cornec, I. Scheider, K.H. Schwalbe, On the practical application of the cohesive model, Eng. Fract. Mech. 70 (14) (2003) 1963–1987.

[83] R.A. Odi, C.M. Friend, An improved 2D model for bonded composite joints, Int. J. Adhes. Adhes. 24 (5) (2004) 389–405.

[84] L.J. Hart-Smith, Design of adhesively bonded joints, in: F.L. Matthews (Ed.), Joining Fibre-Reinforced Plastics, Elsevier Applied Science, London, UK, 1987, pp. 271–311.

[85] C.H. Wang, P. Chalkley, Plastic yielding of a film adhesive under multiaxial stresses, Int. J. Adhes. Adhes. 20 (2) (2000) 155–164.

[86] M. Ignjatovic, P. Chalkley, C. Wang, The Yield Behaviour of a Structural Adhesive Under Complex Loading, 1998. Report DSTO-TR-0728, DSTO, Australia.

[87] W.R. Broughton, L.E. Crocker, M.R.L. Gower, Design Requirements for Bonded and Bolted Composite Structures, NPL Report MATC(A)65, NPL, UK, 2002.

[88] H.L. Groth, Stress singularities and fracture at interface corners in bonded joints, Int. J. Adhes. Adhes. 8 (2) (1988) 107–113.

[89] W.S. Johnson, L.M. Butkus, R.V. Valentin, Applications of Fracture Mechanics to the Durability of Bonded Composite Joints, DOT/FAA/AR-97/56, US Department of Transportation, Federal Aviation Administration, 1998.

[90] R. Kreuger, Virtual crack closure technique: history, approach and applications, Appl. Mech. Rev. 57 (2) (2004) 109–143.

[91] ABAQUS, Technology Brief, TB-05-VCCT-1, October, 2005.

[92] R.C. Alderliesten, Fatigue Crack Propagation and Delamination Growth in Glare, PhD Thesis, TU Delft, 2005.

[93] J.D. Minford, Handbook of Aluminum Bonding Technology and Data, Marcel Dekker, New York, 1993.

[94] K.L. deVries, G.P. Anderson, Analysis and design of adhesive bonded joints, in: Bonded Joints and Preparation for Bonding, AGARD LS 102, NATO, 1979.

[95] Anonymous, Annual Book of ASTM Standards—Section 15—General Products, Chemical Specialties, and End Use Products—Volume 15.06 Adhesives, ASTM International, 2009.

[96] W.L. Schulz, E. Udd, J.M. Seim, I. Perez, A. Trego, Progress on monitoring of adhesive joints using multi-axis fiber grating sensors, Proc. SPIE 3991 (2000) 52.

[97] W. Baker, I. McKenzie, R. Jones, Development of life extension strategies for Australian military aircraft, using structural health monitoring of composite repairs and joints, Compos. Struct. 66 (2004) 133–143.

[98] I. Mckenzie, R. Jones, I.H. Marshall, S. Galea, Optical fibre sensors for health monitoring of bonded repair systems, Compos. Struct. 50 (2000) 405–416.

[99] H.J.K. Lemmen, R.C. Alderliesten, R. Benedictus, J.C.J. Hofstede, R. Rodi, The power of digital image correlation for detailed elastic–plastic strain measurements, in: Proceedings EMESEG 2008, New Aspects of Engineering Mechanics, Structures, Engineering Geology, Heraklion, Crete Island, Greece, July 22–24, WSEAS Press, 2008, pp. 73–89.

[100] R.B. Krieger, Stress analysis concepts for adhesive bonding of aircraft primary structure, in: W.S. Johnson (Ed.), Adhesive Bonded Joints: Testing, Analysis and Design, American Society of Testing and Materials, Philadelphia, 1988, pp. 264–275.

[101] C. Yang, H. Hyang, J.S. Tomblin, D.W. Oplinger, Evaluation and adjustment for ASTM D-5656 standard test method for thick-adherend metal lap shear joints for determination of the stress strain behaviour of adhesives in shear by tensile loading, J. Test. Eval. 29 (1) (2001) 36–43.

[102] R.J. Bucci, Advanced metallic and hybrid structural concepts, in: Proceedings USAF Structural Integrity Program Conference (ASIP 2006), San Antonio, TX, 2006.

[103] G.H.J.J. Roebroeks, P. Hooijmeijer, E. Kroon, M.B. Heinimann, The development of central, in: Proceedings First International Conference on Damage Tolerance of Aircraft Structures, TU Delft, 2007.

[104] R.S. Fredell, J.W. Gunnink, R.J. Bucci, J. Hinrichsen, "Care-free" hybrid wing structures for aging USAF transports, in: Proceedings First International Conference on Damage Tolerance of Aircraft Structures, TU Delft, 2007.

[105] D.E. Packham (Ed.), Handbook of Adhesion, second ed., John Wiley & Sons, Hoboken, USA, 2006.

[106] A.J. Kinloch (Ed.), Durability of Structural Adhesives, Elsevier Applied Science, London, UK, 1986.

[107] W.S. Johnson (Ed.), Adhesively Bonded Joints: Testing, Analysis, and Design, ASTM Special Technical Publications STP 981, American Society for Testing and Materials ASTM, Philadelphia, USA, 1988.

[108] E.W. Thrall, R.W. Shannon (Eds.), Adhesive Bonding of Aluminum Alloys, Marcel Dekker, New York, 1985.

[109] A. Vlot, J.W. Gunnink (Eds.), Fibre Metal Laminates: An Introduction, Kluwer Academic Publishers, Dordrecht, Netherlands, 2001.

Composite to metal bonding in aerospace and other applications

9

R.A. Pethrick[†]

9.1 Introduction

The requirement to reduce weight and increase energy efficiency has produced an increased usage of composites in the aerospace industry. An air-frame is exposed to vibration during take-off and landing, which raises stresses that can result in fatigue damage. Cracking and the effects of corrosion in metals are problems that are well understood and, to some extent, predictable.

Composite materials were initially used in nonstructural components, but are now being used in more flight-critical applications. Glass-reinforced plastic (GRP) and carbon-fiber reinforced plastic (CFRP) are the most widely used materials. In many aerospace applications, composites have to be attached to metals and the problem of bonding dissimilar materials becomes a real concern. Two rather different scenarios are encountered: the first is when the bonding is part of the initial fabrication process and the second is when it is used for repair of structural damage.

The Redux novolac adhesive-bonding system was introduced in the 1950s to construct the Comet aircraft. Bonded structures were subjected to full-scale testing that involved repeated pressurization and depressurization and withstood more than 16,000 cycles, the equivalent of ~40,000 h of airline service. Windows were tested up to a pressure of 12 psi, which is 4.75 psi above the normal service ceiling of 36,000 ft and survived a massive 1250% over-pressurization. The subsequent problems encountered with the Comet airframe were attributed to riveting issues and not adhesive bonding. Redux has subsequently demonstrated impressive durability and was used in the construction of Nimrod aircraft. The Nimrod, being an all weather surveillance aircraft, spends much of its flying time in a moist wet environment that is very testing for most adhesive systems. However, Redux gives off moisture during cure and requires the extensive use of pressurized jigs during fabrication. Epoxy resins are a more user-friendly method of bonding but have introduced concerns about durability of bonded structures. Thermoplastics can be used but are difficult to process as they require high temperatures and the use of jigs during formation of the bond.

Concorde constructed in the late 1960s, contained honeycomb rudder panels to reduce weight. In 1989, a British Airways Concorde experienced rudder damage. New rudders produced in 1993, subsequently failed in 1998. Although the loss of the rudder did not endanger the aircraft it did cause concern, and British Airways undertook extensive nondestructive testing (NDT) of the rudder sections to check their

[†] Deceased.

Welding and Joining of Aerospace Materials. https://doi.org/10.1016/B978-0-12-819140-8.00009-2
Copyright © 2012 Woodhead Publishing Ltd. All rights reserved.

integrity. The problems identified were created during manufacture, and exacerbated during operation. Failures were attributed to problems in the pretreatment and priming stage of the bonding operations.

Recently, composites have been used extensively in the Airbus 380, where the fuselage is aluminum, more than 20% of the airframe has been constructed from composites. CFRP, GRP and quartz-fiber reinforced plastic (QFRP) are used extensively in the wings, fuselage sections; undercarriage and rear end of the fuselage, tail surfaces and doors. The new metal–composite hybrid material, GLARE (GLAss-REinforced fiber metal laminate), is used in the upper fuselage and the wing sections. GLARE is 25% stronger, 20% lighter, better damage tolerance, less tendency to metal fatigue, better corrosion and impact resistance compared with conventional airframe aluminum [1], Fig. 9.1.

GLARE is composed of alternating layers of 0.38 mm thick aluminum sheet and glass-fiber-reinforced bond film. By changing the number of layers in the laminate, the overall thickness and physical properties can be adjusted. As the outer skin is aluminum, surface electrical conductivity to protect the aircraft from lightning strikes is easily achieved. The aluminum acts as an effective crack stopper and the glass fiber reduces the effects of corrosion. The A380–800 has 27 GLARE skin panels covering a total area of 469 m^2 [2]. GLARE is also being used in the C-17 Globemaster III cargo doors.

9.1.1 Peculiarities of composite-metal bonding

Adhesive bonding of composite-composite and metal-metal is a well understood [3, 4]. Adhesive bonding allows the creation of lightweight structures with complex shapes. After over 40 years of sustained research, the crucial issues involved in bonding similar materials are now fairly well understood [3, 5, 6]. The ability to achieve a good bond requires that three criteria must be satisfied:

- A strong bond between the adhesive and substrate must be formed.
- The adhesive must possess sufficient mechanical strength to be able to transfer to load between the substrates.
- The adhesive must be able to accommodate the difference in thermal expansion between itself and the substrate to which it is attached. In the case of a metal to composite bond this is a major issue.

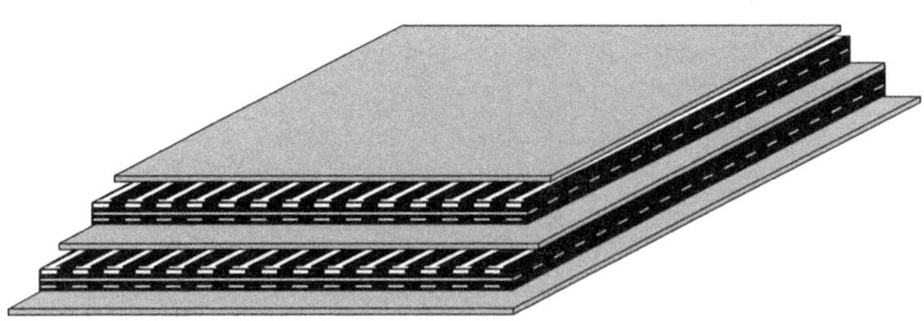

Fig. 9.1 Schematic of a GLARE: aluminum-glass-fiber composite structure.

Failure of a bond may occur either by propagation of a fracture in the adhesive—-*cohesive* failure—or by loss of integrity with the substrate *adhesive* or *interfacial* failure. Failure in the substrate can occur, but this is usually a consequence of poor design. *Cohesive fracture* involves crack propagation in the polymer layer and after fracture both substrates will be covered with adhesive. *Adhesive* or *interfacial* failure involves disbonding between the adhesive and adherent and is characterized by the creation of areas of clean substrate surface on failure.

Fracture is rarely cohesive or adhesive and will often be mixed; the crack path propagating in the adhesive and jumping to the substrate where the failure becomes cohesive. If the adhesive is tougher than the adherent then failure may be restricted to the substrate.

When the substrates are of dissimilar materials, new challenges emerge and temperature cycling can become a very important factor in failure. If the substrates have different thermal expansion coefficients, as in the case of metal and composite materials, shear stresses can create disbonding. In general, adhesives have higher thermal expansion coefficients, α_T, than those of composites or metals. Adhesives and composites can exhibit nonlinear expansion behavior as the material goes through its glass-rubber transition, T_g [3]. Epoxy resins have coefficient-of-expansion values between 50×10^{-6} and $100 \times 10^{-6}\,K^{-1}$, whereas polyester shows values between 100×10^{-6} and $200 \times 10^{-6}\,K^{-1}$, both these values are an order of magnitude larger than that found in metals or composites [7].

9.2 Testing of adhesive bonded structures

Our understanding of failure in adhesive joints is based on mechanical tests that are designed to introduce controlled stresses to the joints. The various modes of failure characterized in Fig. 9.2 are:

- *Mode I* (Fig. 9.2A): the force is applied normally to the adhesive layer and the rate at which the joint opens is monitored and measured in the *double cantilever beam tests* (DCB) [8, 9]. A nonconstant inertia that allows the crack to be arrested is created in the *tapered double cantilever beam* (TDCB) specimen.

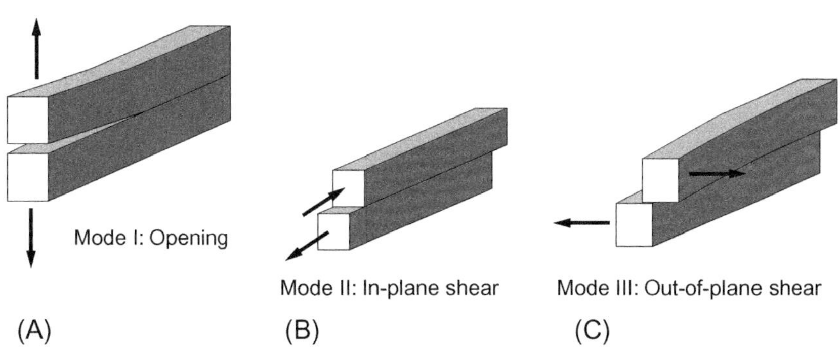

Fig. 9.2 (A–C) Simple failure modes for bonded structures.

- *Mode II* (Fig. 9.2B): the force creates a sliding or in-plane shear mode in a direction perpendicular to the leading edge of the crack and the adhesive will exhibit the highest resistance to fracture.
- *Mode III* (Fig. 9.2C): this involves a tearing motion or antiplane shear mode.

In practice, the loads experienced by a joint are fixed, and good design involves creating a geometry that avoids critical shear stresses being exceeded. Failure can be predicted by considering the stress concentration factor and the strain release rate [10]. The guiding rules for joint design are: (a) the bonded area should be as large as possible; (b) the main loading is carried in mode II; and (c) stable crack propagation follows the appearance of a local failure. Application of these rules to commonly encountered joining situations illustrates good and poor joining practice (Fig. 9.3). In a good design, mode II failure is promoted and the bonded area is maximized. The stress concentration at the edge of a joint can be reduced by use of a tapered spew at the edge of the adhesive layer and reduces the probability of crack initiation.

9.2.1 Characterizing the strength of an adhesive

The three failure modes are characterized in the test methods (Table 9.1). Tests have been developed in which the joint is peeled apart but these require the use of flexible rather than rigid substrates. The double and tapered double cantilever tests are considered the most reliable measurements of adhesive strength. A popular practical test involves the insertion of a wedge into a preformed crack and monitoring the crack force [3, 6] and is widely used in industry for durability studies and probes a combination of mode I and II types of failure [10]. Details of these tests can be found elsewhere [3, 6].

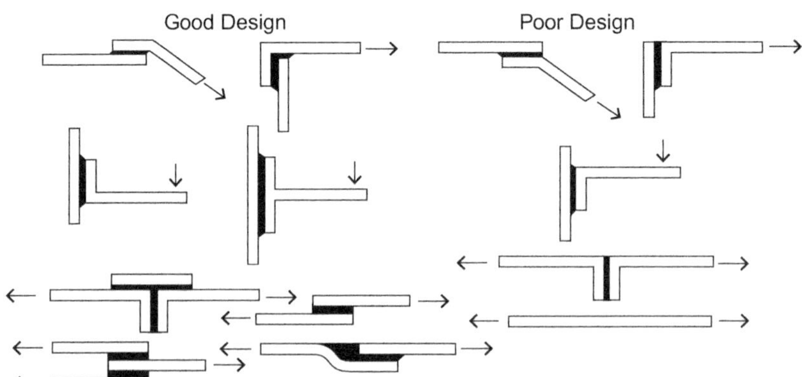

Fig. 9.3 Examples of good and poor joint design.

Table 9.1 Summary of geometries and associated test standards.

Test geometry	Standard	Comment
Axially loaded (tensile) butt joints 	ASTM D897-78 ASTM D 2094-69 and D 2095-72 BS 5350 Part C3	For block form For bar and rod shapes UK version for bar and rod
Lap joints loaded in tension 	ASTM 1002-72 BS 5350 Part C5 ASTM D 2295-72 ASTM D 2557-72 ASTM 3163-73 ASTM D 3983-81	Metal-metal single lap joint UK version Tests at elevated temperatures Test at low temperatures Using rigid plastic substrates Uses thick substrates
Double lap joint 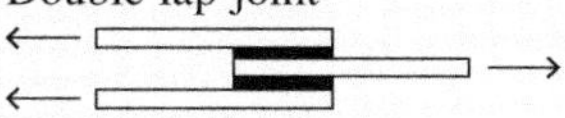	ASTM D 3528-76 BS 5350 Part C5	Double lap joint UK version
Modified double shear 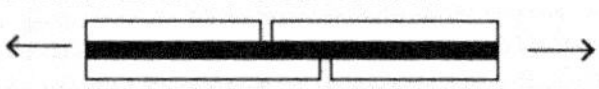	ASTM D 3165-73 ASTM D 906-82	Metal to metal laminate test uses defined bond area
Cleavage		
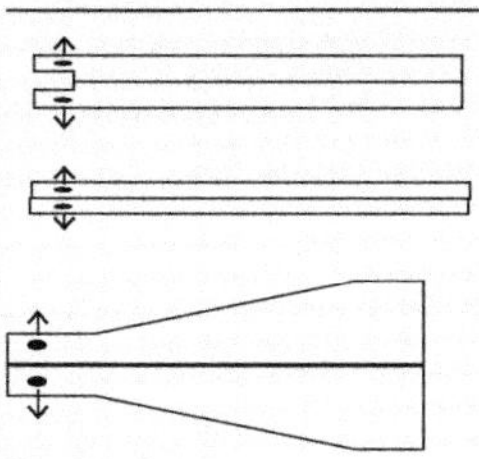	ASTM D 1062-78 ASTM D 3807-79 ASTM D 3433-75	Metal to metal joints Engineering plastics Flat and contoured cantilever beam specimens for adhesive fracture energy G_{Ic}
Peel joints 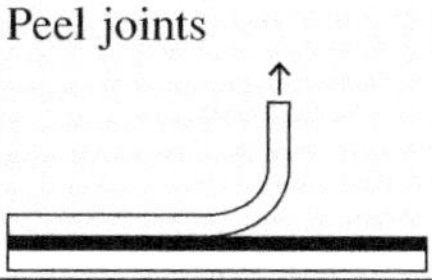	BS 5350 Part C14	90° Peel test for flexible to rigid joints

9.2.2 How is a good adhesive bond created?

For a good bond to be created, the strength of the interaction between adhesive and substrate is critical. Failure of bonds can often be attributed to weakness of the interface either associated with lack of removal of contamination or application of an inappropriate surface treatment [11]. Provided that the adhesive is of sufficient strength then the quality of the interface becomes the critical factor in determining strength and durability.

9.2.3 Importance of the interface

In an ideal situation, the strength of the bond between substrate and adhesive should be comparable with or greater than the strength of the adhesive. For an effective bond to be formed, the adhesive must wet out the substrate and be able to spread across the substrate without application of additional force. The ability to wet out a substrate is determined by a combination of the chemistry of the adhesive and the surface to which it is being applied. The theory of surface wetting has been discussed in many textbooks [3, 6]. For wetting to occur, the adhesive must be able to spread across the surface and have a sufficiently low viscosity for it to enter into the surface microstructure. Polymers such as polyethylene have a low *surface energy* and water that has a high *surface tension* does not wet the surface effectively. Metal oxides have a higher surface energy and hence can be wetted by water, but not by less polar liquids. To increase the compatibility of the adhesive for the substrate, surface treatments can be applied to the substrate. The surface treatment may also allow chemical bonds with the surface to be created.

9.3 Bonding to the metal substrate

In the case of metals, surface treatments are used to promote the growth of oxide layers or enhance surface roughness [12]. The oxide may allow chemical bonding with the adhesive and additionally creates surface roughness that adds a lock and key contribution to the interfacial bond strength. Most aluminum alloys used in aerospace fabrication will be subjected to careful etch and anodisation processes before bonding is undertaken (Fig. 9.4). The oxide has a distinct topography that reflects the growth method used.

Common treatments used for aluminum include phosphoric and chrome/sulfuric acid processes (Fig. 9.4). Whereas both processes promote the growth of an oxide layer, they differ in respect to the depth and characteristics of the 'topography' formed. The egg-box structure created by the etching and anodisation process is very effective at enhancing the lock and key contribution to the adhesive strength. For effective percolation to occur the adhesive must have a low surface tension and viscosity to ensure wetting to the oxide structure [3].

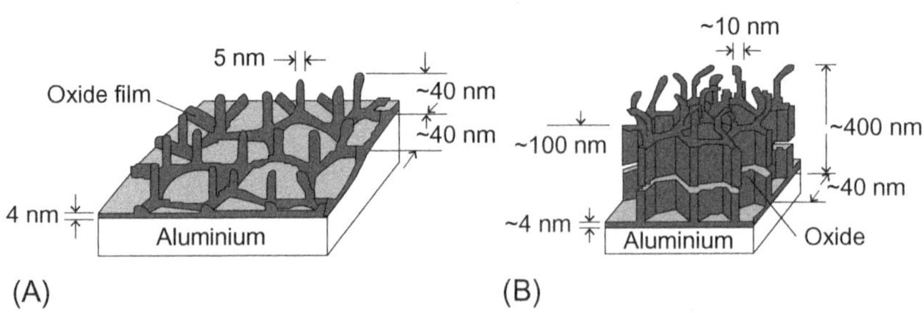

Fig. 9.4 Schematic of the surface structure created using (A) chromic-acid and (B) phosphoric-acid etching of aluminum.

9.3.1 Aluminum pretreatment

The pretreatment aims to develop a well-defined oxide by a combination of etching and anodisation. The delicate oxide produced is protected by a primer that is usually a low-molar-mass version of the adhesive. To aid chemical bonding and to improve the interfacial strength, various treatments involving silanes, zirconium and titanium organic compounds have been developed [3, 6, 13]. These compounds are considered to promote direct coupling between the adhesive and oxide. However, in practice, complex sol-gel reactions occur creating a textured surface that aids bonding, protects the oxide from hydrolysis and toughens the resin [14]. To ensure good bonds are created aerospace aluminum alloys are often clad with an outer layer of pure aluminum. Unclad materials will generate a structure that reflects the micro-segregation of the copper and magnesium alloying components [15].

9.3.2 Titanium-alloy pretreatment

The treatment of titanium alloys recognizes the different types of oxide that are produced on etching and anodisation. Ti-6V-4Al is widely used and has been the subject of a number of investigations [16]. The pretreatments that produce no roughness (macro or micro) yield the poorest bond durability, whereas those that produce significant macro-roughness but little microroughness yield moderate to good durability. Finally those that produce significant microroughness yield the best durability. The best pretreatments involve a combination of abrasion/grit blasting combined with chemical or electrochemical treatment [6]. A typical treatment would involve lightly abrading the metal substrate using 180/220-mesh alumina and cleaning with methylethylketone (MEK) [17–19]. Compressed air must be avoided to aid drying as it is often contaminated with oil or water. Chlorinated solvents must be avoided as they can induce stress corrosion [20, 21]. Detergents must not be used as they can themselves be a source of contamination. Current environmental concerns encourage the reduced use of solvents, and grit blasting is favored. The level of abrasion must be controlled so as to avoid folding of the damaged surface layer. Either the use of high pressures or slow pass rates in the grit-blasting process will cause folding of the surface that may initiate void formation in the bond line [22]. A good process should not remove the surface oxide, which is indicated by a change of color to a dull, dark gray. A range of processes have been developed using alumina and fluorosilic acid as abrasives, followed by a nitric-acid etch, and this process produces good durable bonds [16, 23, 24]. In an attempt to create a more structured oxide, alkaline peroxide etching has been explored [23]. After 36 h at room temperature, or 20 min at 50–70 °C, a 2 μm thick stable grey oxide is produced [25]. Treatment using a combination of trisodium phosphate, disodium tetraborate, potassium fluoride and hydrofluoric acid followed by a 2 min dip in a 5% trisodium phosphate, 2% potassium fluoride and 2.6% hydrofluoric acid mixture produces an anatase oxide layer [23]. The anatase form of titanium dioxide slowly reverts to a rutile form in a moist warm environment, with an 8% volume decrease that can induce interfacial stresses. Stabilization of the anatase form is achieved by adding 0.75% anhydrous sodium sulfate into the etch solution [25–27]. An alternative treatment that involves the use of sodium hydroxide and chromic-acid anodisation (CAA) (TURCO) produces structures with a large amount of macroroughness, 3.4 μm peak to valley, with little or no microroughness, the oxide being ~17.5-nm thick. This process produces no hydrogen embrittlement, which is a problem with the acid-etching process. Pasajell 107 treatment, which involves using 40% nitric acid, 10% combined fluorides, 10% chromic acid, 1% couplers and water to produce a thixotropic paste for brush applications, has good durability [28, 29]. After rinsing, a stable anatase structure is produced that is stable up to 175 °C and converts to rutile at 350 °C.

9.3.3 Primers

Primers are used to protect substrates prior to bonding, increase surface wettability, prevent corrosion and may promote chemical-bond formation with the surface [13]. Silanes are widely used as coupling agents and typically have the structure: R^*–Si(OR′)$_3$, where R^* is a functional group that can react with the adhesive, and may be epoxy or amine groups. R′ is usually an ethyl, methyl or acetyl group. The most commonly used silane is γ-APS (aminopropyl triethoxysilane) [18, 19, 30]. The γ-APS primer is applied as a 1% solution in a volatile solvent and can increase the wet strength of a Ti-6Al-4V/epoxy-resin bond by 50% [18, 19]. In certain situations it is appropriate to use the epoxy γ-GPS (glycidoxypropylmethoxysilane) rather than the amine functionalized primer. The action of the primer is not simply to create chemical bonds, but can generate nanoparticles through a sol-gel process that will toughen the epoxy interface (Fig. 9.5) [14]. The small particles close to the surface significantly increase the toughness and durability of the adhesive bond.

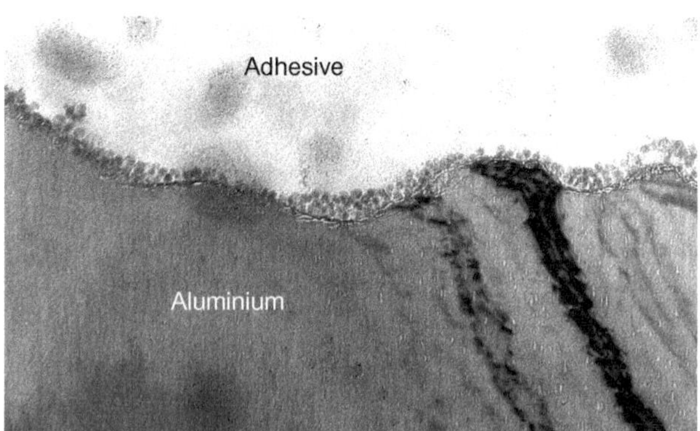

Fig. 9.5 Electron micrograph of a γ-APS-treated clad aluminum surface.

9.4 Composite pretreatment

A roughened surface is desirable to achieve a good bond with the adhesive. Peel ply or tear films can be incorporated into the surface of the composite during manufacture and removed prior to bonding. By using woven ply covered with silicon or fluorinated compounds the films can be delaminated producing a rough surface. However, it is essential that these low-energy contaminants that are essential to achieve the action of the ply do not lead to a weakening of the bond that is produced, and it is therefore essential that they are thoroughly removed usually by abrasion or specialist solvent treatments [6]. Various types of ply are available; nylon mesh requires the use of release agents and polyester requires heat treatment to achieve release but generates a smooth surface that does not produce good bonds [11, 31, 32]. Debris must be removed without the use of detergents. Roughness is best imparted using abrasives [33, 34], grit blasting [35], peel ply [36, 37], tear ply [36], corona discharge [37], plasma treatment [35], flame treatment [38] and laser ablation [37]. Electric corona discharge and plasma or flame treatment create hydroxyl and carbonyl functional groups that enhance bonding between substrate and adhesive.

The surface tension of CFRP and GRP are estimated theoretically to be 560 and 70 $mJ^0\ m^{-2}$, respectively, and are readily wetted by polyester and epoxy resins with surface energies of 35 and 43 $mJ^0\ m^{-2}$, respectively [39, 40]. GRP has a much lower modulus than CFRP, however, the bond integrity is dominated by the characteristics of the matrix rather than the fiber. Thermoset polyester or epoxy composites tend to form a resin-rich surface layer and in some fabrications a gel coat may be used to achieve a smooth surface. Both the gel coat and the resin-rich surface layer are very brittle and can subsequently fail catastrophically when overloaded. It is therefore desirable to use highly compliant adhesives to spread the applied load over as large an area as possible.

Kissing bonds, in which materials appear to be in contact but significant bonds have not been formed, are difficult to find in production and catastrophic if they enter service. These weak bonds are often associated with the incorporation of air during bonding or the migration of species that will either inhibit wetting or create a weak bond [41]. An often unrecognized problem is associated with moisture retained in uncured laminate or condense at the surface when the adhesive is not stored or thawed properly or left out too long in the lay-up room. Water can be absorbed by nylon fibers before the laminates and hence incorporated into the bond structure. The adverse effects of moisture are avoided if good venting is used so that all water vapor is removed during cure. The structure of peel ply is such that it will naturally trap water within the composite. Thoroughly drying all surfaces and components is highly recommended to ensure good joint structures are created [11]. Mixing uncured adhesive films and matrix resins with radically different cure characteristics can produce dry bonds, with virtually zero interfacial strength.

The wedge-crack test is known not to work for composite laminates made from woven fabric layers, because of the tendency of any initial crack to be diverted into 90° fibers on the surface adjacent to the adhesive layer. It is standard design practice not to locate a layer of 90° fibers adjacent to a step on a stepped-lap bonded joint. In adhesive bonding to composite structures, harmful effects of moisture can occur if water is trapped inside the bag during cure.

9.5 Bonding composite to metal

The principle additional problem in bonding a composite to metal is the large difference between the thermal expansion coefficients of the materials involve in bonding. Stresses can be created during thermal cycling that will compromise the joints. Bonding carried out at high temperature may create a situation where the adhesive contracts more than either the composite or the metal. Stresses will be created parallel to the interface and delamination may follow. It is therefore desirable to cure the adhesive close to the working temperature. However, many adhesives are not fully cured and do not develop their full mechanical properties unless they are subjected to a postcure procedure. A variety of materials can be used as adhesives, which adhesive is more appropriate will depend on the particular application.

9.6 Adhesives

The choice of adhesive depends upon the service conditions encountered and the type of application. Sticking a metal tag to a composite structure may not require a structural adhesive, however, uses of a typical pressure-sensitive adhesive-containing solvent, can induce *environmental stress cracking*. Care must always be taken to check that the adhesive used is appropriate for the task. Selection of the adhesive requires consideration of the strength of the joint to be created, temperature range over which it operates and environment it will experience; hot wet, cold dry, contaminated by solvent, moisture, etc. Typically adhesives divide into two classes; structural adhesives that are based on *thermoset* resins and nonstructural adhesives that are based on *thermoplastic* resins. Some high-temperature thermoplastics can be used in structural applications provided their glass to rubber transition, T_g, is sufficiently high to avoid creep. However, thermoplastic adhesives require that the polymer can be applied from the melt phase and this can create its own problems.

9.6.1 Thermoset resins

Thermoset resin, as the name implies, form three-dimensional nonreversible networks during the transformation of the liquid resin to the solid (Fig. 9.6). The dimensions of the resin layer are fixed at the point at which the liquid is transformed to the gel, and continuation of cure will often transform the rubbery gel to a glassy material. The resultant solid has good mechanical properties up to the T_g. Some polymers have T_g values below room temperature and therefore exist as rubbers. These latter materials do not have a sufficiently high modulus for use in structural applications, but can be used for flexible fastenings, fillers and similar applications. Increasing

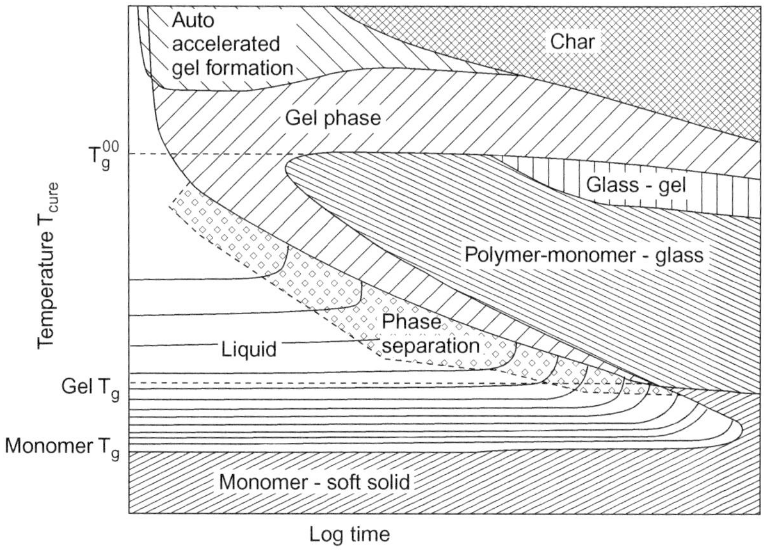

Fig. 9.6 Time temperature transformation diagram for a phase-separating epoxy resin system.

the temperature to T_g^{∞} ensures completion of the cure cycle and usually achievement of good mechanical properties.

In thermosets, the cure process involves chemical reactions transforming reactive monomers into a three-dimensional stable polymer matrix that does not show significant creep [3]. Cure will usually involve the use of heat, radiation, or light (photoinitiation), moisture, activators or catalysts. The main classes of thermosetting polymers are based on the following type of chemistry: (a) phenol-formaldehyde; (b) urea-formaldehyde; (c) melamine-formaldehyde; (d) polyesters; (e) epoxies; and (f) silicones. Many of the adhesives are based on (e) or (f) depending on whether a rigid (structural) or flexible (nonstructural) adhesive is desired. Thermosets are densely cross-linked and have good resistance to heat and solvents and show significantly less elastic deformation under load than the thermoplastics.

9.6.2 Epoxy and polyimide adhesives

Epoxy adhesives are widely used and can be either one-stage (curing agent already mixed in) or two-stage, where the user mixes in the curing agent just before use. One-stage materials can be obtained in film form, very much like a prepreg without the reinforcement, or in a paste [3]. In order to achieve a constant bond line some adhesives are supplied on a nylon mesh. Both room-temperature and elevated-temperature curing systems are used, although often room-temperature curing adhesives require post curing to develop good mechanical properties at elevated temperatures. Cure times can range from a few minutes to more than 12 h for large, critical-performance parts [42–44]. Depending on the curing agent, epoxy resins can be used over the temperature range −50 to 260 °C. Epoxy adhesives based on multifunctional resins can be

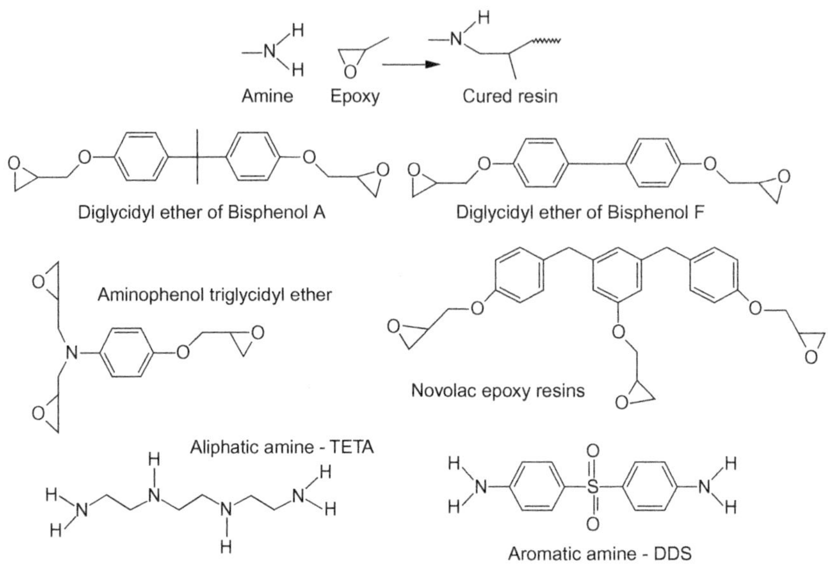

Fig. 9.7 Typical monomers used in the formulation of an epoxy resin.

used up to ~225 °C. Shear strengths will typically be between 35 and 70 MPa. Epoxy resins are generally limited to temperatures no higher than 175 °C. Above this temperature chemical rearrangements occur that will ultimately lead to loss of strength [45]. Some of the structures used to create epoxy adhesives are presented in Fig. 9.7.

The novolac epoxy resins are often solids, but when blended with the bisphenol-based epoxy can form viscous liquids. Mixing the epoxy with an amine initiates cure, however, as in the case of a 1,4-diaminodiphenyl sulfone (DDS), cure reaction will only occur once the melting point of the DDS has been reached. Epoxy resins have relatively low peel strength and flexibility, are sensitive to moisture and surface contamination, can be brittle at low temperatures and have a slow rate of cure. Gelation can occur after 8–10 h, but full cure may require 2–7 days [3]. Epoxy adhesives used in the aerospace industry retain their strength up to 215 °C, but after 3000 h of aging may have lost 20% of their original strength [42].

The FM-R300 film adhesive is typical of an aerospace structural adhesive in which the inactivity of the DDS has been addressed by blending with other more active amines. FM-R300 has been used for bonding the wing-root assemblies, titanium to graphite epoxy in F-18 fighter aircraft and a range of other metal–composite bonding applications in commercial and military aircraft worldwide [46]. Cure requires heating to 120 °C ± 3 °C for 30 min to achieve gelation and then a further 90 min to complete the cure. Typically a pressure of 0.28 ± 0.03 MPa is used to ensure good bonding is achieved. Even this relatively low-temperature cure can lead to significant thermal stresses owing to the difference in coefficient of thermal expansion between metal and composite [46]. A number of variants of the basic FM-R 300 adhesive have been formulated to fully cure at temperatures between 120 and 170 °C [47]. The use of

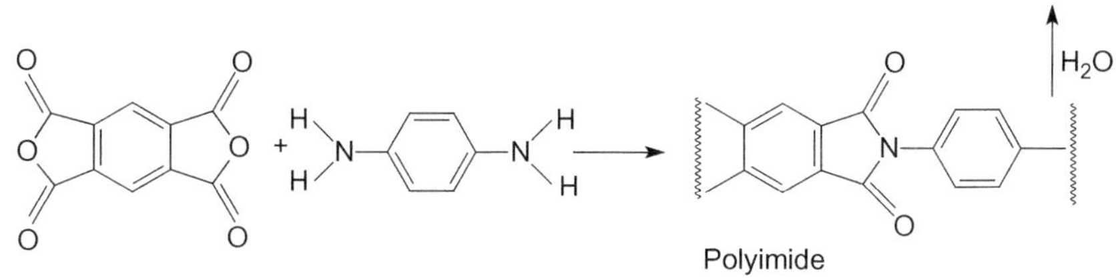

Fig. 9.8 Schematic of polyimide chemistry.

more reactive amines to reduce the cure temperature lowers the T_g of the final matrix material. For high-temperature adhesives required in aerospace applications epoxy phenolics and polyimides (Fig. 9.8) are used and these can operate to above 290 °C.

Polymeric adhesives are insulators and have the potential for creating problems by achieving conducting paths between components crucial to protecting aircraft against lightning strikes. This problem can be overcome by the inclusion of metal particles in the resin and may have the added advantage of improving the mechanical strength. The main disadvantages of using polyimides is their long curing times and the high volatile content, about 12% for filled adhesives and 30% for unfilled adhesives.

FM34-R is a polyimide adhesive that can be used up to 540 °C with mechanical properties similar to epoxy-phenolic adhesive HT-R 424 and does not exhibit any significant drop in strength over 40,000 h at 260 °C.

9.6.3 Hot-melt adhesives

Thermoplastic materials that are solids at room temperature can be used as hot-melt adhesives. Because they are not cross-linked they can creep if held at high temperatures for prolonged periods of time and hence not used for critical structural bonding applications. The molten adhesive is applied to the substrate and cools to a flexible solid film. Thermoplastics generally have good peel and environmental-resistance properties; however, they can melt and flow abruptly when heated close to their melt temperature [3, 6]. Typical hot-melt adhesives include polyamides, poly(ethylene-co-vinyl acetate), nylon, polyacetal "Delrin," vinyl polymers and polycarbonates. The adhesion can generally be improved by priming the surface of the metal substrate with a dilute solution of a phenolic resin [48, 49].

9.6.4 Acrylic-based thermoplastic adhesives

Acrylic adhesives are flexible and are formulated using two or more acrylic monomers, e.g., methyl methacrylate and acrylic acid [49, 50]. The monomers are often liquids, the resultant adhesives are resistant to photo-degradation [51], have good shear strength and will bond to a variety of materials; plastics, metals and composites, even though oily or dirty surfaces [50]. These resins have been used extensively for the production of glass laminate. Acrylic-based thermoplastic adhesives have strengths comparable with the epoxys, at room temperature but lack high-temperature strength.

9.6.5 Cyanoacrylate adhesives

The cyanoacrylates, or super glues, are well known for their rapid curing characteristics. Trace amounts of moisture will often speed the rate of cure [50, 51]. The cyanoacrylates are available as liquids and gels, and have excellent tensile strength but poor peel strength and high-temperature characteristics. Anaerobic acrylic adhesives cure when air is excluded and are often used to secure nuts, etc. [50]. The presence of iron or copper in the substrate catalyzes the cure process. Unfortunately, cyanoacrylates are limited in use to temperatures below 149 °C. Industrial uses include fitting (e.g., bearings into housings), locking (e.g., nuts onto bolts), sealing (e.g., liquid gaskets), retaining (e.g., shafts into hubs) and bonding (e.g., flanged couplings) but are not structural adhesives.

9.6.6 Polyimide adhesives that cure by addition

The original polyimides used condensation chemistry to achieve cure. The volatile emissions can lead to void formation. To avoid this problem polyimide systems have been developed that use addition chemistry to achieve cure (Fig. 9.9). A pressure of 1.4 MPa is usually used to ensure consolidation. Cure will be achieved using a typical radical initiator, such as a peroxide.

9.6.7 Other reactive adhesives

Phenolic and other formaldehyde condensation polymers were some of the earliest systems used in aerospace applications and include the Redux range of adhesives [7, 52]. Redux was originally developed for the Comet aircraft and is based on condensation reactions involving novolac and resol chemistry (Fig. 9.10). The methylene-bridge-formation reaction can be replicated many times and eventually forms a three-dimensional network. The network is produced by the elimination of water and bonding requires the joint to be held in place while this chemistry occurs. The final network is relatively nonpolar and hence has very good environmental stability. Bonds created in this manner have high durability, however, the problems associated with the bonding process have made this type of bonding less attractive than epoxy-resin chemistry.

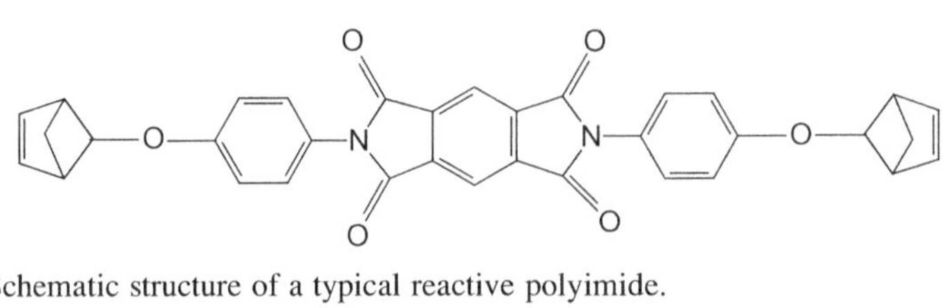

Fig. 9.9 Schematic structure of a typical reactive polyimide.

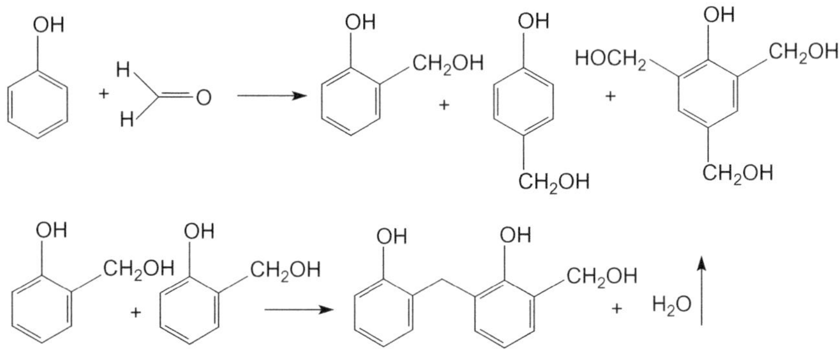

Fig. 9.10 Schematic of novolac methylene-bridge formation.

9.6.8 Bismaleimide (BMI) adhesives

A series of high-temperature adhesives have been developed based on bismaleimide (BMI) chemistry [52, 53]. The BMI resin systems are usually a blend of monomers with properties that have been optimized to give good drape and tack and develop excellent high-temperature characteristics. BMIs can be used between 175 and 230 °C, but can be brittle because of the high level of cross-linking required to achieve the high temperature characteristics. BMI adhesives fit in between high-temperature epoxy and polyimide adhesives. In general, good results can be obtained using a cure of 2 h at 175 °C under 0.28 MPa pressure, followed by a 2–4 h post cure at between 200 and 225 °C [43]. BMI adhesives retain their strength up to about 300 °C, but are not recommended for prolonged use above 210 °C.

There are a wide range of nonstructural adhesives that are discussed elsewhere and will not be reviewed here [3, 54, 55].

9.7 Composite-metal bonded structures

In the previous sections, the factors that influence the selection of pretreatment for the metal-adhesive and composite-adhesive have been considered. Combining these elements in a metal-composite structure creates a number of specific issues. Mixed bonded structures will arise in two distinctly different situations: primary aerospace structure manufacture and service repair. In the former, the possibility of careful control over the conditions used in the preparation is possible. In the latter, many of the pretreatments that use sophisticated etch and anodisation processes are not available, and pretreatment will often be restricted to grit-blasting followed by careful cleaning of the surfaces to be bonded. For the grit-blasting process to be successful the abrasive must cut the composite and metal surfaces and usually dictates the use of metal-oxide powders. The grit must not be recycled, and compressed air must be avoided as it can be contaminated with oil or water. If the processes are carried out on an aircraft, care must be taken that grit particulates do not pass into fuel filters and cause fuel system malfunctions [41, 56].

The discussion of metal-composite adhesive bonds is best divided into consideration of manufacturing processes and repair processes.

9.7.1 Manufacturing of metal–composite and metal/fiber laminate (MFL) structures

Metal/fiber laminates (MFLs) are versatile, high-performance materials with applications in the aeronautical industry. MFLs were developed by Delft University of Technology in the 1980s [57]. These hybrid composites consist of a high-strength alloy and a fiber/epoxy layer. There are three groups, ARALL, GLARE and CARALL, owing to the different fiber-adhesive layers used, i.e., aramid, glass and carbon fibers, respectively [58–60]. CARALL laminates consists of thin layers of carbon-fiber/epoxy (CF/E) prepreg sandwiched between sheets of metal. This class of material offers higher modulus, higher tensile strength and lower density than 2024-T3 alloy, and has good fatigue resistance [57].

Co-cure composites have been used by Airbus in the A340, and more recently in the A380 [2]. Developed by Airbus UK (Bristol, UK) and the moulders, Stork Fokker AESP (Hoogeveen, The Netherlands), a J nose monolithic structure has been created to replace a heavier five-part D-nose flanked by flat composite sandwich panels cored with Nomex honeycomb (DuPont Advanced Fiber Systems, Richmond, VA, USA). This structure demonstrates the possibility of creating other structures using thermoplastic composites and has allowed two-thirds of the wing, a surface of $\sim$55 m^2, to be fabricated from composite.

The A340's tough reinforced thermoplastic composite (RTC) is relatively easy to process. Unlike thermosets, thermoplastics do not cross-link and molding simply requires application of heat. However, for this to be successful it is essential that high T_g thermoplastics are used. Aerospace-grade RTCs closely resemble thermoset advanced composites using continuous-fiber reinforcement and are unidirectional, fabric-based prepregs, narrow tapes, pultruded sheets, tapes and rods. The matrix will be typically polyethersulfone (PES), polyetherimide (PEI), polyphenylene sulfide (PPS), polyetherketoneketone (PEKK) or polyetheretherketone (PEEK). The polymer selected depends on the application and cost. One cautionary note must be added, the outer skin is very easily damaged and because it is formed from very thin aluminum foil is difficult to repair. As a consequence RTCs are likely to be used in areas where impact damage can be restricted. As most surfaces are painted, damage to the paint can often be a good indication of subsurface problems.

Thermoplastics require to be heated to a sufficiently high temperature to achieve the required degree of flow. For PEKK and PEEK, achieving the high temperature is a problem that is solved by resistance heating. The bond is created by passing an electrical current through meshes clamped either side of the bond. The clamp applies pressure to a very thin, open mesh, approximately 0.2 mm thick, which remains in the joint without disrupting the bond. Heating is local to the mesh and reduces the possibility of distortion. This process was used to create the main land gear door for a Fokker 50 regional passenger aircraft and used carbon-fiber/PPS composite. For this process to be successful access to both sides of the bond is essential. Access can be arranged during construction, but is often more difficult in repair situations.

In many applications, fiberglass is used rather than carbon fiber as the former is stiff enough to resist deflection, a desirable property when wishing to maintain a wing's aerodynamic shape. Initially, the fabrication used a compression molding process to produce the fiberglass-reinforced PPS structures. Large structures are usually created in an autoclave and use a semipreg approach [3, 6]. The sized fabric is coated with the thermoplastic and passed through heater rollers to produce a fully impregnated fabric that can still be formed into the desired shape. The cure temperatures are higher than for thermosets, 250–350 °C, and full wet out of the fabric and laminate consolidation are achieved during autoclave cure. Whereas autoclave pressure is similar to that for thermosets, the cycle is much shorter, typically in the order of minutes. Mechanical fasteners are incorporated to avoid problems associated with attempting resistance welding sections, where access to both sides of the bond would be difficult.

The use of RTCs is predicted to increase with potential use in control surfaces such as fuselage, rudders and ailerons [58, 59]. These structures will require integration of composite and fasteners. The use of MFLs requires that an adequate surface treatment is applied to the metal foil. The most common method of surface treatment used is CAA. Concerns over the use of chromate treatments have led to investigations of other approaches, such as sulfuric-boric-oxalic acid anodisation (SBOA) [58, 59, 61–64]. Studies have shown that SBOA is an attractive approach to surface treatment [61]. During controlled manufacture, the MFLs can be stretched after curing in order to reverse the internal stress system in the material arising from the difference in the expansion coefficient differences between the resin and the metal [65].

Fatigue failures of MFLs involve many possible modes: matrix cracking, fiber breakage, fiber–matrix debonds, void growth and delamination. During fatigue, the overall modulus and strength of the material decrease progressively until it is unable to support the applied load and failure ensues. Textile fabrics impart a higher damage tolerance to impact loading, failure in woven fabric composites being initiated by fiber–matrix debonds in fiber bundles oriented transversely to the load direction. In the main directions, weft or warp, the final failure is strongly determined by the fiber bundles oriented parallel to the loading direction. The fatigue properties are influenced by the loading direction, stress concentrations around discontinuities and ductility of the matrix. The mechanical properties of the laminate will depend on the fiber orientation, for instance, a 3/2 lay-up can be made up of multiple cross-plied 0/90 layers, e.g., Al/O°/90°/Al/90°/O°/Al, or may be unidirectional. The fatigue properties of these materials will reflect the way in which the stresses interact with the fiber layer [66].

The MFLs are essentially metal-bonded composites and have either fiber-critical or metal-critical failure. For fiber-critical failure to occur, a crack is observed perpendicular to the outer fiber direction with a simultaneous crack in the outer metal, usually an aluminum layer. In aluminum-critical failure, the crack runs in the rolling direction of the aluminum sheet, which is generally in the fiber direction of the laminate [65]. Owing to their low strain to failure (2%), the aramid (ARALL) and carbon (CARALL) laminates always show fiber-critical-failure behavior. GLARE can exhibit cracking energies that are close to or even better than the monolithic alloys [67].

In CARALL, fatigue-crack growth is often associated with delamination. Fatigue cracks occur in the layers of the laminate and the fibers remain intact and bridge the crack. The fiber bridging reduces the crack-opening displacement and the stress intensity factor at the crack tip. Consequently, the fatigue-crack growth rate is reduced [66, 67]. Delamination buckling is a potential problem with fiber–metal laminates and can occur when a partially delaminated panel is subjected to a compressive force [67, 68]. Interaction of local buckling and extension of the delaminated zone results in a decrease of the residual strength and eventually collapse of the structure (Fig. 9.11). The critical factor that controls the propagation is the interfacial energy and this emphasizes the importance of ensuring a good bond is created between the aluminum and fiber matrix.

The main problem that arises with these materials is that they are usually produced in sheets that have to be joined. A variety of different approaches have been explored [66]. Riveted GLARE joints behave differently from aluminum joints, owing to the different action of the internal stress system. The lower stiffness of the prepreg layers results in the Al layers tending to crack earlier than in a monolithic structure when the same external load is applied. The crack propagation rates in GLARE are orders of magnitudes lower than in monolithic aluminum.

There are as a consequence of these failure modes restrictions on the manufacturing processes that can be used for MFLs. The limits are partly owing to the different

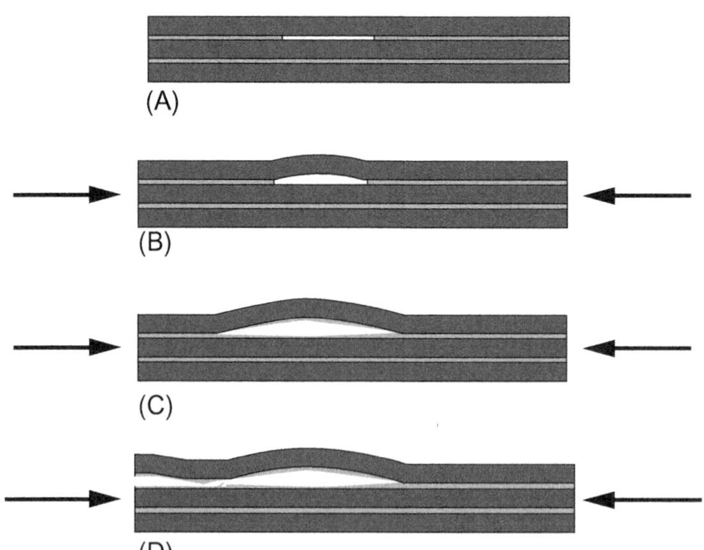

Fig. 9.11 Schematic representation of the delamination buckling failure mechanism: (A) initial delamination; (B) local buckling; (C) delamination growth; (D) failure.

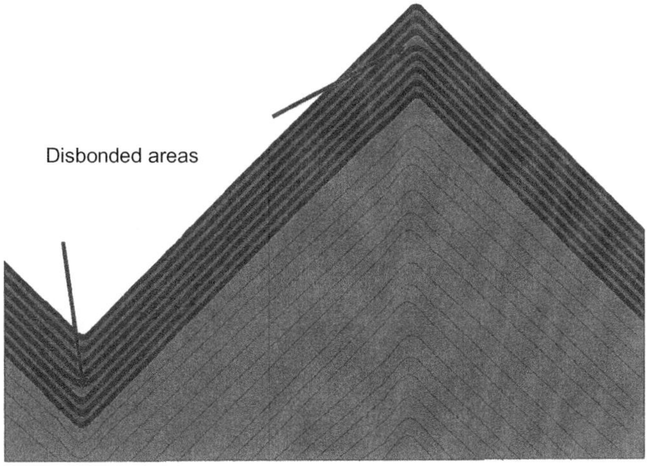

Fig. 9.12 Regions of delamination at the corners formed by forming MFLs.

failure modes and partly owing to the properties of the constituents in the laminate. For machining processes, the wear of the cutting tools during machining operations of GLARE arise from the abrasive nature of the glass fibers. For the forming processes; the limited formability expressed by a small failure strain is related to the glass fibers [67]. Examples of problems that can arise when MFLs are formed are shown in Fig. 9.12.

Delamination can occur at corners when sheets of MFLs are used. The risk of delamination can be reduced by use of appropriate fiber alignment to decrease crack growth. The peel forces are important during milling and drilling processes. During an edge-milling process a large value for the helix angle of the milling tool may result in edge delamination: the top layer of the laminate is peeled off as there is a vertical component of the cutting forces. Similar effects arise when a panel is drilled. Many of the problems can be solved if the unique properties of the materials are taken into account in the fabrication stage. Changes in the way in which the MFLs are processed can reduce, if not eliminate, the delamination problems [68]. The use of MFLs as stringers and joining panels of the material require alternative approaches to be adopted, in which the shapes are created during manufacture [69].

9.7.2 Effects of the environment on aging of MFLs

Exposure of MFLs to high-temperature humid environments depends crucially upon the accessibility of the edges of the laminate. If the edges are sealed then the moisture will not permeate into the resin and the materials are unaffected [58, 59, 61, 70]. However, if moisture is able to enter the laminate, the strength of the material is lowered and the possibility of loss of interfacial integrity can arise. In general, the aluminum-layered structure tends to reduce the moisture uptake compared with conventional resin systems, and the loss in physical properties is correspondingly reduced. Studies of the fatigue failure of MFLs [61] with foil that had been subjected to SBOA and CAA indicated that both treatments produced comparable durability.

Microstructural observations of the fracture surfaces by scanning electron microscopy show hackle formation is the predominant damage mechanism. Pinholes were observed in the bonded structure that reflect the nature of the surface treatment. This insensitivity to moisture is an excellent property but relies on the metal foil not being punctured and the edges of the laminate becoming exposed through damage. The use of thermoplastics has the added advantage of incorporating energy dissipation processes in the resin layer that will improve the damage tolerance of the material [71–73]. If the resin is viscoelastic then the failure becomes rate dependent and increases in toughness are observed relative to brittle materials.

9.7.3 Metal-composite structures in repair situations

The bonding of composite to metal structures is a very important part of the maintenance operation of many aerospace operations. Carbon-fiber patches have been used to reinforce areas where metal has undergone fatigue damage [74–78]. The most effective repair technique is the adhesive bonding of a composite patch on the damaged structure. The advantages of this technique include:

- No stress concentrations
- Good fatigue resistance
- The high specific strength and stiffness of the fiber-reinforced composite enables the patch thickness to be reduced, with concomitant benefits of decreased aerodynamic drag and the relief of the limitation of internal space for patches

A high level of repair safety is provided with good traceability of crack growth underneath the composite patch by eddy-current inspection. One of the issues that arises when assessing the viability of composite bonding is the question of thermal cycling. Good bond performance has been achieved by using a combination of low-temperature and high-temperature adhesives. For a joint with dissimilar adherends, the combination of two adhesives gives a better performance (increased load capacity) over the temperature range than a high-temperature adhesive alone. Thermally cycled mixed adhesive joints were proved to be durable at low temperatures after being exposed to high temperatures, and vice versa [75, 76]. Theoretical modeling of the stress distribution in bonded patches exemplifies the problems that arise as a consequence of the differences in thermal expansion coefficient between the metal and the composite [79]. At high temperatures, a high modulus (brittle) in the middle of the joint retains the strength and transfers the entire load. At low temperatures, a ductile adhesive at the ends of the joint is the load-bearing adhesive. To guarantee that the load is transferred through the low-temperature adhesive, the ends of the overlap can be stiffened. For a joint with dissimilar adherends, the combination of two adhesives gives a better performance (increased load capacity) over the temperature range considered than the use of a high-temperature adhesive alone [75, 76].

It is crucially important that good surface preparation is used in creating the patches, abrasion and grit-blasting being the most appropriate approaches for in-service repairs. As in the case of adhesive-bond design consideration of the edges of the patched areas is critical to avoid high-stress areas. It is common to use tapered patches and use appropriate distributions of fiber orientation to match the underlying load profile [80].

9.7.4 Nondestructive testing (NDT) of metal-composite structures

With the increased use of MFLs, it is desirable to be able to characterize the presence within the structure of hidden delaminations and potential fatigue failures. The most obvious approach to nondestructive testing (NDT) for these composite materials is to use X-ray analysis. Whereas this approach provides a direct visualization of areas of delamination, it is a difficult method to apply within a manufacturing environment and of limited practical application when carrying out in-service repair of structures. A variety of methods have been used to examine MFLs and include: neutron radiography, X-ray tomography, holography/shearography, thermography, laser ultrasonic, ultrasonic phased arrays, ultrasonic with air coupling and acoustic emission [81].

The approaches that are typically adopted include:

- Visual inspection: this is the most commonly used method and often indicates where damage has occurred.
- Ultrasound inspection: a variety of methods can be applied but usually require the component to be removed from the structure and immersed for high-resolution C-scan measurements. Manual inspection can be useful for detection of delamination. The use of ultrasonics for complex MFL structures has been successfully investigated and C-scan measurements are able to provide very useful information on honeycomb types of structure [82].
- Penetrant crack inspection is usually used on metallic components and can be used for the surface of MFLs
- Traditional tapping test—this manually or semiautomated test is often a useful indicator of defective structures.
- Resonance test (Fokker Bondtester): all metal/metal adhesive bonded joints and some sandwich components can be investigated using this approach.
- Eddy-current inspection: This is an electrical conductivity measurement for aluminum alloys to determine the heat-treatment condition and to measure the layer thickness; it can be used for crack inspection only in special cases, e.g., for cold hardening of drill holes. This method can also be used for MFLs in special circumstances. Eddy-current measurements are able to detect damage through changes in the magnetic susceptibility, and the HTS-SQUID magnetometer [83] has been shown to be useful for the study of GLARE.
- X-ray inspection with film and radiography: this technique is difficult to apply in service because of Health and Safety issues relating to the use of X-ray sources.
- Thermography: active—transient thermal NDT and evaluation (E) (i.e., thermography) is commonly used for assessing aircraft composites. Successful studies have been used using pulsed thermography and pulsed phase thermography for the investigation of defects (i.e., impact damage and inclusions for delaminations) in GLARE and GLARE-type composites [84].
- Digital shearography with digital imaging processing, has been used to identifying defects both in small- and large-scale structures [85].

The repair of composite materials is a complex problem that requires a detailed understanding of the nature of a defect and in the case of MFLs the structure characteristics

[68, 74]. NDT and E techniques for assessing the integrity of an aircraft structure are essential to both reduce manufacturing costs and out-of-service time of aircraft owing to maintenance.

Moisture ingress is one of the major causes of failure in adhesive bonds and this is true of MFLs. The application of dielectric NDT to MFLs has been demonstrated in studies of the aging of carbon-fiber-aluminum bonded structures [86].

The dielectric NDT method has significant potential for the study of aging in jointed structures. In carbon-fiber bonded structures and also in GLARE the electric wave propagation is possible down the bond line through the metallic conductors. In the case of GLARE the propagation will follow the metallic layers of the laminate. In the case of the carbon fiber-metal laminate the carbon fiber is sufficiently conductive to allow electric field propagation in the carbon fiber. Many bonded structures are in practice wave guides for the electric field and it is possible to conduct time domain reflectometry (TDR) measurements. The time domain traces as seen in Fig. 9.13 show a peak that corresponds to the reflected signal, which returns to the probe after transversing the joint. As water enters the joint, the dielectric permittivity of the adhesive changes and the peak is moved to longer times. The shift is directly proportional to the amount of moisture that is taken up by the joint. The pattern of peaks within this interval does not change significantly indicating that the integrity of the bond is being

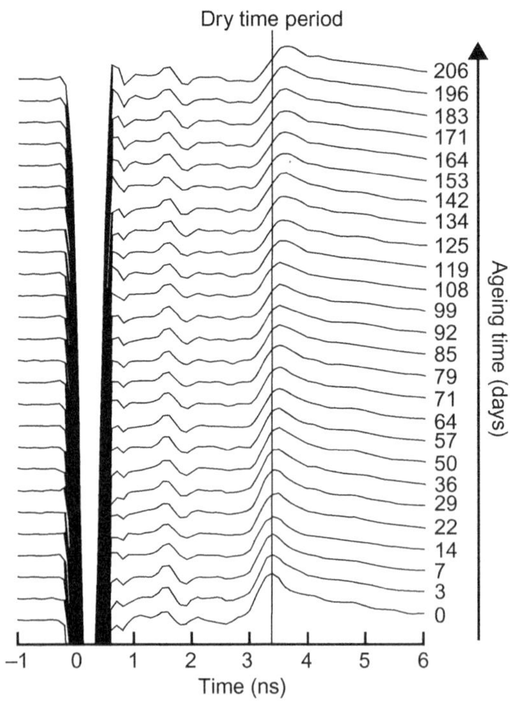

Fig. 9.13 Cascade plot of TDR traces as a function of the aging time for an epoxy-bonded CFRP-aluminum shear joint, aged in water at 50 °C.

maintained. If delamination of disbonding were to occur then additional peaks would be observed. Measurement of the dielectric frequency response of the joint indicates the nature of the moisture in the joint. Moisture entering the joint can either be dispersed in the matrix and plasticise the resin lowering its, and hence its modulus, or collects in microvoids, in which case the water has little effect. In the case of a bond containing a metal, water can interact with the metal surface changing the nature of the oxide interface and leading to lowering of the interfacial strength. In most cases the metal surface is an oxide and water is able to hydrate the oxide. The swelling of the oxide can lead to a loss of integrity of the interface and a weakening of the bond. The ultimate effect of water ingress will be to reduce the interfacial strength and cause adhesive failure. The correlation between the weakening of the bonded interface and the changes in the dielectric frequency spectrum has been demonstrated [86]. Both in the case of MFLs and in other situations, the loss of adhesive strength shows a good correlation with the changes in the dielectric spectrum.

9.8 Conclusions

Successful adhesive bonding of metal-composite structures relies on good surface treatment of both the metal and composite, selection of an adhesive that can compensate for the differences in thermal expansion of the substrates and has good mechanical characteristics. The adhesive must be able to form a good bond with both types of substrate, have good impact and shear characteristics and be able to transfer the load between the substrates effectively. The current commitment to GLARE focuses attention on metal-composite structures and highlights the large potential benefits to be gained in weight savings by the effective use of these materials in aerospace applications.

Acknowledgments

The author wishes to acknowledge the support of the Carnegie Trust for provision of a maintenance grant for a student; British Aerospace, Prestwick, UK, in providing the aluminum alloy and Cytec Aerospace Limited, Wrexham, in providing the prepreg. The support of the Engineering and Physical Sciences Research Council, AFSOR and the Defence Evaluation and Research Agency (Farnborough) are also gratefully acknowledged.

References

[1] L.J.J. Kok, in: N. Coen (Ed.), Around Glare: A New Aircraft Material in Context, Springer, Heidelberg, Berlin, 2002, pp. 99–114.

[2] High Performance Composites, Thermoplastic Composites Gain Leading Edge on the A380, 2006. 3/1/2006.

[3] R.A. Pethrick, Polymer Science and Technology for Engineers and Scientists, Wiley, New York, 2010.

[4] T. Walker, in: S. Henderson (Ed.), The First Jet Airliner: The Story of the De Havilland Comet, Newcastle-Upon-Tyne, Scoval Publishing Ltd., 2000.
[5] K.K. Chawla, Composite Materials, Science and Technology, second ed., Springer, Berlin, 1998.
[6] A.J. Kinloch, Adhesion and Adhesives, London and New York, Chapman and Hall Ltd, 1987.
[7] J.A. Bishopp, The history of Redux® and the Redux bonding process, Int. J. Adhes. Adhes. 17 (1997) 287.
[8] I.A. Ashcroft, D.J. Hughes, S.J. Shaw, Mode I fracture of epoxy bonded composite joints: 1. Quasi-static loading, Int. J. Adhes. Adhes. 21 (2001) 87.
[9] J.M. Whitney, C.E. Browning, W.J. Hoogsteden, A double cantilever beam test for characterizing mode I delamination of composite materials, Reinf. Plast. Comp. 1 (4) (1982) 297.
[10] R.D. Adams, J.W. Cowap, G.M. Farquharson, G. Margary, D. Vaughn, The relative merits of the Boeing wedge test and the double cantilever beam test for assessing the durability of adhesively bonded joints, with particular reference to the use of fracture mechanics, Int. J. Adhes. Adhes. 29 (6) (2009) 609.
[11] L.J. Hart-Smith, A peel-type durability test coupon to assess interfaces in bonded, co-bonded, and co-cured composite structures, Int. J. Adhes. Adhes. 19 (1999) 181.
[12] J.D. Venables, Adhesion and durability of metal-polymer bonds, J. Mater. Sci. 19 (1984) 2431.
[13] D.E. Packham, Surface energy, surface topography and adhesion, Int. J. Adhes. Adhes. 23 (2003) 437.
[14] R. Comrie, S. Affrossman, D. Hayward, R.A. Pethrick, X. Zhou, G.E. Thompson, Study of ageing of adhesive bonds with various surface treatments: Part VI. Dicyandiamide-cured epoxy joints aged at 70°C in a water bath, J. Adhes. Sci. Tech. 20 (16) (2006) 1847.
[15] A. Datta, R.A. Pethrick, S. Affrossman, Surface–analysis of etched and anodized aluminum–copper alloys, J. Adhes. 15 (1982) 13.
[16] P. Molitor, V. Barron, T. Young, Surface treatment of titanium for adhesive bonding to polymer composites: a review, Int. J. Adhes. Adhes. 21 (2001) 129.
[17] J. Comyn, G. Xiao, B. Parker, L. Mascia, Corona–discharge treatment of polyetheretherketone (PEEK) for adhesive bonding, Int. J. Adhes. Adhes. 16 (1996) 301.
[18] G.W. Critchlow, D.M. Brewis, Review of surface pretreatments for titanium alloys, Int. J. Adhes. Adhes. 15 (3) (1995) 161.
[19] G.W. Critchlow, D.M. Brewis, Influence of surface macroroughness on the durability of epoxide-aluminium joints, Int. J. Adhes. Adhes. 15 (3) (1995) 173.
[20] T.G. Gutowski, The Mechanics of Composite Deformation During the Manufacturing Process. First Conference of Composite Materials, American Society for Composites, Dayton, OH, 1986.
[21] R.F. Wegman, Surface Preparation Techniques for Adhesive Bonding, Noyes, Park Ridge, NJ, 1989.
[22] A. Wilson, M. Kindermann, D. Arnott, Void development in an epoxy film adhesive during vacuum bag cure, in: Proc. Int. Aerospace Congress, PICAST 2-AAC 6, Melbourne, 1995, p. 625.
[23] J.K. Park, J.W. Chen, K. Mukherjee, Proceedings of the American Society for Composites 14th Technical Conference, Dayton, OH, 1999, p. 661.
[24] B.M. Parker, Adhesive bonding of fiber reinforced composites, Int. J. Adhes. Adhes. 14 (1994) 137.
[25] A. Mahoon, in: A.J. Kinloch (Ed.), Durability of Structural Adhesives, Applied Science Publishers, London, 1983, p. 255.

[26] Lockheed–California Company, Prepared for National Air Development Centre, 1983, Report No. NADC-84124–60. (US Patent 2 864 732), 1983.
[27] S. Mason, in: J.F. Panday (Ed.), Wetting, Spreading and Adhesion, Academic Press, New York, 1978.
[28] W. Brockmann, O.D. Henneman, H. Kollek, C. Matz, Adhesion in bonded aluminium joints for aircraft construction', Int. J. Adhes. Adhes. 6 (1986) 115.
[29] Pasajell Data Sheets, SEMCO Division, Products Research and Chemical Corporation, 1997. 5454, San Francisco Road, Glendale, CA.
[30] F.J. Boerio, R.G. Dillingham, in: K.I. Mittal (Ed.), Adhesive Joints Formation, Characteristics and Testing, Plenum Press, New York, 1991, p. 541.
[31] F.J. Boerio, C.A. Gosselin, J.W. Williams, R.G. Dillingham, J.M. Burkstrand, Silane primers for improved hydrothermal stability of adhesive bonds to aluminum, in: Proceeding of the Symposium on Polymer Composites, Seattle, Washington, USA, 1983, p. 171.
[32] L.A. Pilato, M.J. Michno, in: L.A. Pilato, M.J. Michno (Eds.), Advanced Composite Materials, Springer-Verlag, Berlin, Heidelberg, New York, 1994, p. 208.
[33] G. Kempe, H. Krauss, Adhesion Bonding Techniques for Highly Loaded Parts of Continuous Carbon-Fiber Reinforced Polyetherketone (CF-PEEK/APC02), 1992. ICAS-92–6.9.1.
[34] G.K.A. Kodokian, A.J. Kinloch, Surface pretreatment and adhesion of thermoplastic fiber-composites, J. Mater. Sci. Lett. 7 (1988) 625.
[35] A. Baalmann, K.D. Vissing, A. Gross, Surface-treatment of polyether ether-ketone (peek) composites by plasma activation, J. Adhes. 46 (1994) 57.
[36] L.J. Hart-Smith, G. Redmond, M.J. Davis, The Curse of the Nylon Ply, 41st International SAMPE Symposium, 1996, p. 303.
[37] J. Wingfield, Treatment of composite surfaces for adhesive bonding, Int. J. Adhes. Adhes. 13 (3) (1993) 151.
[38] J.R. Arnold, C.D. Sanders, D.L. Bellevou, A.A. Martinelli, G.B. Gaskin, Surface treatment of titanium for adhesive bonding, in: Proc. 29th International SAMPE Technical Conference, vol. 1, 1997, p. 345.
[39] A. Baldan, Adhesively-bonded joints and repairs in metallic alloys, polymers and composite materials: adhesives, adhesion theories and surface pretreatment, J. Mat. Sci. 39 (2004) 1–49.
[40] D. Hull, An Introduction to Composite Materials, University Press, Cambridge, UK, 1981.
[41] K.B. Armstrong, Cautionary tales from experiences when working with adhesively bonded composite and metal structures used in aircraft, J. Adhes. Sci. Techn. 23 (4) (2009) 567.
[42] K. Armstrong, Adhesion in hostile environments, Int. J. Adhes. Adhes. 12 (2) (1992) 121.
[43] S.L. Rosen, Fundamental Principles of Polymeric Materials, second ed., John Wiley & Sons, New York, 1993.
[44] M.M. Schwartz, Composite Materials Handbook, second ed., Mc Graw-Hill, New York, 1992.
[45] I.D. Maxwell, R.A. Pethrick, P.K. Datta, Thermal modification of amine cured epoxy–resins—thermal and mechanical studies, Br. Polym. J. 13 (3) (1981) 103.
[46] D.K. Kohli, Improved 121°C curing epoxy film adhesive for composite bonding and repair applications: FM 300–2 adhesive system, Int. J. Adhes. Adhes. 19 (1999) 231.
[47] Cytec web site, 2010. http://www.cytec.com/engineered-materials/index.htm.
[48] P.E. Degarmo, J.T. Black, R.A. Kohser, Materials and Processes in Manufacturing, eighth ed., Prentice Hall, New York, 1999. Chapter 38.
[49] L. Mascia, in: L. Mascia (Ed.), Thermoplastics: Materials Engineering, second ed., Elsevier Applied Science, London, 1989, p. 32.

[50] Loctite web site, 2010. http://www.loctiteproducts.com/questions.asp.
[51] B. George, Y. Grohens, F. Touyeras, J. Vebral, New elements for the understanding of the anaerobic adhesives reactivity, Int. J. Adhes. Adhes. 20 (2000) 245.
[52] Hexel web site, 2010. http://www.hexcel.com/NR/rdonlyres/C30364F7-CA1B-462-F-8B6E-1BDA0E8F8207/0/REDUXSELECTORGUIDE.pdf.
[53] B. Sillion, Polyimides and other heteroaromatic polymers, in: G. Allen, J.C. Bevington (Eds.), Comprehensive Polymer Science, vol. 5, Pergamon, Oxford, UK, 1989, p. 449.
[54] P. Cognard, Handbook of Adhesives and Sealants: Basic Concepts and High Tech Bonding, Elsevier, London, 2005.
[55] H.D. Stenzenberger, Polymer matrices, in: E. Fitzer (Ed.), Carbon Fibres and Their Composites, Springer, Heidelberg, 1985. Chapter 2.
[56] K.B. Armstrong, Efforts to standardize aerospace composite repairs, 1988–2003, Proc. Inst. Mech. Eng. G-J. Aero. Eng. 217 (G5) (2003) 223.
[57] A. Vlot, J.W. Gunnink, Fibre Metal Laminates, Kluwer, Academic Publishers, Delft, The Netherlands, 2001.
[58] E.C. Botelho, A.N. Campos, E. Barros, L.C. Pardini, M.C. Rezende, Damping behavior of continuous fiber/metal composite materials by the free vibration method, Composites B 37 (2–3) (2005) 255.
[59] E.C. Botelho, L.C. Pardini, M.C. Rezende, Hydrothermal effects on damping behavior of metal/glass fiber/epoxy hybrid composites, Mat. Sci. Eng. A 399 (2005) 190.
[60] A. Soprano, A. Apicella, L. D'Antonio, F. Schettino, Application of durability analysis to GLARE aeronautical components, Int. J. Fatigue 18 (4) (1996) 265.
[61] R.S. Almeida, C.A. Damato, E.C. Botelho, L.C. Pardini, M.C. Rezende, Effect of surface treatment on fatigue behavior of metal/carbon fiber laminates, J. Mater. Sci. 43 (2008) 3173.
[62] R.A. Silva, E.C. Botelho, L.C. Pardini, M.C. Rezende, Evaluation of adhesion of continuous fiber–epoxy composite/aluminum laminates, J Appl. Adhes. Techn. 18 (2004) 1799.
[63] R.A. Silva, E.C. Botelho, M.C. Rezende, L.C. Pardini, Evaluation of adhesion of continuous fiber–epoxy composite/aluminum laminates, Mater. Res. 9 (2006) 247.
[64] N.Y. Taranets, H. Jones, 'The influence of surface roughness and pre-oxidation state on the wettability of aluminium nitride by commercial brazes, J. Mater. Sci. 39 (18) (2004) 5727.
[65] A. Vlot, Impact loading on fibre metal laminates, Int. J. Impact Engng 18 (3) (1996) 291.
[66] J.J.C. Remmers, R. de Borst, Delamination buckling of fibre–metal laminates, Comp. Sci. Techn. 61 (2001) 2207.
[67] C.A.J.R. Vermeeren, T. Beumler, J.L.C.G. De Kanter, O.C. Van Der Jagt, B.C.L. Out, Glare design aspects and philosophies, Applied Composite Materials 10 (2003) 257.
[68] J. Sinke, Manufacturing of GLARE parts and structures, Appl. Comp. Mat. 10 (2003) 293.
[69] J. Sinke, Development of fibre metal laminates: concurrent multi-scale modeling and testing, J. Mater. Sci. 41 (2006) 6777.
[70] D.A. De Silva, E.C. Botelho, M.C. Rezende, Hygrothermal aging effect on fatigue behavior of GLARE, J. Reinf. Plast. Comp. 28 (2009) 2487.
[71] A.N. Gent, R.P. Petrich, Adhesion of viscoelastic materials to rigid substrates, Proc. Roy. Soc. London A 310 (1969) 433.
[72] A.N. Gent, J. Schultz, Effect of wetting liquids on strength of adhesion of viscoelastic materials, J. Adhes. 3 (1972) 281.
[73] J. Guillement, B. Bistac, J. Shultz, Relationship between polymer viscoelastic properties and adhesive behaviour, Int. J. Adhes. Adhes. 22 (2002) 1.

[74] W.F. Cole, Technical justification of repairs to composite laminates, Int. J. Adhes. Adhes. 19 (1999) 107.
[75] L.F.M. da Silvaa, R.D. Adams, Adhesive joints at high and low temperatures using similar and dissimilar adherends and dual adhesives, Int. J. Adhes Adhes. 27 (2007) 216.
[76] L.F.M. da Silvaa, R.D. Adams, Joint strength predictions for adhesive joints to be used over a wide temperature range, Int. J. Adhes. Adhes. 27 (2007) 362.
[77] C.-L. Ong, S.B. Shen, The reinforcing effect of composite patch repairs on metallic aircraft structures, Int. J. Adhes. Adhes. 12 (1) (1992) 19.
[78] S.C. Rosalie, M. Vaughan, A. Bremner, W.K. Chiu, Variation in the group velocity of Lamb waves as a tool for the detection of delamination in GLARE aluminium plate-like structures', Compos. Struct. 66 (2004) 77.
[79] A. Deheeger, J.D. Mathias, M. Grediac, A closed–form solution for the thermal stress distribution in rectangular metal/composite bonded joints, Int. J. Adhes. Adhes. 29 (2009) 515.
[80] M. Davis, D. Bond, Principles and practices of adhesive bonded structural joints and repairs, Int. J. Adhes. Adhes 19 (1999) 91.
[81] G. Tober, D. Schiller, 15th World Conf. Nondestructive Testing, in: NDI Reliability Rules used by Transport Aircraft—European View, 2000. Point Roma (Italy), 15–21 October 2000.
[82] S.W. Schramm, I.M. Daniel, W.G. Hamilton, Evaluation of the sensitivity of ultrasonic detection of disbonds in graphite/epoxy–to–metal joints, Int. J. Adhes. Adhes. 1 (1981) 317.
[83] C. Bonavolontà, M. Valentino, G.P. Pepe, Characterization of the damage process in GLARE((R)) 2 using an eddy current technique based on HTS-SQUID magnetometer, Supercond. Sci. Technol. 20 (2007) 51.
[84] N.P. Avdelidis, C. Ibarra-Castanedo, Z.P. Marioli-Riga, A. Bendada, X.P.V. Maldague, A study of active thermography approaches for the non-destructive testing and evaluation of aerospace structures, Proc. Soc. Photo-Opt. Instru. Eng. (SPIE) 6939 (2008) U290.
[85] W. Steinchen, L. Yang, G. Kupfer, P. MaÈckel, Non-destructive testing of aerospace composite materials using digital shearography, Proc. Instn. Mech. Engrs. 212 (1998) G21.
[86] S.T. Halliday, W.M. Banks, R.A. Pethrick, Influence of humidity on the durability of adhesively bonded aluminium composite structures, Proc. Instn. Mech. Engrs. 213 (Part L 1) (1999) 27.

Diffusion bonding of metal alloys in aerospace and other applications

10

Ho-Sung Lee
Korea Aerospace Research Institute, Daejeon, Republic of Korea

10.1 Introduction

The diffusion-bonding process is one of the solid-state bonding processes. During a hot bonding process, two components are typically joined under pressure, with close contact between the surfaces and at a temperature below the melting point of the parent metal. In the case of diffusion bonding, it is important that the fusion temperature is not reached, and filler material is not needed. Diffusion bonding occurs as a result of diffusion of the interface atoms of the bonded materials. Diffusion bonding is an attractive manufacturing method for aerospace applications, where mechanical properties in the bond area and a sound metallurgical bond are important. It is a different process to brazing or the TLP (Transient Liquid Phase) bonding process, in which a foreign metal with a lower melting point is used to weld similar or dissimilar metals together. As a diffusion bond is formed by atomic migration across an interface in a solid state, there is no metallurgical discontinuity at the interface, and the mechanical properties and microstructure at the bonded region are similar to those of the base metal. For example, titanium alloys can easily be joined by diffusion bonding owing to the ability of titanium to dissolve its own oxide in a vacuum. In combination with superplastic forming (SPF), diffusion bonding is a solid-state process that is defined as one in which the components being joined undergo macroscopic deformation by no more than a few per cent, without a liquid phase. The process is dependent on various parameters, in particular time, applied pressure and bonding temperature, to promote microscopic atomic movement and ensure a complete metallurgical bond.

However, in some materials, the diffusion process is not observed and joining with hot pressure results in physical bonding without macroscopic atomic diffusion [1]. This is the case for diffusion bonding or cladding of stainless steel/copper or other dissimilar metals. When bonding stainless steel and copper alloys, structurally sound diffusion bonding is difficult to achieve owing to the formation of a brittle compound layer and a build-up of the oxygen impurities in copper at the bond interface. To avoid this, an insert material is used to improve bonding strength; in most cases a thin layer of Ni-based alloy is applied between the surfaces of the stainless steel and copper alloy.

Welding and Joining of Aerospace Materials. https://doi.org/10.1016/B978-0-12-819140-8.00010-9
Copyright © 2021 Elsevier Ltd. All rights reserved.

The most common diffusion-bonding process in airframe manufacture is utilizing thin sheets with blow forming [2]. This process ensures a uniform pressure profile over the bond area, and the superplastic characteristics of the materials ensure an easy atomic exchange and facilitate diffusion. Lightweight sandwich structures, such as pyramidal trusses, corrugated panels and honeycomb panels, can be manufactured using thin-sheet diffusion bonding and blow forming. Another form of solid-state bonding, massive diffusion bonding, is different from thin-sheet diffusion bonding, which takes place during blow forming of two or three sheets. Massive diffusion bonding [3] is an innovative manufacturing method used to produce heavy-metal components with solid-state bonding of multisheets by low pressure of inert gas. This process makes near-net-shape forming of components possible from presized multisheets of titanium, enabling significant weight and cost savings. One major advantage of massive diffusion bonding is its ability to produce heavy-metal sections with less material waste than conventional methods, such as mechanical machining from solid bulk. It is also possible to produce closed sections, which could not be manufactured by other machining, extrusion or forging techniques.

The main mechanism of solid-state diffusion bonding is based on creep behavior [4]. Both are thermally activated processes and thermal energy provides the necessary activation energy required to overcome the potential energy barrier that prevents atomic movement. Diffusion bonding begins with the process of creep, which is then followed by vacancy diffusion to allow for complete void closure and bonding across the joint interface [5, 6]. This process of void closure is the opposite to the cavitation process, which occurs during the final stage of creep rupture. The cavitation model for nucleation and growth of cavities demonstrates the critical size of a stable cavity, based on surface energy and applied stress [7]. In other words, there is a critical size that needs to be reached by voids for enclosing and complete bonding to take place. The whole mechanism is governed by surface diffusion, volume diffusion, grain boundary diffusion and power-law creep. Some theoretical models predict the kinetics of diffusion-related processes [8–10].

The proposed mechanism will describe the solid-state diffusion-bonding model and its application in explaining the mechanism of solid-state bonding of titanium alloys in a nonvacuum environment. The whole process can be divided into six stages. In the first stage (see Fig. 10.1A), the two surfaces must be in contact. The amount of initial contact between the surfaces depends on surface conditions such as irregularity, clearness, roughness, the time from initial chemical etching and surface treatment. For example, the rougher the surface, the higher the pressure and the longer the bonding time required. In the second stage (see Fig. 10.1B), diffusion bonding begins with microplastic deformation at the interface, where ridges on the surface are deformed plastically in such a way that there is no macroscopic deformation in the parts to be bonded. During this process, voids will be produced and aligned at the interface. The voids become isolated and the gas pressure inside the voids is equal to the pressure in the furnace. At this stage, it is important to increase the temperature. During the third stage (Fig. 10.1C), the surfaces start to absorb the gas and because the voids are isolated, the pressure inside the void decreases. A recent study of the kinetics of decreasing gas pressure in the closed volume at high temperatures [11] shows that

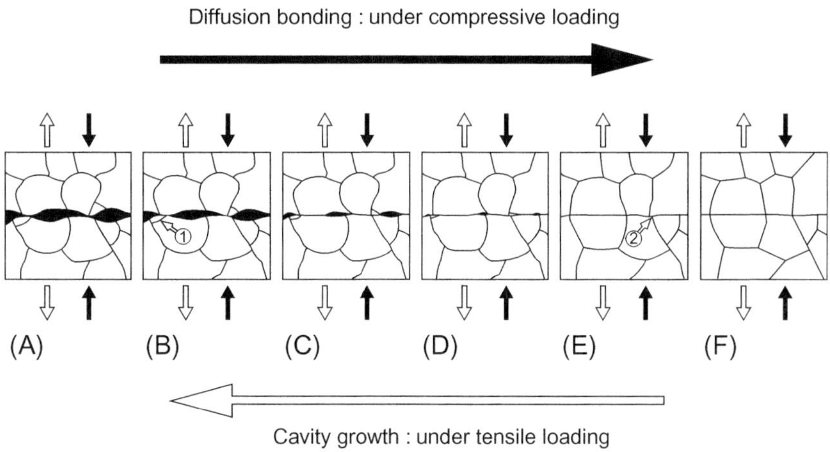

Fig. 10.1 (A–F) Schematic view of the formation mechanism of the diffusion bonding process.

at 550 °C the gas pressure inside a void of 100 μm is reduced from 7.3 to 3×10^{-4} Pa within several minutes. At higher temperatures, e.g., 900 °C, it is expected that the vacuum will form in several seconds. The stability of the titanium oxides in the surface region has been investigated in ultra-high-vacuum conditions [12], and it has been shown that at temperatures of more than 400 °C all the oxides decompose and the oxygen is diffused into the bulk. The fourth stage consists of void shrink-age owing to diffusion creep and the chemical potential difference between surface energy and volume energy. At this stage, the grain boundaries start to migrate to accommodate the shrinkage, in order to maintain a constant volume. Some of the smaller grains, such as that represented by ① in Fig. 10.1B, are in the process of shrinking in this stage. In the fifth stage, voids that are smaller than the critical size are unstable and will disappear rapidly. The collapse of such voids has been explained using a mathematical model [13], which takes into account the chemical potential gradient between the stressed interface and the stress-free surface of the void. This is just the opposite of the cavity nucleation process that occurs during elevated temperature cavitation [14]. In the final stage, there is no void or discontinuity at the welded interface and there is no clear difference between it and the grain boundaries. The welded interface can only be distinguished because it is a straight line. The grain boundary migrates to find the minimum surface energy as indicated by ② in Fig. 10.1E, and eventually the grain boundary becomes a smooth and continuous line to reduce surface energy.

In this chapter, diffusion bonding of titanium, steel and copper alloys is discussed in terms of its applications to the manufacture of several aerospace components with various configurations. These components include a high-pressure tank for attitude control of spacecraft, a combustion chamber with copper cooling channels and other lightweight structural panels. It should be noted that this chapter focuses only on the aerospace applications of diffusion bonding, and the presentation material is limited to the author's own work, as many applications are proprietary and their details are not available to the public.

10.2 Diffusion-bonding process

10.2.1 Titanium alloys

Ti-6Al-4V alloy is a dual-phase alloy, with fine equiaxed alpha phases mixed with a transformed beta phase. The chemical composition is 6.05Al-3.89V-0.21Fe-0.11O-0.01C-0.007N with less than 0.005 Y. β transus of this alloy was determined by DTA (Perkin Elmer DTA7) at a cooling rate of 10 °C min. This is because the ratio of the two phases is important for flow stress in high-temperature deformation. It is well known that the phase portion of the two-phase titanium alloy is related to the superplastic condition [15]. This phase ratio is affected by the beta transus temperature, which depends on the alloy composition. The average grain size at room temperature was about 5 μm, and the original sheets of this alloy were 2.04 mm thick and were cut to 100 mm in diameter. The flow stress behavior at high-temperature flow was obtained from a series of tensile tests with strain rates ranging from 10^{-4}/s to 10^{-2}/s and at temperatures from 800 to 950 °C. The material was shown to have high elongations above 800 °C and at strain rates ranging from 10^{-4}/s to 10^{-3}/s as shown in Table 10.1. An example of the stress-strain behaviour of this alloy with a strain rate of 10^{-3}/s at different temperatures is shown in Fig. 10.2. The material was shown to have good formability above 800 °C and at strain rates ranging from 0.001/s to 0.0001/s (Table 10.1).

The diffusion-bonding fixture consisted of an austenitic Fe-Cr-Ni heat-resistant alloy steel (25Cr-20Ni-0.3C-1.0Mn), which was designed to permit simultaneous inert-gas pressurization on one side of the specimen. As the product was required to withstand gas pressure, it was necessary to sustain the sealing condition at high temperature and high pressure. When using diffusion bonding of multisheet metal to manufacture a structural part, as in this instance, a complex-shaped heat-resisting structure can be manufactured by changing the shape of the insert.

Diffusion bonding of multiple sheets was performed in a furnace under the conditions summarized in Table 10.2. As a good surface condition is vital for complete bonding, the surfaces of the blanks were carefully prepared for bonding by a special procedure [16]. The surfaces to be bonded were machined and grinded with 600 grit SiC paper before being chemically cleaned. The rinsed surfaces were then air dried with clean filtered air. Prior to diffusion bonding, the specimens were immediately

Table 10.1 Elongation at various temperatures and strain rates.

	Strain rate (%)		
Temperature (°C)	**0.01/s**	**0.001/s**	**0.0001/s**
800	727	1229	1232
850	904	1898	1304
900	900	1458	1270
950	950	760	510

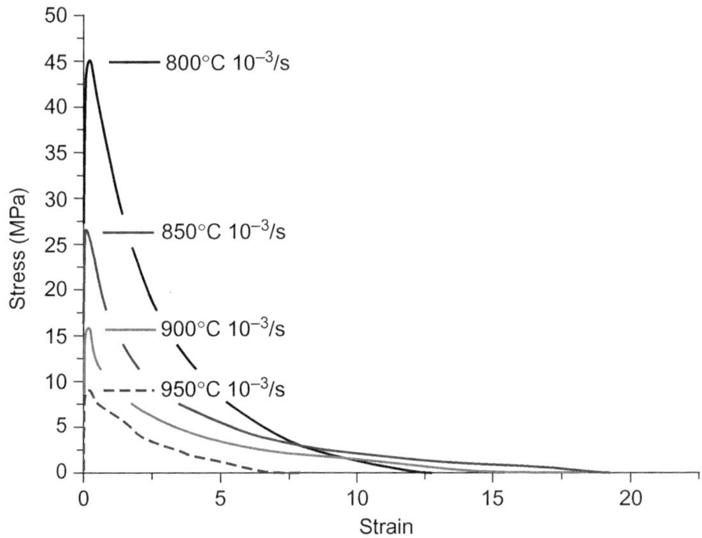

Fig. 10.2 Stress-strain behavior of Ti-6Al-4V at a strain rate of 10^{-3}/s at different temperatures.

Table 10.2 Diffusion-bonding conditions.

Specimen	Bonding temperature (°C)	Applied pressure (MPa)	Number of sheets
A	850	4	4
B	850	3	12
C	875	4	4
D	875	4	11
E	900	4	40
F	920	4	3

cleaned again with a high-purity solvent and dried. Next, multiple sheets of the titanium alloy were sequentially stacked in the tool according to the present embodiment. After stacking, the tool was sealed and heated to about 800 °C and 4 MPa of inert gas was injected into the top of the tool. When the target temperature was reached (i.e., about 875 °C) the 4 MPa of pressure was sustained by supplying inert gas. After 60 min, the tool was depressurised and the temperature was slowly reduced.

Diffusion-bonding processes of between 3 and 40 sheets of Ti-6Al-4V alloy were conducted in a furnace under different conditions (see Table 10.2). The set time of 1 h was chosen in order to simplify the procedure and reduce the parameters. After stacking, the tool was sealed and heated to a target temperature. Following this, the pressure was applied to the top of the tool for a given amount of time. Using gas pressure as a loading medium prevents nonuniform pressure application on the bonding area.

Diffusion bonding of Ti-6Al-4V was carried out at a superplastic temperature (850–920 °C) where diffusional transport of atoms is substantially enhanced. The sequence of micrographs in Fig. 10.3 illustrates the microstructural change occurring at each different stage of Fig. 10.1. The bonding time was 1 h and the pressure applied was 4 MPa. The optimum solid-state bonding condition was obtained at 875 °C. It is interesting to note that each stage proposed in Fig. 10.1 was in fact observed in the experiment, as shown in Fig. 10.3. At a bonding temperature of 850 °C, it is shown that the bonding is in the initial stage and not yet completed, with evidence of bridging two surfaces and presence of voids, as proposed in Fig. 10.1B. The initial stage of diffusion bonding begins when the two surfaces come into contact, oxide layers meet and some of the oxide surfaces begin to break down. Even though there is no macroscopic deformation, it is inevitable that microscopic deformation will occur, owing to the physical contact of the surfaces under pressure. As some surfaces are in contact

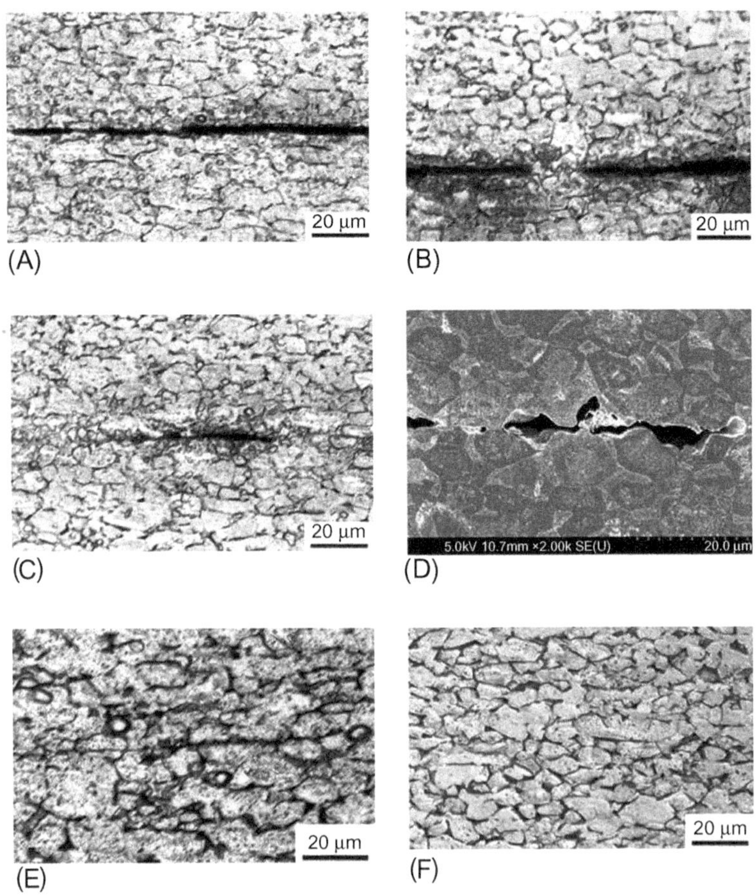

Fig. 10.3 Microstructure of welded region at 850 °C (A–D) and 875 °C (E, F) with 4 MPa for 1 h.

with each other, voids will be isolated, as shown in Fig. 10.3C. Fig. 10.3D illustrates the point just before the fifth stage of the diffusion-bonding process, in which void closure is almost completed. The results of diffusion bonding at 875 °C in Fig. 10.3E and F show that complete solid-state bonding can be obtained under this condition.

Fig. 10.4 shows the initial stage of the process: it is not yet complete, but there is evidence of bridging between the surfaces and voids are present, as described above. The results of diffusion bonding at 875 °C show that a complete diffusion bond can be obtained at this temperature, as shown in Fig. 10.4B. This is one of the examples of thin-sheet diffusion bonding combined with superplastic blow forming. It is clear from the micrograph that under these conditions the microstructure of the bonded interface cannot be distinguished from the matrix, so bonding must be complete, with atomic diffusion and grain-boundary migration. There is no evidence of oxide residue or other foreign phases in the bonded region. It can be concluded that diffusion bonding of the titanium alloys tested in this study produces a diffusion-bonded microstructure at optimum conditions in an inert environment.

The microstructure bonded at 920 °C exhibits an oxygen-enriched alpha phase at the bonded interface, which is typically needle-shaped. One of the reasons for the formation of this phase at the higher temperature is the higher diffusion rate of oxygen in the presence of oxygen. The needle-like phase is the typical shape of the product of nucleation and shows typical growth from beta to the lower-temperature allotrope in

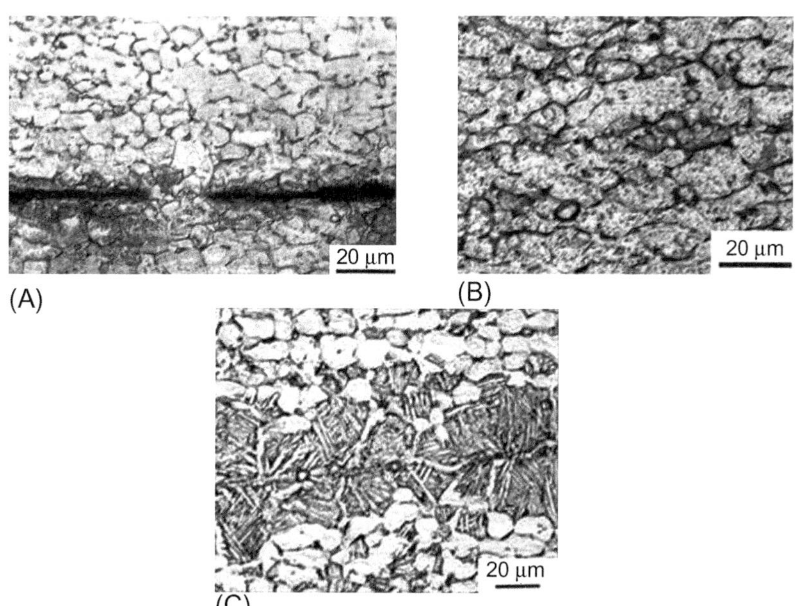

Fig. 10.4 Microstructure of interface of specimens bonded at (A) specimen A; (B) specimen C; and (C) specimen E.

the presence of oxygen, which is observed in Fig. 10.4C. This phase is not desirable owing to its brittle nature.

Typical diffusion-bonding conditions used by Petrenko et al. [17] are as follows: 950 °C, 2 MPa and 30–120 min. Mutoh et al. [18] reported that the optimum conditions for bonding Ti-6Al-4V were a temperature of 900 °C and a pressure of 10 MPa for 1 h, in which the bonding process was performed in a vacuum with mechanical pressure loading. Zhang et al. [19] selected 1010 °C, 3 MPa and 3 h as the best conditions for diffusion bonding of Ti-6Al-4V. In this study, the bonding temperature of 875 °C was used based on the result of the β-transus-temperature measurement.

The final two stages of diffusion bonding were clearly shown for Ti-15V-3Cr-3Sn-3Al alloy, which is welded under a pressure of 6 MPa for 3 h at 900 and 925 °C [20]. Fig. 10.5A shows that an oxide layer formed at 900 °C. However, in Fig. 10.5B there is no sign of oxide residue or other foreign phases at the interface welded at 925 °C. The microstructure is observed with higher magnification in Fig. 10.6, which indicates that the bonding quality is about the same as that of the grain boundary. The process of shrinkage of small grains at the welded interface is clearly visible in Fig. 10.6A. Some of the grain-boundary migration to accommodate grain-boundary formation at the boned interface is shown in Fig. 10.6B. From the micrograph it is clear that under these conditions, the microstructure of the bonded interface cannot be distinguished from the matrix, showing that bonding must be complete, with atomic diffusion and grain-boundary migration. Atomic flow and grain rearrangement occur with the help of grain-boundary migration. Some grains can merge and coalescence of grains can be observed during the last stage. Eventually, it is impossible to distinguish the welded interface from the other grain boundaries and the microstructure becomes completely homogeneous.

10.2.1.1 Titanium lightweight honeycomb panels

In an airframe structure, a lightweight honeycomb panel design is important to achieve complete stabilization of the component faces, as the desired compressive strength can then be attained with thinner gauge materials. The honeycomb core provides buckling resistance for the structure, while at the same time transmitting shear forces. In addition, properties inherent to sandwich structures include remarkable stiffness, vibration damping, thermal, acoustic and insulation properties. The application of sandwich structures with honeycomb cores for use at elevated temperatures is limited only by the properties of the materials and the type of bonding method used to manufacture them. Han et al. [21] fabricated a three-sheet titanium sandwich panel under bonding conditions of 7.5 MPa for 30 min at 1030 °C.

The schematic configuration of the stop-off application profile is shown in Fig. 10.7. The thickness of the outer skin is 3 mm and that of the inner web is 1.6 mm. The panel consists of five cells. After applying BN powder as a stop-off material to both surfaces of the core sheet, the four sheets were carefully mounted in the apparatus. Using hydrostatic gas pressure as a loading medium ensures that the pressure on the bonding area is uniform. After diffusion bonding four sheets under the conditions described earlier, the gas pressure inside the specimen was increased according to the pressure profile from the finite element analysis. The application

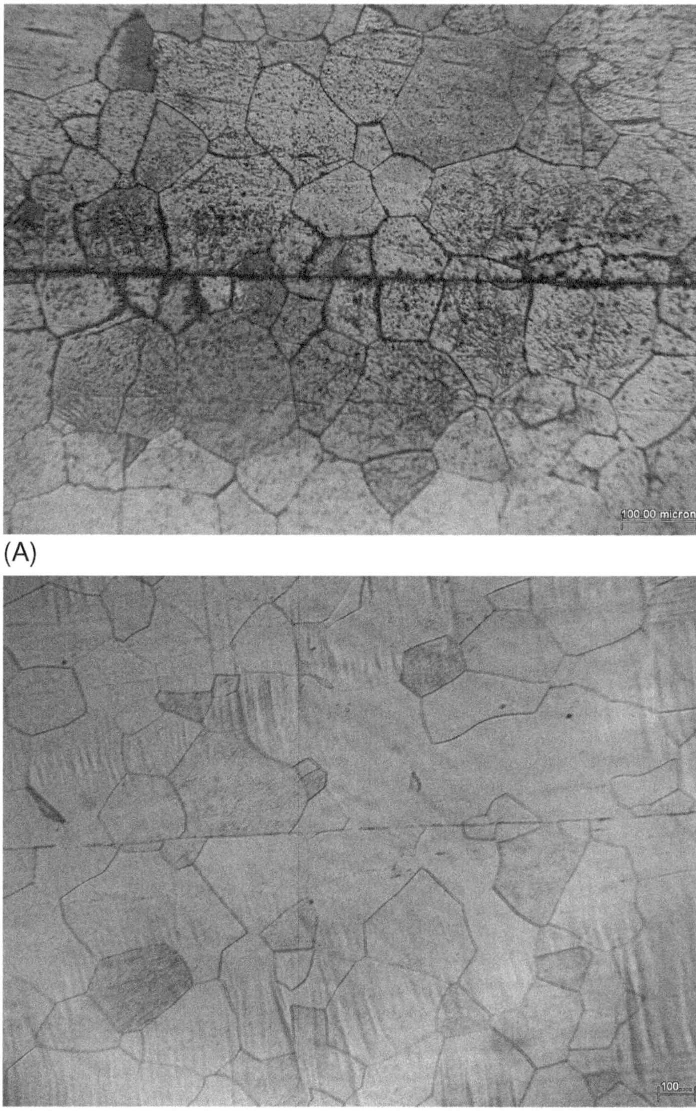

Fig. 10.5 Micrographs of Ti-15V-3Cr-3Sn-3Al alloy bonded under a pressure of 6 MPa for 3 h at (A) 900 °C and (B) 925 °C.

of inert-gas pressure to the envelope in a fixture produces the sandwich structure. In the analysis produced using MARC™, the diffusion-bonded sheet was discretised by a four-node quadrilateral element with axi-symmetric solid properties, and the upper and lower dies were treated as rigid bodies. Forming gas pressure was imposed on the lower and upper surfaces of a core sheet. The maximum forming pressure was limited to 4.0 MPa. The incremental configuration from finite element method (FEM)

Fig. 10.6 Higher magnification of micrographs of Ti-15V-3Cr-3Sn-3Al alloy bonded under a pressure of 6 MPa for 3 h at 925 °C.

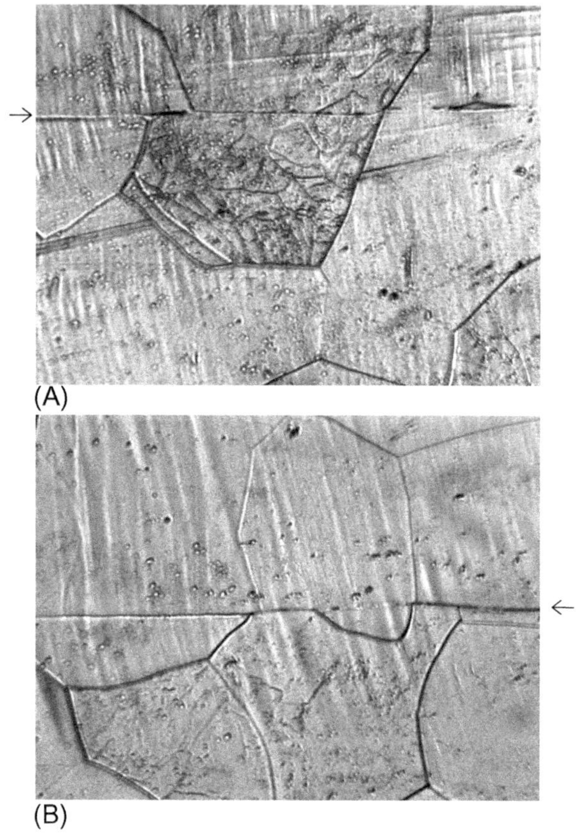

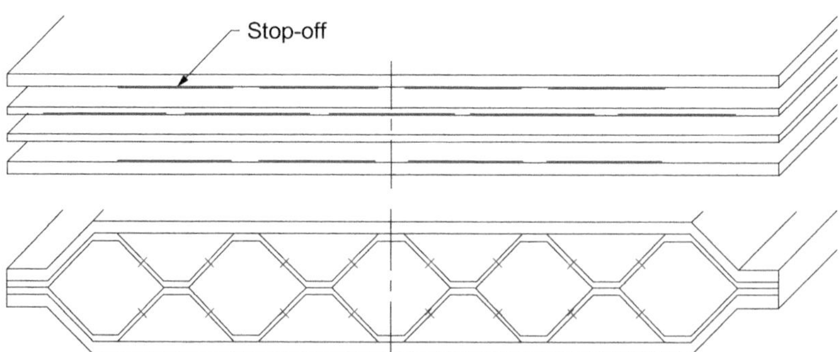

Fig. 10.7 Schematic profile of stop-off application for a four-sheet panel.

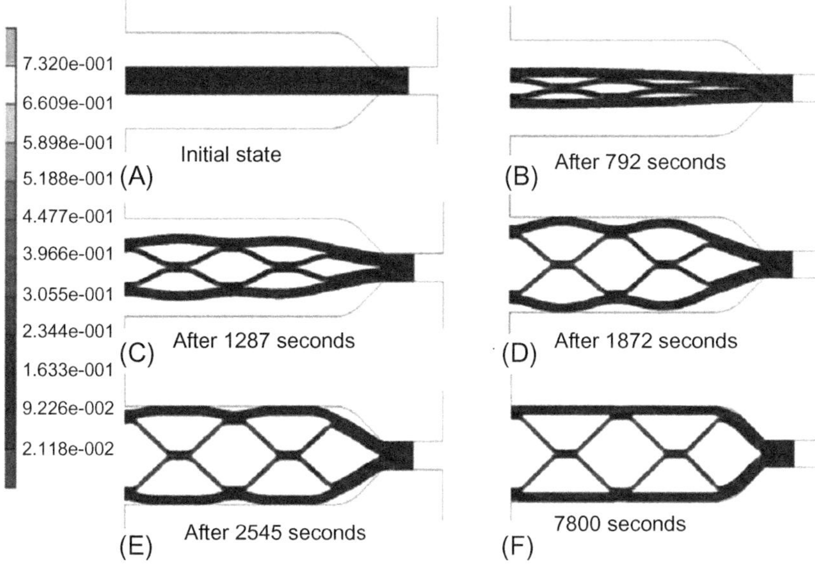

Fig. 10.8 Results of effective strain distribution during blow forming of a four-sheet bonded panel.

analysis during blow forming is shown in Fig. 10.8 [22]. The result showed that when $t = 1872$ s, the skin surface began to come into contact with the die contour and the metal flow accommodated the excessive materials around the groove. This took a relatively long time. At $t = 7800$ s, it was possible to obtain a flat and groove-free surface, and this was what the actual part underwent during the blow-forming process, as shown in Fig. 10.9.

After fabrication of the sandwich panel, the microstructure and thickness measurement were analysed and compared with the analysis result. Fig. 10.9 shows a schematic view of the apparatus for a four-sheet sandwich panel and the microstructure of the pressure-welded region. It was clearly shown that diffusion bonding of this alloy at 875 °C with 4 MPa for 1 h produces a metallurgically homogeneous bonding microstructure.

10.2.1.2 Attitude control pressurant vessel

The typical application of titanium alloys is in pressurized vessels for attitude control, which store relatively high-pressure gas or fuel. Examples include various commercially available tanks for spacecraft and launch vehicles, pressurized tanks for attitude control, KSR-III, PMD (Propellant Management Device) fuel-supply tank of ARIANE [23] and the teardrop-type fuel tank of the Japanese MUSES-A spacecraft [24]. Titanium tanks can be fabricated by spin forming, blow forming or machining, although the first two methods are used most often in order to reduce manufacturing costs.

In this study, two sheets of Ti-6Al-4V were circumferentially diffusion bonded and then blow formed to manufacture a spherical vessel [25]. The diffusion-bonding

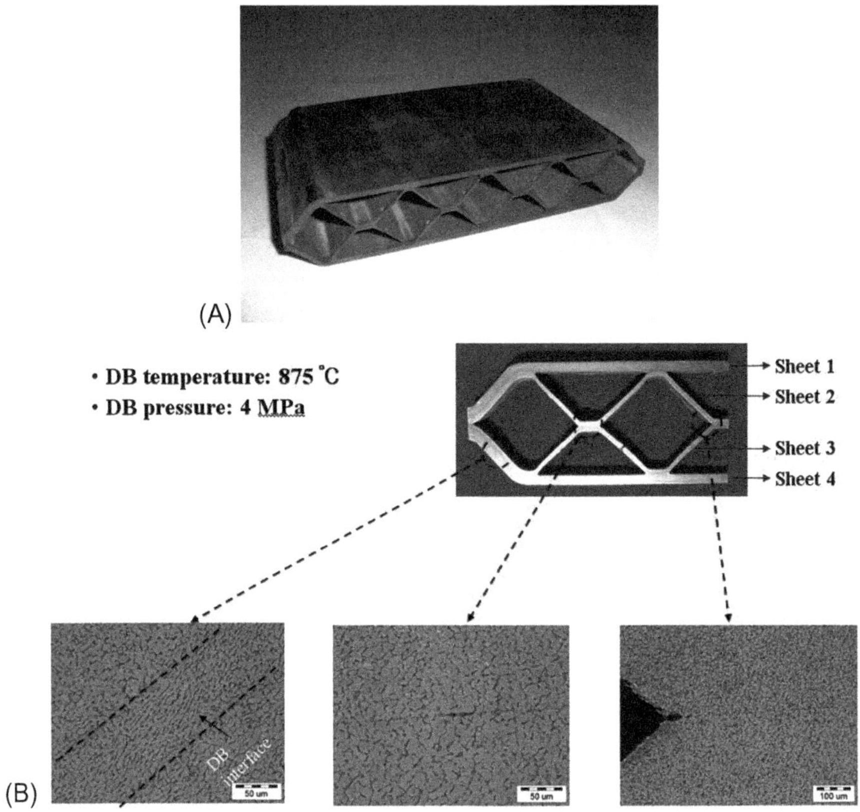

Fig. 10.9 (A) Lightweight honeycomb panel article (B) and micrographs of the bonded region.

conditions were the same as those described in Section 10.2.1.1. Fig. 10.10 shows the gas-pressure profile obtained from both the finite element simulation and the actual applied gas pressure. It is worth noting that the initial slope of the pressurizing rate is a little lower than the analyzed value in order to accommodate the time delay for maintaining uniform pressure inside the vessel before blow bulging. The maximum gas pressure was 3.0 MPa.

This is an example of free blow forming, different to conventional bulging of circular diaphragms in which the circumferential edge of the circular sheet is usually clamped [26]. As there is no clamping or fixed boundary, the equatorial diameter of the forming sheet is not equal to the initial diameter. Kruglov et al. [27] developed a mathematical model of the free-forming process, taking into account the varying equatorial diameter of the sheet during blow forming. According to this model, the upper limit of the ratio of the envelope's diameter to the spherical shell diameter is about 1.25 and independent of the envelope's dimension. Therefore, this is the model used in this study.

The initial dimension of the circular specimen was 210 mm, but without a fixed boundary the equatorial diameter of the specimen during free forming is not constant.

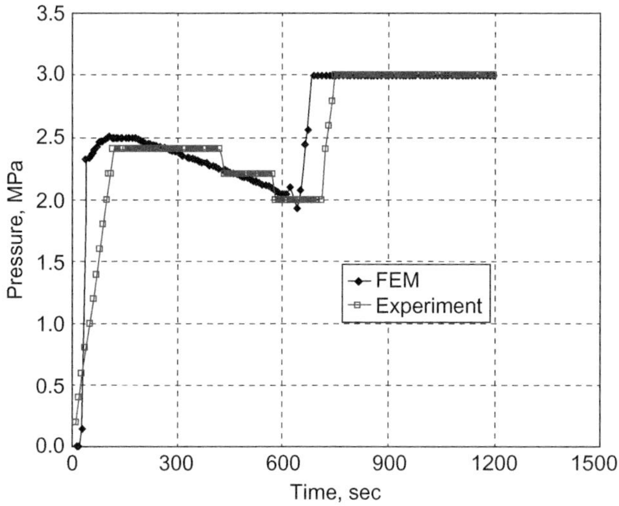

Fig. 10.10 Gas-pressure profile for a forming spherical vessel.

Initially the major diameter is large and becomes smaller. This is shown in Fig. 10.11, using the conditions according to Fig. 10.10 (after 23 min up to 1.1 MPa). This process is necessary to stretch the pressure-bonded zone so that a symmetric spherical shape is possible. Otherwise, the thickness of the pressure-bonded zone is not uniform and therefore hard to deform. The free-forming process was performed for 33 min at 1.7 MPa, so that the major diameter of the specimen was less than 180 mm in order to fit it into a sizing tool. Finally, the specimen was installed in the sizing tool and formed for 20 min to obtain the final dimensions. Fig. 10.12 illustrates each stage of the blowing process of a pressure-welded vessel. The thickness distribution of the spherical vessel is shown in Fig. 10.13.

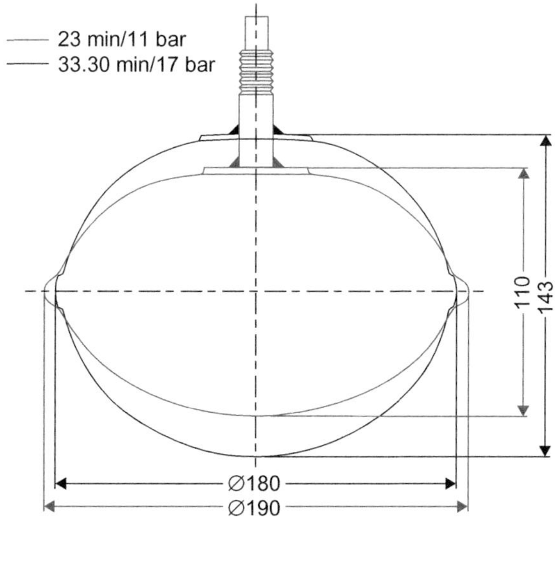

Fig. 10.11 Change of dimension of the vessel after forming.

Fig. 10.12 Series of blow-forming processes for spherical vessel.

Fig. 10.13 Thickness distribution of the spherical vessel.

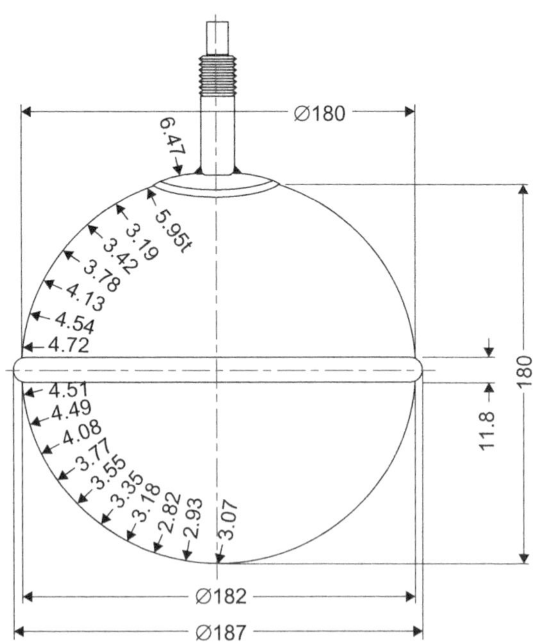

Fig. 10.14A shows the hydraulic pressurizing test. The strain and acoustic emission signals were observed to investigate the effect of the solid-state bonding method on the failure mode and performance of the spherical vessel. The result of the pressurizing test of the spherical vessel is depicted in Fig. 10.14B. This shows that failure did not occur after the proof pressure of 7000 psi. It is found that the solid-state bonded region does not provide a path for fracture crack, and structural integrity was demonstrated up to proof pressure.

10.2.1.3 Hollow fuel tank

The forming profile of hollow configuration was analysed, and the finite element model was successfully demonstrated to predict forming behaviour [25]. In this process the end of two sheets were diffusion welded together, and dome and cylinder parts were formed in a single step from a double-layer tube, as shown in Fig. 10.15. The diffusion-bonding conditions were 875 °C at 4 MPa for 1 h, as described above. The gas-pressure profile was obtained from the finite element simulation and the

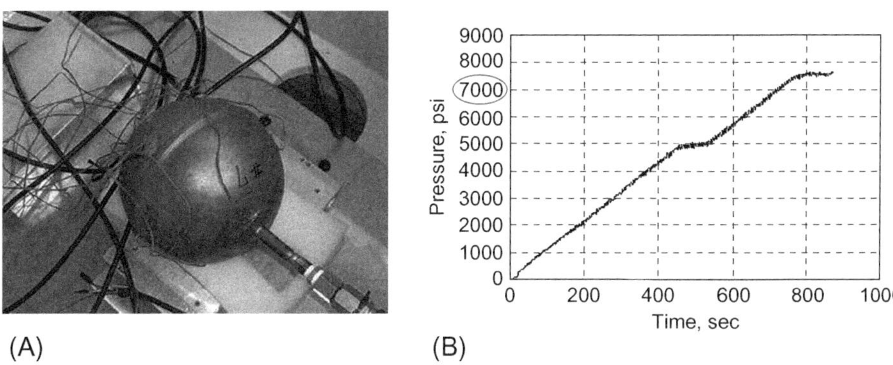

Fig. 10.14 Photograph during a hydraulic pressure test (A) and pressurizing profile (B).

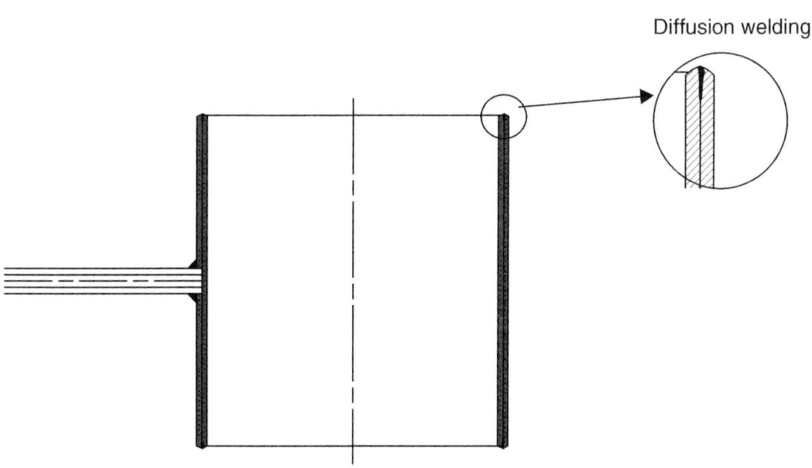

Fig. 10.15 Schematic view of the diffusion-welded region of a double-layer titanium tube.

gas pressure increased almost linearly. Based on the gas-pressure profile obtained from the finite element simulation presented in Fig. 10.16, the forming of the hollow vessel was performed at 875 °C at the maximum pressure of 1.5 MPa. The final shape was obtained at 4000 s. The forming tool and formed article are shown in Fig. 10.17. The result shows that the forming of a complicated shape, such as a hollow tank, has been successful with diffusion-bonding technology.

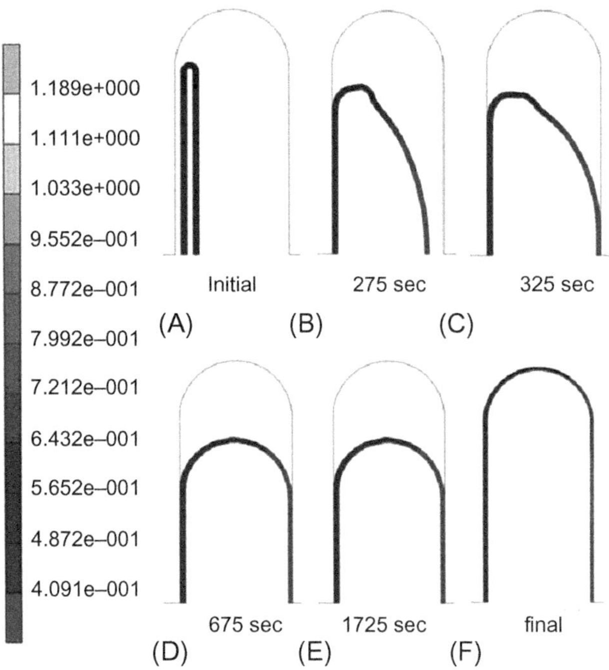

Fig. 10.16 Analysis results of effective strain distribution during blow forming of a hollow cylinder.

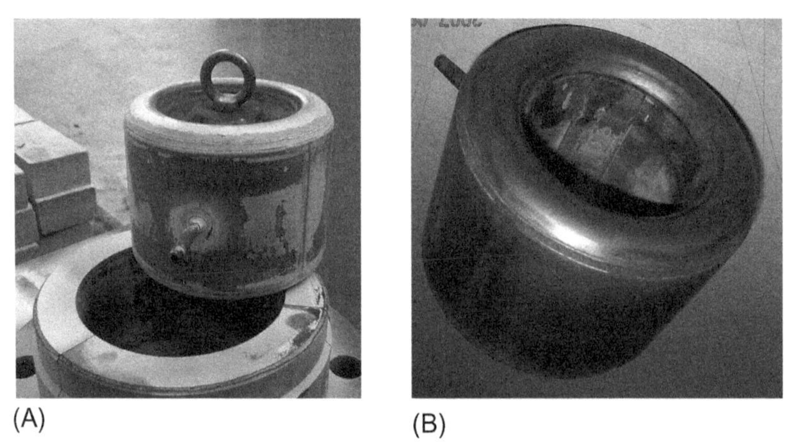

Fig. 10.17 Article with forming tool (A) for blow forming and a titanium hollow vessel (B).

10.2.2 Steel and copper alloy

It is well known that joining steel and copper alloys is not easy because dissimilar alloys have different diffusion coefficients and may result in micro voids under some bonding conditions. To avoid this effect, most methods involve low-melting alloy brazing or casting of molten copper onto steel material. During heating in a hydrogen environment, a low-melting alloy begins to form at the gold interface from the inter-diffusion between copper and gold. The pressure built between the steel and copper spreads the gold–copper liquid metal uniformly, and the gold atoms diffuse into solid copper and steel with a concentration gradient in each metal. Another method is to apply a copper-alloy coating to a steel surface first and then heat the material to allow atomic diffusion. The coating can be applied with an aluminothermic-reduction reaction process. However, this method is not appropriate for large surface areas because the significant difference between the melting temperatures of the reduced metal and slag phases can produce gas entrapment between the solid-slag phase and the molten-metal phase.

In this study, it is necessary to join copper and steel to produce a combustion chamber for a liquid-propellant launch vehicle. The inner shell of the chamber is made of copper, with cooling channels for the regenerative engine, and the outer shell is dual-phase steel to keep high pressure inside the chamber. In order to characterise the flow strength of these materials at high temperatures, several tensile tests were performed at temperatures ranging from 800 to 950 °C. An example of high-temperature flow stress for SUS329J1 is shown in Fig. 10.18 for a strain rate of 10^{-4}/s at 900 °C. The shape of the curve indicates that this material is superplastic under these conditions. This information is used to estimate the optimum conditions for diffusion bonding and blow forming. Diffusion bonding of copper and steel was performed at three different pressures and at temperatures of 850 and 900 °C. Fig. 10.19 shows an example of testing a welded specimen, and the results are shown in Fig. 10.20. It is shown that the optimum condition for diffusion bonding is 4 MPa at 850 °C for 1 h. Hydraulic pressure tests

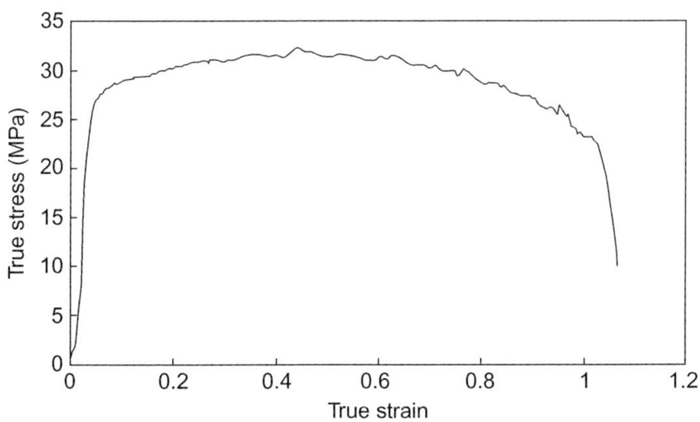

Fig. 10.18 Stress-strain behavior of SUS329J1 for a strain rate of 10^{-4}/s at 900 °C.

Fig. 10.19 Typical shape of a pressure-welded specimen and failure behavior for lap shear test.

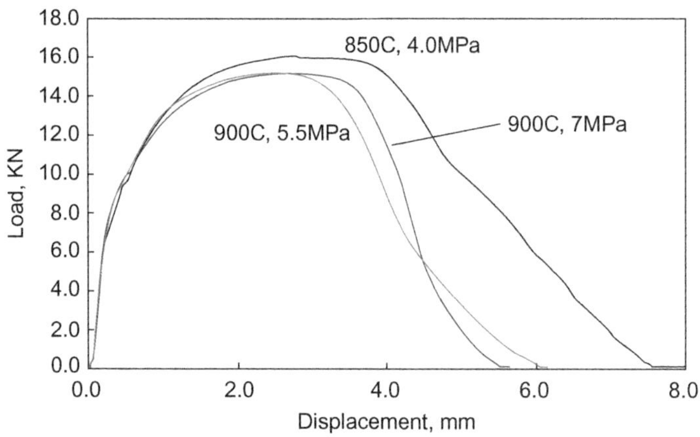

Fig. 10.20 Results of lap shear test of pressure-welded specimen at three different conditions.

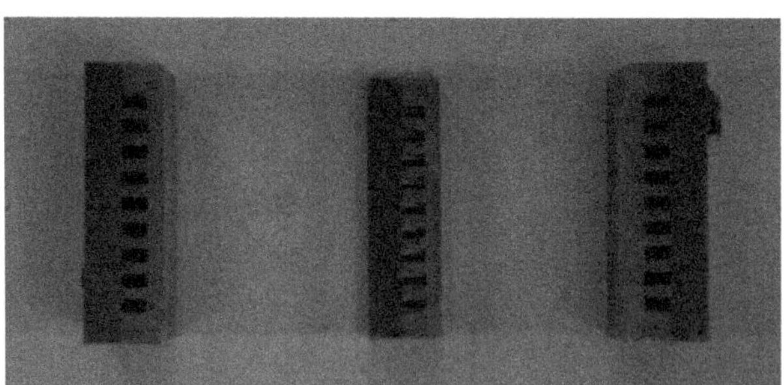

Fig. 10.21 Typical cooling-channel specimens.

were also performed for welded cooling-channel specimens. Some of the cooling-channel specimens are shown in Fig. 10.21.

The result of EDX analysis and a scanning electron microscope (SEM) micrograph are shown in Fig. 10.22. It can be concluded that the atomic diffusion process occurs at the interface and copper atoms diffuse into steel, whereas iron and chrome atoms diffuse into copper.

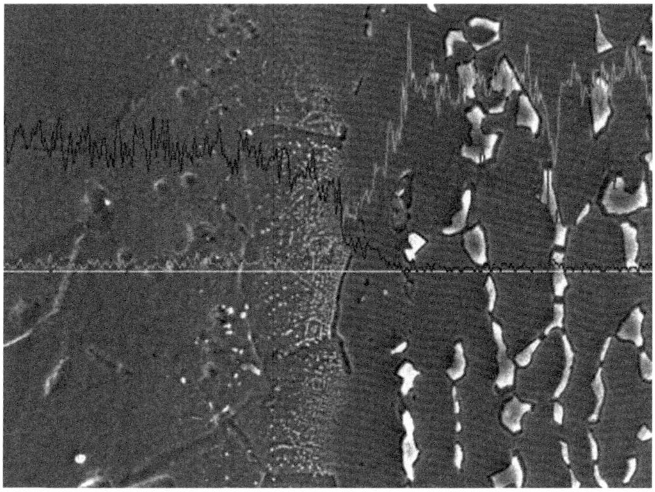

Fig. 10.22 SEM micrograph with EDX analysis at the diffusion-bonded line.

10.2.2.1 Combustion chamber with cooling channels

The combustion chamber of a liquid-propellant launch vehicle is one of the most important components that determines the performance of the launch vehicle. It consists of two major parts: the outer skin made of stainless steel to sustain the internal pressure and the inner shell made of copper alloy for regenerative cooling channels. Two outer skins are prepared using the blow-forming process, one for each half of the chamber as divided along the symmetric axis. The inner shell with cooling channels is then pressure welded to the outer steel jacket.

A concave fixture was fabricated for diffusion bonding of the copper inner layer and steel outer skin as shown in Fig. 10.23. The steel outer-skin article was blow formed and the maximum gas pressure was 7 MPa, according to FEM analysis [28]. Fig. 10.24 shows a forming die and examples of formed steel outer skins.

For the inner layer, the copper alloy was machined to create cooling channels and then placed in the diffusion-bonding fixture. It is important at this stage to match the contour with that of the steel outer skin for a complete match. Fig. 10.25 presents photographs of copper alloy and a steel outer skin, with the diffusion-bonding fixture. The pressure was supplied inside the inner layer using the tubes shown in Fig. 10.24, and the whole fixture was clamped in a hydraulic press.

The results of this experiment show that using diffusion bonding to join steel and copper alloy has been successful for near-net-shape forming of the chamber of a launch vehicle, as shown in Fig. 10.26.

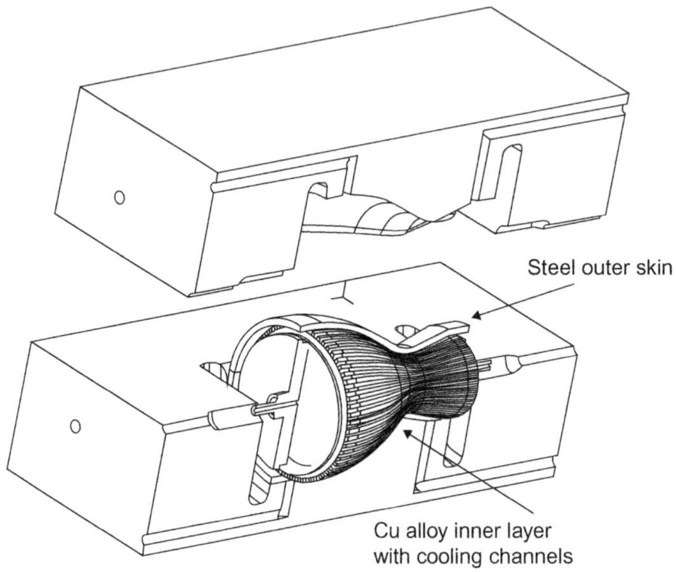

Fig. 10.23 Schematic view of fixture for diffusion bonding of the copper inner layer and steel outer skin.

Fig. 10.24 Steel outer-skin article after blow forming.

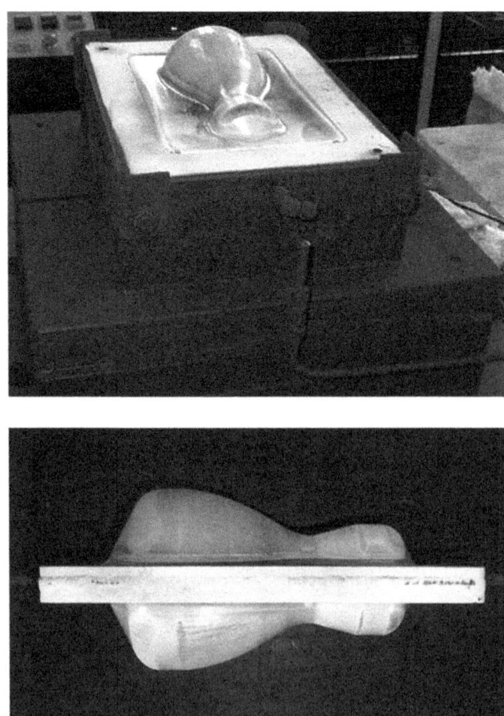

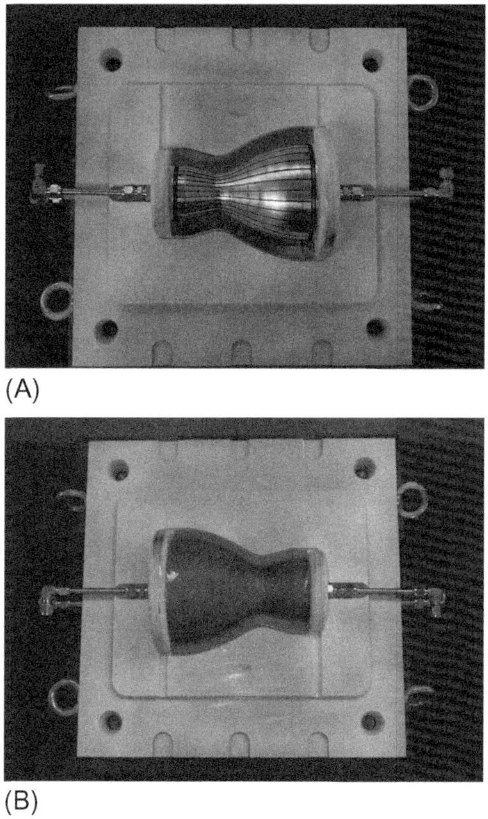

Fig. 10.25 Photographs of pre-machined copper inner layer (A) and steel outer skin (B) before diffusion bonding.

Fig. 10.26 Photographs of combustion chamber pressure welded with copper cooling channels.

10.3 Conclusions and future trends

The present work demonstrates that diffusion bonding of titanium, steel and copper alloys in an inert environment has been successful for near net-shape forming of products with complex configurations. It is notable that the micrographs of diffusion-bonded titanium alloys show grain-boundary migration and diffusion bonding, demonstrating that a homogeneous micro-structure can be obtained. Evidence of atomic diffusion can also be found in the bonding of steel and copper alloys.

The result shows that the diffusion-bonding method has been successfully applied to the manufacture of aerospace components, including light-weight titanium sandwich panels, high-pressure vessels for attitude control of spacecraft, hollow fuel vessels and combustion chambers of steel with copper cooling channels.

Recent research into diffusion bonding has focused on the important issues of reducing the weight of aerospace components and providing cost-effective manufacturing. Quantitative non-destructive testing methods are required to inspect diffusion-bonding integrity. Diffusion bonding of light-weight alloys such as aluminum [29] and magnesium [30] is currently being investigated.

References

[1] D.J. Stephenson, Diffusion Bonding 2, Elsevier Applied Science, London, UK, 1991.

[2] R. Pearce, Diffusion Bonding, School of Industrial Science, Cranfield, UK, 1987.

[3] D. Stephen, Superplastic forming and diffusion bonding of titanium, in: Designing with Titanium, Institute of Metals, London, UK, 1986, pp. 108–124.

[4] N.F. Kazakov, Diffusion Bonding of Materials, Mir Publishers, Moscow, Russia, 1985.

[5] B. Derby, E.R. Wallach, Theoretical model for diffusion bonding, Met. Sci. 16 (1982) 49.

[6] B.P. Kashyap, A.K. Mukherjee, Review: cavitation behavior during high temperature deformation of micrograined superplastic materials, Res. Mechanica 17 (1986) 293–355.

[7] A.H. Chokshi, A.K. Mukherjee, The role of cavitation in the failure of superplastic alloys, in: C.H. Hamilton, N.E. Paton (Eds.), Superplasticity and Superplastic Forming, TMS-AIME, Warrendale, PA, 1988, pp. 149–159.

[8] M. Kajihara, T. Takenaka, Kinetic features of solid-state reactive diffusion between Au and Sn base solder, Mater. Sci. Forum 539–543 (2007) 2473–2478.

[9] J. Pilling, The kinetics of isostatic diffusion bonding in superplastic materials, Mater. Sci. Eng. A100 (1988) 137–144.

[10] J. Wang, Y. Li, H. Ma, Y. Yin, Microstructure and diffusion kinetics at the bonded Fe-16Al/Cr18-Ni8 interface, React. Kinet. Catal. Lett. 87 (2006) 67–75.

[11] L.V. Usacheva, I.L. Bataronov, V.R. Petrenko, V.F. Selivanov, A kinetic model of the development of the volume interaction in diffusion welding, Svar. Proizvod. 57 (2004) 11–15.

[12] Y. Mizuno, F.K. King, Y. Yamauchi, T. Homma, A. Tanaka, Y. Takakuwa, T. Momose, Temperature dependence of oxide decomposition of titanium surface in ultrahigh vacuum, J. Vacuum Sci. Tech. A 20 (5) (2002) 1716–1722.

[13] J. Pilling, N. Ridley, M. Islam, On the modelling of diffusion bonding in superplastic materials: superplastic super alpha-2, Mater. Sci. Eng. A205 (1996) 72.

[14] A.H. Chokshi, A.K. Mukherjee, T.G. Langdon, Superplasticity in advanced materials, Mater. Sci. Eng. RI0 (1993) 237–274.

[15] M.L. Meier, A.K. Mukherjee, Superplasticity of microduplex Ti-6Al-4V, in: T.R. McNelley, C.H. Heikkenen (Eds.), Superplasticity in Aerospace II, TMS-AIME, Warrendale, PA, 1990, pp. 317–332.
[16] ASTM D2651-01, Standard Guide for Preparation of Metal Surfaces for Adhesive Bonding, 2016.
[17] V. Petrenko, A.V. Peshkov, V.Y. Polevin, Increasing the service characteristics of titanium diffusion welded laminated structures, Welfd Int. 19 (12) (2005) 995–998.
[18] Y. Mutoh, M. Kobayashi, Y. Mae, T. Satoh, Fatigue properties of SPF/DB joints in titanium alloys, in: S. Hori, M. Tokizane, N. Furushiro (Eds.), Super-plasticity in Advanced Materials, The Japan Society for Research on Superplasticity, Osaka, Japan, 1991, pp. 405–410.
[19] Z.Y. Zhang, C.H. Cai, Z.R. Lin, The diffusion bonding of Ti-6Al-4V titanium alloy under superplastic condition, in: S. Hori, M. Tokizane, N. Furushiro (Eds.), Superplasticity in Advanced Materials, The Japan Society for Research on Superplasticity, Osaka, Japan, 1991, pp. 693–698.
[20] H.S. Lee, J.H. Yoon, Y.M. Yi, Solid state diffusion bonding of titanium alloys, Solid State Phenom. 124–126 (2007) 1429–1432.
[21] W. Han, K. Zhang, G. Wang, X. Zhang, Superplastic forming and diffusion bonding for sandwich structure of Ti-6Al-4V alloy, J. Mater. Sci. Technol. 21 (1) (2005) 60–62.
[22] H.S. Lee, J.H. Yoon, Y.M. Yi, Lightweight sandwich structures from solid-state bonded titanium sheets, in: 60th International Astronautical Congress, October 12–16, 2009, Daejeon, Korea, 2009.
[23] W. Beck, L. Duong, H. Rogall, Titan 6–4 hemispheres for SCA system of Ariane 5, Mat. Wiss. U. Werkstofftech. 39 (4) (2008) 293–297.
[24] E. Sato, S. Sawai, K. Uesugi, T. Takami, K. Furukawa, M. Kamada, M. Kondo, Superplastic titanium tanks for propulsion system of satellites, Mat. Sci. Forum 551–552 (1) (2007) 43–48.
[25] H.S. Lee, J.H. Yoon, Y.M. Yi, Manufacturing of titanium spherical and hollow cylinder vessel using blow forming, Key Eng. Mater. 433 (1) (2009) 57–62.
[26] Y.S. Yang, A.K. Mukherjee, An analysis of the superplastic forming of a circular sheet diaphragm, Int. J. Mech. Sci. 34 (4) (1992) 283–297.
[27] A.A. Kruglov, F.U. Enikeev, R.Y. Lutfullin, Superplastic forming of a spherical shell out a welded envelope, Mater. Sci. Eng. 323 (1) (2002) 416–426.
[28] H.S. Lee, J.H. Yoon, Y.M. Yi, A study on high temperature blow forming of duplex stainless steel, in: SAE Paper 2009–01–3260, SAE AeroTech Congress, November 10–12, 2009, Seattle, WA, 2009.
[29] M. Muratoglu, O. Yilmaz, M. Aksoy, Developing diffusion bonding characteristics of aluminum metal matrix composites (Al/SiCp) with pure aluminum for different heat treatments, J. Mater. Proc. Tech. 178 (1) (2006) 211–217.
[30] G. Mahendran, V. Balasubramanian, T. Senthilvelan, Developing diffusion bonding windows for joining AZ31B magnesium-AA2024 aluminium alloys, Mater. Des. 30 (4) (2009) 1240–1244.

High-temperature brazing in aerospace engineering

11

A. Elrefaey
Dortmund University of Technology, Dortmund, Germany

11.1 Introduction

Brazing, using low-temperature filler metals, has been used for joining materials to produce decorative articles and engineering construction since before the birth of Christ. Nowadays, the revolutionary change in industrial applications of materials has made the need for brazing at high temperatures routine in most industrial sectors. High-temperature brazing is a joining process that takes place in a vacuum furnace or controlled atmosphere at above 900 °C to create high-strength bonds with good corrosion and oxidation resistance. As the brazing process is carried out at high temperature, much care should be directed to the furnace or controlling atmosphere that will keep the base metal clean and enhance flow of the brazing alloy into the joints. Additionally there is particular emphasis on the possibility of corrosion of the base metal by the brazing alloy and the probability that the base metal may also require a heat-treating process different from the brazing cycle to gain maximum strength. These factors need to be addressed prior to the selection of filler metal and consequently add more challenges to the brazing process designer.

There are plenty of components in aerospace engineering constructions that should only be brazed at high temperatures, such as aircraft engines, honeycomb sandwich structures and many pipes for the transfer of fuel within the aircraft. Filler metals employed in such joints vary in their physical and mechanical properties, in addition to the cost criterion. The challenge in brazing technology is to find solutions that meet today's multiple business priorities and creating bonds that fulfill performance needs in the most cost-effective and environmentally effective way.

This chapter is an attempt to explore the most common filler metals that have the greatest impact in the field of high-temperature brazing in aerospace engineering. Furthermore, the chapter highlights the novel innovations and the new trends in brazing at high temperature. The aim of the chapter is to provide technicians, designers and engineering students with a basic knowledge and to suggest how the potential of the process can be realized.1

11.2 Filler metals

Brazing filler metals (BFM) are metals that are placed between the two or more base metals in the form of a thin layer for the purpose of filling the gap and therefore joining them. The filler metal must be compatible with the base metal, joint clearance and the

Welding and Joining of Aerospace Materials. https://doi.org/10.1016/B978-0-12-819140-8.00011-0
Copyright © 2012 Woodhead Publishing Ltd. All rights reserved.

brazing procedure to be used. When designing a brazed joint for a specific service application, it is important to consider the properties and compatibility of the base metal and filler metal in the brazing operation, as well as the final brazed joint in the environment in which they will operate. Filler metals are available in a variety of forms, namely foil, paste, sheet, powder, wire and rod. The basic characteristics of these filler metals, that are prerequisites for the successful base metals bonding, are strength, workability at high temperatures and corrosion resistance.

There are a large number of metals and alloys that are used as filler metals and all of them have different compositions, which make it crucial to carefully compare and choose from them. This section will highlight the most common filler-metal systems used in aerospace engineering.

11.2.1 Nickel-based filler metal

Nickel-based filler metals are used to braze ferrous and non-ferrous high-temperature base metals. These BFMs are generally used for their strength, high-temperature properties and resistance to corrosion, but are also used for subzero applications down to the liquefaction temperatures of oxygen, helium or nitrogen. Pure nickel is not widely used as a filler metal because of its high melting point, except for some certain applications, such as joining of molybdenum and tungsten components intended for subsequent operation at high temperatures. Table 11.1 lists some important nickel-based BFMs [1].

The main melting-temperature depressants in these systems are phosphorus, boron and silicon. Because of the diffusion of the filler-metal temperature depressant into the parent metal, the remelt temperature of the solidified braze is raised. This phenomenon is employed in the manufacture of engines with superalloy turbine blades because it allows the brazed joint to be used at a higher temperature than its brazing temperature. However, care must be taken to select an appropriate braze/parent metal combination because the diffusion of certain elements into structural engineering alloys can result in intergranular embrittlement. Conversely, the formation of borides, carbides, phosphides and similar compounds in the joint-gap through reaction with the constituents of the parent metals can cause a reduction in joint ductility and impact strength [2].

The basic filler metals are modified by the addition of other elements such as chromium, manganese and iron. For example, the oxidation and corrosion resistance of the basic chromium/phosphorus filler metal has been modified by increasing the chromium content from 12% to 23% without significant change in application temperature, although the alloy is a little less fluid [3].

One of the main applications of nickel brazes is joining of stainless steels and other heat-resistant alloys. Nickel and high-nickel alloys are embrittled by sulfur and many low-melting-point metals, including zinc. The braze alloy, therefore, must not contain such elements. They are also susceptible to stress cracking in the presence of molten brazes so considerations of stress relief are important during both the heating and cooling stages of the brazing process.

Table 11.1 Nickel-based BFMs.

AWS designation	Composition (% weight)													
	Cr	B	Si	Fe	C	P	S	Al	Ti	Mn	Cu	Zr	Ni	Other
BNi-1	13.0–15.0	2.75–3.50	4.0–5.0	4.0–5.0	0.6–0.9	0.02	0.02	0.05	0.05	–	–	0.05	Bal	0.05
BNi-1a	13.0–15.0	2.75–3.50	4.0–5.0	4.0–5.0	0.06	0.02	0.02	0.05	0.05	–	–	0.05	Bal	0.05
BNi-2	6.0–8.0	2.75–3.50	4.0–5.0	2.5–3.5	0.06	0.02	0.02	0.05	0.05	–	–	0.05	Bal	0.05
BNi-3	–	2.75–3.50	4.0–5.0	0.5	0.06	0.02	0.02	0.05	0.05	–	–	0.05	Bal	0.05
BNi-4	–	1.5–2.2	3.0–4.0	1.5	0.06	0.02	0.02	0.05	0.05	–		0.05	Bal	0.05
BNi-5	18.5–19.5	0.03	9.75–10.50	–	0.10	0.02	0.02	0.05	0.05	–	–	0.05	Bal	0.05
BNi-6	–	–	–	–	0.10	0.0–12.0	0.02	0.05	0.05	–	–	0.05	Bal	0.05
BNi-7	13.0–15.0	0.01	0.10	0.2	0.08	9.7–10.5	0.02	0.05	0.05	0.04	–	0.05	Bal	0.05
BNi-8	–	–	6.0–8.0	–	0.10	0.02	0.02	0.05	0.05	21.5–24.5	4.5–5.0	0.05	Bal	0.05
BNi-9	13.5–16.5	3.25–4.0	–	1.5	0.06	0.02	0.02	0.05	0.05		–	0.05	Bal	0.05
BNi-10	12	2.5	3.5	3.5	0.5	–	–	–	–		–	–	Bal	16W
BNi-11	10	2.5	3.5	3.5	0.4	–	–	–	–		–	–	Bal	12W

Nickel-based filler metals are one of the widely used brazing systems in aerospace engineering. It has a good compromise of the needed physical and mechanical characteristics such as melting-point range, fluidity, joint remelt temperature, vapor pressure, strength, ductility, corrosion resistance, cost and availability. American Welding Society (AWS) BNi-1 finds application for high-strength, heat-resistant joints in assemblies like turbine blades and jet-engine parts. Gas turbine engines, such as the one shown in Fig. 11.1, usually require high strength to withstand considerable vibration in service. Fit-up of the joint edges is not well controlled and may vary from contact to 0.254 mm joint clearance. This situation calls for a sluggish filler metal that gives good strength in the braze. BNi-1 and BNi-1a, the low-carbon version of the same alloy, are employed for this job. Nickel-chromium-boron-silicon filler that contains 17% tungsten gives strength to the braze metal in joints with wide joint clearances. It is also sluggish at brazing temperatures, making it ideal for wide joint clearances [4].

Certain critical parts of aircraft engines, such as honeycomb air seals (see Fig. 11.2), are in service at high temperatures. Therefore, the filler metals used for brazing these parts should have high solidus temperatures. AWS BNi-5 is commonly used in such applications because it possesses the highest solidus temperature in the entire BNi family of alloys. This filler metal also provides adequate strength

Fig. 11.1 Gas turbine engine brazed with BNi-1.

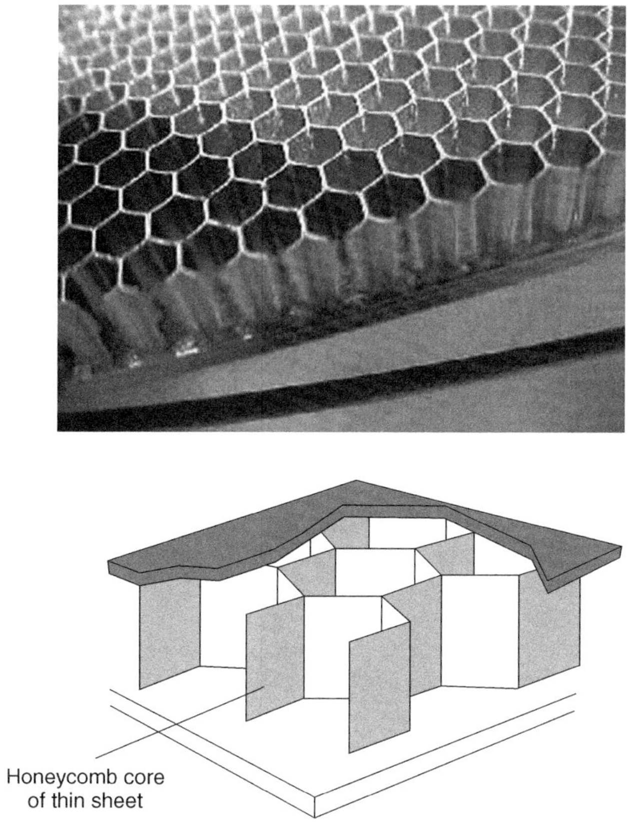

Fig. 11.2 Honeycomb air seals used in aircraft engines.

and corrosion resistance owing to its high chromium content (about 19% weight). The quantity of Ni-based BFM used is probably the most important factor for a successful honeycomb brazing. Brazing temperature and times also are very important, as the honeycomb material is very thin and can be easily eroded by the aggressive BFM. In the brazing cycle, the honeycomb assembly should be heated to an intermediate holding temperature about 55 °C below the solidus temperature until the temperature is uniformly distributed, and then quickly raised to a brazing temperature about 55 °C above the liquidus temperature, unless experience shows that a slightly lower brazing temperature can be used. A short holding time (1–5 min) should be used at the brazing temperature to minimize BFM/honeycomb interaction, as many manufacturers tend to use more BFM than is actually required to braze the honeycomb to its substrate. The ideal amount of BFM required for brazing honeycomb is just enough to fill the volume of space in the nodes between each cell, as well as the space between the bottom of the honeycomb and the substrate to which it is being joined, which means only a tiny amount of BFM is actually needed. Any extra BFM applied to the honeycomb will either build-up unneeded filets in each cell, or could erode the honeycomb [5].

One of the fastest growing and economically important areas of application is repairing cracks that appear in blades and vanes of aircraft and stationary turbines during service. The brazing gaps have vastly different dimensions along the crack length, requiring high BFM flowability and wetting. Additionally, the joint should have a structure similar to that of the BFM parts and possess good mechanical properties. Traditional Ni/Cr-based alloys with B and Si as melting-temperature depressants have been used for a long time together with NiCr/Zr/Hf compositions. Ni-Zr/Hf alloys were specifically developed for, and implemented in, brazing repairs. Recently, new Ni—Ge and Ni-(20–60)Mn BFM were successfully applied for repairing. The latter yield strong epitaxially crystallized joints owing to azeotropic nature of Ni-Mn-alloy system with unlimited mutual solubility of Ni with Mn in the solid state. The major advantage of Mn as a melting-point depressant is its complete solubility in superalloys without formation of secondary intermetallic phases that appear in joints made of B- and Si-containing BFM [6].

Nickel-based filler metals not only have the advantages discussed previously, but also disadvantages and limitations that must be recognized to ensure proper selection. Some react severely with heat-resistant structural alloys. If the brazing conditions promote prolonged reactions of this nature, erosion and/or penetration of thin metal sections may occur. Another important consideration is the diffusion of certain alloying elements, such as boron and silicon to a lesser degree, into the grain boundaries of the base metal that can result in the reduction of joints with poor mechanical properties. Also, some nickel-based filler metals produce joints that lack ductility. To a large degree, the unfavorable characteristics of some nickel-based filler metals can be minimized by proper control of brazing variables [1].

11.2.2 Silver-based filler metal

These filler metals are used for joining most ferrous and non-ferrous metals, except aluminum and magnesium. Silver brazes combine high tensile strength, ductility and thermal conductivity, with unusual wettability to most metals plus the added value of being bactericidal. Silver brazing alloys are used widely in applications ranging from air-conditioning and refrigeration equipment to power distribution equipment in the electrical engineering sector. It also has wide applications in the aerospace industry. Silver-based filler metals are not recommended to braze joints subjected to high-temperature applications. Tests have shown that joints brazed with silver-based filler alloys have good long-term oxidation resistance at temperatures up to 427 °C. Joints with silver-based filler alloys are being used successfully in engine components that have service environments up to 427 °C [7]. In combination with other metals, silver-based alloys provide a melting range of up to 1000 °C.

Silver-based filler metals include a range of silver-based filler-metal compositions that may have various additions, such as copper, zinc, cadmium, tin, manganese, nickel and lithium. The brazing system could roughly be classified into pure silver, silver-copper, silver-copper-zinc and silver-based filler alloyed with different alloying elements.

Pure silver BFM is usually used when high-temperature brazing is needed and when the alloying elements in silver-based alloys can harm the joint properties. In this regard, it is reported that pure silver has been used in brazing Ti-6Al-4V to molybdenum alloys and to niobium alloys for high-temperature applications [8, 9]. Infrared brazing of these joints at 1000 °C for 60 s effectively inhibits the interfacial reaction between the braze alloy and substrate owing to its very rapid thermal cycle. The average shear strength of the infrared-brazed Ti-6Al-4V/Ag/Nb joint was 173.0 MPa, and it is much higher than the furnace-brazed joint (117.5 MPa).

Silver-copper BFM is based on eutectic equilibrium between 72% silver and 28% copper at 779 °C. This alloy is malleable and can be readily worked into a wide variety of perform geometries. Silver-copper alloys that contain additions of phosphorus are known as self-fluxing filler metals in that they can be used in air without the addition of flux to the joint area, provided the parent metal is not too refractory. The silver content improves both the brazed joint strength and elongation of this group. Another important member of this group is the active braze based on the silver-copper eutectic with addition of up to 5% titanium. The purpose of adding titanium is to introduce a highly reactive braze that is capable of wetting and bonding to non-metallic material, particularly various ceramics and graphite [2]. Indium, as an alloying element, is added to the previous system to greatly decrease the melting temperature of the filler metal, as low-temperature filler metals are considered an advantage in many cases, especially when brazing titanium and titanium alloys.

It is worth noting that the use of silver-based BFM has found many applications in brazing titanium and titanium to dissimilar metals that are widely used in aerospace engineering. Commercially pure titanium has been successfully brazed using Incusil-ABA (Ag-27.2Cu-12.5In-1.25 Ti, % weight) in a high-vacuum furnace, at a temperature lower than the so called "beta transus," the critical temperature for the α to β phase transformation. Brazing temperature and holding time have been found to control the shear strength of the brazed joints, as is shown in Fig. 11.3. A brazing temperature of 750 °C provided the best results in terms of joint strength. At the same time, the shear strength of joints revealed a general tendency to increase with an increase in holding time. The change of joint strength with the brazing parameters was to a large extent dependent on the thickness of the reaction layer in between the substrate and the brazed alloy. The best result was achieved when the thickness was the minimum to achieve good wetting and interaction at the titanium/braze alloy interfacial area [10].

Among the promising silver-based BFM commonly used in applications that are subjected to high-temperature service and/or corrosion-resistance requirements, is the 81Ag-10Pd-Ga filler (Gapasil-9). This filler has a brazing temperature in the range of 789–921 °C, and is available as wire, foil and powder. It has been used in brazing thin walls of commercially pure titanium tubes by an innovative electron-beam (EB) vacuum spot-brazing process and vacuum-furnace process. Preliminary observations indicated that Gapasil-9 showed excellent filler-metal fluidity and complete penetration by capillary action through the entire length of the braze joint brazed by the EB brazing process. In vacuum-furnace brazing, Gapasil-9 is somewhat sluggish. The EB brazing demonstrated that the capillary action was so vigorous that Gapasil-9 filler

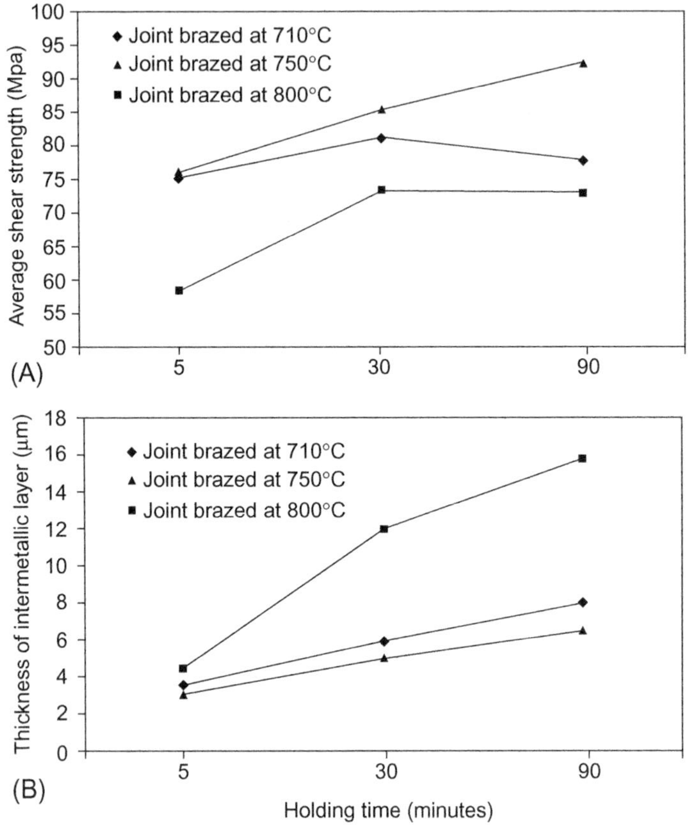

Fig. 11.3 Effect of brazing parameters on strength of the joint. (A) Relation between average shear strength of the joint, brazing temperature and holding time; (B) relation between thickness of intermetallic layer, brazing temperature and holding time.

metal penetrated through the braze joint with practically zero gap. Additionally, the interfacial reaction layer (Ti—Pd phase) appeared to be much thinner in the case of EB brazing than vacuum-furnace brazing [11].

One of the key technologies that concerns the manufacturing of hypersonic engines and liquid rocket-propulsion systems worldwide is the development of ceramic composites C_f/C (Carbotex). Advances in joining science and technology are important to realize the benefits of these advanced materials. A general requirement for aerospace applications is the joining of the C_f/C to a high-temperature alloy that could be nickel-based alloys or nimonic. C_f/C composites are joined to a nimonic alloy using TiCuSil filler metal (Ag-26.7Cu-4.6Ti % weight). The wetting of the filler was enhanced with the deposition of a Cr film on the composite by magnetron sputtering and heat treatment before brazing, as schematically shown in Fig. 11.4A. A brazing regime was conducted as it is presented in Fig. 11.4B. At the C_f/C-filler interface a layered structure of the metallic wetting elements was observed. Crack-free joints

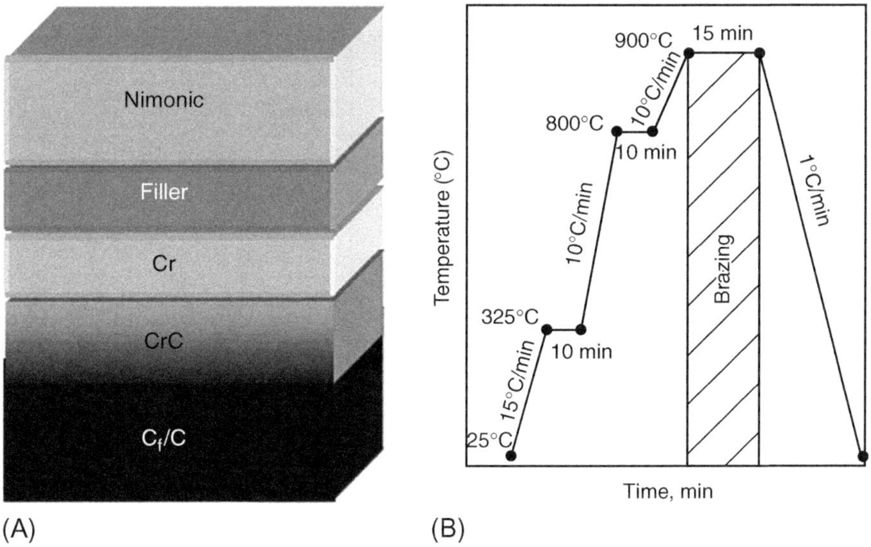

Fig. 11.4 Schematic of the metal-ceramic joint (A) and the thermal cycle of brazing (B).

have been produced, and shear tests showed that failure occurred within the composite that was considered an indication of very strong joint [12].

Although silver-based alloys are used for general-purpose applications, there is limitation for use the alloys at high temperature together with the unsuitability in some circumstances, such as their susceptibility to chloride ion-induced corrosion that cause troublesome problems during commercial use.

11.2.3 Titanium-based filler metal

Titanium-based filler metals are mainly used in brazing titanium alloys and titanium to other metals. Because of the variety of titanium alloys and corollary heat treatments, the brazing alloys must be selected to ensure that the brazing temperature will be within the recommended heat treatments of the alloys being brazed. Commercially pure and alpha alloys should be brazed below the beta transus to prevent a coarse structure that reduces ductility. BFM selection should also occur within the consideration to the environment to which the brazement will be exposed. Of particular concern is the possibility of galvanic corrosion [7].

Nowadays, Ti-Cu-Ni and Ti-Zr-Cu-Ni alloy systems should be considered as the best choice for filler metals, especially for joints that should operate at high temperatures and in a highly corrosive environment. Brazed joints made with these fillers have a heterogeneous microstructure containing a number of intermetallics such as Ti2Cu, Ti2Ni, TiCu and Ti3Cu4. In spite of that, the fine-grained strong microstructure can always be obtained if holding time at the brazing temperature is short ($<$10 min, preferably $<$5 min) and cooling is $>$35 grad/min [13]. In addition, Ti-Zr-Cu-Ni filler

Table 11.2 Ti-based and Ti-Zr-based BFM.

Alloy	Composition (% weight)	Temperature (°C)		
		Solidus	Liquidus	Brazing
BTi-1(Ticuni®,70, MBF-5003)	Ti-15Cu-15Ni	902	950	980–1050
BTi-2(Ticuni®, 60, MBF-5006)	Ti-15Cu-25Ni	901	914	930–950
BTi-3(Ti Braze 260®)	Ti-26Zr-14Cu-14Ni-0,5Mo	–	–	860–890
BTi-4 (MBF-5005)	Ti-20Zr-20Cu-20Ni	848	856	870–890
BTi-5(MBF-5002) (Ti Braze 375®)	Ti-37.5Zr-15Cu-10Ni	839	843	850–880
MBF-5001	Zr-17Ni	982	986	–
MBF-5004	Ti-25Zr-50Cu	842	848	–
MBF-5011	Ti-18.5Cu-27.5Ni	910	920	970–980
MBF-5012	Ti-20Cu-20Ni	915	936	–
VPr16	Ti-Zr-Cu-Ni system	–	900	920–970
VPr28	Ti-Zr-Cu-Ni system	–	860	880–920
Stemet®1201	Ti-12Zr-24Cu-12Ni	–	–	900–1000
Stemet®1406	Zr-11ZrTi-13Cu-14Ni	–	–	900
–	Ti-12Zr-12Cu-12Ni-14Cr	–	930	–
–	Ti-(31–49)Zr-(2.5–38.5)Ni	–	–	870–885
–	Ti-(37–46)Zr-(4–24)Ni-(2–6)Be	–	–	870–885
–	Ti-21Cu-21Ni-16Ag	–	–	930–960
–	Ti-45Zr-4.7Be-5Al	–	–	–
–	Ti-49Cu-2Be	–	–	–
–	Ti-48Zr-2Be	–	–	930

metals such as Ti-26Zr-14Cu-14 Ni-0.5Mo (BTi-3) provide lower intermetallic formation; that is why they are recommended for thin-wall brazed structures like heat exchangers and honeycombs. Table 11.2 shows the Ti-based and Ti-Zr-based BFMs in powder and foil state [14].

The brazing of Ti-6Al-4V and Ti-15–3 titanium alloys using Ti-15Cu-15Ni and Ti-15Cu-25Ni filler foils have been studied to investigate the effect of composition of filler metals on brazing performance [15]. Infrared brazing was performed at brazing temperatures ranging from 970 to 1060 °C, and two holding times of 180 and 300 s. Effects of post-brazing annealing at 900 °C for 3600 s were also used to evaluate the effect on brazed joints. It was concluded that brazing temperature and soak time control the amount of the Cu-Ni-rich Ti phase in the brazed joints. The higher the Cu and/or Ni contents in the braze alloy, a higher brazing temperature and/or longer brazing time are required to homogenize the microstructure of the joint. A post-brazing annealing is also shown to be an effective way to homogenize the microstructure of brazed joint for improved joint strength. Moreover, the average shear strength increases with increasing infrared brazing temperature and when a post-brazing annealing is applied. Similar results were achieved when brazing TiAl intermetallic compound, which has been regarded as a potential replacement for titanium alloys in the aircraft compressor, to Ti-6Al-4V alloy [16].

Another interesting application of titanium filler metal, Ti-15Cu-15Ni, has been reported for vacuum brazing of niobium C103, a refractory alloy widely used for rocket and aerospace applications, to Ti-6Al-4V alloy. The employed brazing temperature was 960 °C and the holding time was 15 min. The high-temperature shear strength of the joint was investigated to evaluate its limit service temperature. The results reveal that the fracture load of the joint reaches a maximum of 31,076 N, and the joint fractures in the C103 parent metal, when the overlap length of the joint increases to 3.5 times the thickness of the parent metal (7 mm), as is shown in Fig. 11.5A. However, the joint shear strength decreases with the overlap length of the joint because of the non-uniform stress distribution in the lap joint during the shear

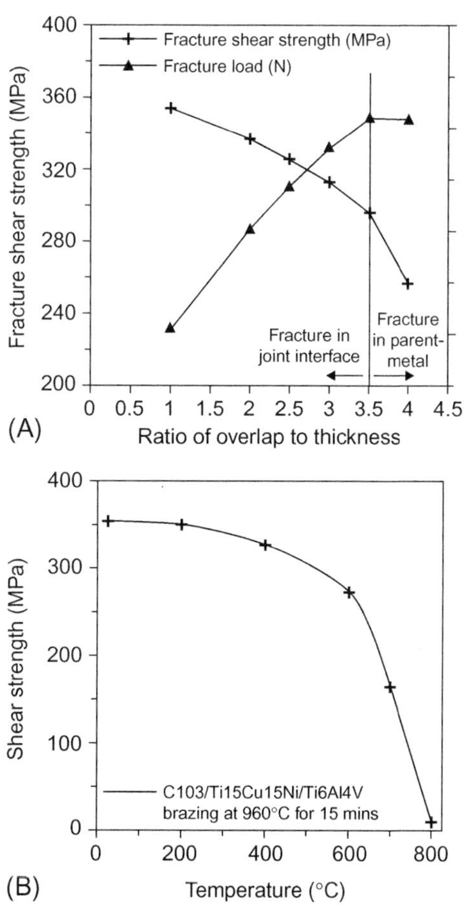

Fig. 11.5 Correlation between shear strength and overlap distance of the C103/TiCuNi/Ti-6Al-4V joint brazed at 960 °C for 15 min (A) and the high-temperature shear strength of the joint (B). 900 °C for 10 min showed fatigue strength of about 600 MPa, which is more than 80% of that of the Ti-6Al-4V base metal. All the brazed joints exhibited excellent corrosion resistance.

test. Post-brazing treatment at 880 °C for 4 h increases the shear strength of the joint by 16%, from 352 to 411 MPa. However, pro-longing the post-brazing time to 6 h reduces the shear strength of the joint to 299 MPa. The results of the high-temperature shear test reveal that the shear strength of the joint maintains at around 300 MPa, at the test temperature below 600 °C (see Fig. 11.5B). However, as the test temperature exceeds 600 °C, the shear strength of the joint declines greatly [17].

Amorphous Ti-Zr-Cu-Ni-based filler metals were used in Ti-6Al-4V brazing [18, 19]. It was found that joints with tensile strength close to those of the base metals could be obtained when the brazing temperature was above 900 °C but below the beta transus. Tensile strength decreased sharply to less than 50% of the base-metal strength when brazing was performed at temperatures well above beta transus. Joints brazed at.

Traditional Ti-15Cu-15Ni filler metal and the majority of Ti-Zr-Cu-Ni-based filler metals are suitable for brazing titanium aluminides and Ti-matrix composites with the joint strength about 0.4–0.5 of that of base metals. Both TiAl alloys and Ti-matrix composites have higher strength than traditional alloys and apparently require a higher strength of joints, especially at high service temperatures. The potentialities of BFMs based on the Ti-Zr-Cu-Ni can provide higher strength of joints after brazing at the temperature above 1000 °C. A promising method of increasing hot strength in this case is to braze the joints at a temperature that is significantly higher than the standard brazing temperature of this filler metal. The structure of brazed joints was fully dense, without voids, but with a few pores on the interface (see Fig. 11.6) that were caused by residual porosity of hot-pressed titanium aluminide base metal. Mechanical testing has proved that these pores do not affect shear strength of brazed joints. Surprisingly, brittle intermetallic layers were not observed in the joint microstructure, although intermetallics were always present in titanium joints brazed with the same filler metal,

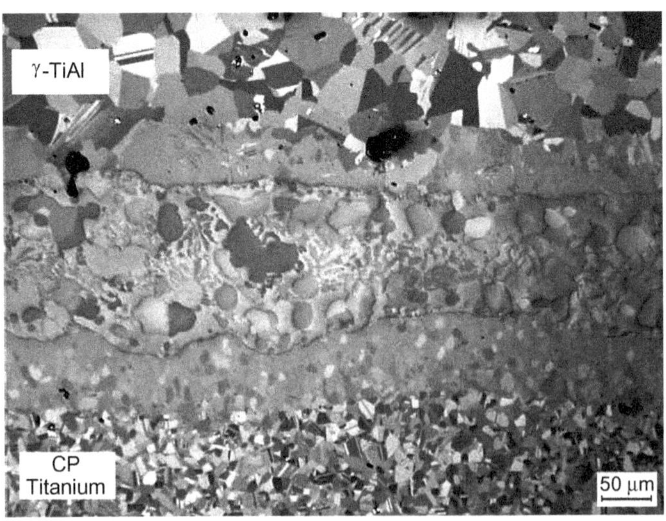

Fig. 11.6 Microstructure of γ-TiAl-to-titanium joint brazed with TiBraze375 (Ti-37.5Zr-15Cu-10Ni-0.1Cr/Fe).

but at 860–900 °C. The absence of intermetallics resulted in better plasticity and shear strength of brazed joints than that of titanium joints brazed at lower temperatures [20].

It is worth noting here that brazing using titanium filler metals needs special care because of the high affinity of titanium to most elements and gases. A high-vacuum or inert-gas atmosphere should be used during vacuum or induction brazing. It is also important to ensure that the vacuum chamber is free of contaminates from pervious brazing cycles. Also, the choice of fixture material must be carefully considered. Nickel and the high nickel-content material generally should be avoided, as low-melting eutectic is formed at 942 °C. Coated-graphite or carbon-steel fixtures may be used.

11.2.4 Gold-based filler metal

Gold is the most malleable and ductile of all the metals. It readily creates alloys with many other metals and it is unaffected by air, moisture and most corrosive reagents, and is therefore well suited for use in many applications. Gold has been used for many years in brazing of jewellery, but in response to technological demand, particularly from the electronics, nuclear power and aerospace industries, many gold-bearing brazing alloys have been developed. These brazing alloys are particularly suited for use in corrosion-resistance assemblies endowed with enhanced mechanical properties. Operating temperatures for the gold-based BFM are below 538 °C and the fillers are characterized by very little erosion attack to base metals, such as stainless steel, super alloys and refractory metals. Gold-based brazing alloys in which gold is the major constituent are listed in Table 11.3 [21].

Most of the commercially available brazing alloys contain gold, copper, nickel and palladium as their principal components. Gold-copper alloys form a continuous series of solid solutions, the melting point of each of these metals is reduced by the addition of the other. Moreover, the liquidus and solidus curves converge near the 80% gold–20% copper composition, which means that this single-phase binary alloy has no melting range. This property is often important from the brazing point of view because it is usually associated with high fluidity and ability to fill narrow joint gaps and to form

Table 11.3 Chemical composition for gold filler metals.

AWS classification	UNS number	Composition (% weight)				
		Au	Cu	Pd	Ni	Other elements
BAu-1	P00375	37.0–38.0	Remainder	–	–	0.15
BAu-2	P00800	79.5–80.5	Remainder	–	–	0.15
BAu-3	P00350	34.5–35.5	Remainder	–	2.5–3.5	0.15
BAu-4	P00820	81.5–82.5	–	–	Remainder	0.15
BAu-5	P00300	29.5–30.5	–	33.5–34.5	35.5–36.5	0.15
BAu-6	P00700	69.5–70.5	–	7.5–8.5	21.5–22.5	0.15

small radius fillets. Other useful properties of gold-copper brazing alloys include the ability to wet copper, nickel, iron, cobalt, molybdenum, tantalum, niobium, tungsten and to produce ductile joints without excessive inter-alloying. The latter factor is important in cases when excessive erosion of the work piece by molten brazing alloy could affect the dimensional accuracy of brazed parts or unduly weakened thin-walled structures [22]. The addition of nickel to gold-copper brazes helps to improve their ductility. A typical composition is Au-16.5Cu-2Ni, which has a melting point of 899 °C. The nickel in the filler metal considerably improves its resistance to creep at elevated temperature, and the filler becomes far less brittle than the conventional 80 Au-20 Cu alloy.

The next important group of gold-brazing alloys is the gold-nickel BFM. 82 Au-18 Ni is a very high-strength alloy, possessing excellent corrosion resistance. Owing to its favorable melting point and exceptional wetting and flow characteristics on tungsten, molybdenum and stainless steel, it is widely used for brazing these metals to copper, nickel and the glass sealing alloys, Kovar and Rodar. Because its solidus and liquidus are at the same temperature, this alloy is of particular value for brazing metals with widely different thermal expansions, for example copper to Kovar or molybdenum. The alloy wets a wide range of high-temperature iron- and nickel-based alloys. It does not alloy excessively with these materials, nor does it produce the severe intergranular penetration normally associated with the nickel-base brazing alloys containing boron [23].

High-temperature properties of gold-nickel alloys can be modified by the introduction of alloying elements. For instance, chromium added to these alloys increases their oxidation resistance (at the expense of their free-flowing characteristics) and confers on them the ability to wet graphite. Alloying can be used also to reduce the noble-metal content (and therefore the intrinsic cost) of materials of this kind without losing any of their useful properties; this, incidentally, was the reason for developing the alloy Au-Cu-Ni-Cr-B, which was developed in the Johnson Matthey Research Laboratories to provide a less-expensive equivalent of the standard 82.5 Au-17.5 Ni alloy used in relatively large quantities in the aircraft industry, the largest consumer of heat-resistant brazing alloys [22].

Another interesting application of brazing with the 82.5 Au-17.5 Ni alloy is the brazing of different parts of the Apollo spacecraft system, such as brazing of a large number of stainless-steel tubing carrying fuel, oxidizer (N204) and helium at high pressures. The gold-nickel alloy is used because of its resistance to N204. Additionally, the alloy is one of the three filler materials used in the brazing of the thrust chamber of the F-l engine powering the first-stage Saturn V launch vehicle of the Apollo spacecraft system [24]. Each thrust chamber has 178 primary tubes and 356 secondary tubes and is constructed mainly of a high-temperature nickel alloy (Inconel X-750). The joints are made by a three-step furnace brazing process.

Materials recently used in aerospace and turbine blades, such as silicon-nitride ceramic (Si3N4), with high strength and hardness, as well as good thermal and oxidation resistances, have been brazed by using Au58.74Ni36.50V4.76% weight filler alloy. The addition of V to the Au—Ni system improves the wettability of filler alloy

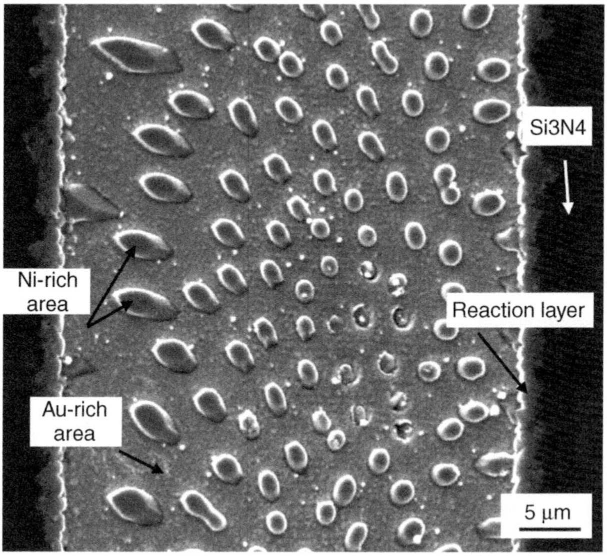

Fig. 11.7 SEM image of the Si3N4/Si3N4 joint brazed at 1150 °C for 60 min using Au58.74Ni36.50V4.76 filler alloy.

on Si3N4 ceramic. Fig. 11.7 shows a scanning electron micrograph image of the joint brazed at 1150 °C for 60 min. It is shown that the joint consists of three main phases: a continuous interfacial wetting reaction layer (VN compound) with a thickness of 1–2 μm between the filler alloy and Si3N4 ceramic; a Au-rich solid-solution phase as the matrix of the central joint; and Ni-rich solid-solution particles distributed homogeneously in the Au-rich solid-solution matrix homogeneously. Air annealing for 100 h was found to sharply decrease the bending strength of the joint at temperatures more than 700 °C owing to the oxidation of the filler alloy at the annealing temperature [25].

The addition of palladium to the gold-copper and gold-nickel alloys improves their resistant oxidation at elevated temperatures. These alloys are therefore used mostly for joining superalloy and refractory metal components that need to serve in relatively aggressive environments, such as in modern jet engines [26]. Commercially available brazes of this type have melting temperatures that reach approximately 1200 °C [2].

Gold-based BFMs are being used at increasing rates every year. They have better properties than silver brazing alloys and are without some of the deficiencies associated with nickel-alloy brazing filler. In situations where high strength and excellent oxidation resistance are required at a temperature around 870 °C there is a strong interest in gold alloys. The high gold content makes gold alloys expensive on a per-ounce basis, but in most cases the value of the assembly and the extreme importance of reliable high-temperature performance compensate the cost of the small amount of alloy needed to make the joint.

11.2.5 Palladium-based filler metal

Palladium is the lightest and and has the lowest melting point of the platinum group metals. It is a silver-white metal and does not tarnish in air. When annealed, it is soft and ductile. Cold working greatly increases its strength and hardness. It resists high-temperature corrosion and oxidation, but it is attacked by nitric and sulfuric acid. Palladium was found to produce considerable improvements in the flow characteristics of copper/silver filler metals, when small quantities are added. Therefore, a series of alloys have been developed containing this element, such as nickel, manganese and gold-based filler metals, to produce alloys with a progression of melting temperature. Table 11.4 shows palladium-bearing brazing alloys. These filler metals possess many of the beneficial properties of the gold-bearing filler metals but are less expensive. They allow joints which have the following characteristics [1]:

- Good mechanical integrity and freedom from brittle intermetallics, a consequence of the fact that palladium forms solid solutions with most common metals
- Enhanced mechanical strength at elevated temperatures; they tend to be superior to the family of gold filler metals that do not contain palladium or other platinum group metals
- High-oxidation resistance at elevated temperatures, especially in the case of the palladium-nickel filler metals
- Good corrosion resistance, although not as good as the gold-bearing filler metals
- Low vapor pressure at typical brazing temperatures, comparable to that of the gold-bearing metals
- A consistently narrow melting range, in most cases no more than 25–50 °C

Ultra-high-temperature materials, such as hot-pressed ZrB2-SiC (ZS) and composites with carbon (ZSC) or SCS-9a SiC fibers (ZSS), were joined using Pd-based brazes, Palco (65% Pd-35% Co) and Palni (60% Pd-40% Ni). Joint integrity was exhibited by the defect-free ZS/Palco/ZS joints with large interaction zones (~200 μm), as is shown in Fig. 11.8. Meanwhile, the ZSS/Palco and ZSC/Palco joints revealed interfacial cracking owing to residual stresses. It is also reported that the joints with Palni exhibited either poor wetting/bonding or cracking because the filler achieved poor wettability to the base materials in addition to problems associated with the large

Table 11.4 Palladium-bearing brazing alloys.

Composition (% weight)					
Pd	**Ag**	**Cu**	**Ni**	**Other**	**Melting range (°C)**
65	–	–	–	35Co	1230–1235
60	–	–	40	–	1273
54	–	–	36	10Cr	1232–1260
25	54	21	–	–	900–950
21	–	–	48	31Mo	1120
15	65	20	–	–	850–900
5	68.5	26.5	–	–	805–810

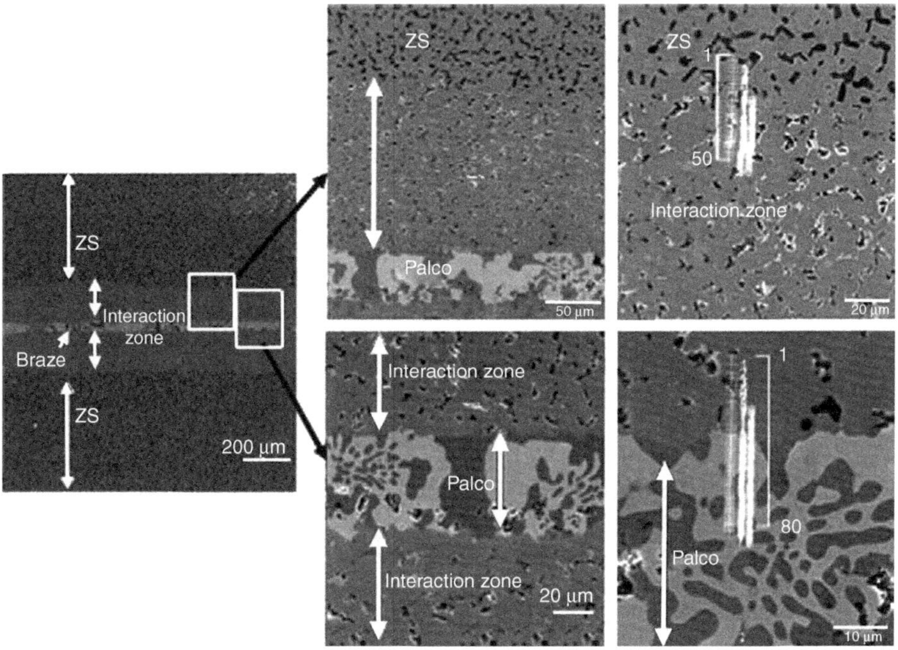

Fig. 11.8 SEM images of a ZS/Palco/ZS joint.

difference in thermal expansion coefficient [27]. Extension of this study has shown that brazing ZS, ZSC and ZSS to titanium and Inconel is possible by the same filler metals at 1240–1260 °C. The tendency for the ceramic to fracture was greater in the ceramic to titanium joints made using Palni than Palco, consistent with the larger strain energy and lower ductility of Palni than Palco joints [28].

The use of palladium in different industrial fields is limited by the availability of the element as it is a comparatively rare and very expensive metal, and to justify its use as a significant improvement in general brazing properties must be demonstrated.

11.2.6 Cobalt-based filler metal

The cobalt filler metals are used for their high-temperature properties and their compatibility with cobalt-based metals. Brazing in a high-quality atmosphere or diffusion brazing gives optimal results. Special high-temperature fluxes are available for torch brazing. Cobalt-based brazes are often substituted for high-temperature gold-based brazes (e.g., Co-20Cr-l0Si, melting point 1180 °C) because they have similar application temperatures, coupled with good corrosion resistance and, of course, substantially lower cost. Several of these alloys contain small percentages of palladium. Because cobalt-based brazes have inferior workability in bulk form, an economic method of fabricating is by rapid solidification, as is the case of all compositions listed in Table 11.5 [2].

Table 11.5 Cobalt-based brazing alloys produced as foils by rapid solidification.

Composition (wt%)							Melting range (°C)	
Co	**Cr**	**Ni**	**W**	**B**	**Si**	**Pd**	**Solidus**	**Liquidus**
Bal	21.0	–	4.5	2.15	1.6	–	1136	1163
Bal	21.0	15.0	4.5	4.4	4.2	–	1078	1139
Bal	21.0	15.0	4.5	4.4	4.4	3.0	1068	1156
Bal	21.0	15.0	4.5	4.4	4.4	5.0	1068	1152

Cobalt-based alloys are used in the brazing of light-weight high-temperature-resistant honeycomb structures for leading edges of wings and other body parts of supersonic jets and reusable shuttles. In these applications, the base metals to be joined are mostly nickel, cobalt-based superalloys and high chromium iron-based alloys. These alloys contain some aluminum, titanium and sometimes yttrium to improve their high-temperature and high-oxidation resistance. The latter is achieved owing to the intrinsic formation of an oxide alumina/titania surface-protecting film on base-metal parts. The presence of active metalloid elements such as boron and silicon in the filler metal causes a partial or even complete dissolution of the protective oxide films in the base metals, which promotes catastrophic oxidation. Additionally, boron, owing to its very small atomic radius, diffuses extensively out of the joint area into the base alloys, particularly in those containing chromium, because of its tendency to form chromium borides. These borides are formed preferentially at grain boundaries resulting in alloy brittleness and excessive oxidation. Therefore, it is preferable that cobalt-based BFM contain less boron, silicon and elements that stop born diffusion to the base metal. Palladium forms a predominant layer of a high oxidation-resistant AlPd intermetallic phase at joint interfaces thus preventing boron diffusion to the base metal. Further, palladium improves the oxidation resistance of the filler and facilitates producing of the foils in amorphous form [29].

11.3 Trends in brazing at high temperature

This section is an attempt to present the recently developed techniques and the novel innovations in high-temperature brazing in the aerospace industry. It is worth noting that owing to the limited chapter size, brief overviews of the techniques are presented, with some applications to clarify the processes as much as possible.

11.3.1 Transient liquid phase bonding process

Transient liquid phase (TLP) bonding is a diffusion-bonding process currently used for joining several types of heat-resistant alloys, for example nickel and cobalt-based superalloys. It has the advantage of not requiring the rather high pressures needed in typical solid-state diffusion-bonding processes. Like brazing, it employs an interlayer material as a bonding agent. Unlike brazing, however, the process occurs isothermally. The interlayer melts and solidifies as a result of interdiffusion with the base material, all at a constant temperature. As a result, the solidified bond region consists of a primary solid solution similar in composition to the base material itself. Thus, the possibility of the formation of brittle phases in the bond region, as in brazing or other joining processes, is largely avoided. Alternatively, because the latter stages of TLP bonding involve solid diffusion the process is very slow. The TLP bonding process has been described in detail by a number of sources. Following the approach applied by Tuah-Poku et al., [30], the mechanism of the TLP bonding process in a binary alloy system is usually illustrated by a phase diagram, as shown in Fig. 11.9. For simplicity, an interlayer of pure element B is used to join two pieces of pure element A; element B serves as the melting-point depressant for the matrix element, A. As a practice, in most cases an interlayer with a eutectic composition is used, rather than a pure element, to shorten the overall bonding time. The TLP bonding process can be broken down into six stages: (a) initial bonding assembly (T_B is the bonding temperature); (b) dissolution of the inter-layer and interdiffusion of A and B during heating up to and holding T_B; (c) widening and homogenizing of the liquid interlayer to maximum width; (d) shrinkage

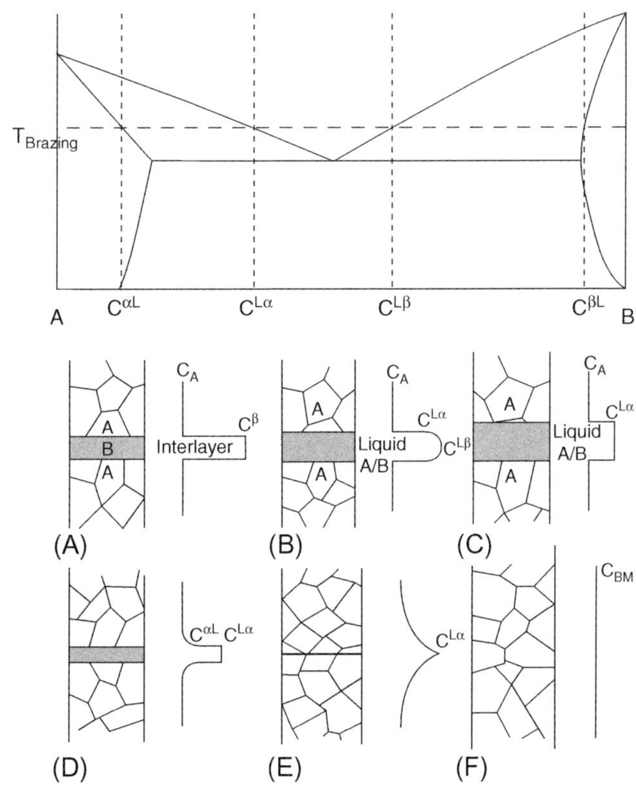

Fig. 11.9 The mechanism of the TLP bonding process in a binary alloy system.

and solidification of the liquid interlayer owing to a loss of solute B; (e) complete solidification at T_B; and (f) homogenization of the bond region.

It should be noted that in TLP bonding, the solidification process occurs isothermally at T_B (Fig. 11.9D and E). The driving force for the bonding to proceed is the concentration gradient of solute B between the interlayer and the parent metal. Theoretically, a homogenous microstructure can be obtained by holding the composite for a sufficient time at a high enough temperature during the TLP bonding process. Some grain growth may well be expected during the homogenization stage. Ideally, no residue of the melting-point depressant or the interlayer phase remains after bonding, as shown schematically in Fig. 11.9F. It has also been noted that the dissolution, widening and homogenization of the interlayer usually take only a few minutes (because the liquid-state diffusion is rapid), whereas the controlling stages of the TLP bonding, isothermal solidification of the liquid and homogenization of the bond region, take hours to complete (under solid-state diffusion control) [31].

The commonly used melting-point depressants boron, silicon and phosphorus are an essential part of TLP brazing, because they can diffuse rapidly out of the liquid filler metal during the brazing process. The loss of melting-point depressants from the filler metal raises the melting temperature of the subsequent joint to a level similar to that of the parent metal. This melting-point shift differentiates TLP bonding from high-temperature brazing and makes it attractive for applications requiring elevated service temperatures. In traditional brazing processes solidification is induced by cooling, resulting in a heterogeneous bond. Isothermal solidification proceeds via epitaxial growth of the substrate, and, when followed with an appropriate homogenization will result in a bond microstructure similar to the bulk material [32]. Since the microstructure and, hence, mechanical properties of the bond tend to match the base material, TLP bonding shows great potential for joining materials that are not easily joined by conventional fusion welding processes.

Successful application/optimisation of the TLP bonding process involves a proper control of parameters that include bonding temperature, time, filler alloy thickness and composition. This is to obtain a joint whose micro-structure and mechanical properties are comparable with those of the base alloy. Several studies have tried to predict the relation between TLP brazing parameters and microstructure features of the joint with particular emphasis on the isothermally solidified zone. The effects of bonding temperature and brazing time on isothermal solidification rate during the TLP bonding of IN 738LC superalloy were investigated [33]. Nickel-based filler alloy, Ni-Cr-B filler alloy 100 μm thick, was used at brazing temperatures of 1130, 1145, 1160 and 1175 °C. The isothermal solidification rate increased as the bonding temperature increased from 1130 to 1145 °C, such that complete isothermal solidification was achieved after bonding for 6 h at 1130 °C and 5 h at 1145 °C, as shown in Fig. 11.10. This attributed to the increase in solid-state diffusion rate of boron with an increase in bonding temperature. At higher bonding temperatures 1160 and 1175 °C, isothermal solidification of the residual liquid was not complete. As seen in Fig. 11.10, a complete isothermal solidification of the joint could not be achieved until after 8 h of bonding at 1160 and at 1175 °C, however, even after 12 h of bonding, a complete isothermal solidification of the residual liquid did not occur. The change in solidification rate

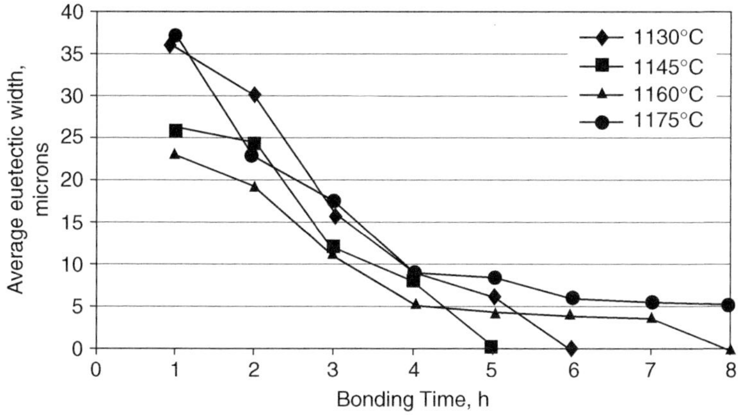

Fig. 11.10 Variation in average eutectic width with bonding time and temperature.

was attributed to the substantial enrichment of the liquid interlayer with the base alloy solute elements, particularly Ti and Cr, and its continuous modification during isothermal solidification. These factors also influence the nature of the phases formed in the centreline eutectic constituents subsequent to the incomplete isothermal solidification.

An enrichment of the liquid interlayer with base-metal solute elements, and the subsequent modification of its concentration has been reported by several authors during the TLP bonding process [34–36]. Recently, this phenomenon was also reported in the case of TLP bonded GTD-111 nickel-based superalloy using a Ni-Si-B interlayer. At low bonding temperatures the microstructure of the joint centreline was found to be controlled by B diffusion. However, at high bonding temperatures the effect of base-metal alloying elements on the joint micro-structure development was more pronounced [37].

The effect of gap size or the thickness of brazed layer on the time required for complete isothermal solidification was found to have a proportional relationship [38]. At a given constant temperature and holding time, the width of the centreline eutectic is expected to decrease with the decrease in the initial filler gap size. This implies that a longer time will be required to achieve complete isothermal solidification in large gap size samples. Fig. 11.11 shows this relation in case of bonding Waspaloy superalloy with MPF-51 foil at 1140 °C for 60 min. Alternatively, increasing the holding time at TLP brazing temperature has shown to decrease the average eutectic centreline width, as clearly shown in Fig. 11.10. Some authors reported that the eutectic width decreased linearly with the square root of holding time at brazing temperature [39, 40].

TLP bonding is a widely used technique for joining and repairing of heat-resistant alloys. This is because a complete isothermal solidification of the liquid interlayer results in the formation of a solid-solution phase. When this phase is homogenized, it produces joints that have microstructure and mechanical properties that are comparable with those of the base alloy. However, there will be a limitation to the commercial application of this technique if a complete isothermal solidification could not be achieved within a reasonable time, as this will increase the cost of repairing damaged

Fig. 11.11 SEM micrograph of joints bonded at 1140 °C for 60 min with a gap size of (A) 50 μm; (B) 100 μm; and (C) 150 μm.

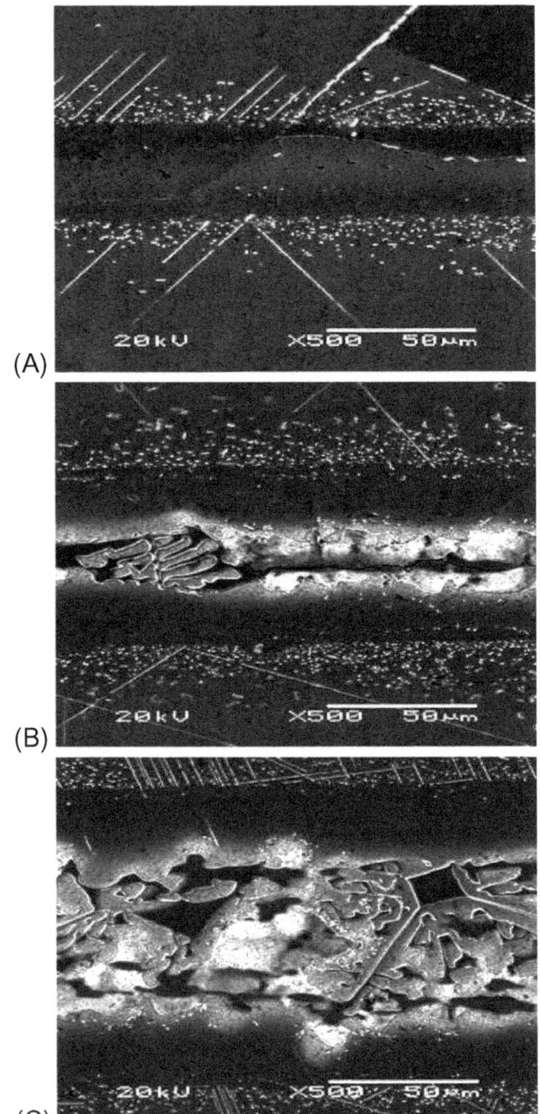

components. In order to shorten the time that is required for a complete isothermal solidification, one of the factors that is commonly controlled is the composition of the filler alloy. Melting-point depressants such as boron, which is an interstitial element with high diffusivity, are usually incorporated into the filler alloys employed for TLP bonding of Ni-based heat-resistant alloys. However, certain elements, such as Al and Ti, are deliberately excluded or restricted in amount from the interlayer because they can form deleterious phases during bonding, and also prolong the time required for the completion of isothermal solidification [33].

11.3.2 Rapidly solidified amorphous filler metal

Rabinkin [6] states that it is not an exaggeration to attribute the appearance of rapidly solidified amorphous ductile BFM as the major step forward in the development of high-temperature brazing technology over the last 40 years. Indeed, most BFMs with metalloids are inherently brittle in the conventional crystalline form and cannot be produced in continuous shapes, such as foil and wire. Therefore, they are available only as powders or their derivatives. Alternatively, the presence of metalloids at or near eutectic concentration promotes the rapid solidification (RS) conversion of such filler metal into a ductile amorphous foil. The most important advantages of RS-amorphous and microcrystalline BFM alloys versus BFM powder forms are their flexibility and ductility, even for compositions that are exceedingly brittle when produced by conventional metallurgical methods. Because a ductile amorphous alloy brazing foil, usually called METAGLAS Brazing Foil (MBF), maybe used as a preplaced preform, there is no need for large brazement gaps, as those used with powder pastes, to achieve a complete filling of the braze cross-section. Moreover, MBF, for example, has a particular advantage over powder and polymer-bonded tape forms because of its superior flow characteristics. MBF flows more freely upon melting than any powder form [41]. A smaller clearance also promotes improved retention of base-metal properties because of curtailed erosion by the use of a smaller volume of filler material in the MBF form. From an economic point of view, only about one third of the filler metal weight is needed per unit joint area when using MBF as a filler metal versus using sprayed or pasted powder. For all these reasons, a preplaced self-fluxing thin MBF preform is superior to the powder-containing paste [42].

MBF alloys are mostly Ni/Cr-based compositions, with boron and silicon as melting-point depressors and eutectic-forming components. Table 11.6 lists commercial and developmental BFM for joining steel and metal alloys at temperatures higher than 900 °C. Recently, new alloys have appeared on the world market with modified compositions. These alloys contain phosphorus as the major metalloid added to Ni-, Fe- or Cr-based. Phosphorus depresses the alloy's melting characteristics more strongly than boron and silicon. The partial or complete replacement of boron and silicon with phosphorus permits the increase of the amount of iron, chromium and molybdenum without increasing the melting temperature above 1150–1200 °C. As a consequence, these alloys have a large chromium concentration providing good corrosion resistance [6].

Amorphous filler metal alloys are mostly used for joining high-performance structural steels and alloys in applications that require high-strength joints that exhibit high-corrosion/oxidation resistance under high temperatures. The prime application area of Ni-based fillers was confined to the aerospace industry, where these types of filler metals were used in brazing of aircraft structural parts, acoustical tail pipes (J-pipes), thrust reversers, turbine blades, seals and honeycomb parts of space vehicles

Table 11.6 Commercial and developmental BFM for joining steel and metal alloys at temperature more than 900 °C.

Alloy system	Alloy class	Major alloying elements	Forms	Brazing temperature (°C)	Service temperature (°C)
Transition metal-metalloid alloy system					
Ni/Fe/Co-(B)-(Si)-(C)-(P)	Eutectic	Cr, Mo, W, Ti, Al	Powder, Paste Tape, RS Foil	950–1200	≤1200
Ni/Pd-(Si)-(B)	Eutectic	Cr, Co, W, Mo	Powder, Paste, Tape, RS Foil	900–1000	400–800
Transition metal-metal alloy system					
Ni-(20 – 23) Ge	Eutectic	–	Powder, Paste	1200	>1200
Ni/Zr/Hf	Eutectic	Cr	Powder, Paste	1200–1250	>1150
Noble metal system					
Au/Pd/Ag	Solid Solution	Cu, Ni, Cr	Powder, Foil, Wire	900–1300	≤1200

for many years. Brazed plates/sinusoidal fin specimens with designs similar to that of industrial turbine heat exchangers were extensively studied by Rabinkin [29]. Plates and fins were AISI 436 stainless-steel sheets with 100 and 50 μm thickness, respectively. Effects of varying foil thicknesses on joint performance was evaluated. MBF-20 foil with 25, 37 and 50 μm thickness were used in brazing at temperature ranging from 1050 to 1150 °C and holding time of 10 min. Details of joint configurations are found in the paper authored by Rabinkin [29]. It was concluded that samples brazed using 25 and 37 μm average thickness foils showed failure in the brazements (Fig. 11.12A). Also, in some samples brazed using 25 μm foil, large unbrazed spots were observed. These spots had formed because an insufficient amount of brazed filler metal needed to fill occasional dents or other defects existed in the fins. In the case of the 50 μm foil sample, failure occurred in the middle of the fins, as depicted in Fig. 11.12B, and no unbrazed spots were observed. Therefore, the strength of the brazed structure in this case was ideally determined by the strength of the base metal. The high strength of joints brazed with 50 μm foil was attributed to the reinforcement of joint by large filets with larger joint cross-sections as depicted in Fig. 11.13.

In another study related to brazing-cycle optimisation of AISI 316L stainless steel, brazed by MBF-51 25 and 50 μm thick foils, brazing cycles of various temperature/time were applied. It was found that optimal combination of strength, ductility and fatigue resistance of joints is obtained when brazing is combined with high-temperature annealing in one extended cycle. The joint after such treatment is free of the brittle central eutectic line, and the base metals adjoining the braze area have

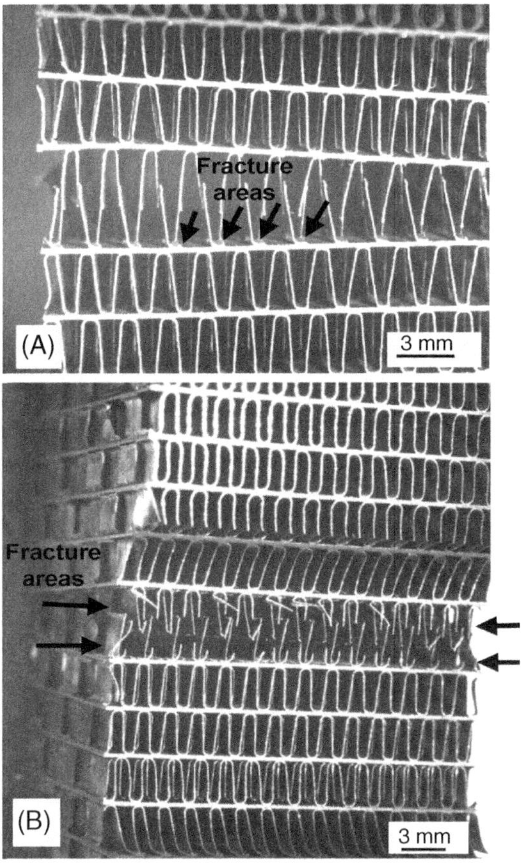

Fig. 11.12 Samples of a 436 stainless-steel plate/fin structure after failure under tensile mechanical testing at 25 °C. (A) Joint brazed using 25 μm thick foil and (B) joint brazed using 50 μm thick foil.

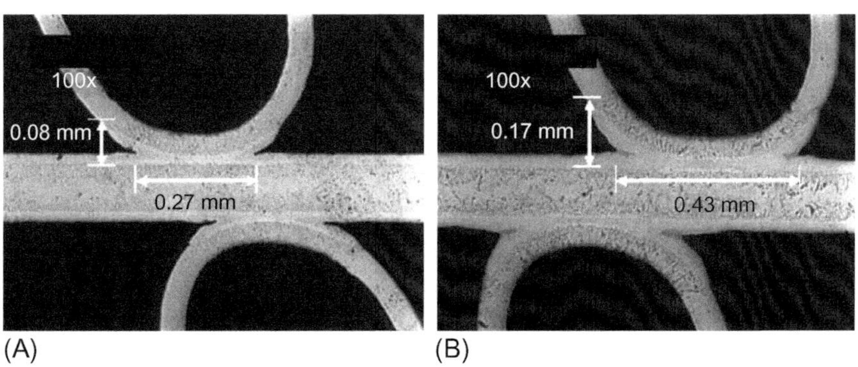

Fig. 11.13 Gas turbine brazed with BNi-1.

only limited and localized segregation of chromium borides. The joint itself is uniform and has a strong and ductile microstructure with only one nickel chromium-based solid-solution phase. Fig. 11.14 shows the joint brazed at 1175 °C with 25 and 50 μm foil [43]. The chromium boride is absent if the amount of filler metal per joint area is small, i.e., a thin foil or when the foil is thick, after long-term brazing or brazing combined with post-brazing annealing are applied (Fig. 11.14A and C). This is owing to the practically complete diffusion of boron into the base metal.

Amorphous active filler metal based on Zr-Ti-Ni-Cu is widely used in brazing joints that need comparatively low brazing temperature and difficult-to-wet materials. Therefore, these MBF have more potential in brazing of titanium and titanium to dissimilar metals. The titanium-to-stainless-steel joint is one of the most important

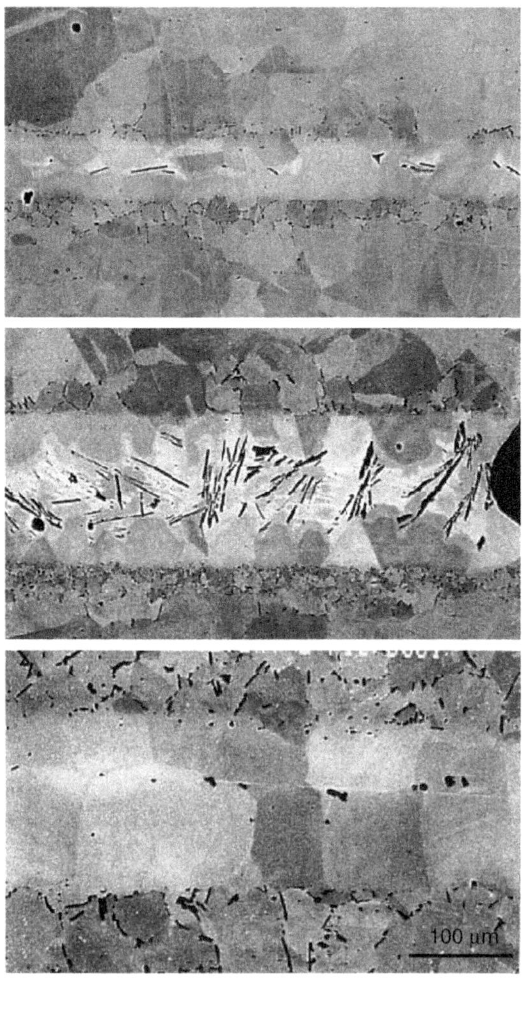

(A) (B) (C)

Fig. 11.14 SEM micrograph of 316/MBF-51/316 joints brazed using (A) 25 μm; (B) 50 μm thick MBF-51 ribbon and short brazing cycle; and (C) 50 μm foil and a long brazing/annealing cycle.

examples in this regard. Brazing at low temperatures protects titanium from β transformation and decreases the thickness of intermetallic formation as well [44]. The filler-metal activity is very beneficial, especially when brazing ceramics and the hard-to-wet materials. It can accelerate atomic diffusion and surface reaction during the high-temperature brazing process. In addition, it is expected to decrease brazing temperature so as to reduce residual stress developed in the joint and hence to increase the joint strength. An interesting study of brazing of Si3N4 ceramic using Ti40Zr25Ni15Cu20 amorphous foil has been performed in which the effects of brazing temperature and time on joint strength and interfacial microstructure were investigated. The joint strength first increased then decreased, with the increasing brazing time and temperature, which was significantly affected by reaction layer thickness. Under the same brazing conditions, the strength of the joint brazed with amorphous Ti40Zr25Ni15Cu20 filler alloy was much larger than that bonded with Ti40Zr25Cu20Ni15 crystalline filler. Improved joint strength was attributed to the positive wettability improvement using amorphous filler metal [45].

Current efforts are directed toward development of new amorphous filler-metal compositions finely compatible and attuned to the specific needs of each of many metallic material groups. The application of new amorphous filler-metal compositions can also be augmented by further advancement of RS-casting technology in order to decrease MBF production costs. Such cost reduction will alleviate their partial cost disadvantage and improve their economic competitiveness versus gas-atomized powders. There is no doubt that amorphous brazing foil will dominate the high-temperature brazing area very soon because its current production growth rate is very high [46].

11.3.3 Self-propagating high-temperature systems

Many methods of bonding require a heat source. The heat source may be external or internal to the structure to be joined. An external source is typically a furnace that heats the entire unit to be bonded. An external heat source often presents problems because the bulk materials can be sensitive to the high temperatures required. Materials can also be damaged by the mismatch in thermal contractions. Internal heat sources often take the form of reactive powders that are typically mixtures of metals or compounds that will react exothermically to form a final compound or alloy. Such powders foster bonding by self-propagating high-temperature synthesis (SHS) [47]. Reactive powder has an advantage of achieving high heat input in the brazed area, which is suitable for joints that require brazing at high temperature. Moreover, the powder could be used in most brazing applications as both a heat source and a brazing material. However, powder compacts have several limitations, such as the green density of the powder and the size of the powder particles is often large compared with the characteristic diffusion distance for a given system, making it difficult to achieve full intermixing. Additionally, particles of highly reactive components will often have a passivating outer coating that acts as a barrier to diffusion with other reactants.

With modern thin-film deposition techniques, fully dense multilayer materials with similar exothermic reactions can be fabricated. Such reactive multilayer foils consist of hundreds or thousands of nanometer-scale alternating layers of two or more materials, known as reactants, which can mix exothermically. When heat is applied locally to the foil, it undergoes SHS in a fashion similar to powder compacts. Structures of reactive multilayer foils offer several potential improvements over powder compacts. The individual layers in a reactive foil are usually on the scale of tens of nanometers, which significantly decreases the diffusion distances involved in interatomic diffusion between the reactants. The thickness of the individual reactant layers can be controlled during fabrication, allowing significant control over the properties of the foil. The individual layers are in intimate contact with each other, which increases both thermal and atomic diffusion in the system while also eliminating voids in the system. Moreover, having thin layers of reactants that are in intimate contact with each other makes the SHS reaction propagate faster in most reactive foils than in a powder compact with the same components [48]. So far, self-propagating reactions have been reported in Ti/B, Ni/Si, Zr/Si, Rh/Si, Ni/Al, Monel (7Ni,3Cu)/Al, Ti/Al, Pd/Al, Pt/Al and CuO_x/Al multilayer materials.

Reactive multilayer materials are most commonly fabricated using physical vapor deposition (PVD) methods, such as magnetron sputtering and EB evaporation. The PVD methods involve creating a vapor of a material, known as the source material, and then depositing the vapor onto a substrate. The rate at which the vapor is deposited is controllable, allowing the growth of layers with a thickness ranging from nanometers to micrometers. Alternatively, multilayer foils can also be made by cold rolling.

In practice, the nanofoils are combined with additional brazing foils placed between two components to be joined. A small applied pressure allows the braze to flow and wet all surfaces. The reaction can be triggered electrically or thermally. For the ignition of the short-circuit current, a 9 V battery is sufficient. By varying the thickness and composition of the nanofoils the maximum temperature, the speed and the absolute heat input of the joining process can be controlled. Fig. 11.15 shows a schematic drawing of a self-propagating reaction in a multilayer foil [49]. Reactive multilayer foils provide a unique opportunity to dramatically improve conventional soldering and brazing technologies by using the foils as local heat sources to melt solder or braze layers and thereby join components. The only limitation of this process until now is its suitability to be used as a heating source for soldering applications than its use as a BFM, in addition to its original function as a heat source. Additional applications are being developed, such as a multilayer system that is optimized for use as a filler metal as well as higher-energy foils. The chemistry of this foil will produce more heat and energy to melt braze layers with higher melting temperatures.

Self-propagating-high-temperature reactive powders have been used successfully in brazing TiAl intermetallic alloys using different mixtures of Ti-Al-C powder. The highest joint strength was achieved using Ti50-Al35-C15% weight [50]. It was also noted that the porosity formed in direct SHS joining decreased the strength of joints. The wettability, density and joint strength of joints were improved by inserting Ag-based brazing foil together with the powder compacts [51]. Reactive powders have also been used in repairing turbine-blade materials using the high-pressure

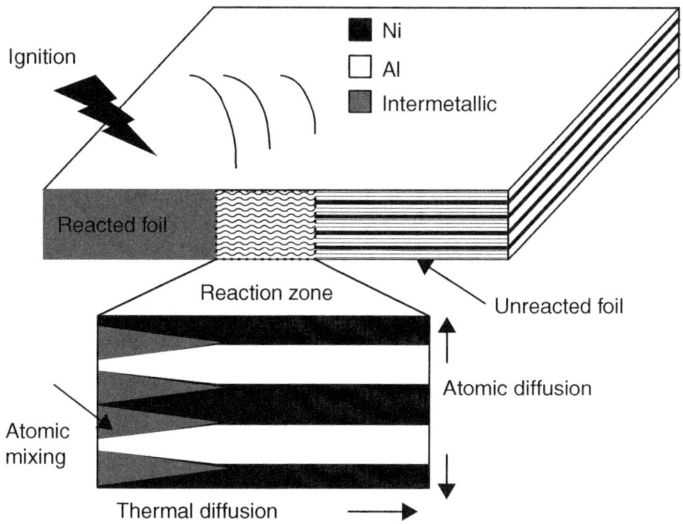

Fig. 11.15 A schematic drawing of a self-propagating reaction in a multilayer foil.

combustion synthesis of the NiAl compound. In this process, a nickel-based braze was inserted between a substrate of nickel-based superalloy and a powder compact (Ni—Al). The heat released during the combustion synthesis gave rise to interdiffusion of the elements and consequently to formation of a metallurgical bond. The advantages of this process are the speed and the low cost. Moreover, unlike the conventional repair techniques, it allows limitation of the thermally affected zone [52].

Alternatively, nanofoils have been used in brazing titanium alloy (Ti-6Al-4V) in air by using Al/Ni multilayer foils to melt a silver-based braze [53]. Fig. 11.16 shows a schematic of the joint and backscattered scanning electron micrograph of a cross-sectioned joint. The foils are capable of undergoing self-sustaining exothermic reactions. By systematically controlling the properties of the foils and by numerically modeling the reactive joining process, good wetting and adherence between all interfaces are achieved. A joint characteristic is a crack in the reactive foil that is filled with braze material during the joining process. It is also concluded that the temperatures reached by the foils during reaction are crucial in determining the success of joining when using higher-melting-temperature braze layers. The maximum temperature is directly dependent on the foil's heat of reaction. Other factors, such as reaction-velocity foil thickness and total heat, have less impact on joint strength.

In addition to the previous-mentioned application of nanofoils, it can also be used to reduce the temperature and/or the pressure needed for solid-state diffusion bonding. The multilayer can, simultaneously, improve the diffusivity, owing to their nanocrystalline nature and high density of defects, and act as a local heat source. In this respect, a variety of metallic, ceramic and intermetallic materials have been bonded with the help of nanofoils. These applications are not within the scope of this chapter as the process used is mainly diffusion bonding.

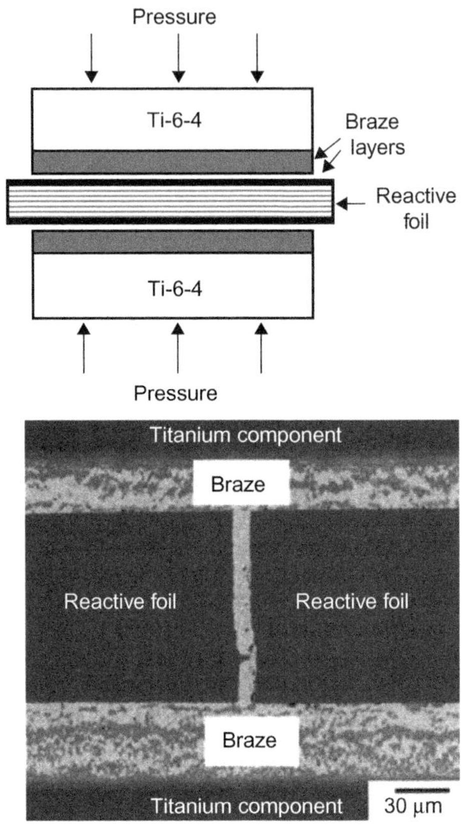

Fig. 11.16 A schematic joint configuration (top) and backscattered electron SEM micrograph of a joint cross-section.

11.4 Conclusion and future trends

In all branches of industries that use metals, the application of brazing is increasing. The origin of those technologies is assumed to be 6000 years ago and, for a long time, they were limited to a few industrial fields. In recent years, many new advanced materials are appearing and brazing technologies have to fulfill the brazing requirements of these materials. Developments and innovations in brazing technology are mainly focused on the two main branches of brazing, which are BFMs and brazing techniques.

During the last few years, new alloys with modified compositions have appeared on the world market. The motivation has been to either improve the characteristic properties of the brazed joint, such as strength, wetting and corrosion, especially at high temperatures, or to reduce the cost of materials. It is worth noting that all of the BFMs used in the field of brazing at high temperature are based on the most expensive metals. Therefore, efforts have to be made to reduce some of the material costs by developing tailor-made low-cost alloys. Examples of low-cost BFMs with excellent overall properties are the 19Au-7Ni-6Pd-5Mn-43Cu and 30Au-10Ni-10Pd-6Mn-34Cu fillers. These alloys are less expensive than normal gold-based alloys and are characterized by a brazing temperature range from 960 to 1010 °C, in addition to high strength, excellent wetting, good oxidation and salt-spray resistance. These filler metals were developed for vertical tube-to-tube brazing in the space-shuttle main-engine flight nozzle.

The cost of BFMs could also decrease by using the right quantity of filler metal. Foil produced by conventional techniques usually has a limited minimum thickness, which in many cases exceeds the actual needed quantity of filler metal in a specified application. The application of a filler metal by using the PVD process is considered a novel trend in this regard. PVD technology offers a wide range of depositable metal coatings suitable for brazing. PVD allows the deposition of BFM alloys, metal thin films as wetting surfaces, as well as reactive intermediate layers for materials that are difficult to braze by conventional means. Metallic PVD coatings show extremely high purity, and thus guarantee high-quality filler metal coatings and metallization with excellent substrate adhesion. Also, the amount of liquid filler metal affects the joint microstructure and especially the thickness of the intermetallic layer at the interface. The relatively thick joints (joint clearance $\geq$100 μm) usually have a continuous intermetallic layer that generates cracks at any thermal stresses. Besides, spattering cleaning of the substrate in vacuum prior to film deposition can be done. Therefore, oxide layers, that represent one of the major problems in brazing, and more precisely wetting alloys are overcome by this technique. After spatter cleaning, a clean, fairly unoxidized and ideally undisturbed surface of the material is ready for coating. The possibility of using PVD technology in brazing opens many new avenues for difficult-to-join materials. By depositing one or more layers of filler metals on the joint surfaces, traditionally difficult-to-join materials can be brazed together.

One of the promising technologies that is expected to influence the science of brazing in the near future is the application of nanotechnology. In many industrial sectors, nanotechnology is considered a driving force of innovation. As mentioned earlier, an interesting approach in brazing is based on the combination of size effects with exothermic reactions. As all chemical reactions are connected to energy transformation, in this case the energy, released in the form of heat in an exothermic reaction, is used for joining. The essential advantage when simultaneously using the size effects and exothermic reactions for joining is the fact that the heat needed is released directly at the joining area. Thereby, the heat input is reduced significantly in comparison with conventional joining methods. The joint is thus relieved from thermally induced mechanical stresses. Moreover, the joining process does not take much time because of the high velocity of propagation. Wide-spread applications of this technique are expected when the nanofoil is optimized for the use as a filler metal as well as a source of heat generation.

It is unfair to conclude this chapter without mentioning the TLP process, which is widely used in bonding varieties of difficult-to-join materials, especially those used at an elevated-temperature service, such as nickel-based superalloys, oxide-dispersion strengthened alloys, metal-matrix composites, structural intermetallics and dissimilar materials. This process is expected to continue its proportional and successful application in honeycomb structures, engine frames for aero gas turbines and other applications. The capability of producing parent-metal-like microstructures and consequently high mechanical-properties joints is the main point of attraction in this process. Besides, the process is particularly suitable when a high working temperature is needed, as the loss of melting-point depressants from the filler metal raises the melting temperature of the subsequent joint to a level similar to that of the parent metal.

References

[1] M. Schwartz, Brazing, ASM International, USA, 2003.
[2] D.M. Jacobson, G. Humpston, Principles of Brazing, ASM International, Ohio, USA, 2005.
[3] G. Sheward, High-Temperature Brazing in Controlled Atmospheres, Pergamon Materials Engineering Practice Series, Oxford, 1985. http://www.amazon.com/High-Temperature-Controlled-Atmospheres-Materials-Engineering/dp/0080261701.
[4] M. Weinstein, R.L. Peaslee, F.M. Miller, How to choose nickel-based-filler metals for vacuum brazing—Selection of the right filler metal can lead to more economical and functional brazements, Weld. J. 88 (2009) 59–61.
[5] D. Kay, Honeycomb-brazing essentials for successful use as turbine seals, Ind. Heating 70 (2003) 33–37.
[6] A. Rabinkin, High temperature brazing development since the time of the 'Bow-Tie Generation: In memory of Robert L Peaslee, in: 9th International Conference Brazing, High Temperature Brazing and Diffusion Bonding, LÖT 2010, Aachen, Germany, 2010.
[7] ASM International, Metals Handbook, Vol. 13, Corrosion: Materials, 1987. Park, Ohio, USA.
[8] H.Y. Chan, R.K. Shiue, Study of brazing Ti-6Al-4V and TZM alloy using pure silver, J. Mater. Sci. Lett. 22 (2003) 1659–1663.
[9] D.W. Liam, R.K. Shiue, Brazing of Ti-6Al-4V and niobium using three silver-base braze alloys, Metall. Mater. Trans. A 36 (2004) 2415–2427.
[10] A. Elrefaey, W. Tillmann, Effect of brazing parameters on microstructure and mechanical properties of titanium joints, J. Mater. Process. Technol. 209 (2009) 4842–4849.
[11] Y. Flom, Electron beam brazing of titanium for construction in space, in: 3rd International Brazing and Soldering Conference (IBSC), San Antonio, Texas, April 24–26, 2006.
[12] N.V. Moutis, C. Jimenez, X. Azpiroz, T. Speliotis, C. Wilhelmi, S. Messoloras, K. Mergia, Brazing of carbon–carbon composites to Nimonic alloys, J. Mater. Sci. 45 (2010) 74–81.
[13] O. Botstein, A. Rabinkin, Brazing of Ti-based alloys with amorphous Ti-25Zr-50Cu filler metal, Material. Sci. and Eng. A 188 (1994) 305–315.
[14] A. Shapiro, A. Rabinkin, State-of-the-art of titanium-based brazing filler metals, Weld. J. 82 (2003) 36–43.
[15] C.T. Chang, Y.C. Du, R.K. Shiue, C.S. Chang, Infrared brazing of high-strength titanium alloys by Ti-15Cu-15Ni and Ti-15Cu-25Ni filler foils, Material. Sci. Eng. A 420 (2006) 155–164.
[16] R.K. Shiue, S.K. Wu, Y.T. Chen, C.Y. Shiue, Infrared brazing of Ti50Al50 and Ti–6Al–4V using two Ti-based filler metals, Intermetallics 16 (2008) 1083–1089.
[17] I.T. Hong, C.H. Koo, The study of vacuum-furnace brazing of C103 and Ti–6Al–4V using Ti–15Cu–15Ni foil, Mater. Chem. Phys. 94 (2005) 131–140.

[18] T. Onzawa, A. Suzumura, M. Ko, Structure and mechanical properties of CP Ti and Ti-6Al-4V alloy brazed joints with Ti-based amorphous filler metals, Quart. J. Jpn. Weld. Soc. 5 (1987) 205–211.
[19] T. Onzawa, A. Suzumura, M. Ko, Diffusion brazing of Ti-21V-4Al b-alloy using Ti-based filler, Quart. J. Jpn. Weld. Soc. 7 (1989) 460–467.
[20] A.E. Shapiro, Heat-resistant brazing filler metals for joining titanium aluminide and titanium alloys, in: 3rd International Brazing and Soldering Conference (IBSC), San Antonio, Texas, April 24–26, 2006.
[21] AWS committee, Brazing Handbook, American Welding Society, Miami, Florida, 1991.
[22] M.H. Sloboda, Industrial gold brazing alloys, Gold Bull 4 (1971) 2–8.
[23] M.M. Schwartz, Applications for gold-base brazing alloys, 8 (1975) 102–110.
[24] F.X. De Carlo, Brazing gives apollo a lift, Am. Machinist. 111 (1967) 84–87.
[25] J. Zhang, Y. Sun, C.F. Liu, W. Zhang, Interfacial microstructure of Si3N4/Si3N4 joint brazed using Au–Ni–V filler alloy, J. Mater. Sci. 45 (2010) 2188–2193.
[26] Y.K. Yu, D.Y. Liaw, R.K. Shiue, Infrared brazing Inconel 601 and 422 stainless steel using the 70Au-22Ni-8Pd braze alloy, J. Mater. Sci. 40 (2005) 3445–3452.
[27] R. Asthana, M. Singh, Joining of ZrB2-based ultra-high-temperature ceramic composites using Pd-based braze alloys, Scr. Mater. 61 (2009) 257–260.
[28] M. Singh, R. Asthana, Joining and integration of ZrB_2-based ultra-high temperature ceramic composites using advanced brazing technology, J. Mater. Sci. 45 (2010) 4308–4320.
[29] A. Rabinkin, Cobalt-Chromium-Palladium-Based Brazing Alloys, 2000. United States patent number 6165290, Dec. 26, 2000.
[30] K. Tuah-Poku, M. Dollar, T.B. Massalski, A study of the transient liquid phase bonding process applied to an Ag/Cu/Ag sandwich joint, Metall. Trans. A. 19 (1988) 675–686.
[31] M.W. Tseng, D.B. Williams, K.K. Soni, R. Levi-Setti, Microstructural evolution during transient liquid-phase bonding in a Ni-base superalloy/sapphire fiber composite, J. Mater. Sci. 34 (1999) 5187–5197.
[32] Y. Zhou, W.F. Gale, T.H. North, Modeling of transient liquid phase bonding, Int. Mater. Rev. 440 (1995) 181–196.
[33] O.A. Idowu, N.L. Richards, M.C. Chaturvedi, Effect of bonding temperature on isothermal solidification rate during transient liquid phase bonding of Inconel 738LC superalloy, Mat. Sci. Eng. A 397 (2005) 98–112.
[34] W.F. Gale, E.R. Wallach, Influence of isothermal solidification on micro-structural development with Ni–Si–B Filler Metals, Mater. Sci. Technol. 7 (1991) 1143–1148.
[35] S.V. Orel, L.C. Parous, W.F. Gale, Transient liquid phase bonding of NiAl using Ni–Si–B interlayers, in: Proceeding of Advanced Joining Technologies for New Materials II, American Welding Society, Florida, 1994. pp. 5–19, 2–4.
[36] S.V. Orel, L.C. Parous, W.F. Gale, Diffusion brazing of nickel aluminides, Weld. J. 74 (1995) 319s–324s.
[37] M. Pouranvari, A. Ekrami, A.H. Kokabi, Effect of bonding temperature on microstructure development during TLP bonding of a nickel base superalloy, J. Alloys Compd. 469 (2009) 270–275.
[38] O. Adu, N.P. Wikstrom, O.A. Ojo, M.C. Chaturvedi, Transient liquid phase bonding of nickel-based superalloys, in: Fourteenth International Conference on the Joining of Materials and The 5th International Conference on Education in Welding, JOM-Institute, Helsingor, Denmark, April 29–May 2, 2007.
[39] M.C. Chaturvedi, O.A. Ojo, N.L. Richards, Diffusion brazing of cast inconel 738 superalloy, AZo-OARS, J. Mater. Online 1 (2005) 1–12.

[40] M. Pouranvari, Diffusion brazing of a nickel based superalloy Part 1-Solidification behavior, in: Association of Metallurgical Engineers of Serbia (AMES), 2010. Scientific paper UDC: 621. 792. 3:669. 248., 241–247.

[41] N. DeCristofaro, D. Bose, Rapidly solidified filler metals in brazing and soldering applications, in: Proceedings of the Fifth International Conference on Rapidly Quenched Metals, Amsterdam, North-Holland, 1985, pp. 415–424.

[42] A. Rabinkin, New brazing development in the automotive industry, in: Proceedings of the International Brazing and Soldering Conference, IBSC 2009, Orlando, 2009, pp. 380–386.

[43] A. Rabinikin, E. Wendski, A. Ribaudo, Brazing stainless steel using a new MBF-series of Ni-Cr-B-Si amorphous foil, Weld. J. 77 (1998) 66s–75s.

[44] J.G. Lee, J.K. Lee, S.M. Hong, M.K. Lee, C.K. Rhee, Microstructure and bonding strength of titanium-to-stainless steel joints brazed using a Zr–Ti–Ni–Cu–Be amorphous filler alloy, J. Mater. Sci. 45 (2010) 6837–6840.

[45] J. Zou, Z. Jiang, Q. Zhao, Z. Chen, Brazing of Si3N4 with amorphous Ti40Zr25Ni15Cu20 filler, Mater. Sci. Eng. A 507 (2009) 155–160.

[46] A. Rabinkin, Brazing with (NiCoCr)–B–Si amorphous brazing filler metals: alloys, processing, joint structure, properties, applications, Sci. Technol. Weld. Join. 9 (2004) 181–199.

[47] T.P. Weihs, M. Reiss, Method of making reactive multilayer foil and resulting product, 2001. US patent No. 6534194 B2, 18 March 2003.

[48] X. Qiu, Reactive Multilayer Foils and Their Applications in Joining, 2007. M.Sc. Thesis, Tsinghua University.

[49] J.S. Kim, T. LaGrange, B.W. Reed, M.L. Taheri, M.R. Armstrong, W.E. King, N.D. Browning, G.H. Campbell, Imaging of transient structures using nanosecond in situ TEM, Science 321 (2008) 1472–1475.

[50] J. Cao, J.C. Feng, Z.R. Li, Effect of reaction heat on reactive joining of TiAl intermetallics using Ti–Al–C interlayers, Scr. Mater. 57 (2007) 421–424.

[51] J. Cao, J.C. Feng, Z.R. Li, Joining of TiAl intermetallic by self-propagating high-temperature synthesis, J. Mater. Sci. 41 (2006) 4720–4724.

[52] C. Pascal, R.M. Marin-Ayral, J.C. Tedenac, C. Merlet, Combustion synthesis: a new route for repair of gas turbine components—principles and metallurgical structure in the NiAl/RBD61/superalloy junction, Mater. Sci. Eng. A 341 (2003) 144–151.

[53] A. Duckham, S.J. Spey, J. Wang, M.E. Reiss, T.P. Weihs, 'Reactive nanostructured foil used as a heat source for joining titanium, J. Appl. Phys. 96 (2004) 2336–2342.

Quality control and nondestructive testing of self-piercing riveted joints in aerospace and other applications

12

P. Johnson
Liverpool John Moores University, Liverpool, United Kingdom

12.1 Introduction

This chapter discusses the need for nondestructive testing (NDT) for self-pierce riveting [1] (SPR), as well as monitoring systems and sensors that can be used to provide this testing. Joining techniques in such sectors as the automotive and aerospace industries are predominantly driven by advances in materials, working with dissimilar materials and the call for increased automation owing to a decline in the skilled labor force [2].

Riveting machines apply rivets to materials in a wide variety of configurations, from manually operated, handheld riveting guns to multihead automated riveting tools that are electrically, pneumatically or hydraulically actuated. There are three main types of riveting machinery: compression riveting, nonimpact riveting (also known as orbital riveting) and impact riveting. Compression riveting forms the head as a result of squeezing or pulling the rivet shank, whereas nonimpact riveting forms the head of the rivet by performing a spinning or rolling action to the end of the top of shank. Impact riveting—as used in SPR—forms the head by impacting the top of the shank, using riveting hammers/punches.

Henrob Ltd. are industry leaders in SPR joining and monitoring systems. Their client list includes eight leading automotive manufacturers—BMW, Audi, Jaguar, Volvo, Chrysler, Mercedes, Freightliner and Hyundai. In 1994 Audi used Henrob's preclamping SPR system for the majority of the A8s single-point joints [3]. In 1999 the same SPR system was used for various parts of the Audi A2, which featured a lightweight all-aluminum body [4]. Jaguar also used self-piercing rivets in 2001 for their X350 [5]. The engineers decided to use SPR because significant parts of the X350 body shell were made from aluminum. Jaguar implemented a SPR system developed by Henrob, and used Kawasaki robots to apply the rivets. Some 4 years later in 2005, Jaguar reevaluated their manufacturing processes and concluded SPR still presented the best solution for joining aluminum [6].

Henrob along with Bollhoff—another prominent supplier of SPR machinery—offer monitoring systems to accompany their joining machinery. Like Henrob,

Welding and Joining of Aerospace Materials. https://doi.org/10.1016/B978-0-12-819140-8.00012-2
Copyright © 2012 Woodhead Publishing Ltd. All rights reserved.

Bollhoff operates in volume manufacturing plants in the automotive industry [7]. These companies offer a variety of hydraulic and electric servo joining solutions tailored to their customers joining specification. The monitoring system may be purchased at an additional cost. Henrobs monitoring system is called RivMon, and can be purchased at an additional cost for its hydraulic and electric servo systems, whereas Bollhoffs monitoring system is an optional extra on its RIVSET systems [5, 8–10]. The main feature of these monitoring systems are their dedicated sensors that monitor the setting force and punch movement throughout the riveting process, resulting in a force displacement curve. This curve is compared with a taught reference curve. If the process curve fits within a predefined tolerance of the reference curve, the joint is passed; otherwise the joint may be flagged for attention or even halt the process. The sensors employed depend on the variant of SPR system. Hydraulic systems have a positional sensor to locate the punch and up to two pressure sensors to monitor the clamp and punch pressures. Electric servo systems monitor the punch location, punch velocity and torque (motor current).

Orbitform offers a process monitoring system called Watchdawg [11]. Although this system is designed for nonimpact (orbital) riveting, it works using the same force displacement technology. The screen shows numerical force and positional information, as well as the force displacement graph. The graph contains three plots; the solid line shows the force displacement curve, with force measured on the x-axis in pounds and displacement on the y axis measured in inches. Here the upper and lower dashed plots identify the tolerance bands around the joint and it can be clearly seen that this example joint is well within tolerance.

12.1.1 Nondestructive testing (NDT) techniques

The most prominent alternative NDT techniques for SPR not based on force displacement-curve technology are real-time visual inspection and ultrasonic testing. The General Engineering Research Institute (G.E.R.I.) at Liverpool John Moores University [12] is currently researching real-time visual inspection for various aspects of the SPR process. Ultrasonic testing appears to be another promising method of NDT, under evaluation by Warwick Manufacturing Group at the University of Warwickshire [13] and Uppsala University [14].

12.2 Computer vision

12.2.1 Rivet status

The machine jaws are an ideal location to begin preprocess monitoring as this is where the rivets rest prior to setting. It is possible to detect the status and orientation of a loaded rivet by visual inspection of the machine jaws. It is possible to prevent a number of incidents from occurring, such as a punch imprint on the joining material or a multiple rivet piercing, by identifying the rivet status preprocess. A punch imprint is caused when a rivet has failed to load into the machine jaws and the SPR process

Fig. 12.1 A punch imprint in aluminum joining materials.

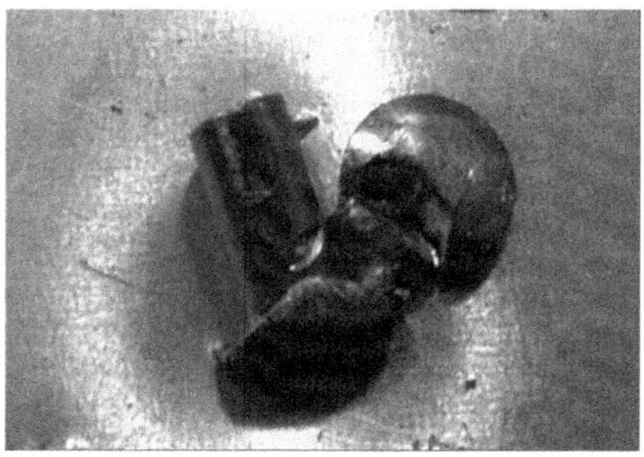

Fig. 12.2 Two rivets set into aluminum joining materials.

triggers regardless. Without a rivet present the punch makes direct contact with the joining materials causing an indentation, as shown in Fig. 12.1. If multiple rivets are loaded into the rivet jaws it is also possible for the SPR process to trigger. These rivets may still pierce the joining materials, but the resulting joint will be unacceptable, as shown in Fig. 12.2.

These incorrect rivet settings not only compromise joint strength and aesthetics, but often require replacement of the joining materials. If these materials are a car panel on an automotive production line, any stoppage means spiralling costs owing to decreased production and redundant staff.

The problem of incorrect rivet settings may be prevented by monitoring image amplitude across the machine jaws. To do this a camera must be positioned with a clear view of the jaws, which on the riveting machine (Doidge 120) [15] used in G.E.R.I is only accessible at 90 degrees anticlockwise from the front of the machine (Fig. 12.3).

This view allows the rivet jaws and feed track to be monitored (Fig. 12.4). The rivets travel down the feed track and sit in the jaws until they are set. If multiple rivets are present they will cause a backlog starting at the rivet in the jaws and regressing up the rivet feed track.

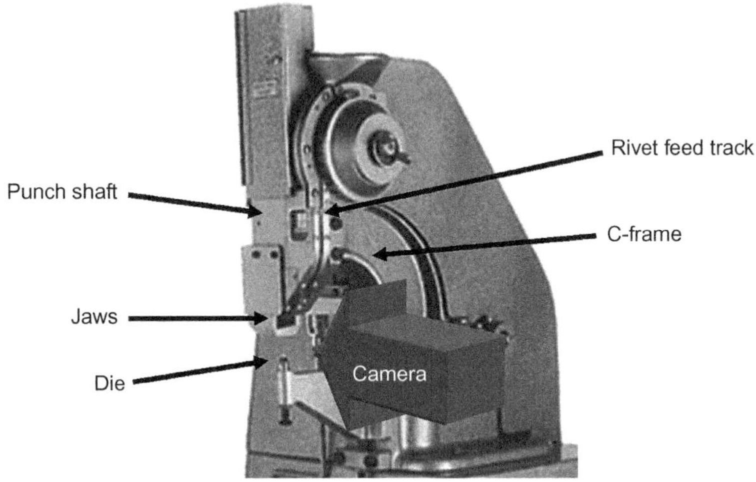

Fig. 12.3 The camera position for checking the rivet status.

Fig. 12.4 The camera view of the jaws from the position shown in Fig. 12.3.

The G.E.R.I. algorithm works by finding and comparing the amplitude of x values on the y axis (Fig. 12.5). The white line plots the amplitude, where $y = 120$, whereas the green lines highlight the three points of interest (50 pixels each) on the x axis where $x = 70–110$ (labeled as 1), $x = 150–200$ (labeled as 2) and $x = 230–280$ (labeled as 3).

The amplitudes of point 1 and 3 are used as a brightness control and are compared with that of point 2. The difference in amplitude is used to identify whether a rivet is missing, present or if multiple rivets are present. The differences vary depending on ambient lighting, which is why the bespoke software includes a calibration feature. Once the software has been calibrated for a particular joint, the algorithm will determine the status of the loaded rivet (Fig. 12.6). A single rivet will give the highest amplitude value, with multiple or missing rivets achieving a much lower intensity. Multiple rivets are identified as having intensity below the lower tolerance of a single rivet but above zero, whereas amplitude of a missing rivet will be below zero when subtracted from the amplitude of the machine jaws.

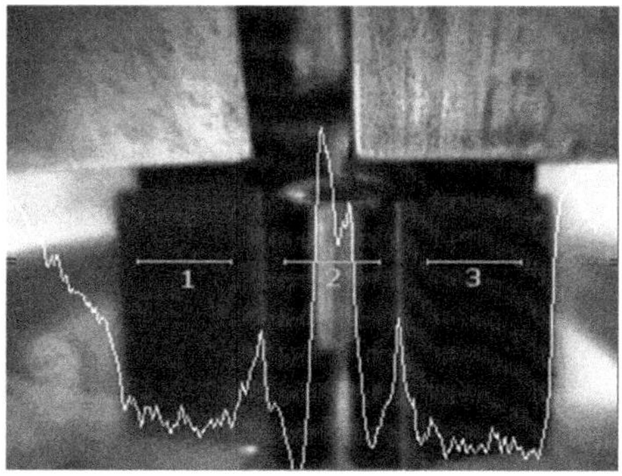

Fig. 12.5 Amplitude values of a single rivet plotted across the x-axis.

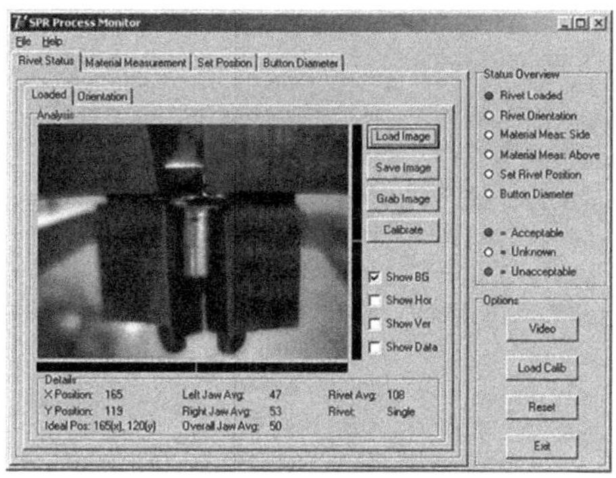

Fig. 12.6 The bespoke process monitoring software showing the details of the loaded rivet status tab.

The software displays values for each of the individual amplitude measurements as well as a final decision on the rivet status. The "Rivet" label will provide an indicator of "missing," "single" or "multiple" depending on the outcome of the algorithm. A simplified color indicator in the "Status Overview" produces a red or green light to let the operator know if the rivet was set correctly. The indicator will appear red for missing or multiple rivets, and green for a single rivet. For users requiring further analysis there are checkboxes to display horizontal and vertical amplitude overlays, as well as numerical data for both axes.

12.2.2 Rivet orientation

When a rivet is loaded into the machine jaws the shaft should run parallel to the jaws (which is vertical on the camera view) to ensure the punch hits the rivet head in the center. If the rivet is not vertical before setting (as shown in Fig. 12.7) the punch can cause the rivet to rotate (or tumble), creating a poor joint that is unlikely to completely

Fig. 12.7 Nonvertical rivet orientation.

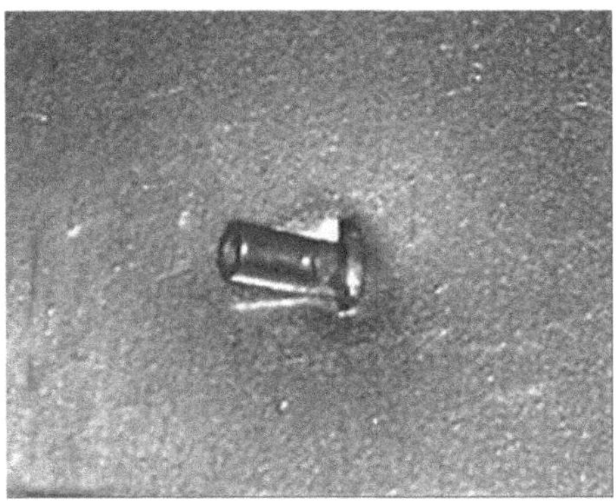

Fig. 12.8 Tumbled rivet set into aluminum.

pierce the top material (Fig. 12.8). The camera is situated in the same location as the "Rivet Status" technique discussed previously in Fig. 12.3.

The rivet in Fig. 12.7 is orientated only a few degrees from vertical, but this misalignment often causes the rivet to tumble when it is impacted by the punch. The rivet in Fig. 12.8 has tumbled approximately 110 degrees clockwise resulting in the rivet head piercing the joining materials, rather than the rivet shaft and setting incorrectly. The rivet head is neither designed nor long enough to pierce the joining materials and results in a substandard joint that damages the top material, to be repaired or replaced before being reset.

12.2.3 Material measurement

The ability to measure the joining materials presented to the SPR machinery can solve a number of problems associated with incorrectly configured machinery. Incorrect triggering can cause damage to the joining materials and or riveting machinery. The vision system is used to compare the presented materials against the expected materials, which is determined by the machines setting force. For example if a rivet

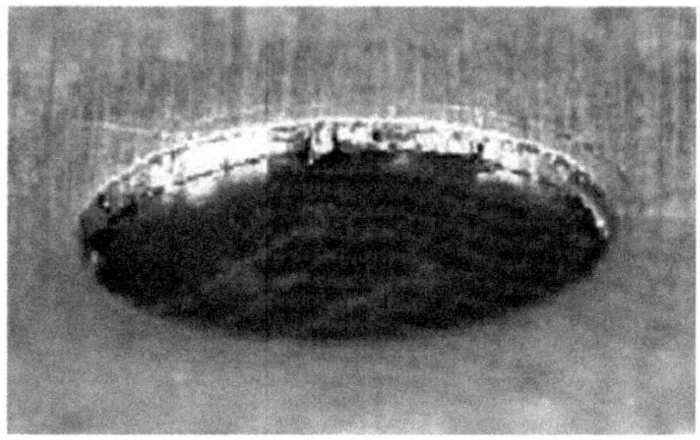

Fig. 12.9 Deep-set rivet head.

Fig. 12.10 High-set rivet head.

head is set too deep it could mean the materials were too thin or the setting force was too great. Similarly a high-set rivet head could mean the materials were too thick or the setting force was too little (Figs. 12.9 and 12.10).

Such instances will compromise the joint strength and aesthetics. Monitoring the presented materials and comparing them against the expected materials can therefore facilitate a change in setting force or halt the joining process if necessary. Two measurement techniques have been developed, each of which will be discussed in turn.

12.2.3.1 Side material measurement

The measurement technique uses a laser line and optical filters in conjunction with the camera to provide a laser based measurement system. The laser and camera are fixed to the back of the c-frame at 180 degrees to the front of the machinery (Fig. 12.11). These locations on the rear of the c-frame provide the greatest joint clearance. This is of paramount importance as the system is designed to operate on a robot arm where joint access is the single-most limiting factor for SPR joining applications.

The camera is situated perpendicular to the side of the joining materials. The laser can be positioned at any point on the c-frame that enables a clear line of sight to the joining material, as the laser stripe must fall vertically on the stacked joining materials (Fig. 12.12).

Neutral density and red filters are placed directly in front of the camera lens. The neutral density filters reduce the overall brightness, and the red filter ensures only red

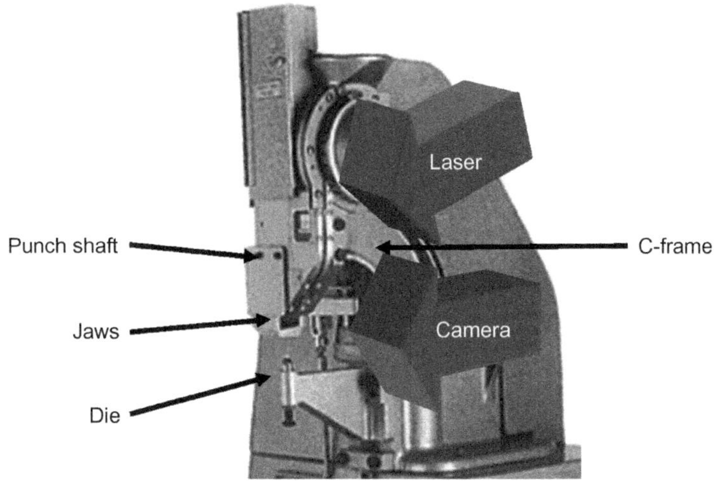

Fig. 12.11 The camera and laser position for material measurement from the side.

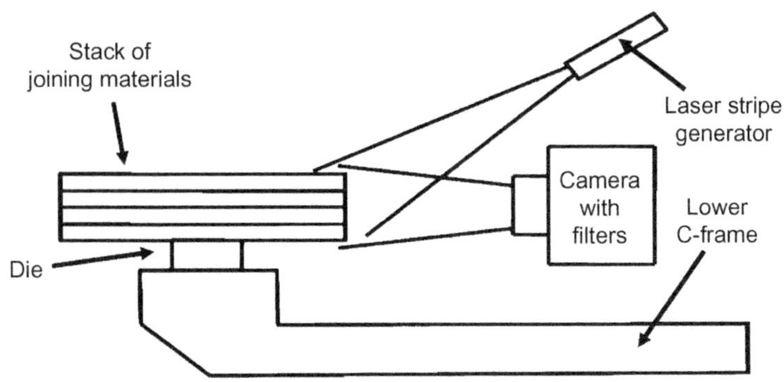

Fig. 12.12 Diagram of the side laser measurement set-up.

light that matches the wavelengths generated by the laser (~670 nm) reaches the camera (compare Figs. 12.13 and 12.14). The filters help ensure light reflected from the metallic measuring materials is presented to the camera in a way that minimizes corrective image preprocessing.

Fig. 12.13 shows a stack of four sheets of 1.5 mm aluminum without any optical filtering. As can be seen there are many bright areas that would require substantial preprocessing before feature extraction could begin. Fig. 12.14 on the other hand shows the same stack of aluminum with the laser line using the optical filters. The filters provide a crude segmentation and minimize further preprocessing. The next step is to perform a threshold, generating a black and white image that highlights the laser stripe in white as it appears on the side of the measurement materials. The center point of the highlighted pixels is marked red, creating a read line along the *y*-axis (Fig. 12.15).

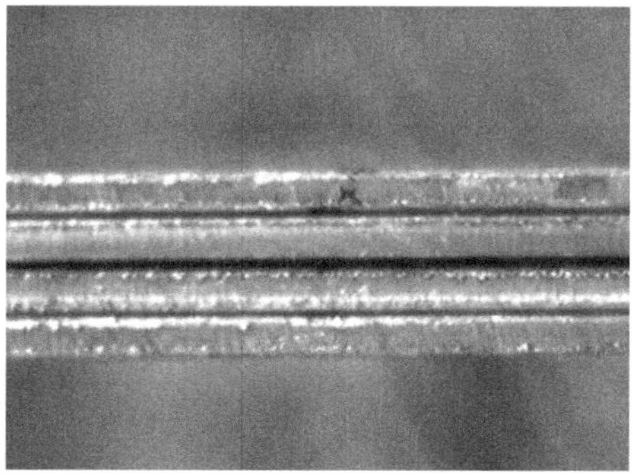

Fig. 12.13 Stacked aluminum without optical filtering.

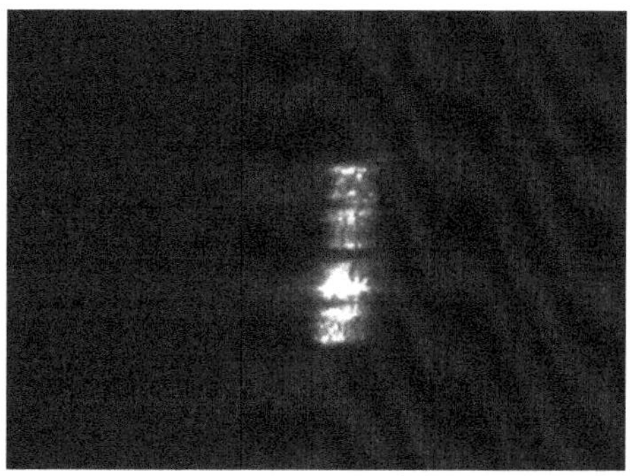

Fig. 12.14 Stacked aluminum with neutral density and red filters.

Fig. 12.15 The measurement materials with a laser line threshold and threshold midpoint.

It is important to notice the threshold and red midpoint pixels do not create a continuous line, owing to the gaps between the stacked materials. These gaps have no noticeable effect when joining soft metals such as aluminum, but it is important the measurement technique is able to distinguish between the overall width that includes the air gaps, and the joined width that excludes the air gaps. The most useful measurement is exclusive of air gaps as this enables the material thickness to be checked against the current setting force for suitability.

To obtain a real-world measurement from the digital image the number of red pixels are counted up and calculated according to the camera distance from the measurement materials.

12.2.3.2 Top material measurement

The top measurement technique also uses a laser line and optical filters in conjunction with the camera for the laser based measurement system. The camera is directed parallel to the rivet shaft and the laser is positioned at 45 degrees clockwise on the c-frame as shown in Fig. 12.16.

When the measuring materials are stacked prior to joining, the position of the bottom material is known because it rests upon the die. This is known as the reference height and can be used to derive a thickness measurement for stacked joining materials. To achieve this, the camera observes the joining materials from above, whereas the laser stripe is directed into the cameras field of view (FOV) from the back of the c-frame at a 45 degree angle (Fig. 12.17).

The bottom of the stacked joining materials is known as the reference height, whereas the top is known as the measured height. Owing to the laser stripe angle of emission, the position where the stripe intersects the material stack is dependent on the stack height and can be used to measure the stack. Several examples of what would be observed by the camera in the previous diagram are shown in Figs. 12.18–12.21.

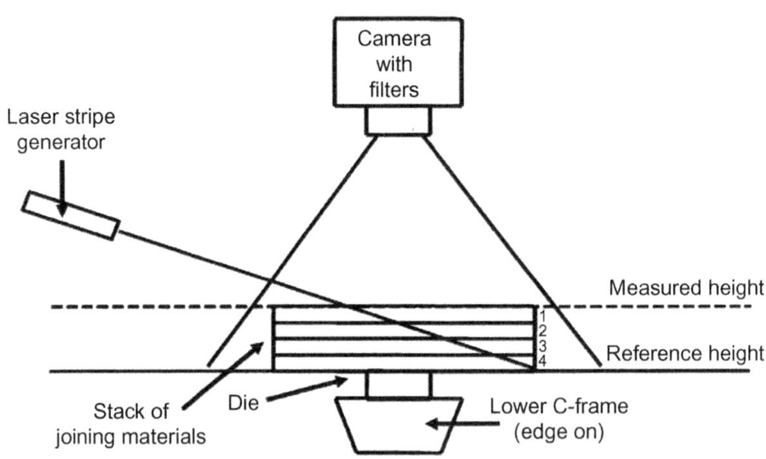

Fig. 12.16 The camera and laser position for material measurement from the top.

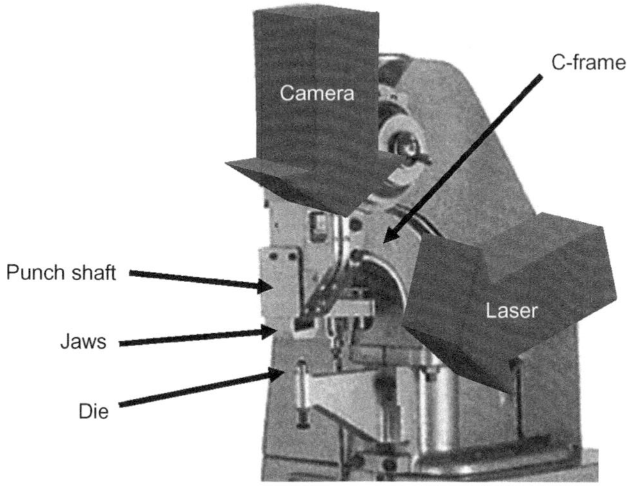

Fig. 12.17 Diagram of the top laser measurement setup.

Fig. 12.18 Bottom material (labeled as 4 in Fig. 12.16).

Fig. 12.18 shows how the bottom material (labeled as 4 in Fig. 12.16) would be observed by the camera. As the thickness of the stack increases, the laser stripe will be observed moving further down in the captured images. Fig. 12.21 shows how the top material (labeled as 1 in Fig. 12.16) would be observed by the camera.

The location of the laser stripe as observed by the camera is converted into a pixel measurement. However, variances in measurement height introduces errors into the calculations for scaling pixels to real-world measurements (e.g. mm). Therefore the measurements are only accurate at a set distance. This effect can be minimized by

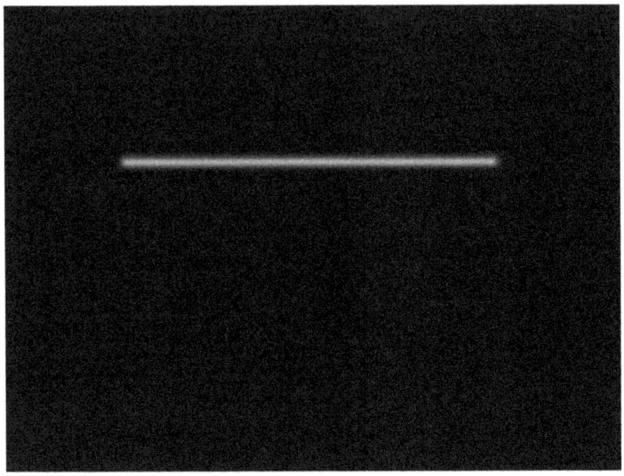

Fig. 12.19 Third material (labeled as 3 in Fig. 12.16).

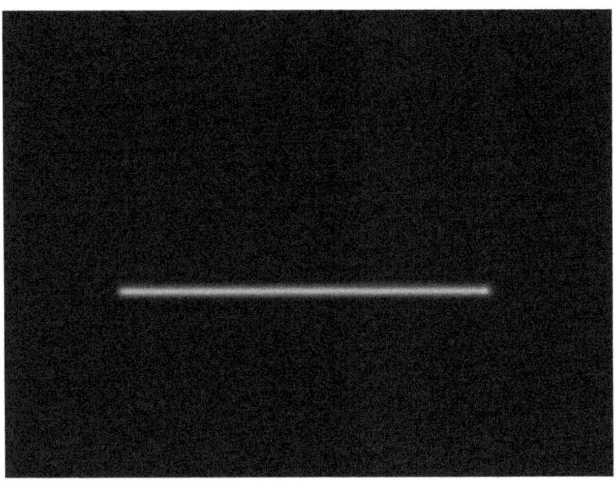

Fig. 12.20 Second material (labeled as 2 in Fig. 12.16).

changing the camera lens and therefore changing the cameras FOV [1, 2, 16] (Fig. 12.22) where θ_a is a wide FOV and θ_o is a narrow FOV. Each of the lenses available for the camera was tested for pixel-to-millimeter scaling accuracy to determine which lens is best suited to a particular range.

The lenses and corresponding FOV are shown in Table 12.1. Figs. 12.23 and 12.24 show an identical scene viewed through 16 mm and 3.6 mm lenses respectively. Although these figures show the stacked materials being measured from the side, the same is true of any top down measurement.

The section of stacked joining materials visible through the 16 mm lens is highlighted (in red) in the image captured through the 3.6 mm lens. The 3.6 mm lens

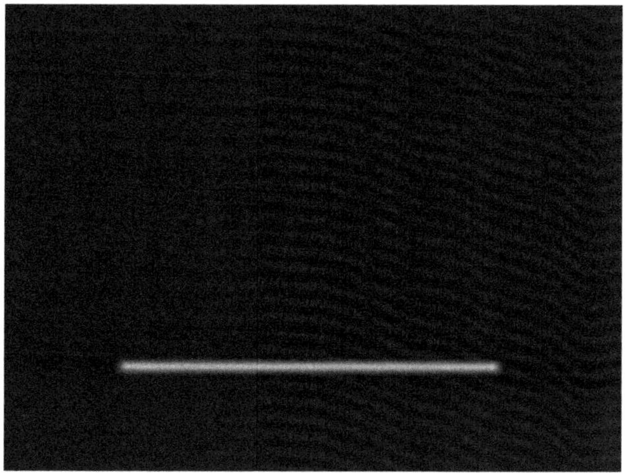

Fig. 12.21 Top material (labeled as 1 in Fig. 12.16).

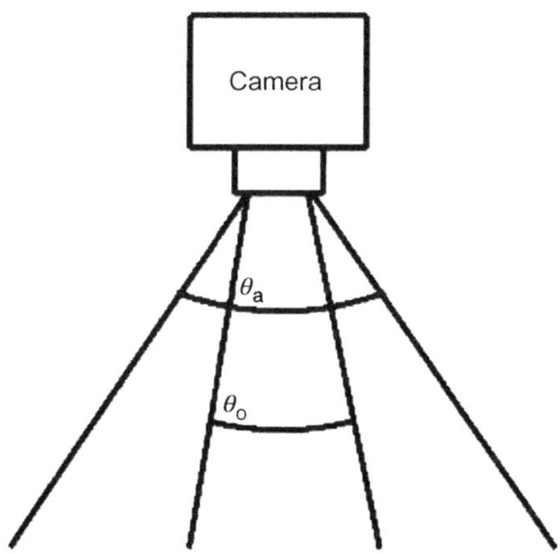

Fig. 12.22 Camera FOV.

Table 12.1 The lenses available for this experiment.

Lens (mm)	Field of view (°)
3.6	92
6	53
8	40
12	28
16	19

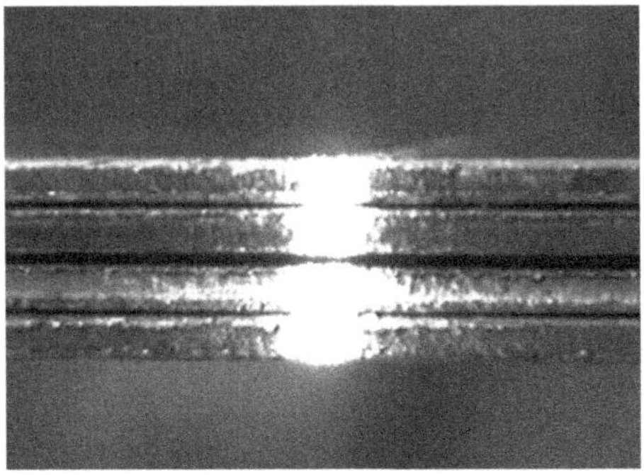

Fig. 12.23 Stacked measuring materials using the 16 mm lens.

Fig. 12.24 Stacked measuring materials using the 3.6 mm lens.

offers such a wide FOV that part of the camera housing that holds optical filters in place is visible.

As the FOV of the 16 mm lens is much smaller than the 3.6 mm lens, the camera has to be moved back further from the measuring materials to gain focus. By increasing the distance between the camera and measuring materials, it allows for the angular error to be reduced closer to the ideal of finite distance and 0° FOV, where the trace lines of points of interest enter the camera in a parallel beam, no matter where they originate from in the image. In an ideal case, an object of known size would measure the same number of pixels, no matter what distance from the camera. By decreasing the FOV and increasing the distance, this assumption still approximately holds, albeit over a short range of distances.

12.2.4 Rivet head position

One of the problems highlighted through discussions with companies involved in automotive SPR was the need to visually measure the head of a set rivet (Fig. 12.25) relative to a defined point.

In order for the rivet head position to be of maximum use the defined point should be a feature of a known location, which can be used as a reference point. Through

Fig. 12.25 The head of a rivet after setting into aluminum sheets.

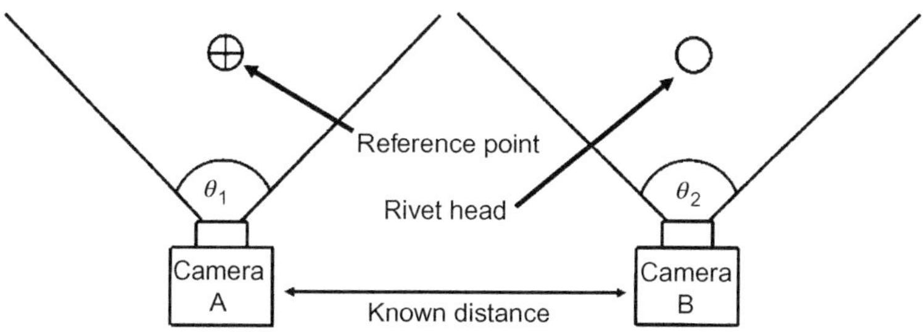

Fig. 12.26 Rivet head position measurement diagram.

careful camera positioning and image processing (Fig. 12.26) it is then possible to calculate the position of the rivet head relative to the feature. Assuming this feature is a known location, the calculation provides an absolute measurement of the rivet head position.

Fig. 12.26 shows two cameras; camera A observes the reference point and camera B observes the rivet head position. Although the points of interest should roughly appear central to the cameras FOV, their positions may move slightly over time as a result of machine wear and vibrations. The FOV of camera A (θ_1) and B (θ_2) depends on the lens used, which depends upon the size of the objects being observed. A 16 mm lens with a narrow 19° FOV is used for this technique as the points of interest are relatively small. The narrow lens also reduces any distortions caused by the "fish eye" effect that can interfere with scaling pixels to real-world measurements.

The algorithm uses a two-dimensional array (x, y) to calculate an absolute measurement of the rivet head position. For example if the reference point is expected to be observed at 160 × 120 pixels (the center of a 320 × 240 image) in the captured image, but after image processing it is found to be observed at 152 × 122 pixels, it means the center of the reference point is 8 pixels to the left and 2 pixels higher than originally expected. The difference between the expected and actual pixel location must be

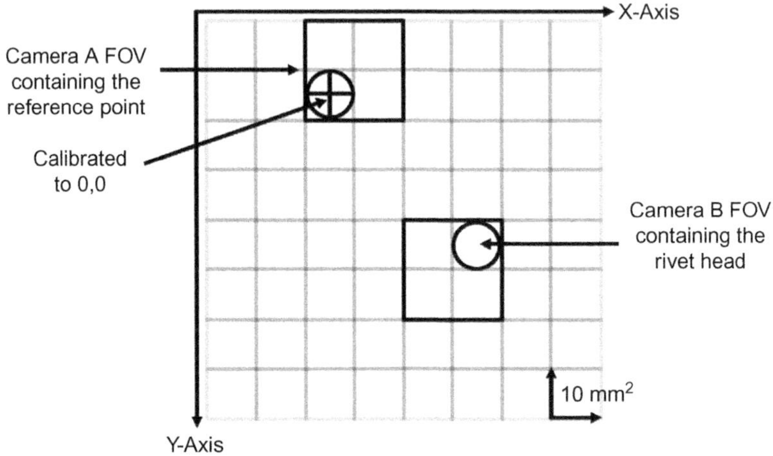

Fig. 12.27 Simplified rivet head position array.

included in the algorithm and used to fine tune the reference point location. After these adjustments have been made the reference point location is designated as 0, 0 on the two-dimensional array (Fig. 12.27). As the distance between camera A and B is known, this only leaves fine-tuning adjustments of the rivet head location in order to gain an absolute measurement for the rivet head. The same adjustment procedure is applied to camera B, by comparing the expected pixel values for the location and the actual values.

If the above array is used as an example with the upper left corner of the reference point calibrated as 0, 0, the rivet head would be 3, 3. The rivet head is therefore 30 mm below and 30 mm to the right of the reference point before any fine-tuning adjustments are made. Using Pythagoras theorem (Hypotenuse [2] = Opposite [2] + Adjacent [2]) for this simplified example, the distance between the two locations is 4.24 mm to two significant figures. Any pixel adjustments can then be scaled to real-world measurements to give an absolute rivet head location. The complete calculation is as shown below where RP is the reference point and RH is the rivet head location.

$$RP + RP_{\text{Adjustment}} = 0,0$$
$$\text{Absolute measurement} = \left(RP + RP_{\text{Adjustment}}\right) - \left(RH + RH_{\text{Adjustment}}\right)$$

12.2.5 Rivet button diameter

The Welding Institute states a fair visual indication of SPR joint quality can be determined by performing dimensional checks of the rivet button formation [17]. The button formation is created as a consequence of the rivet pushing through and displacing the joining materials.

Computer vision can perform dimensional checks on this formation using the edge detection and thresholding techniques. This segmentation in conjunction with other simple image-processing techniques to remove noncircular features and random noise identifies the button formation.

The easiest way to do this is to measure the diameter of the button in pixels across an axis. As the button formation is spherical it has an infinite number of axes, any of which can be chosen for measurement. It is important that the camera is situated central and perpendicular to the button formation, as this removes the need for corrective depth-perception processing and reduces the complexity of the algorithm.

Once the diameter has been calculated across the axes, it can be scaled to real-world measurements (e.g. mm) and compared with known good dimensions for that joint. Scaling is dependent on how far the camera is from the button formation in addition to the cameras FOV, which is determined by the lens.

12.3 Ultrasonic testing

Evaluation of ultrasonic testing as a means of NDT for the SPR process provided some interesting results. Experimental procedures have focussed on the quality of the self-piercing rivet [14], as well as the completed SPR joint [18]. These locations are inspected by an ultrasonic continuous wave through a differential piezoelectric transducer operated at a predetermined frequency. The quality is evaluated by monitoring the variations in the electrical impedance of the transducer presented by both the phase and the amplitude displayed in the complex impedance plane.

12.3.1 Self-piercing rivet

To test the quality of a self-piercing rivet a spring-loaded transducer is built into a probe that can easily be centered on the rivet head. The tip of the probe is matched to the diameter of the rivet head. Transducer design is key to the procedure's success, and therefore such transducer features as its center frequency, diameter, as well as the configuration of piezo-electric elements, have to be carefully selected to achieve satisfactory test results.

The experimental assessment of self-piercing rivets using ultrasound presented a new technique known as narrowband ultrasound spectroscopy (NBUS) [19]. The difference between the impedance corresponding to the inspected rivet and the predetermined scatter values, corresponding to a sound rivet, is monitored. The technique has confirmed that the specially designed spring-loaded probe is capable of detecting defects encountered in self-piercing rivets.

12.3.2 Riveted joint

During the NDT of a completed SPR joint, the probe is positioned above the rivet head. The SPR transducer excites elastic waves that propagate in different directions and are refracted from the interfaces of the joint. The mechanical load of the joint is transformed into electrical impedance and is measured as the output of the transducer.

The experimental work determined that differences can be identified between aluminum joints; however the success was dependent on operator skill and therefore raises concerns regarding the systems robustness. Furthermore, the sensitivity of the transducer means the differences in joints were undetectable when working with very thin aluminum.

12.4 Conclusion

It is clear computer vision is a valuable addition when determining the quality of the mechanical interlock for SPR joints. The status of a rivet can be identified using common image-processing techniques and provide an indication of whether a rivet is missing, present or if there are in fact two rivets loaded into the machine jaws. The material measurement techniques provide a means of determining the material thickness from above or from the side depending on joint constraints. By employing these techniques a SPR system can prevent a number of common failures that may compromise the mechanical interlock in terms of strength and aesthetics. Future work will focus on trialling image subtraction techniques for identifying the rivet status, and circular feature recognition to better identify the rivet head position.

References

[1] J. Sprovieri, The spectrum of riveting, Assembly 1050-8171, 50 (4) (April, 2007) 46–55.

[2] R.W. Messler Jr., Joining technologies for the next century: drivers and directions, Assem. Autom. 17 (1) (1997) 55–65.

[3] H. Brochure, Automotive Applications, Accessed 09/07/2008. http://www.henrob.co.uk/languages/english_uk/downloads.htm.

[4] A. Kochan, Audi moves forward with all-aluminium cars, Assem. Autom. 20 (2) (2000) 132–135.

[5] J. Mortimer, Jaguar uses X350 car to pioneer use of self-piercing rivets, Ind. Robot. 28 (3) (2001) 192–198.

[6] J. Mortimer, Jaguar ‘roadmap’ rethinks self-piercing technology, Ind. Robot. 32 (3) (2005) 209–213.

[7] Bollhoff: Self Pierce Riveting Systems for Automatic Processing. Flexible, economic and safe—ABF/MBF/MTF Product Catalogue. Bollhoff Fastenings Ltd, Midacre, Willenhall, West Midlands, WV13 2JW, United Kingdom.

[8] H. Brochure, Introduction to SPR, Accessed 09/07/2008. http://www.hen-rob.co.uk/languages/english/downloads.htm.

[9] H. Brochure, General Industry Applications, Accessed 09/07/2008. http://www.henrob.co.uk/languages/english_uk/downloads.htm.

[10] J. Mortimer, Mix of robots used for Jaguar’s aluminium-bodied XJ luxury car, Ind. Robot. 30 (2) (2003) 145–151.

[11] Orbitform Product Catalogue 1600 Executive Drive, Jackson, MI 49203, USA.

[12] P. Johnson. “Online visual inspection of self-piercing riveting to determine the quality of the mechanical interlock” General Engineering Research Institute (G.E.R.I.), Liverpool John Moores University, Byrom Street, Liverpool L3 3AF.

[13] L. Han. "Comparison of self-pierce riveting, resistance spot welding and spot friction joining for aluminium automotive sheet", Warwick Manufacturing Group, University of Warwick, Coventry, CV4 7AL, UK.
[14] T. Stepinski, Assessing Quality of Self-Piercing Rivets Using Ultrasound, Uppsala University, Sweden, 2006.
[15] Doidge Fastenings Ltd, Type 120 Machine, Accessed 16/07/2008. http://doidge.com/original/english/machine120.html.
[16] X. He, I. Pearson, K. Young, Self-Pierce Riveting for Sheet Materials: State of the Art, Warwick Manufacturing Group, International Manufacturing Centre, University of Warwick, Coventry CV4 7AL, UK, 2008.
[17] TWI TWI World Centre for Materials Joining Technology, High Speed Sheet Joining—by Mechanical Fastening, January/February 1996 Bulletin.
[18] L. Han et al. 'An Evaluation of NDT for Self-Pierce Riveting', Warwick Manufacturing Group, University of Warwick, Coventry, CV4 7AL.
[19] T. Stepinski, M. Engholm, Narrowband Ultrasonic Spectroscopy for Inspecting Multilayered Aerospace Structures, in: 9th European Conference on NDT, Berlin, September 2006.

Assessing the riveting process and the quality of riveted lap joints in aerospace and other applications

13

Gang Li, Guillaume Renaud, and Min Liao
Aerospace Research Centre, National Research Council Canada, Ottawa, ON, Canada

13.1 Introduction

The following sections are covered in the chapter to study the riveting process and its effect on riveted lap joints. This section will first review the riveting process and quality assessment of the rivet installation, including solid rivets, joint design rules, and joint quality assessment. It then goes on to review several experimental and numerical methods to determine the residual stress/strain and interference induced by the riveting process, followed by a review summary and recommendations for the riveting process research. At the end of the chapter are three case studies using experimental and finite element methods (FEMs). The first case study looks at the effect of the riveting process on the residual stress/strain and interference, the second case studies the stress condition in the riveted lap joints during the joint tensile loading stage, and the third case assesses fatigue life of a three-row countersunk riveted lap joint using an analytical methodology. Finally brief concluding remarks and future work are provided on potential future development trends for riveting tools, rivets, and riveted-bonded attachment method aimed at achieving greater fatigue performance in riveted lap joints. Acknowledgments and references are then provided.

The main objective of this chapter is to identify and justify the recommended feasible and reliable research approaches for an accurate study of the stress/strain condition in the riveted lap joints, from the riveting process to the joint tensile loading stage. The obtained knowledge can be used to: (i) identify weak locations in a joint, (ii) ensure that the riveted joints have high fatigue performance, (iii) accurately assess joint fatigue life, and (iv) improve the fatigue life by developing new riveting tools and rivets.

13.2 Riveting process and quality assessment of the rivet installation

13.2.1 Solid rivets

Solid rivets are one-piece fasteners consisting of a smooth cylindrical shaft with a head on one end, which are installed by mechanically upsetting on the shaft tail side.

Welding and Joining of Aerospace Materials. https://doi.org/10.1016/B978-0-12-819140-8.00013-4
Copyright © 2021 Elsevier Ltd. All rights reserved.

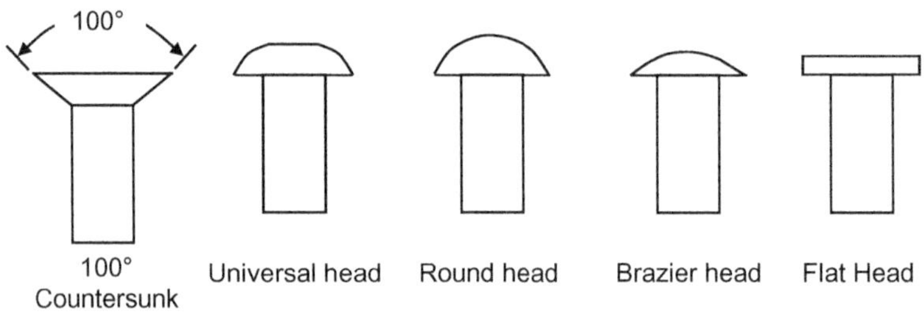

Fig. 13.1 Solid rivet head shapes.

Table 13.1 Some rivet information [1–4].

Material	Aluminium			Aluminium alloy	
	1100-F	2117-T4	2017-T4	2024-T4	5056-H32
Material code	A	AD	D	DD	B
Markings	Plain	Dimple	Raised dot	Raised double dash	Raised cross
	○	◦	•	--	+

Solid rivets can have flush or protruding heads. These two types of solid rivets are extensively used in aircraft structures [1]. Flush head rivets are also called countersunk head rivets. Protruding head rivets include brazier head, flat head, round head, and universal head rivets, as shown in Fig. 13.1. Countersunk head rivets are used where high aerodynamic efficiency is required, for instance, at longitudinal lap joints in a fuselage. Brazier head rivets are often used on external surfaces of noncombat aircraft. Other protruding head rivets, like round and flat head rivets, are used on structures based on major factors of integrity and/or aerodynamic efficiency. Protruding head rivets can be replaced by universal head rivets.

Typical materials for aircraft rivets include aluminium alloys, titanium, and nickel based alloys. The aluminium alloys are usually 1100, 2017, 2024, 2117, 7050, 5056, and V-65. Markings on the manufactured rivet head, such as small raised or depressed dimples, raised cross, raised dot, raised double dash, or raised ring, indicate the rivet's material, as shown in Table 13.1. This material coding is used to identify the rivet material.

Rivet diameters are measured in 1/32 in. (25.4/812.8 mm) increments and lengths in 1/16 in. (25.4/406.4 mm) increments. The shank diameter and rivet length are labelled after the material code in the rivet designation. As an example, the rivet designation of a 100° countersunk head rivet is presented in Fig. 13.2 where "MS" stands for "Military Standard specification". Similarly, "AN", if used to replace "MS", stands for "Air Force/Navy Standards specification", and "NAS" for "National

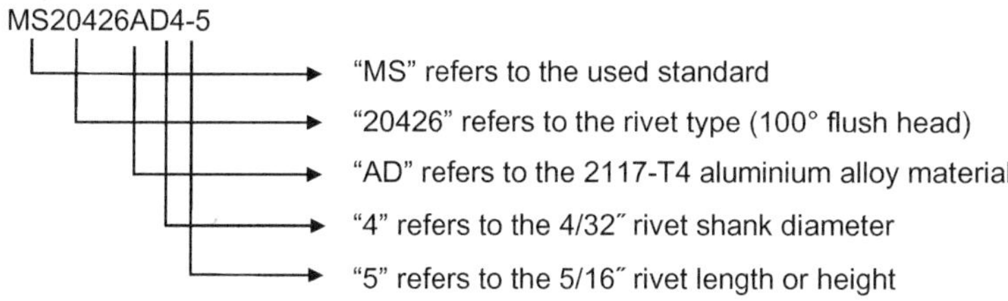

Fig. 13.2 A rivet designation.

Aerospace Standard specification". The rivet type number "20,426" refers to a 100° countersunk head rivet; material code "AD" refers to the 2117-T4 aluminium alloy material for the rivet, the rivet shank diameter is 4/32″ (101.6/812.8 mm), and the rivet length is 5/16″ (127/406.4 mm).

The installation of these solid fasteners requires access to both sides of the riveted structures. Solid rivets are driven using a hydraulically, pneumatically, or electromagnetically driven squeezing tool. A riveted joint using the same rivets is the most efficient, since all the rivets could reach capacity simultaneously.

13.2.2 Preparations of riveting process and quality assessment

Several factors are assessed to identify the suitable solid rivets for joints. These factors include the aerodynamic efficiency, the sheet material and thickness, the rivet strength, the rivet type including shank diameter and height, the joint working requirements, etc. Some design rules for good fatigue life capability can be summarized as: [1–4] (i) the rivet spacing should be four rivet shank diameters or more; (ii) the minimum edge margin should be at least two rivet shank diameters; (iii) the rivet shank diameter should be equal or greater than three sheet thicknesses; (iv) before riveting, the rivet height out of the other sheet surface should be 1.5 rivet shank diameters; (v) the hole should be 3 to 6 thousandths of an inch greater than the rivet shank diameter; (vi) for flush head rivets, the countersunk depth should not penetrate the entire outer sheet thickness and/or the sheet thickness should be at least 1.5 the countersunk depths; (vii) the recommended maximum driven head diameter should be within the range of 1.3 to 1.8 shank diameters; usually, 1.5 shank diameters is suggested; and (viii) the driven head height beyond the sheet surface should be 0.3 shank diameters or more.

Good hole filling and hole expansion are important to achieve high fatigue resistance of the joint holes and a high yield stress of the rivet is preferred. After riveting, rivets fill the hole completely to establish an interference fit at the hole edge region. To avoid defects being introduced during the riveting process, the rivet driven head deformation should not be beyond the rivet material deformation limits to avoid cracking in the rivets. The riveted joint structures must be inspected visually to ensure good rivet installations. Unacceptable defects must be identified based on related specifications [1, 2] and should be discarded. The defects can include poorly filled holes, tilted rivets (usually cannot be beyond 2 degrees [5]), loose rivets, cracks in rivet flush and driven heads, bulging skin, etc.

13.3 Determination of residual strains and interference in riveted lap joints

Due to plastic deformation, the total elastic deformation energy cannot be completely released after the riveting process. Thus, interference and compressive residual stresses and strains are created at the hole edge region. The residual stresses and strains, as well as the interference, have a significant effect on the joint fatigue behaviour [1–16]. Because of the complexity associated with the joint riveting process, there are no accurate theoretical solutions to handle the three dimensional residual stress and strain fields in the vicinity of the hole. Experimental and finite element (FE) methods are usually adopted to estimate the induced residual stress/strain condition and interference [4, 7–28].

13.3.1 Experimental measurements

Several methods can be used to measure residual strains [4, 9, 10, 17, 18, 21–28]. The corresponding residual stresses are then determined using the measured strains. These methods include strain gauges, neutron diffraction, X-ray diffraction, ultrasonic, optical, magnetic based on Barkhausen noise, local melting, and partially destructive techniques like hole drilling. Several measurement methods are briefly summarized in the following.

Entire in-situ strain variations in both the radial and hoop (tangential) directions, during the riveting process, at specific surface positions, can be measured using micro-strain gauges [21–28]. Gauges, mounted on some area experiencing large deformations, will be damaged if the actual strains are greater than the gauges' strain limits. Residual strains in the hole vicinity and inside the panels cannot be measured using this method. The measured strains are average values over the strain gauge area. This method is convenient, reliable, economical, and nondestructive.

Neutrons can penetrate through many centimeters of aluminium alloy sheets; but X-rays can penetrate only a few atomic layers [10, 24]. Thus, the neutron diffraction technique and X-ray can measure residual strains inside the material, which is superior to the strain gauge method. Both methods are non-destructive. Stress/strain free (unriveted open hole) coupons are used as references to estimate the residual strains in the tested/riveted coupons. The provided strains are the average values of each measured gauge volume. The gauge volume is determined by material properties, grain size, statistical quality, spatial resolution of the instrument, experienced personnel, etc.

The ultrasonic method is based on the well-established behaviour of echo signal amplitudes [9], which are a function of interference or residual strain. The ultrasonic method is non-destructive. However, to establish the correlation between the interference or residual strain and an ultrasonic signal amplitude, destructive actions have to be carried out on other coupons riveted under the same conditions to get the hole edge expansion data. The measured amplitude obtained using an open hole coupon results in the zero interference fit signal.

Photoelastic stress measurements, rivet-sheet springback measurements, and micro-hardness measurements are also used to measure the residual stress at the faying surface in riveted lap joints. However, as pointed out by Müller [4] in 1995; experimental techniques for the direct measurement of residual stresses are far from simple.

13.3.2 Finite element methods

FEMs are powerful tools that are used to simulate many problems, including the study of residual stresses/strains in riveted joints since the 1990s. Initially, some researchers [11, 12, 19] used simplified methods to introduce the assumed interference fit in riveted lap joints. These simplified methods include the misfit method, thermal expansion method, and rivet squeeze force method. With these methods, variations in the assumed residual stresses/strains and interference through the hole thickness direction cannot be accurately introduced. For the misfit method, the radial and clamping interference stresses are generated by forcing radial and rivet height conformity between the initially oversized ratio of rivet shank over the sheet hole dimension, and the undersized ratio of the shank height value over the package depth of the two joint sheets. For the thermal expansion method, the rivet shank is given orthotropic properties with different coefficients of thermal expansion in the shank axial and lateral directions. The temperature of the rivet shank is changed to reduce its length and to expand it in the radial direction. Thus, the two sheets are gripped by the rivet and initial interference stresses are generated [11, 12]. For the squeeze force method, the squeeze force is directly applied and maintained at the top and bottom of each rivet. The drive head shape of these rivets is assumed. Stresses/strains generated in the joints are treated as residual stresses/strains. Following the use of these three methods, the effect of the residual stress/strain can be further qualitatively analyzed by applying a joint remote tensile load [11, 12, 19].

Current FE models can analyze the complicated riveting process either using the displacement- or the force-controlled riveting methods. Nonlinearities in material behaviour, structural geometry, and contact boundary conditions can be taken into account. For the displacement- or force-controlled riveting process, two load steps are needed. One step is to use displacements or forces to control the rigid pusher to squeeze the rivet, and the second step is to move the pusher back and away from the rivet. The created stresses and strains in the riveted joint are the residual stresses and strains. Rivet driven head shapes are determined by the applied compressive displacement or rivet squeeze force, and are therefore not assumed as in the above simplified methods.

Some researchers used velocity to control the rivet installation [21, 22]. The squeezing velocity was 2 mm/min during the testing and 10 m/s was used for the three-dimensional (3D) FE analysis. A countersunk 7050 aluminium alloy rivet, with a 4 mm shank diameter, was used along with two 2024-T351aluminium alloy sheets, 1.6 mm thick. The researches found that inertia and wave propagation effects were limited if the crushing velocity was on the order of 10 m/s. Ryan and Monaghan [14] in 2000 used an elastoplastic two-dimensional (2D) axisymmetric FE model to simulate the residual stress fields for both fiber metal laminate and typical aluminium alloy riveted countersunk joints. One rigid die was stationary and the other rigid pusher moved at a constant velocity of 17 mm/s. The sheet materials were 2024-T3 aluminium alloy. Fiber metal laminates were made of two aluminium alloy layers, with a layer of glass fiber in between. The rivet was made from 2117-T3 aluminium alloy.

A large number of aircraft manufacturers use rivet-tail shape parameters, such as tail height and tail diameters, as a means to assess the quality of the riveted joints. Rivets are usually installed using displacement-controlled method, which results in an unknown rivet squeeze force. Thus, the relation between the rivet squeeze force and the induced residual stresses/strains and/or interference cannot be identified. To establish this relation and provide the necessary knowledge to enhance the joint fatigue life, the superior force-controlled riveting method is used. A 2D axisymmetric FE model can be employed [4, 15, 23, 24]. After the rivet squeeze force is removed, the residual stress/strain and interference, induced by the riveting process, can then be analyzed by extracting the corresponding data at specific locations from the FE model. Because the finite element model needs only a one-quarter joint geometry, a very fine mesh can be used in the high stress areas. A 2D axisymmetric FE model usually has three deformable contact bodies, two circular sheets and one rivet, and two rigid contact surfaces, one stationary at the rivet flush head side and a rigid pusher contacting the rivet tail. To eliminate edge effects, the sheet should extend approximately 5 rivet shank diameters from the axis of symmetry. It has been found from these FE results that an increasing maximum squeeze force not only pushes the zone of tensile hoop stress away from the hole periphery, but also results in a larger driven head size. The relation between the rivet driven head deformation and squeeze force can be identified, interference magnitude and variation through thickness can be determined at each specific squeeze force, and full-field contours for both the residual stresses and strains can be provided. However, the 2D axisymmeteric FE model cannot simulate the final stress conditions when joints are in tension. A 3D FE model is required for this situation [25–28].

13.4 Summary and recommendations for the riveting process research

An order of 2–3% interference fit should be enough to ensure that a joint is waterproof and has a prolong service life [7]. The compressive residual hoop stresses in the area near the hole periphery can greatly retard crack nucleation, as well as slow the growth of a crack in its early stage. Experimental results showed that fatigue life increases with larger squeeze forces, and decreases with an increase in fatigue load [15]. Due to the large nonlinear variation through the joint thickness, there can be large differences present in the interference fit and residual stress between the inner and outer sheets [25, 26]. For instance, large interference and residual stresses are created in the inner sheet neighboring the rivet drive head side. On the other hand, the corresponding interference and residual stresses are very small in the joint outer sheet. Accurate understanding of the magnitude and variation of the interference fit and residual stresses/strains is critical to achieve a better understanding of the joint fatigue life. To achieve this goal for improving the riveted joint fatigue performance, high fidelity FE models are required. Currently, there are no theoretical solutions for the complicated 3D residual stress/strain induced by the riveting process. Thus, experimental and FE methods are usually adopted. Experimental data are used to validate the FE models, and the validated FE models are then further used to carry out parametric studies. A force-controlled riveting method in the FE methods, combined with the necessary experimental testing, is highly recommended. A quasi-static force-controlled riveting method offers better control of the rivet installation, and provides the consistent conditions at and around the rivet/sheet hole interface in the riveted lap joints [4, 6, 15]. Accurate elastoplastic stress-strain curves of the joint materials are required for a good finite element analysis. Experimental data like the measured residual strains, driven head deformations, and/or interference are needed to validate the finite element results at specific squeeze force levels. Two-dimensional axisymmetric FE methods are only appropriate to study the riveting process [4, 15, 23, 24]. When the joints are remotely loaded in tension, the final stress condition in the joints, considering the effects of the squeeze force and other factors, can be studied using 3D FE methods [25–27]. Further fatigue life of a three-row countersunk riveted lap joint is assessed using a method that combines a global/local 3D FE model coupling approach and the Smith-Watson-Topper (SWT) strain life equation [28]. For clarity, three case studies are provided in the following section.

13.5 Case study using the force-controlled riveting method

Case studies will address three issues: (i) the effect of the riveting process, (ii) the stress condition during the joint tensile loading stage, and (iii) a methodology for assessing fatigue life of the countersunk riveted lap joints. Joints riveted using the countersunk head rivets of the MS20426x-x type are studied.

13.5.1 Case study 1: Effect of the riveting process on residual stress/strain in joints

A two stage approach was used: (1) validate the 2D axisymmetric FE method using corresponding experimental data, and (2) study the effect of squeeze force on the residual stress/strain using the validated 2D axisymmetric FE model [23, 24]. For the first stage, three joint specimens were fabricated, one with micro-strain gauges and the other with no gauges. The assembly process used a 53.38 kN squeeze force. Micro-strain gauges and neutron diffraction were used to determine the strain variations in the joint during and after the riveting process. The experimental strain data and rivet driven head deformations were used to verify the numerical model. In the second stage, six additional nonstrain-gauged joints, two joints at each squeeze force level, were riveted using the rivet squeeze forces of 26.69 kN, 35.59 kN, and 44.48 kN. Only the rivet driven head deformations were used to validate the corresponding FE results.

The residual stress/strain conditions at any position within the lap joint, as well as the full-field stress and strain contours, can be extracted from the validated FE results. Details can be found elsewhere [25, 26] and some information is provided in this chapter.

13.5.1.1 Joint and materials

The specimen configuration used in this study is shown in Fig. 13.3 and consisted of two 76.20 × 76.20 mm bare 2024-T3 aluminium alloy sheets, each 2.03 mm thick, and one 2117-T4 aluminium alloy 100° countersunk type rivet MS20426AD8-9. The material properties used for the bare sheets were $E = 72.4$ GPa, $v = 0.33$, and $\sigma_y = 310$ MPa (initial yield stress), whereas those used for the 2117-T4 aluminium alloy rivet [15] were $E = 71.7$ GPa, $v = 0.33$, and $\sigma_y = 172$ MPa. The rivet had a total length of 14.29 mm and shank diameter, D, of 6.35 mm. The mean inner sheet hole diameter was 6.45 mm and the rivet mean protruding height above the inner sheet surface was 9.95 mm based on the optical measurement results of 20 specimens. An isotropic hardening behaviour was assumed for both the sheet and rivet materials. When the true stress was beyond the initial yield stress, the material constants C and m were determined using the curve fitting method by substituting the uniaxial tensile test data into Eq. (13.1):

$$\sigma_{true} = C(\varepsilon_{true})^m \tag{13.1}$$

where σ_{true} is the true stress, ε_{true} is the true strain. The hardening parameters used for the 2.03 mm thick sheet were: $C = 765.67$ MPa and $m = 0.14$ when $\varepsilon_y < \varepsilon_{true} \leq 0.02$ (ε_y was the initial yield strain), $C = 744.62$ MPa and $m = 0.164$ when $0.02 < \varepsilon_{true} \leq 0.10$, and the slope of the linear hardening curve was 1.034 GPa when $\varepsilon_{true} > 0.10$. The hardening parameters [15] used for the rivet were: $C = 544$ MPa and $m = 0.23$ when $0.02 < \varepsilon_{true} \leq 0.10$, and $C = 551$ MPa and $m = 0.15$ when $0.10 < \varepsilon_{true} \leq 1.0$.

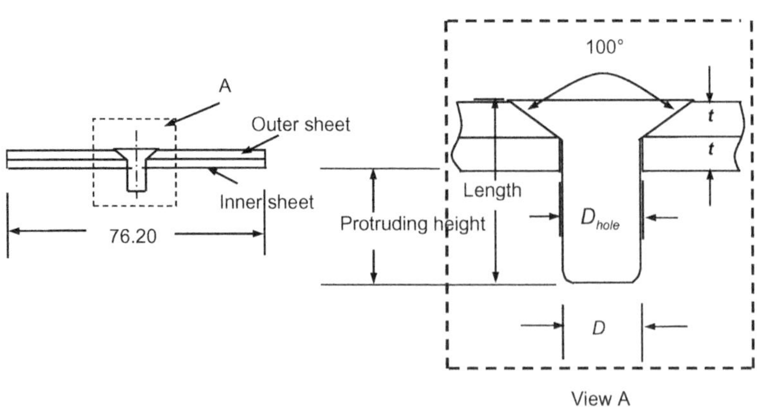

Fig. 13.3 Specimen geometry (dimensions: mm).

In order to avoid both the thermal and inertial influence on the specimen, a small constant loading ramp of 111.2 N/s was chosen for the force-controlled riveting [23, 24], which is very slow compared to the actual rivet loading rates.

After riveting, one riveted and one unriveted baseline specimen (strain/stress free joint) were sent to the Chalk River Laboratory of the Steacie Institute for Molecular Science, National Research Council Canada, to measure the residual strains in the riveted lap joint using neutron diffraction. Since neutrons penetrate several centimeters through aluminium alloys, neutron diffraction has been used as a non-destructive technique to measure the internal strains present in aluminium components.

13.5.1.2 Measurements

In-situ measurements of the strain variation in both the radial and hoop (tangential) directions were obtained during the riveting process using the micro-strain gauges. Fig. 13.4 shows the micro-strain gauge arrangement on the joint inner sheet surface. By defining the radial coordinate value of r to be zero at the joint hole centre, the strain gauge locations can be identified. Gauges R1 to R4 were located at $r = 7.85$ mm, 10.55 mm, 13.35 mm, and 16.15 mm, while Gauges H1 to H4 were at positions $r = 7.88$ mm, 10.35 mm, 12.86 mm, and 15.38 mm. Gauges R1 to R4 (Micro-Measurements EA-13-031EC-350 with gauge factor of 2.09 ± 1.0%) were used to measure the radial strain values and gauges H1 to H4 (Micro-Measurements EA-13-031DE-350 with gauge factor of 2.06 ± 1.0%) were used to measure the hoop direction strain values.

By aligning the specimen in the proper direction, three normal strains, at specific locations inside the sheets, can be determined using the neutron diffraction technique. The three normal stresses can be estimated from these three strain components. The goal was to obtain data close to the rivet/sheet and sheet/sheet interfaces using the neutron diffraction technique. However, the fine spatial resolution required a small instrumental gauge volume. Aluminium sheet has a relatively small coherent cross section for neutron measurements, imposing a lower limit of approximately 1 mm^3

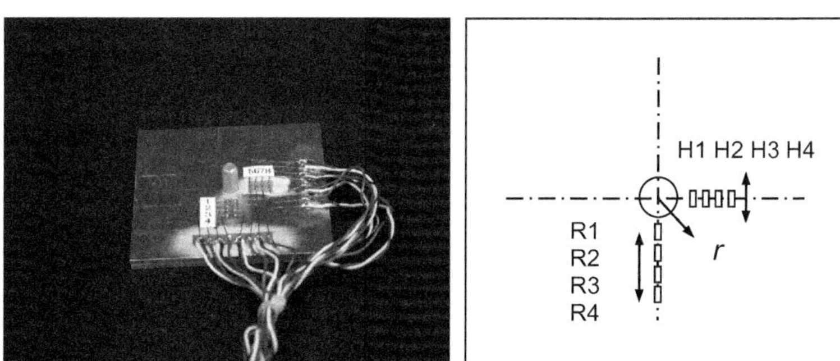

(A) (B)

Fig. 13.4 Micro-strain gauge arrangement on the inner sheet surface before riveting for (A) an experimental coupon setup, and (B) a schematic representation.

on the gauge volume to obtain data of sufficient statistical quality. Two gauge geometries were therefore required. For the measurements of the x and y components of strain, the gauge dimensions were 0.5 × 0.5 × 5 mm with the 5 mm dimension parallel to z and a diamond cross-section in the x-y plane. For the z (hoop) component, the dimensions were 1 × 1 × 1 mm with a rectangular cross section in the x-z plane. The setup of the riveted lap joint for neutron diffraction measurement is shown in Fig. 13.5. The neutron diffraction measurement locations were relatively far from the sheet surface for accuracy reasons.

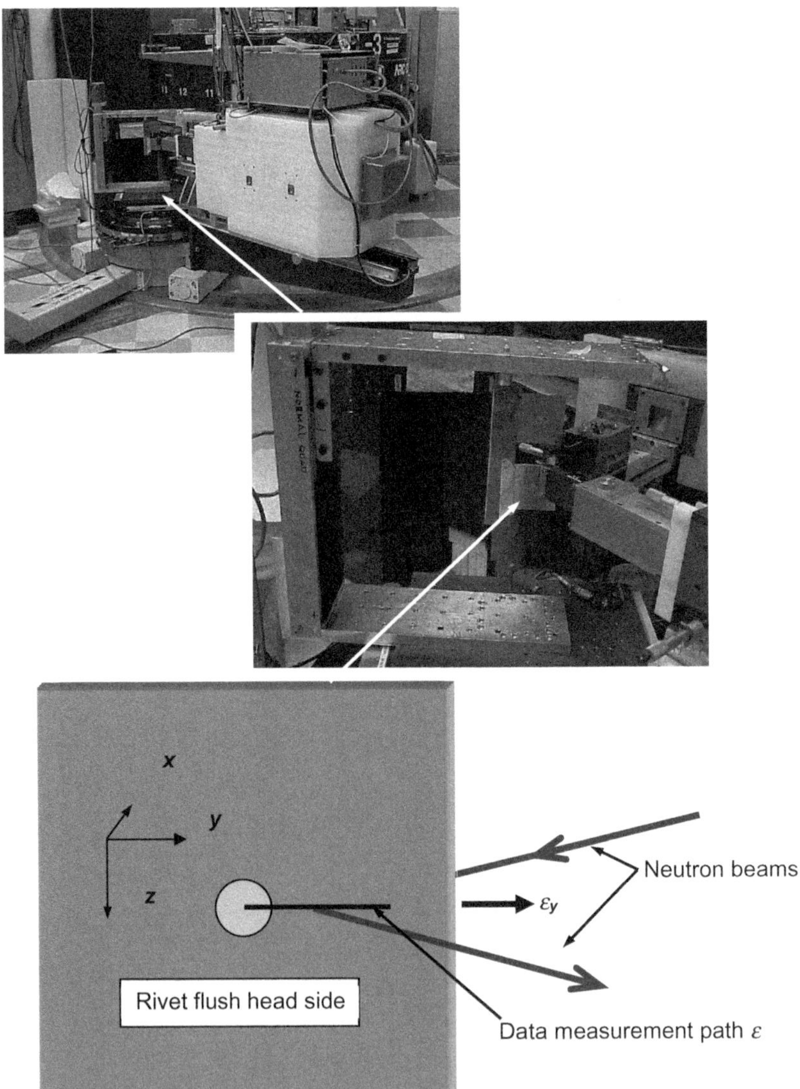

Fig. 13.5 Riveted lap joint setup for neutron diffraction measurement.

13.5.1.3 Numerical simulation

A 2D axisymmetric FE model, using MSC.Patran (pre- and post-processor) version 2001r1 and MSC.Marc (solver) version 2001, was developed to study the riveting process [23, 24]. Material and geometry non-linear properties, as well as the contact situations, were considered in the numerical simulations. Different mesh sizes were examined to determine the convergence of the rivet deformation and the maximum rivet shank diameter deformation ratio of D_{max}/D, where D is the rivet shank diameter before riveting and D_{max} is the maximum rivet shank diameter after riveting. A very fine mesh was chosen with a total of 5410 nodes and 5099 4-node axisymmetric elements with reduced integration. The mesh for the sheets extended to 5 rivet shank diameters (5D) from the joint symmetric axis, **x**, as shown in Fig. 13.6. The squeeze force was applied to the rigid pusher, which compressed the rivet driven head edge. A coefficient of friction of 0.2 was used in the Coulomb model [15, 23–28]. A tabular listing of the stress and plastic strain values were entered into a table provided by the MSC.Patran interface, which used linear interpolation for values between the points to implement the hardening behaviour of the model. Isotropic hardening behaviour was assumed for both the rivet and sheet materials.

During the riveting process, the contact areas are: (i) the faying surface between the inner and outer sheets, (ii) the area between the inner sheet and rivets, (iii) the area between the outer sheet and rivets, (iv) the area between the rivets' driven heads and the rigid pushers, and (v) the area between the rivets and the rigid supporting surface. Evident penetrations between the contact pairs should be avoided in the FE analyses. Two different sets of boundary conditions were used in the 2D axisymmetric FE model: [25]

Set 1: Radial displacement $U_y = 0$ was applied at the rivet axis and the far ends of the sheets. A vertical displacement $U_x = 0$ was applied at the two corner points of the outer side edge. One rigid set supported the rivet bottom and one rigid pusher was used to squeeze the rivet driven head;

Set 2: A third rigid body was introduced to support the external surface of the outer sheet on the basis of the boundary conditions in set 1, as shown in Fig. 13.6.

The outer sheet surface did not touch the bottom rigid set during the riveting process if the finite element model used the boundary conditions in set 1, which was only an idealized case. To more accurately simulate actual testing conditions and to improve set 1, the joint outer sheet touched the third rigid body located at the joint bottom during the riveting process in set 2.

13.5.1.4 Results and summary

Very little difference was found for the rivet deformations obtained from the numerical model using the displacement boundary conditions from sets 1 and 2. Variations in the rivet driven head compressive displacement versus the squeeze force during the riveting process are presented in Fig. 13.7. It can be seen from this figure that the experimental results and the finite element predictions agreed very well. The rivet diameter deformation ratio of D_{max}/D was 1.70 for both the experimental and numerical results. Fig. 13.8 shows the unriveted and riveted specimens, as well as the simulated rivet deformation.

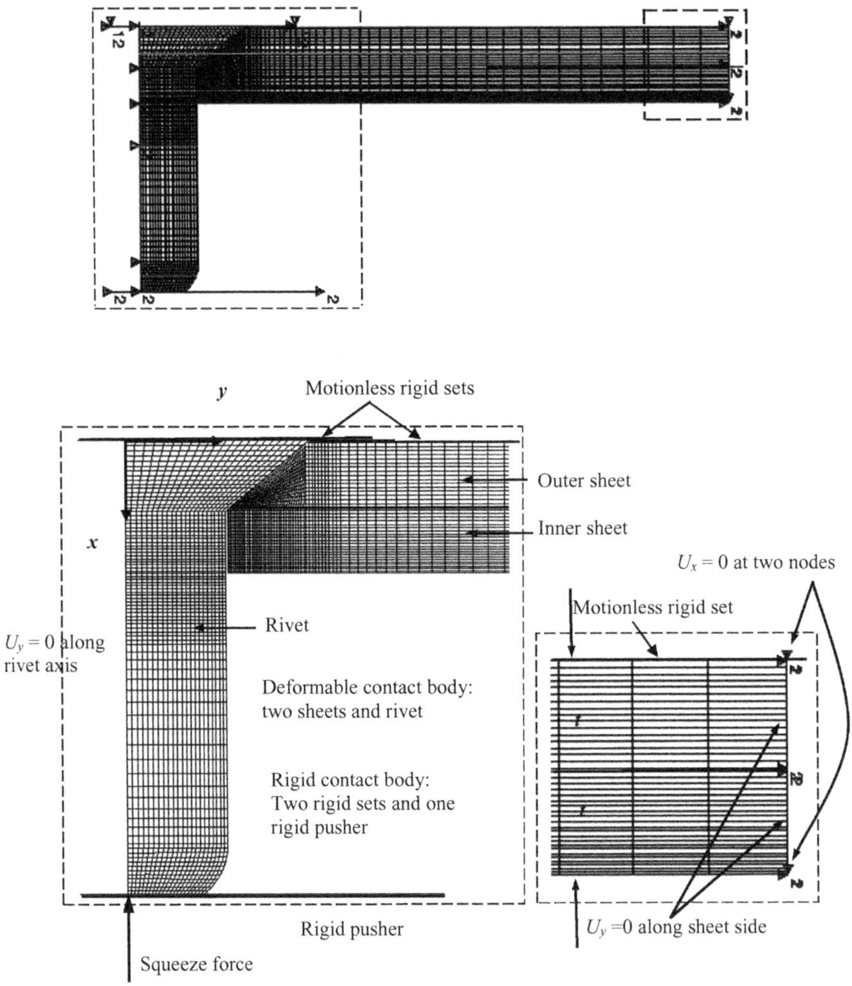

Fig. 13.6 Schematic diagram of the 2D axisymmetric finite element model with three deformable contact bodies, three rigid contact bodies, and the FE mesh and boundary conditions of set 2 for the simulation of a lap joint.

Comparisons of the strain variations obtained from the experimental tests and FE analyses for the radial and hoop directions are presented in Figs. 13.9 and 13.10, respectively. As can be seen from these figures, during the riveting process compressive strains were present in the radial direction while tensile strains were present in the hoop/tangential direction. Comparisons between the strain variation trends show good agreement between the FEM and experimental results for all the strain pairs except for a small portion during the entire riveting period. The numerical simulation resulted in excellent predictions for the hoop strain but was less accurate for the radial strain. The set 2 boundary conditions gave more accurate results than the boundary condition in

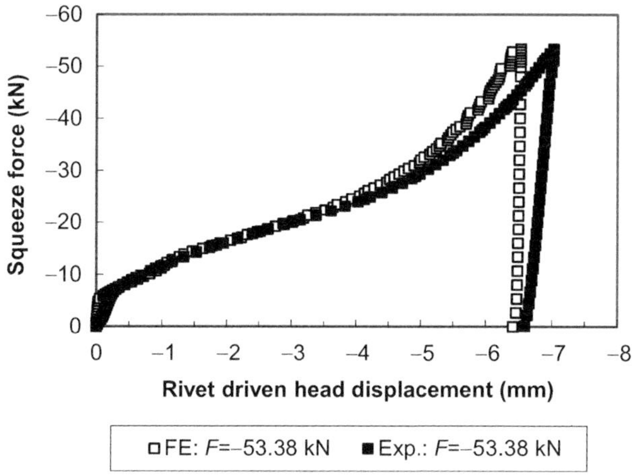

Fig. 13.7 Comparison of the rivet driven head displacement during the entire riveting period using the 53.38 kN rivet squeeze force determined by the experimental test and finite element method using the displacement boundary conditions of set 1 or set 2.

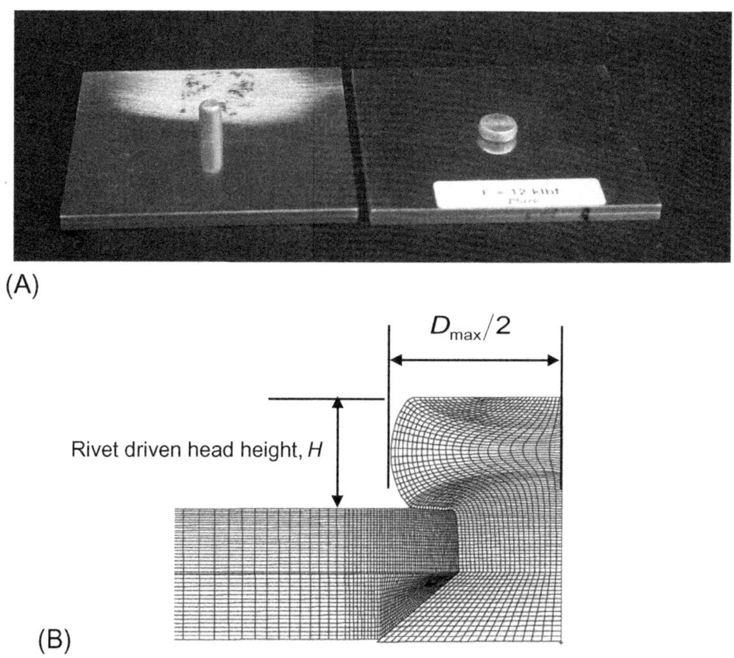

Fig. 13.8 Rivet driven head shape obtained from the experiment and the FE analysis. (A) Photographs of the joint before riveting and after riveting using the 53.38 kN squeeze force, (B) FE prediction of rivet driven head shape riveted at the 53.38 kN squeeze force.

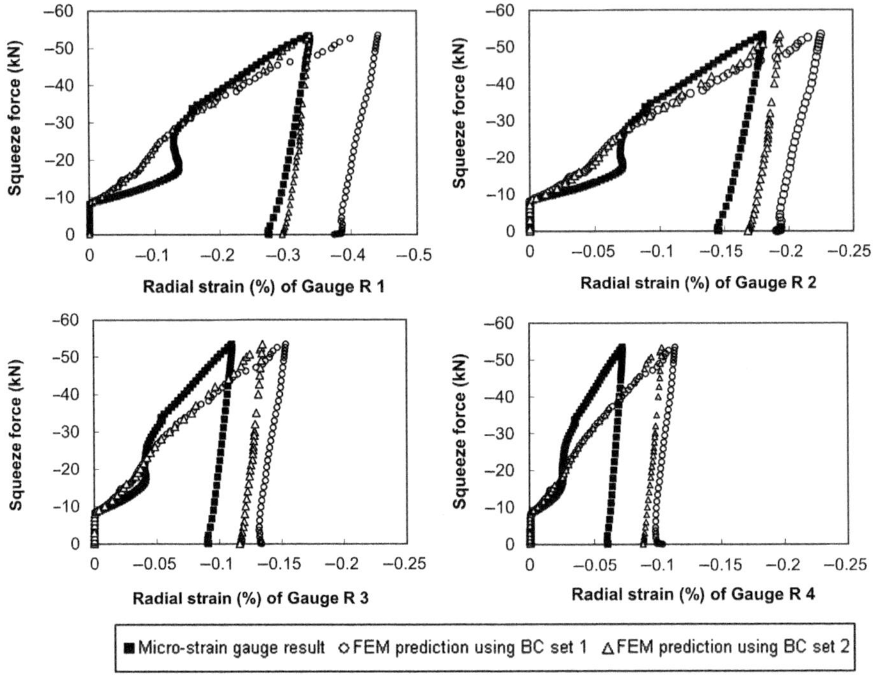

Fig. 13.9 Comparison of the radial strain variations in gauges R1 to R4 on the inner sheet surface during the riveting process under the 53.38 kN squeeze force obtained from micro-strain gauge and 2D FE results.

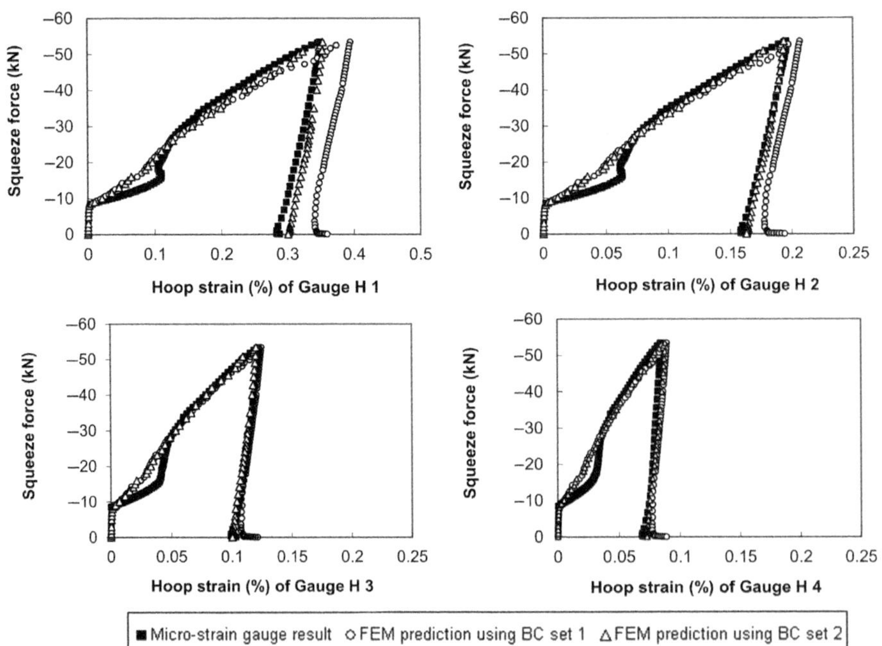

Fig. 13.10 Comparison of the hoop strain variations in gauges H1 to H4 on the inner sheet surface during the riveting process under the 53.38 kN squeeze force obtained from micro-strain gauge and 2D FE results.

set 1. A total of six contact bodies were used in set 2 while five contact bodies were used in set 1. The three rigid contact bodies were two rigid sets and one rigid pusher, while the three deformable contact bodies were the two sheets and one rivet. From the experimental point of view, the boundary conditions of set 2 were more representative than that of set 1.

Comparisons between the strain distributions in the radial, hoop and clamping directions from the neutron diffraction technique and the FE model, using the boundary conditions of set 2, are presented in Figs. 13.11 and 13.12 for the outer sheet and

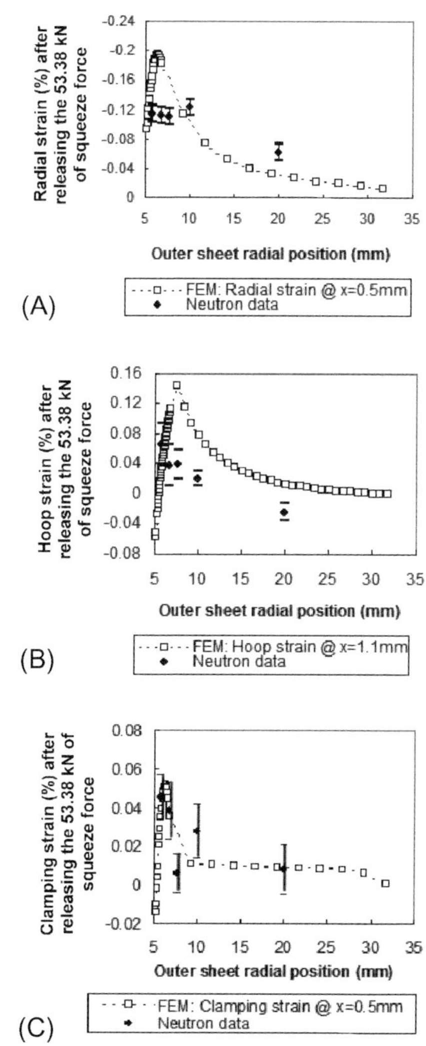

Fig. 13.11 Comparison of residual strains in the outer sheet along the radial position predicted by the numerical method and neutron diffraction after unloading the 53.38 kN squeeze force. (A) Radial strain, (B) Hoop strain, and (C) Clamping strain.

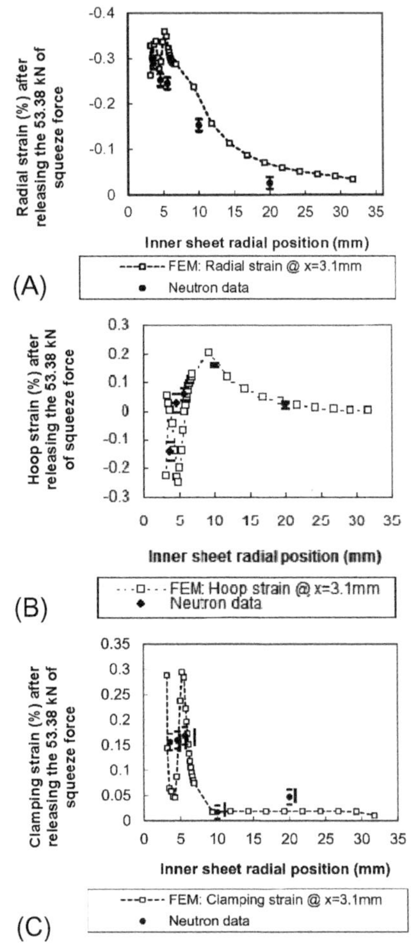

Fig. 13.12 Comparison of residual strains in the inner sheet along the radial position predicted by the numerical method and neutron diffraction after unloading the 53.38 kN squeeze force. (A) Radial strain, (B) Hoop strain, and (C) Clamping strain.

the inner sheet, respectively. Upper and lower bounds for the strain values obtained from the neutron diffraction technique are also presented in these two figures. The depths, x direction, of the data points for ε_x (or ε_y) and ε_z inside the outer sheet were different due to the measurement adjustment of the gauge centre. Reasonable agreement was achieved between the neutron diffraction results and the numerical predictions. The reasons of the discrepancy in the results are briefly explained elsewhere [23, 24]. These comparisons show that the 2D axisymmetric FE model can be effectively used to simulate the riveting process. Therefore the residual stresses could be obtained using the developed 2D FE model for any location within a lap joint or at any loading.

Large squeeze forces induced big rivet driven head deformations. The final rivet deformations obtained from both experimental and finite element methods after

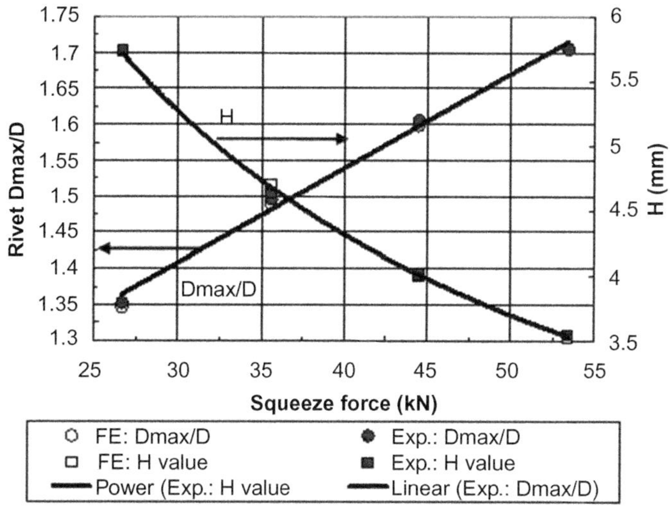

Fig. 13.13 Comparison of rivet driven head deformation and the rivet driven head height, *H*, obtained from experimental and FE results under four different rivet squeeze forces.

releasing the squeeze force are plotted in Fig. 13.13. The maximum relative error of the finite element predictions to the experimental results was around −0.5% for the ratio of the rivet shank diameter deformation D_{max}/D and 1.5% for the rivet head height *H*. Good agreement was achieved for the rivet driven head deformations between the experimental and finite element results.

Interference is an important parameter to evaluate the quality of the joint riveting process [4, 5, 7–9, 16], which should be a nonzero positive value and induced by a large rivet plastic deformation. A tight connection between fastener and sheets can be achieved if a large interference is introduced during riveting. The non-linear radial displacement and the interference distribution profiles at the sheet hole edge obtained from the numerical simulation results are shown in Fig. 13.14. In this figure, interference is defined to be the sheet hole radial expansion strain. The maximum and minimum interference at the sheet hole edge was approximately 5.1% and 0.27% respectively, after releasing the 53.38 kN squeeze force. A large interference is achieved only in the inner sheet. Variation in the outer sheet interference is not sensitive to the applied squeeze force and is small when the squeeze force is up to 53.38 kN. Results show that the connection between the outer sheet and rivet was weaker than that with the inner sheet.

Full-field residual contours of Von Mises stress, radial stress, hoop/tangential stress, and hoop strain after riveting are presented elsewhere [23, 24]. Full-field contours show that the magnitude of the compressive residual stress increased with the squeeze force, and a large expansion of the rivet against the sheet hole interface was achieved. Non-linear distributions for the radial stress in both the radial and sheet thickness directions were observed.

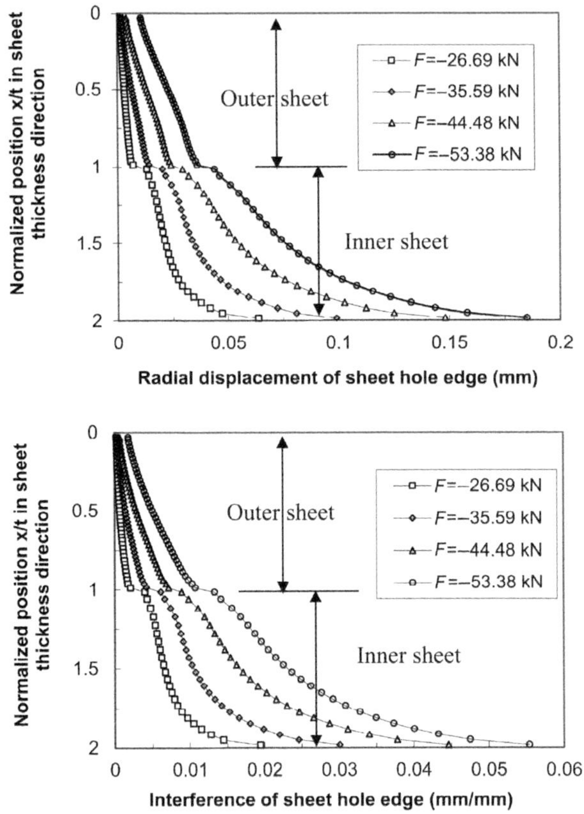

Fig. 13.14 Variations in the interference at the hole edge through the joint sheet thickness.

13.5.2 Case study 2: Stress condition in three-row countersunk riveted lap joints

Three-row countersunk riveted lap joints are typically used for fuselage structures. Therefore, the study of the stress condition in this kind of lap joint is of practical importance. In this joint secondary bending is generated due to the eccentricity of the load path when the riveted lap joints are loaded in tension. The residual stress/strain field, joint configuration, and secondary bending make the stresses/strains in the hole vicinity very different from the remote tensile stress/strain conditions. The main objective of this case study was to demonstrate that 3D FE methods with proper settings can handle this complicated stress analysis. This study covered the entire loading sequence from the riveting process to the joint remote tensile loading stage [26, 27].

13.5.2.1 Experimental aspect

Three joint specimens were tested, which consisted of two bare 1.60 mm thick 2024-T3 aluminium alloy bare sheets, riveted with three 2117-T4 aluminium alloy countersunk-type MS20426AD5-6 rivets. The material properties used for the sheets were $E = 72.4$ GPa, $\nu = 0.33$, and $\sigma_y = 310$ MPa (initial yield stress), whereas the ones used for the 2117-T4 aluminium alloy rivet [15] were $E = 71.7$ GPa, $\nu = 0.33$, and $\sigma_y = 172$ MPa. Referring to Eq. (13.1), the hardening parameters used for the sheets were: $C = 676$ MPa and $m = 0.14$ when $\varepsilon_y < \varepsilon_{true} \leq 0.02$ (ε_y was the initial yield strain), $C = 745$ MPa and $m = 0.164$ when $0.02 < \varepsilon_{true} \leq 0.10$, and the slope of the linear hardening curve was 1.034 GPa when $\varepsilon_{true} > 0.10$. The hardening parameters [15] used for the rivet were the same as in case study 1. Three rivet squeeze forces of 10 kN, 14 kN, and 18 kN were used to install the rivets in each joint. After riveting, the lap joints were loaded in tension to a maximum remote stress of 98.6 MPa. Joint configuration and dimensions are given in Fig. 13.15 and Table 13.2, respectively. The mean radius clearance between the rivet and hole was 0.06 mm. Tabs with dimensions of 50 × 25.4 × 1.60 mm were bonded to the

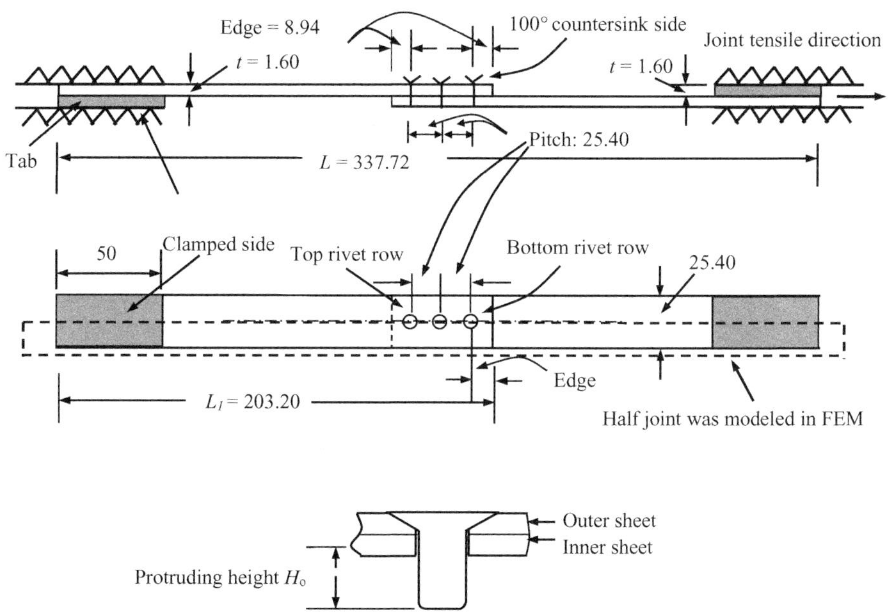

Fig. 13.15 Diagram of the lap joint with three countersunk rivets (dimensions: mm).

Table 13.2 Joint dimensions (mm) based on optical measurements of the three specimens.

Sheet dimensions	**Distance between the panel side edge and the nearby hole centre**	**Mean inner sheet hole diameter, D_{hole}**	**Mean rivet protruding height, H_o**	**Rivet shank diameter, D**
203.2 × 25.4 × 1.60	8.94	4.09	6.06	3.97

ends of each joint, to eliminate the initial secondary bending moment that would be induced when the joints were installed in the load frame. Micro-strain gauges, MM EA-13-031DE-350 and MM EA-13-031EC-350, were used to capture strain variations during the riveting process and then during the joint tensile loading stage. Details of the strain gauge mounting sequence, and positions on the joints can be found elsewhere [26, 27].

Lap joints were riveted and then loaded in tension using a 250 kN MTS Load Frame with serial number 455 and model number 311.11. A small constant load ramp of 111.2 N/s was chosen for all the rivet installations [29]. Three rivets were installed one at a time, starting from the middle rivet. To keep the riveting condition the same, each coupon was held firmly to a mobile heavy metal plate. Gauges were reset to zero before the next riveting process to avoid any potential disturbances to both gauge and coupon. After riveting, the joints were loaded in tension to a maximum remote stress of 98.6 MPa. The tensile loading rate was 0.5 mm/min. These strain data and rivet driven head deformations were used to validate the numerical predictions.

13.5.2.2 Three-dimensional FE modelling

Owing to the joint symmetry, only half of the joint, as shown in Fig. 13.15, was modelled. Symmetric boundary conditions were applied to the joint centre plane along the longitudinal x direction. The FE model was generated in accordance with the experimental joints using the FE software packages MSC.Patran (pre- and post-processor) version 2004r2 and MSC.Marc (solver) version 2001. A total of 12,362 nodes and 9096 8-node three-dimensional reduced integration brick elements (type 117) were used, as shown in Fig. 13.16. Element type 117 is an 8-node isoparametric arbitrary hexahedral for general three-dimensional applications using reduced integration. This element uses an assumed strain formulation written in natural coordinates that insures good representation of the shear strains in the element and is preferred over high-order elements when used in a contact analysis [30].

Five deformable contact bodies, two sheets and three rivets, four rigid contact bodies, three pushers and one rigid set, were defined in the FE model. The rigid set contacted the joint bottom surface. Three rigid pushers were used to squeeze the three rivet driven heads. Specific contact pairs for both surface-to-surface and point-to-point were not needed since the current finite element software package could handle this particular contact situation if it occurred. A friction coefficient of 0.2 was used in the Coulomb model for all contact surfaces. A force-controlled riveting method was used in this numerical study. The same squeeze force was used to install all three rivets in one joint [23, 24].

The multiple load steps, with their specific boundary conditions, were defined in a single loading sequence. Load step 1 applied the squeeze force to the centre pusher to squeeze the centre rivet; load step 2 released the squeeze force back to 0; load step 3 applied the squeeze force to the top rivet pusher, which squeezed the top rivet; load step 4 released the squeeze force back to 0; load step 5 applied the squeeze force to the lower rivet pusher, which squeezed the lower rivet; load step 6 released the squeeze force back to 0; and load step 7 applied the in-plane loading to the joint, up to a

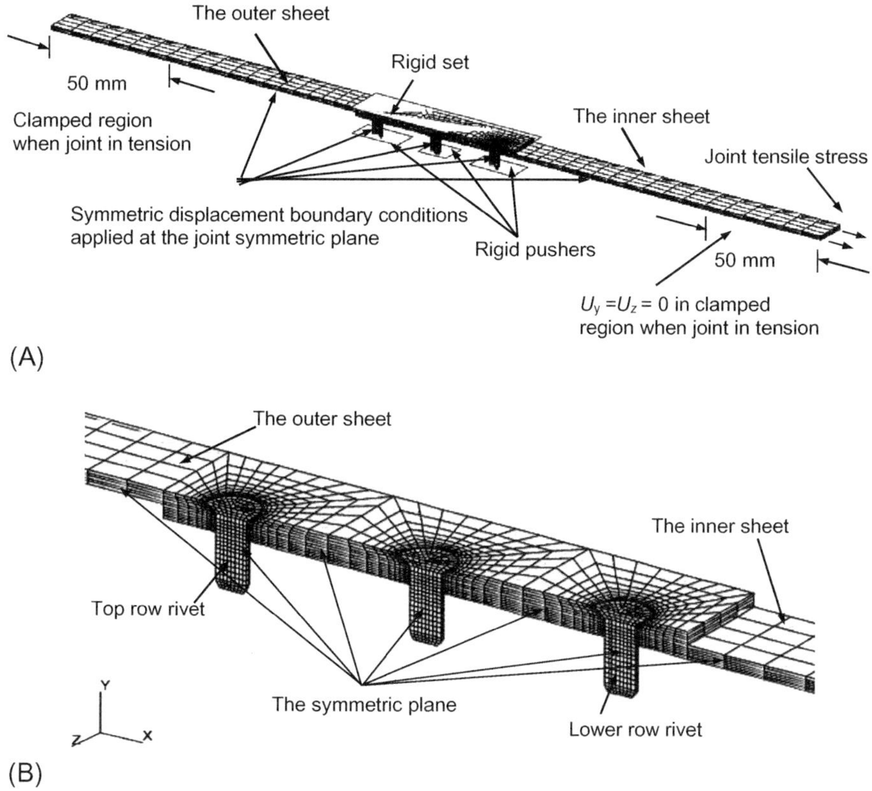

Fig. 13.16 The 3D FE model for the lap joints meshed using a total of 9096 8-node reduced integration brick elements and 12,362 nodes. (A) Three-dimensional FE model, (B) Overlap region after removal of four rigid bodies.

maximum stress of 98.6 MPa. In the final load step, the three deformable bodies contacted each other while the rigid bodies were deactivated. Details of the displacement boundary conditions and contact pairs used in each step can be found elsewhere [28, 29].

13.5.2.3 Comparisons of the experimental and FE results

Rivet driven head deformations are summarized in Table 13.3 and shown in Figs. 13.17 and 13.18. The relative difference between the experimental and numerical results for the driven head deformation ratio D_{max}/D was within 2%. During the riveting process, micro-strain gauges 1 to 3 were used to capture the hoop strains of the three holes, and gauge 4 was used to capture radial strains at one hole. Their positions are shown in Fig. 13.17. Due to the distance between the rivets, the riveting process did not influence the adjacent holes [16, 17]. To have a clear comparison of hoop strain variation during the riveting process, the experimental average hoop strains for gauges 1 to 3 ($\varepsilon_{avg} = \frac{\varepsilon_{G1} + \varepsilon_{G2} + \varepsilon_{G3}}{3}$) were used because they were located at similar positions

Table 13.3 Rivet deformations of D_{max}/D obtained from the experimental and the 3D finite element analysis results.

Rivet squeeze force	Test D_{max}/D	FEA D_{max}/D
10 kN	1.31	1.32
14 kN	1.50	1.52
18 kN	1.63	1.66

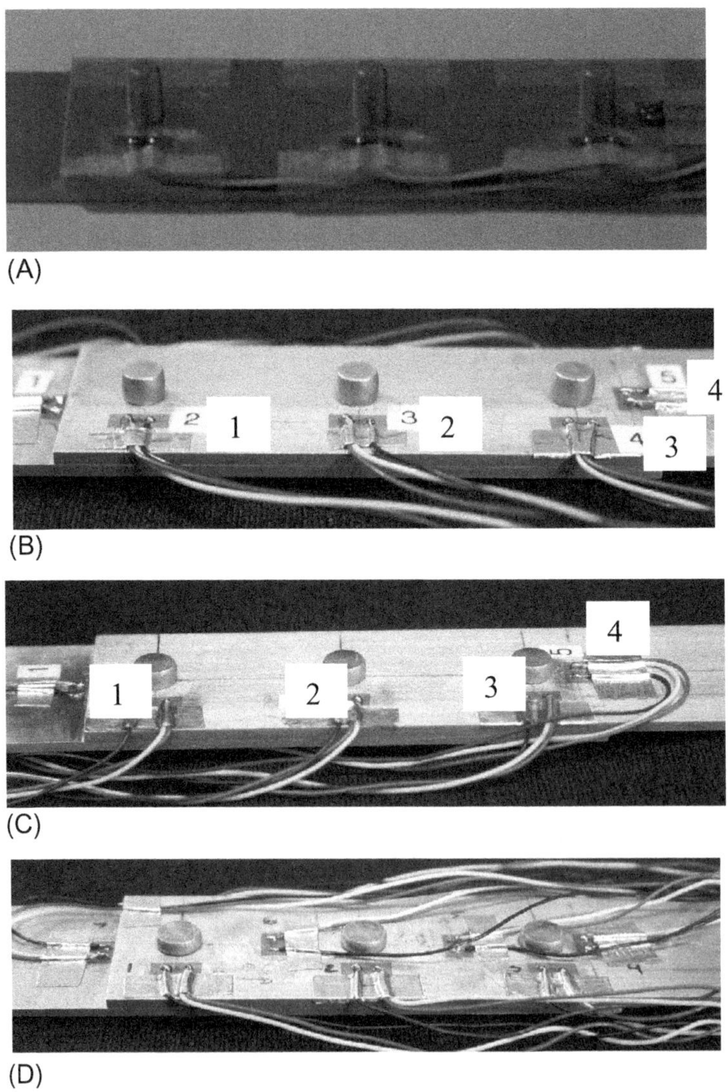

Fig. 13.17 Photos of the MS20426AD5-6 rivet driven head deformations in the lap joints, after the riveting process, using different rivet squeeze forces. (A) Before riveting, (B) Riveted by rivet squeeze force = 10 kN, (C) Riveted by rivet squeeze force = 14 kN, and (D) Riveted by rivet squeeze force = 18 kN.

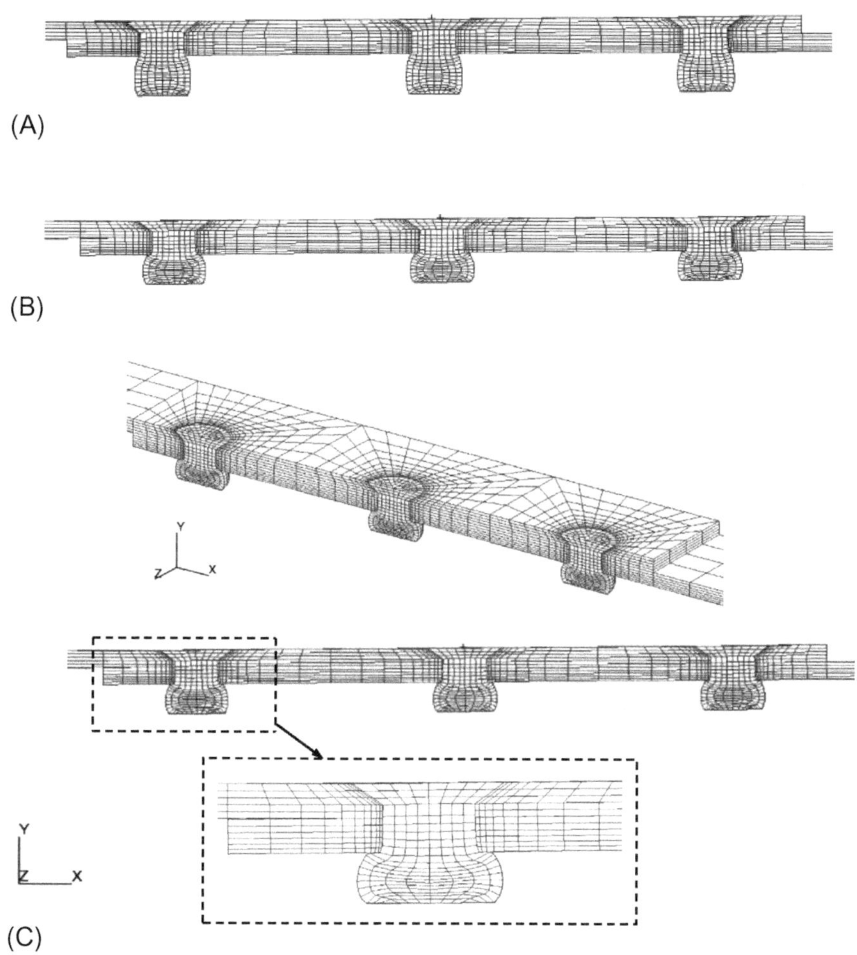

Fig. 13.18 Lap joint deformations during the tensile loading stage, after the riveting process. No penetrations occurred. (A) Joint was loaded in tension to 98.6 MPa after the riveting process using the 10 kN squeeze force, (B) Joint was loaded in tension to 98.6 MPa after the riveting process using the 14 kN squeeze force, (C) Joint was loaded in tension to 98.6 MPa after the riveting process using the 18 kN squeeze force.

relative to their neighboring holes. Comparisons of the average hoop strain from gauges 1 to 3 and the radial strain from gauge 4, during the riveting process, are presented in Fig. 13.19A. Quantitative strain comparisons, during the tensile loading stage, were further carried out to validate the numerical results. To be consistent, the average residual strain, ε_{avg}, were again used as the initial value for the gauges 1 to 3 during the tensile loading stage. Comparisons of the strain variations during the joint tensile loading stage are presented in Fig. 13.19B. Discussions on these comparisons can be found elsewhere [26, 27]. Generally, good agreement was achieved for

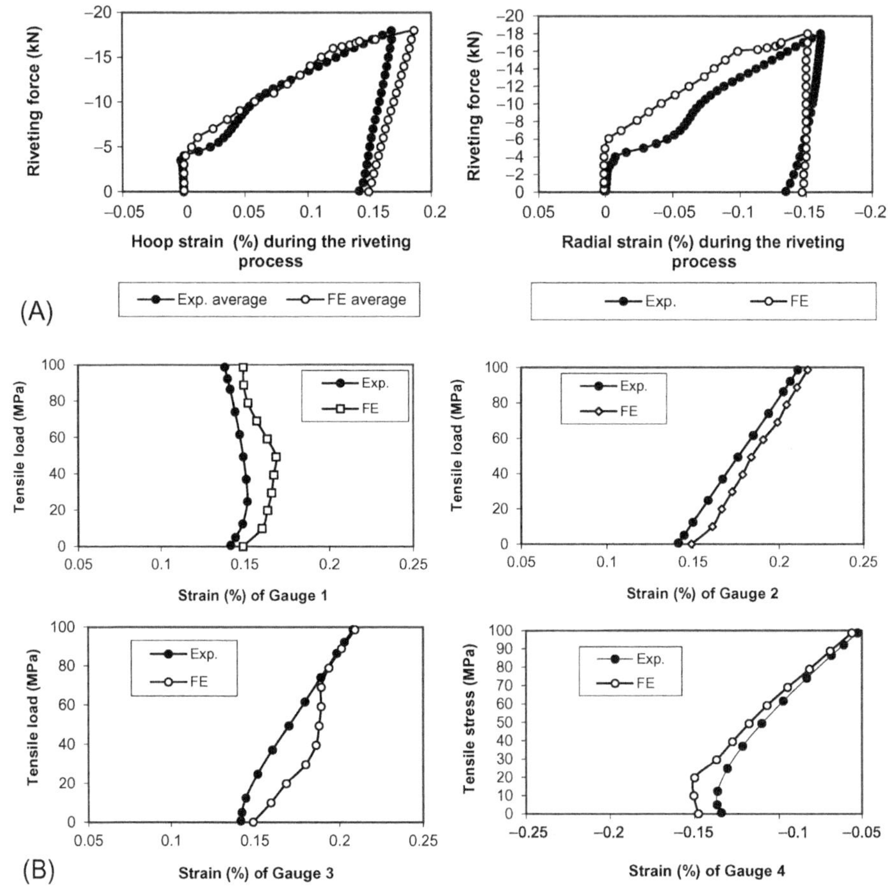

Fig. 13.19 Comparison of the strain variations in gauges 1 to 4 joints during: (A) Strain variations during the riveting process using the 18 kN squeeze force, (B) Strain variations during the tensile loading stage after releasing the 18 kN squeeze force.

both the rivet driven head deformation and strain variations between the experimental and FE results. It could be drawn from these comparisons that the residual stress, induced by the riveting process and stress conditions during the joint tensile loading stage, could be analyzed using the current numerical model with reasonable accuracy.

13.5.2.4 Parametric study using the validated 3D FE model

Two more factors were numerically studied: [27] (i) the clearance between the sheet/rivet interface and (ii) the friction coefficient. The different tasks are listed in Table 13.4. It can be seen that two additional clearances of 0.12 mm and 0.18 mm were considered along with the previous value of 0.06 mm, and that two additional friction coefficients of 0.4 and 0.6 were considered besides the previous value of 0.2 in the Coulomb friction model.

Table 13.4 Tasks carried out in the case study 2 for the three-row riveted lap joints.

Condition	Mean clearance	Rivet squeeze force	Notes
Task 1 Validation of 3D FE model using corresponding test data: the friction coefficient was 0.2 used in the FE analysis			
Clearance 1	0.06 mm	10 kN 14 kN 18 kN	Both Experimental and FE carried out
Task 2 Parametric study using the validated 3D FE model			
Task 2.1 Effect of clearance on the stress condition			
Clearance 2 Clearance 3	0.12 mm 0.18 mm	18 kN 18 kN	3D FE only using the 0.2 friction coefficient
Task 2.2 Effect of friction coefficient on the stress condition			
Coefficient 2 Coefficient 3	0.4 0.6	18 kN 18 kN	3D FE only under the Clearance 1 of 0.06 mm condition

13.5.2.5 Results and summary

The magnitude of the hoop stress in the top fastener hole vicinity along a transverse path on the outer sheet faying surface, were used as an indicator of the effects of rivet squeeze force, clearance, and friction coefficient on the stress condition in the riveted lap joints during the joint tensile loading stage. This path was selected because the section in the top rivet row area of the outer sheet is relatively weak and usually fails first.

The hoop stress variations along the prescribed path, considering the influence of the rivet squeeze force, are presented in Fig. 13.20. The following characteristics can be observed from this figure: (i) the large rivet squeeze force generates a large compressive hoop stress at the hole edge and (ii) nonlinear variations in the hoop stress occurred during the tensile loading stage. This figure shows that the residual stresses induced by the rivet squeeze force have considerable effect on the stress variations during the tensile loading stage. When the joint remote tensile stress was 98.6 MPa, the corresponding increments of the hoop stress magnitude was approximately 128%, 58%, and 50% using the three rivet squeeze forces, compared to the residual hoop stress at the hole edge induced by the 10 kN rivet squeeze force.

Provided that the joint was riveted using the 18 kN rivet squeeze force and, assuming a 0.2 friction coefficient, the effect of the clearance between the sheet to rivet interface on the hoop stress variations is shown in Fig. 13.21. Three different clearances of 0.06 mm, 0.12 mm, and 0.18 mm were studied. It can be seen from this figure that the clearance had a significant influence on the hoop stress in the hole vicinity. A small clearance led to large compressive hoop stresses in the hole vicinity and a large

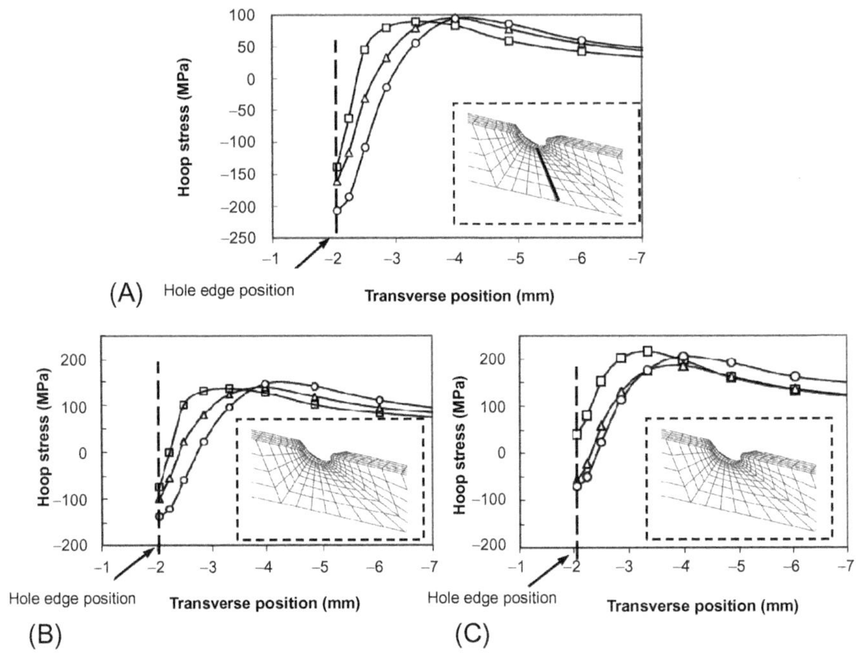

Fig. 13.20 Hoop stress variations along the transverse path during the tensile loading stage, after releasing the three different rivet squeeze forces, □: 10 kN, Δ: 14 kN and O: 18 kN. (A) Remote tensile stress of 0 MPa, (B) Remote tensile stress of 49.3 MPa, (C) Remote tensile stress of 98.6 MPa.

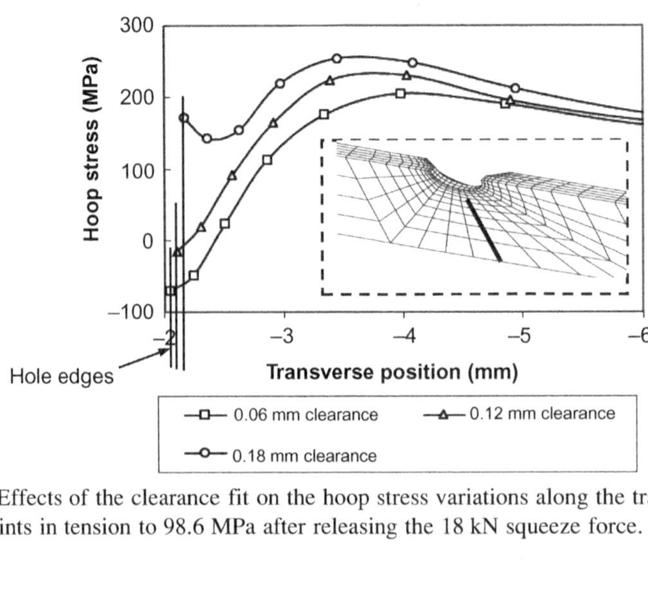

Fig. 13.21 Effects of the clearance fit on the hoop stress variations along the transverse path when the joints in tension to 98.6 MPa after releasing the 18 kN squeeze force.

Table 13.5 Increment (%) in the hoop (longitudinal) stress as compared to the joint with the 0.06 mm clearance.

Condition	During the tensile loading stage				
Clearance	Stress at the hole edge	Increment at the hole edge	Maximum stress in the hole vicinity	Distance to the hole edge	Increment of the maximum stress
0.06 mm	−69.2 MPa	0	205.9 MPa	1.94 mm	0
0.12 mm	−14.5 MPa	+79%	231.3 MPa	1.93 mm	+12%
0.18 mm	172 MPa	+349%	254.6 MPa	1.28 mm	+24%

clearance resulted in tensile stresses in the hole vicinity. For clarity, hoop stress information in the hole vicinity is summarized in Table 13.5. When the clearance was changed from the 0.06 mm to 0.18 mm, it can be observed in this table that: (i) the stress increment at the hole edge could be up to +349%, (ii) the increase in the maximum stress was as high as +24%, and (iii) the distance between the location of the maximum stress and the hole edge decreased from 1.94 mm to 1.28 mm, when the large clearance of the 0.18 mm was used. One clear finding is that the large clearance would significantly increase the hoop stress magnitude at the hole edge, which strongly suggested that the fatigue strength of the lap joints would be seriously reduced, which is consistent with available experimental results [31].

When the 0.06 mm clearance joints are loaded in tension after releasing the 18 kN squeeze force, the effects of the friction coefficient on the hoop stress variations were examined and plotted in Fig. 13.22. Little difference was observed in the results for the three different friction coefficients used.

The full-field stress distributions of the maximum principal stress can be found elsewhere [27]. The effect of rivet squeeze force is presented in Fig. 13.23. The maximum principal stress is of great importance in the study of crack nucleation during the tensile loading stage. These full-field stress contours show that large maximum principal stresses mainly occurred at the top rivet row hole region. A large rivet squeeze force moved the high stressed region away from the holes' edge vicinity. The largest maximum principal stress in the top rivet region increased with the increment in the clearance, which intersected the top fastener hole for both the 0.12 and 0.18 mm clearance fits. A large friction coefficient increased both the stress magnitude and high stress area in the top rivet hole region; however, its impact was less than that of the clearance. The large maximum principal stress distribution area and shape were consistent with the experimental fatigue testing results [18]. In the fatigue tests of non-corroded joints, cracks typically originate in the outer sheet heavily fretted area around the upper rivet hole a short distance away from the top rivet hole edge [18].

It is possible to study the entire loading stages for the three-row countersunk riveted lap joints using 3D FE methods. The effect of rivet squeeze force, clearance, and friction coefficient on the stress/strain condition in the lap joints can be studied.

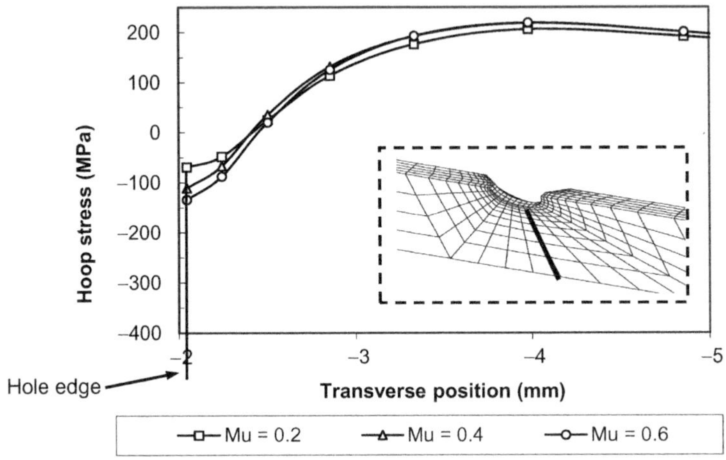

Fig. 13.22 Effects of the friction coefficient in the Coulomb model on the hoop stress variations along the transverse path when the joints were loaded to 98.6 MPa after releasing the 18 kN squeeze force.

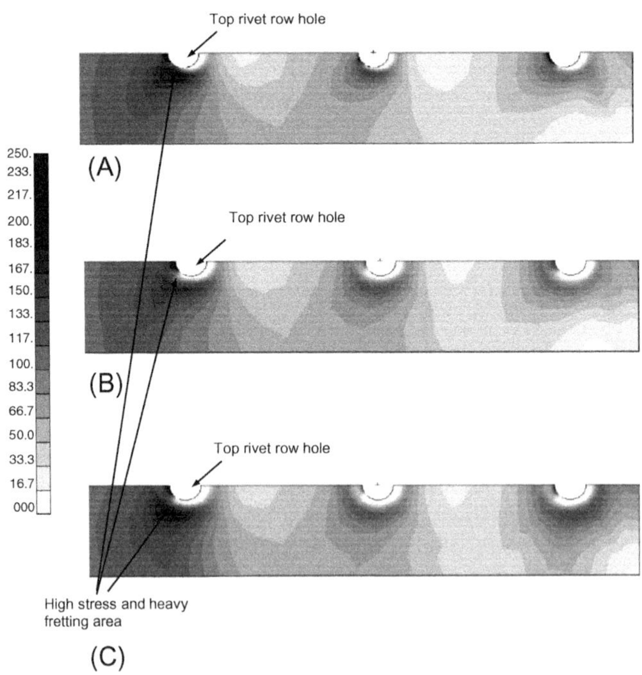

Fig. 13.23 Maximum principal stress (MPa) on the outer sheet faying surface when the joints were loaded in tension by the 98.6 MPa stress. The 0.06 mm clearance and 0.2 friction coefficient were used in the FE analysis. (A) Tensile load of 98.6 MPa after releasing the 10 kN squeeze force, (B) Tensile load of 98.6 MPa after releasing the 14 kN squeeze force, (C) Tensile load of 98.6 MPa after releasing the 18 kN squeeze force.

The numerical results suggest that high strength lap joints could be fabricated using: (i) a relatively large squeeze force to install rivets, and (ii) a small clearance between the sheet hole and rivet shank interface.

13.5.3 Case study 3: Fatigue life assessment for three-row countersunk riveted lap joints

Details of this study can be found elsewhere [28]. An analytical methodology is applied to assess the fatigue "initiation life" (to an approximately 0.5 mm crack length) using a riveted flat panel test structure, considering the combined effects of joint manufacturing and external cyclic loading conditions. Joint manufacturing factors such as the rivet-hole clearance and rivet driven head deformation were considered, since they directly affect the joint integrity condition [4, 15, 23–28]. The stresses and strains induced by the manufacturing and external cyclic loading conditions were obtained using contact FE models through a global/local FE model coupling approach (MSC/Patran).

This work was conducted using the Smith-Watson-Topper (SWT) strain-life method [32]. Displacement fields obtained from the global FE model were used as the remote loading applied to a local 3D FE model after the simulation of the riveting process. Each simulation consisted of multiple loading steps, including rivet squeezing and releasing, and then remote cyclic loading. A cyclic stress-strain curve was used in the local model [33]. Stresses and strains extracted from the local FE model at the third cycle were then used to calculate the joint fatigue crack "initiation life" via the SWT equation. The effects of the rivet head deformation and rivet-hole clearance on calculated fatigue life were studied. The NRC in-house tool, CanGROW, was then used to carry out a preliminary Monte Carlo simulation for the fatigue life distribution to the first link-up between cracks. It involved the development of correction factors to include the countersunk geometry and the post-riveting residual stresses effects in the simulations.

13.5.3.1 Experimental aspect

Information about the experimental lap joint geometry, configuration, and strain gauge (SG) locations can be found elsewhere [28]. Only one joint was tested for fatigue life evaluation. Based on the JAXA panel joint drawing, the rivet hole diameter was 4.039 ± 0.127 mm (max: 4.166 mm, min: 3.912 mm, note the minimum size could cause an interference fit). The riveted lap joint consisted of two 1.27 mm thick 2024-T3 aluminium alloy panel sheets and one 1.6 mm thick and 20 mm wide 2024-T3 aluminium alloy strap attached to the middle row to simulate a fuselage panel stringer. This joint was riveted by 3 rows of 20 fasteners, including 9 button head rivets installed on both sides of the panel and MS20426AD5-5 42 countersunk rivets in the remaining central locations, where the joint multiple-site damages (MSD) was evaluated. The rivet driven head deformation range, D_{max}/D ratio, measured on the test article, was between 1.50 and 1.52 for most of the rivet holes, with an average of 1.516. The pitch spacing was 20 mm in the longitudinal (perpendicular to loading)

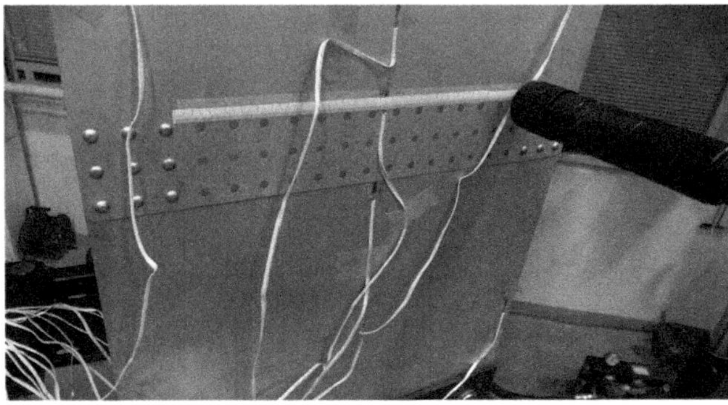

Fig. 13.24 Fatigue test setup [28].
Courtesy of JAXA and Techno Press.

direction and 23 mm in the "hoop" (parallel to loading) direction. A total of 24 unidirectional micro-strain gauges with 5 mm gauge length, KFG-5-120-C1, and 12 rosettes with 1 mm gauge length, KFG-1-120-D17, were used for strain measurements with a 2.05–2.11 $\pm$ 1% gauge factor. These gauges were categorized in three groups: (i) far field gauges: five pairs at A- and B—12S to 17S, located far away from the overlap section; (ii) near field gauges: six pairs at A- and B—1S, 2S, 6S to 8S, 11S, near the overlap section; and (iii) rosette gauges within the overlap section: seven rosette pairs at A- and B-3R to 5R, 9R, 10R, 18R, and 19R. In this nomenclature, "A" refers to the SGs mounted on the outer surface and "B" refers to the SGs mounted on the inner surface of the joint.

A static tensile test was conducted first. The tensile load was applied from 0 to 42.42 kN (83.5 MPa stress) and then unloaded back to 0 kN for strain data collection. Then tension-tension fatigue testing at a 10 Hz frequency was carried out with a maximum stress of 65 MPa and a stress ratio of 0.069 for other panel manufactured at same period. During the fatigue test, visual inspection was conducted around each fastener using a 50 $\times$ magnification charge-coupled device (CCD) camera to detect the locations and lengths of the cracks generated. Fig. 13.24 shows the setup of the fatigue test specimen and a camera. The crack lengths were measured every 10,000 cycles after the first fatigue crack had been detected.

13.5.3.2 Prediction of the fatigue life and further crack growth

Fatigue life prediction

The MSC/Patran FE package (pre- and post-processor) and the MSC/Marc (solver) version 2014r1 were used for the global/local FE model coupling analyses. Static tensile loading analyses were conducted for validation of both the global and local FE models using experimental strain data.

A. *Global/local FE model coupling*
  B.1. *Global FE model*

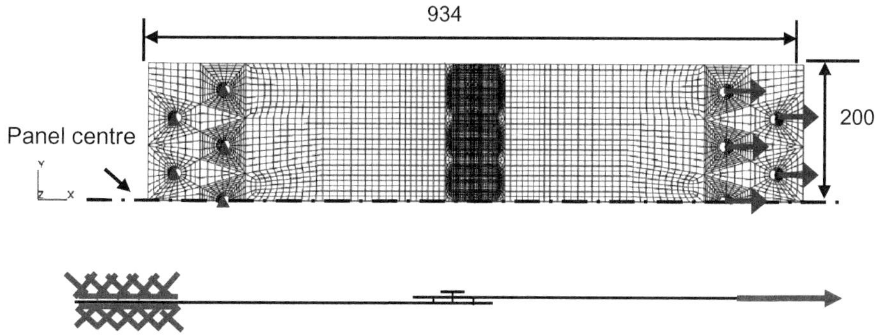

Fig. 13.25 Global finite element model using shells for the riveted panel joint structure and loaded through pins at the two end sections (dimensions: mm).

Due to the symmetric joint configuration, only half of the joint structure was simulated with the global FE model. The generated global FE model is shown in Fig. 13.25 with 8713 4-node shell elements and 10,080 nodes, in which fasteners and panel holes were not considered. Shell elements were created at the mid-planes of the panels and strap. The linkages between the fasteners and the panels were modelled using rigid link MPCs (multipoint constrains) which allowed secondary bending to occur under the tensile loading. Only geometric nonlinearity (large displacements) was considered in the global model. A tensile loading step, from 0 to 21.21 kN, was applied to the global FE model and MPCs were set for all the right side pins ensuring they all had the same tensile displacement.

B.2. *Local FE model and material parameters*

The local model covered nine rivets to avoid edge effects on the stress condition at the top-row of rivets, where the riveting process was simulated. For simplicity, eight out of nine rivets were modelled as "welded" to the panels in the local FE models. The riveting simulation was only performed on the central top-row of rivets. The local FE joint model is presented in Fig. 13.26 with 48,820 8-node reduced-integration contact brick elements and 59,643 nodes. This local joint model had a length of 216 mm in the loading direction (corresponding to the fuselage hoop direction) and a width of 60 mm in the transverse direction (corresponding to the fuselage longitudinal direction). The length of the overlap section was 60 mm. The strap, used to simulate a stringer structure, was assumed welded to the middle row rivets in the local model. The rivet shank diameter, *D*, was 3.968 mm. Three local models, with diameters of 4.064, 4.115, and 4.166 mm for the hole protruding part, were created. The corresponding radial clearances between the rivet shank and protruding hole part were 0.048, 0.0735, and 0.099 mm, respectively. The countersunk depth in the outer panel was set to be 0.94 mm, leaving a 0.325 mm thick edge for the hole protruding part.

A fine mesh was applied to the rivet, overlap section and nearby region, while a coarser mesh was used for the panel areas away from the overlap ends. A total of 120 elements were created along the countersunk hole edge perimeter, 6 elements along the inner panel thickness, 4 elements along the protruding edge, transformed

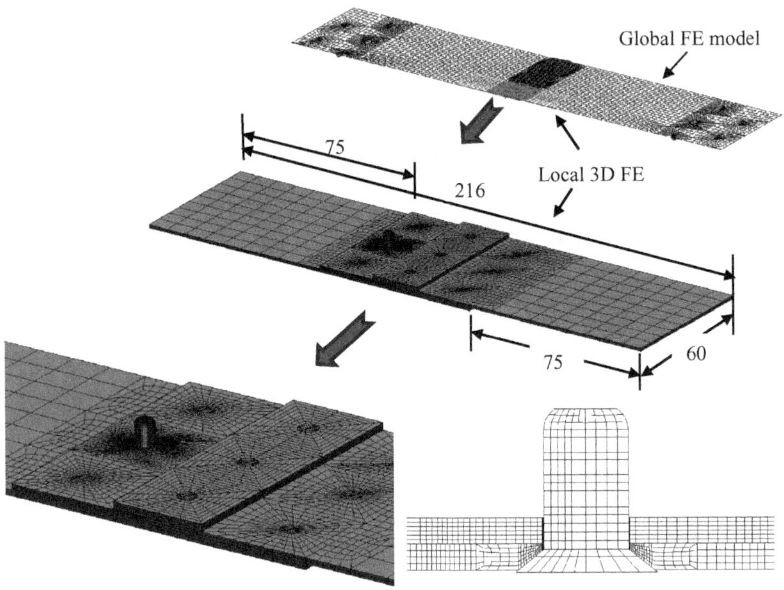

Fig. 13.26 Schematic presentation for the local 3D model (dimensions: mm).

to 6 elements in the outer panel hole region. This mesh strategy was determined through a mesh sensitivity study based on the first model (size 1), which had a 0.048 mm rivet-hole clearance. Multiple loading steps from the riveting process to a three-cycle remote loading stage were applied to the local FE mode, in which nonlinearity in material, geometric deformation, and contact boundary condition was considered. It was assumed that the cyclic stress-strain curve would be stabilized in the third cyclic loading stage [33]. Displacement fields generated at the maximum and minimum remote stress levels were applied to the mid plane nodes around the perimeter of the local 3D FE model to induce the cyclic loading condition. Explanations in the contact pair definition and solution procedure for multiple load cases can be found in author's previous publications [23–28].

An isotropic hardening behaviour was assumed for both the rivet and sheet materials. The material parameters of the 2117-T4 Al alloy MS20426AD5-5 rivet [15, 23–28] are presented in Table 13.6. The material constants C and m were calculated by substituting uni-axial tensile test data into Eq. (13.1).

Table 13.6 Elastic and plastic properties for MS20426ADx-x rivet material.

Parameter of rivet	Value
Young's modulus, E	71.7 GPa
Poisson's ratio, v	0.33
Initial yield stress, σ_y	172 MPa
Hardening parameters when $0.02 \leq \varepsilon_{true} \leq 0.1$	$C = 544$ MPa and $m = 0.23$
Hardening parameters when $0.1 < \varepsilon_{true} < 1.0$	$C = 551$ MPa and $m = 0.15$

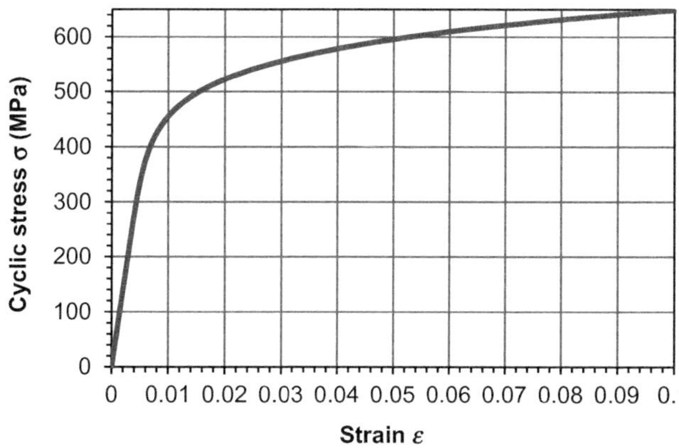

Fig. 13.27 Cyclic stress-strain curve for 2024-T3 aluminium alloy (transverse direction).

A cyclic stress-strain curve for 2024-T3 aluminium alloy in the transverse direction [33] was used for the skins during the application of the three load cycles. The cyclic stress-strain curve (transverse direction) is shown in Fig. 13.27 with a yield stress of 430 MPa. A Young's modulus of $E = 70.3$ GPa and a Poisson's ratio of $\nu = 0.33$, were used for the sheet material parameters in the elastic deformation stage.

B. *Fatigue life assessment*

B.1. *Smith-Watson-Topper (SWT) equation*

The strain-life method was used to estimate the fatigue "initiation life" associated with each combination of hole sizes and rivet deformations. The calculation was performed using the SWT equation [32] considering mean stress effect:

$$\sigma_{\max}\frac{\Delta\varepsilon}{2} = \frac{\left(\sigma_f'\right)^2}{E}\left(2N_f\right)^{2b} + \sigma_f'\varepsilon_f'\frac{1}{2}\left(2N_f\right)^{b+c} \tag{13.2}$$

where $\sigma_{\max}\frac{\Delta\varepsilon}{2}$ is calculated at every node of the local 3D FE model in all directions. The maximum calculated value, referred to as the SWT parameter, was used to calculate N_f, the number of cycles required to meet a predefined "failure" criterion. The SWT equation was derived using the Basquin and Manson-Coffin equations for covering the elastic and plastic strain amplitude effects [32]. The Basquin and Manson-Coffin equations are shown in Eqs. (13.3) and (13.4) respectively:

$$\left(\frac{\Delta\varepsilon}{2}\right)_e = \frac{\sigma_f'}{E}\left(2N_f\right)^b \tag{13.3}$$

$$\left(\frac{\Delta\varepsilon}{2}\right)_p = \varepsilon_f'\left(2N_f\right)^c \tag{13.4}$$

where: $\sigma_f^{'}$ = the tensile fatigue strength coefficient;
$\varepsilon_f^{\prime}$ = the tensile fatigue ductility coefficient;
c = the fatigue ductility exponent;
b = the fatigue strength exponent;
N_f = the number of cycles to obtain, in this case, a 5% decrease in the tension load of a cylinder coupon (transverse direction in the coupon fatigue test). This was back-calculated to be approximately equivalent to a 0.5 mm deep thumb nail crack;
E = the Young's modulus; and
$\Delta\varepsilon = \varepsilon_{max} - \varepsilon_{min}$ the strain range.

For the 2024-T aluminium alloy sheets, the transverse-direction (TL) strain-life constants were: [33] $\sigma_f^{\prime} = 835$ MPa, $\varepsilon_f^{\prime} = 0.174$, c = -0.644, b = -0.096, and E=70.28 GPa. The relation between the calculated SWT parameter and the "initiation life" is presented in Fig. 13.28.

Crack growth analysis

Fatigue crack growth from a 0.5 mm crack length to the first link-up of two adjacent cracks between two holes was evaluated using the NRC in-house tool, CanGROW, which includes capabilities to perform Monte-Carlo simulations for MSD problems [34, 35]. Correction curves on the part-through crack in a countersunk hole, expressed as functions of crack length, were developed to consider the effects of the riveting induced residual stresses, obtained from the local FE model, and the countersunk geometry. These correction curves were calculated using AFGROW 5.2, which includes a countersunk geometry stress intensity factor solution. Using the AFGROW COM interface, the applied remote load (spectrum) was automatically adjusted, based on the crack tip position, to consider the stress fields under the maximum and minimum cyclic loads, including the post-riveting residual stresses. This strategy was used because the residual stress option was not available for the countersunk geometry in

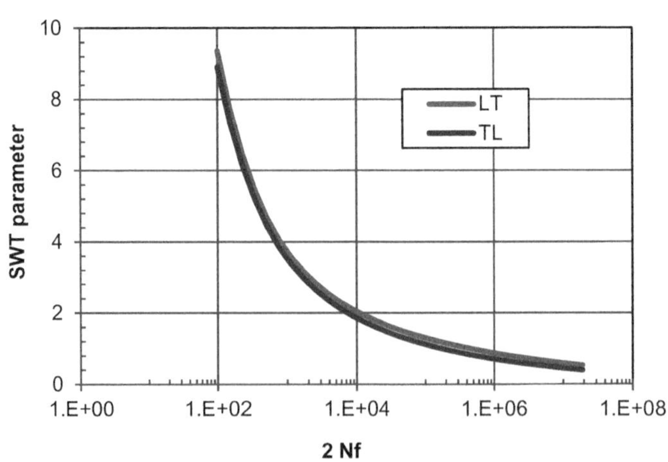

Fig. 13.28 SWT parameter (MPa)—life relationship.

that version of AFGROW. The stress intensity factor curves obtained from AFGROW were then converted to correction curves applicable to MSD cracks located at straight holes in a CanGROW model, which could then be used to estimate the life to the first link-up of the two adjacent cracks.

13.5.3.3 Results and discussion

Joint deformation and strain comparison

Due to the tensile load path eccentricity, secondary bending occurred in the single-lap joint under tension. Fig. 13.29 shows this deformation feature in both global and local FE models. Rivet driven head deformation was obtained from the local FE model

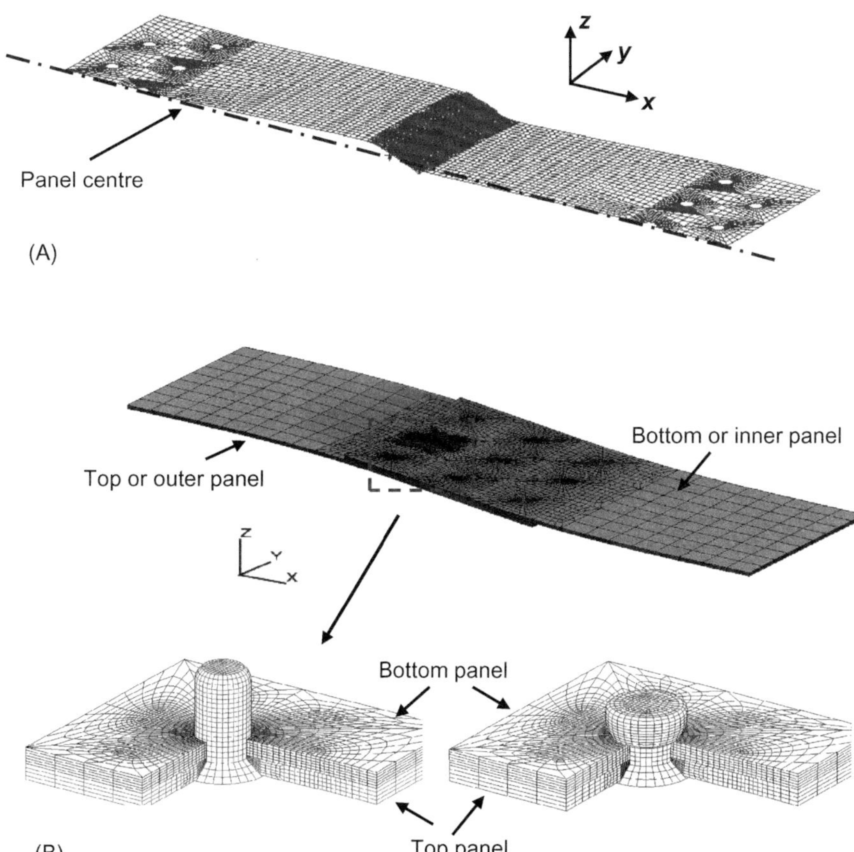

Fig. 13.29 Joint secondary bending predicted by both global and local FE models. The riveting process was conducted in the local FE model prior to the fatigue loading analysis. (A) Enlarged secondary bending in the global FE joint model in tension. (B) Joint secondary bending (enlarged) and rivet head original and deformed shape predicted by the local FE model.

through the riveting process simulation steps and no penetration was observed between the deformable contact pairs. Good agreement in the rivet deformation was obtained between the experimental and FE results. The rivet head deformations obtained from the FE analysis were from 1.501 to 1.527, whereas the provided experimental data were 1.50, 1.51 and 1.52.

A. *Strain comparison*

Good correlation was confirmed for most of the strain gauge locations. Strains and stress values in the SG locations obtained from three local models were similar because they were away from the fine meshed area in a radial hoop region equal to 0.4D from the studied hole edge. Strain comparisons at selected key locations were presented only to show the FE model validity. Close agreements between the experimental and FE strains were obtained on all the selected locations. This comparison clearly shows that the local FE model is valid because it closely predicted the strains for both the bending and nonbending sections, especially in the overlap section at the top-row region. Failure is expected to occur in the highly stressed region of the outer panel, within the top fastener row. Comparison details can be found from previous work [28].

B. *Crack nucleation sites prediction*

Fig. 13.30 shows the crack nucleation sites predicted by the three local FE models through the SWT parameter. These nucleation sites were at very similar positions; they were on the top panel faying surface, at the hole transverse edge area. These predicted crack nucleation sites are consistent with the high stress locations observed in the full-field stress contours, as well as the joint failure observed in the test.

Estimated fatigue life and test results

Table 13.7 lists the fatigue lives to a 0.5 mm crack length estimated by the SWT equation using the stresses and strains extracted from the three local models.

Rivet driven head deformations of 1.50, 1.51 and 1.52 were referenced to select proper riveting displacement in the local 3D size 2 and 3 models for obtaining the appropriate rivet deformations [23–28]. The size 1 model, although still within the specification with a radial rivet-hole clearance of 0.048 mm, resulted in lives exceeding 85 million cycles even at the 1.459 rivet deformation. Therefore, further fatigue life assessment in this size 1 model with higher rivet deformations such as 1.50 and above was not conducted. Numerical results showed that the effects of hole size and rivet deformation on the fatigue life were significant. The SWT parameter was found to be very sensitive to the local stress and strain states that are induced by the rivet-hole clearance and rivet head deformation levels in this study. The fatigue lives predicted from the size 2 and 3 models, with radial rivet-hole clearances of 0.074 mm and 0.099 mm, ranged between 2.15 and 4.59 million cycles. The estimated lives obtained from the size 2 and 3 models were slightly higher but relatively close to the JAXA experimental fatigue life of about 1.95 million cycles. The predicted high

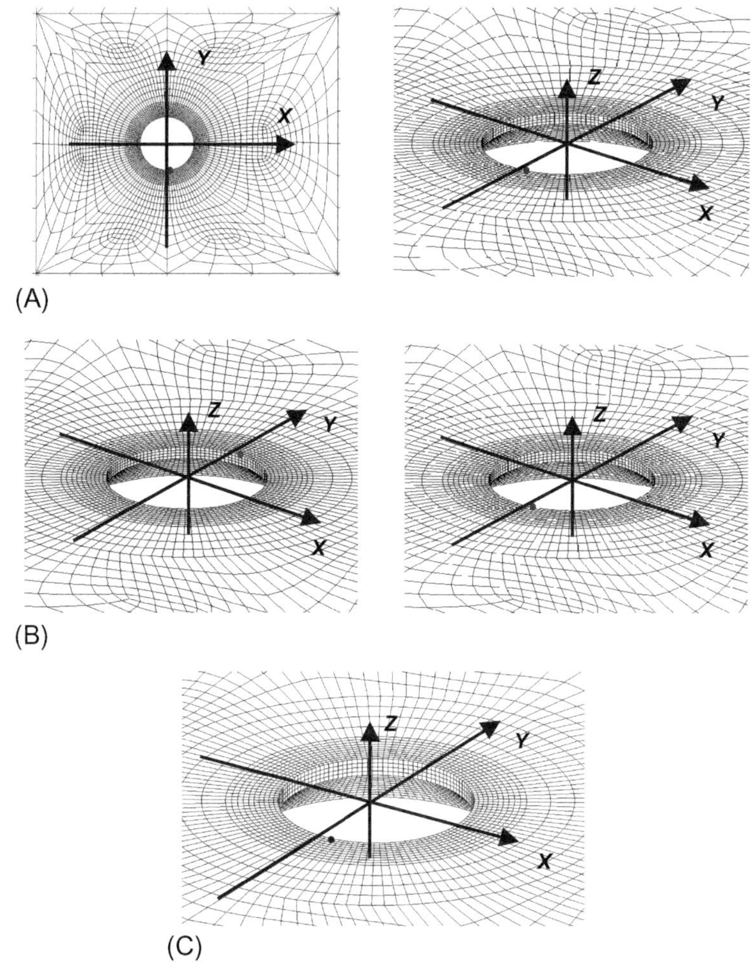

Fig. 13.30 The predicted crack nucleation positions (dotted points) on the top panel faying surface under different driven head deformations for size 1 to 3 models. The *x*-axis is in the global planar remote tensile direction. (A) Crack nucleation position (in red) in the size 1 model with the 1.459 rivet head deformation, (B) Crack nucleation position (in red) in the size 2 model with the 1.507, 1.517 (left), and 1.527 (right) rivet head deformations, (C) Crack nucleation position (in red) in the size 3 model with the 1.501, 1.509, and 1.519 rivet head deformations.

stress location, shown in Fig. 13.31, and fatigue life in Table 13.2 are in relatively good agreement with the failed joint results shown in Fig. 13.31. The fatigue fracture occurred at the top-row position of the outer panel, and fretted areas were found in the hole edge vicinity on the faying surface and fracture surface. The observed cracks nucleated in the fretted area in the hole edge vicinity area and then propagated along a transverse path. The numbering of the rivet holes starts on the left side of the panel in

Table 13.7 Predicted lives to 0.5 mm crack length.

	D_{max}/D for the countersunk MS20426AD5-5 rivets			65 MPa max stress $R = 0.069\ \sigma_{max}$	
Hole diameter (protruding part)	**FE**	**Experimental**	**Dif. (%)**	σ_{max} **(Δε/2)**	N_f **(cycles)**
Size 1: 4.064 mm	1.459	–	–	0.2608	85.11E+6
Size 2: 4.115 mm	1.507	1.50	0.47	0.5176	2.4337E+6
	1.517	1.51	0.46	0.4851	3.4020E+6
	1.527	1.52	0.46	0.4578	4.5895E+6
Size 3: 4.166 mm	1.501	1.50	0.07	0.5300	2.1526E+6
	1.509	1.51	−0.07	0.5170	2.4471E+6
	1.519	1.52	−0.07	0.5147	2.5031E+6

Fig. 13.31. Ten through cracks and 19 part-through cracks were identified by fractography. During the fatigue test, the first crack was observed at about 1.95 million cycles at the right side of rivet no. 4, and the second crack was found at about 1.955 million cycles at the left side of rivet no.12. The first link-up of two fatigue cracks was at about 2.11 million cycles between rivet no. 4 and 5 and the lap joint finally fractured at 2,133,534 cycles.

Correlation between the crack growth analysis and the test result

A preliminary assessment of the fatigue life to first link-up was carried out using the size 2 hole diameter and the 1.517 driven head deformation. These values were assumed to be, amongst the cases presented in Table 13.2, the closest to the test nominal rivet diameter and measured deformation. The corresponding analytical life, 3.402 million cycles, was assumed to be the median life to a 0.5 mm crack length. Also, a standard deviation of 0.1296 for Log10 (life), used in DEF STAN 00-970 (PART 1/5, SECTION 3, LEAFLET 35) recommended for metallic airframe full scale tests, was assumed in order to generate a time to a 0.5 mm crack length size (TTCS) distribution. The stresses predicted by the local model were then used in AFGROW to generate a correction curve that included the effect of the riveting residual stresses and countersunk geometry. This correction curve was then used in CanGROW [34, 35] to generate an equivalent initial flaw size (EIFS) distribution by regressing the predicted TTCS distribution to time zero. It should be noted that, EIFS used here is not for life prediction; it is just used in CanGROW to conveniently grow a crack to 0.5 mm and then continue to grow the crack to the first linkup. A MSD Monte Carlo simulation of 45,000 trials, each consisting of different random initial cracks sampled from the EIFS distribution and positioned at two adjacent rivet holes, was then performed to estimate the distribution of life to crack link-up between the two holes. When considering the

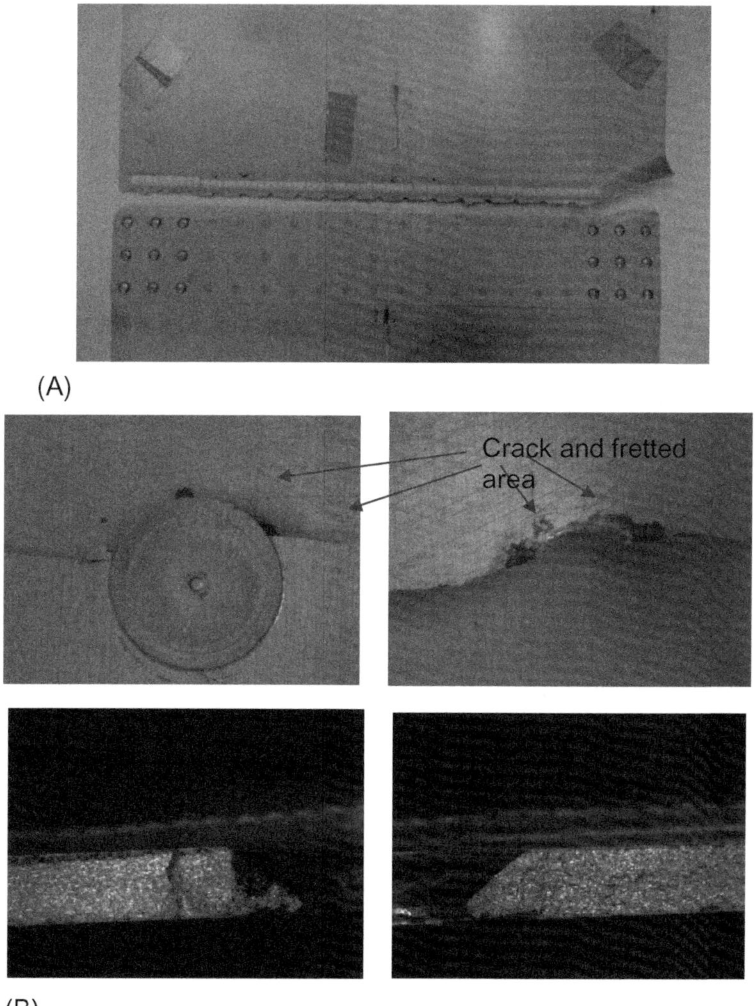

Fig. 13.31 Failure information observed from the failed joint under fatigue testing [28]. (A) Fracture occurred at the top-row position of the outer panel, (B) Fretted areas of the outer panel. Courtesy of JAXA and Techno Press.

13 potential link-ups between the 14 adjacent holes, the median of the combined probability of first link-up was calculated to be approximately 2.5 million cycles, as shown in Fig. 13.32. Good correlation was obtained between first link-up assessment and experimental result in the fatigue life. It should be noted that so far, only one lap joint test was completed and used for comparison. The comparison can be further improved when more results become available.

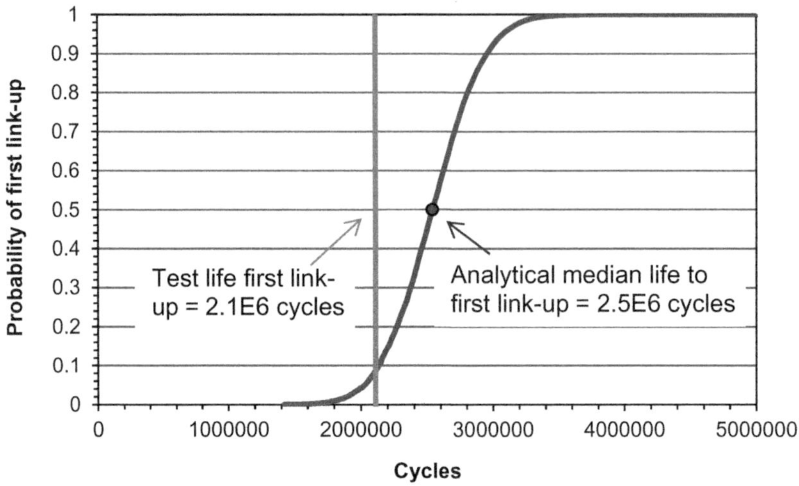

Fig. 13.32 Stress probability of first link-up within 14 holes.

13.5.3.4 Commentary on the study of fatigue life assessment

Fatigue life estimation of a multirow countersunk riveted lap joint was performed analytically. The stress and strain conditions in a highly stressed substructure of the joint were investigated using a global/local FE model coupling approach. After validation of the FE models using experimental strain measurements, the stress/strain condition in the local 3D FE model was simulated under a fatigue loading condition. This local model involved multiple load cases with nonlinearity in material properties, geometric deformation, and contact boundary conditions. The resulting stresses and strains were used in the SWT strain life equation to estimate the fatigue "initiation life", defined as the life to a 0.5 mm deep crack. Effects of the rivet-hole clearance and rivet head deformation on the predicted fatigue life were identified, and good agreement in the fatigue life was obtained between the experimental and the numerical results. Further crack growth from a 0.5 mm crack length to the first linkup of two adjacent cracks was evaluated using the NRC in-house tool, CanGROW. Good correlation in the fatigue life was also obtained between the experimental result and the crack growth analysis. The calculated life distribution was based on an assumed life standard deviation. If more hole measurements became available, a more accurate fatigue life distribution, with a higher degree of confidence, could be developed. This life distribution could then be used for evaluating the widespread fatigue damage behaviour, and develop a special inspection program. The study shows that the selected methodology is promising for assessing the fatigue life for the riveted lap joints, and can improve research efficiency by reducing test quantity and cost.

13.6 Concluding remarks and future work

The weak location of riveted lap joints is the outer sheet faying surface at the top row hole region. To improve the joint fatigue performance, a certain amount of interference and compressive residual stress/strain should be introduced to that region. However, current rivets and installation tools cannot achieve this target. It can be noticed from Fig. 13.14 that the higher level of interference is only present in the inner sheet and very low interference is generated in the outer sheet. Only the rivet driven head experiences a large compressive displacement, whereas the flush head or protruding head side does not experience enough compressive deformation. Two potential developments could be carried out to both the rivets and the corresponding rivet installation tools, to change this situation in the outer sheet. For a new generation of rivets aimed at introducing a certain amount of interference in both the inner and outer sheets, two aspects should be considered. These requirements could be that: (i) rivet heads could also be driven heads so that both sides of the rivet would experience large compressive displacement during the rivet installation, and (ii) the induced interference magnitude could be controlled and its distribution through the joint thickness should be made relatively uniform. The corresponding rivet installation tools could therefore be designed to install these new rivets in metallic joints using one or two strokes to squeeze both rivet ends.

A methodology was used to estimate the fatigue life of a riveted lap joint based on a global/local FE model coupling approach and the Smith-Watson-Topper (SWT) equation. The estimation compared reasonably well with the experimental fatigue life, when accurate stress, strain, and material fatigue behaviour conditions were employed. To further improve the fatigue life prediction, other information on fretting damage, material anisotropic hardening behaviour and its fatigue parameters, in-service working history, etc. should be studied and effectively considered in a practical methodology for the life assessment.

Presently, carbon fiber-reinforced laminated composite structures in many configurations can be cost-effectively manufactured using automated fiber placement technology. Reducing weight and improving performance in the new generation aircrafts can be accomplished by the extensive use of composites to replace traditional metallic materials. To this end, relevant joint structures, metal-to-composite and/or composite-to-composite, with targeted strength and fatigue performance are needed. Also, attachment using the riveted-bonded method can significantly improve the joint integrity condition [36–38]. Hence, knowledge of damage tolerance, including the fail-safe behaviour of the riveted-bonded metal-to-composite joints should be explored to support the emerging engineering applications. Typical aspects, such as hole drilling and preparation of the composite structures, galvanic and fretting corrosion, hygrothermal aging, bonding and riveting process technique, on the joint performance also need to be studied.

Acknowledgments

This chapter mainly is a summary of our previous work on "Residual Stress in Riveted Lap Joints" project (2001–2006) and its extension work, present in case study 3, through a NRC-JAXA joint collaborative project (2013–2015). Sincere acknowledgment to JAXA peers, Mrs. Takao Okade and Shigeru Machida, for their experimental work referenced in the case study 3. Deep appreciation to our colleagues Dr. G. Shi, Mrs. N.C. Bellinger, J.P. Komorowski, G. Eastaugh, and Y. Bombardier for their valuable discussion, suggestion, and support on the research in forming the chapter. Deep appreciation also to Mrs. Marko Yanishevsky and Robert S. Rutledge for their proof reading and editing. Many thanks to those people who have, in one way or another, contributed to the work.

References

[1] 'Metallic Material and Elements for Aerospace Vehicle Structures'. MIL-HDBK-5H, Department of Defense, USA, Dec. 1998.

[2] 'Rivets, buck type, preparation for and installation of', *MIL-R-47196A-Notice 1*, Department of Defense, USA, 1986.

[3] M.C.Y. Niu, Airframe Structural Design, Hong Kong Conmilit Press Ltd, Hong Kong, 1997.

[4] R.P.G. Müller, An Experimental and Analytical Investigation on the Fatigue Behaviour of Fuselage Riveted Lap Joints, Ph.D. Thesis, Delft Univ. of Technology, Delft, The Netherlands, 1995.

[5] L. Hazlehurst, M. Woods, Automated Rivet Interference Inspection System, SAE International, Warrendale, PA, 1994. paper 941842.

[6] D.R. Donaldson, K.J. Kenworthy, Fatigue design and test program for the American SST, in: J.Y. Mann, I.S. Milligan (Eds.), Aircraft Fatigue Design, Operational and Economics Aspects, Rushcutters Bay, Pergamon, 1972, pp. 437–476.

[7] R.I. Nepershin, V.V. Knigin, V.V. Klimenov, Interferences and residual stresses in riveted joints, J. Mach. Manuf. Reliab. 5 (1992) 47–51.

[8] T.B. Ryzhova, Estimation of the reliability of ultrasonic quality control of riveted joints with clearance, Russ. J. Nondestruct. Test. 30 (1994) 418–421.

[9] T.B. Ryzhova, Ultrasonic assessment of the radial clearance of riveted joints in aircraft structures, in: ICAS, Congress, 21st Melbourne, Australia, 1998, pp. 1–7. ICAS-98-R, 4.

[10] T.J. Fitzgerald, J.B. Cohen, Residual stresses in and around rivets in clad aluminium alloy plates, Mater. Sci. Eng. A188 (1994) 51–58.

[11] C.-P. Fung, J. Smart, An experimental and numerical analysis of riveted single lap joints, J. Aerosp. Eng. 208 (2) (1994) 79–90.

[12] C.-P. Fung, J. Smart, Riveted single-lap joints. Part 1: a numerical parametric study, J. Aerosp. Eng. 211 (1) (1997) 13–27.

[13] S.A. Fawaz, J. Schijve, A.U. de Koning, Fatigue crack growth in riveted joints, in: ICAF 97 – Fatigue in New and Aging Aircraft; Edinburgh, UK, 1997, 1, 1997, pp. 553–574.

[14] L. Ryan, J. Monaghan, Failure mechanism of riveted joint in fibre metal laminates, J. Mater. Process. Technol. 103 (2000) 36–43.

[15] M.P. Szolwinski, T.N. Farris, Linking riveting process parameters to the fatigue performance of riveted aircraft structures, J. Aircr. 37 (1) (2000) 130–137.

[16] A. Trego, D. Cope, Evaluation of damage tolerance analysis tools for lap joints, AIAA J. 39 (12) (2001) 2250–2254.

[17] N.C. Bellinger, J.P. Komorowski, T.J. Benak, Residual life prediction of corroded fuselage lap joints, Int. J. Fatigue 23 (2001) S349–S356.
[18] G.F. Eastaugh, P.V. Straznicky, D.V. Krizan, A.A. Merati, J. Cook, Experimental study of the effects of corrosion on the fatigue durability and crack growth characteristics of longitudinal fuselage splices, in: Proceedings of the Fourth DoD/FAA/NASA Aging Aircraft Conference, St. Louis, MO. 15–18 May 2000, 2000.
[19] G. Shi, N. Bellinger, Y. Xiong, Three-dimensional nonlinear finite element modelling and analysis of riveted lap joints with corrosion pillowing, in: S.N. Atluri, F.W. Brust (Eds.), Advances in Computational Engineering & Science, vol. II, Tech Science Press, 2000, pp. 1542–1547.
[20] J. Mackerle, Finite element analysis of fastening and joining: a bibliography (1990–2002), Int. J. Press. Vessel. Pip. 80 (2003) 253–271.
[21] E. Markiewicz, B. Langrand, E. Deleotombe, P. Drazetic, L. Patronelli, Analysis of the riveting forming mechanisms, Int. J. Mater. Prod. Technol. 13 (3–6) (1998) 123–145.
[22] B. Langrand, L. Patronelli, E. Deleotombe, E. Markiewicz, P. Drazetic, An alternative numerical approach for full scale characterization for riveted joint design, Aerosp. Sci. Technol. 6 (2002) 343–345.
[23] G. Li, G. Shi, Effect of the riveting process on the residual stress in fuselage lap joints, Can. Aeronaut Space J. 50 (2) (2004) 91–105.
[24] G. Li, G. Shi, N.C. Bellinger, Study of the residual strain in lap joints, J. Aircr. 43 (4) (2006) 1145–1151, https://doi.org/10.2514/1.18125.
[25] G. Li, G. Shi, N.C. Bellinger, Study of residual stress in single-row countersunk riveted lap joints, J. Aircr. 43 (3) (2006) 592–599, https://doi.org/10.2514/1.18128.
[26] G. Li, G. Shi, N.C. Bellinger, Residual stress/strain in three-row countersunk, riveted lap joints, J. Aircr. 44 (4) (2007) 1275–1285, https://doi.org/10.2514/1.26748.
[27] G. Li, G. Shi, N.C. Bellinger, Stress condition in triple-row riveted lap joints under the influence of specific factors, J. Aircr. 48 (2) (2011) 527–539, https://doi.org/10.2514/1.C031129.
[28] G. Li, G. Renaud, M. Liao, T. Okada, S. Machida, A methodology for assessing fatigue life of a countersunk riveted lap joint, Adv. Aircr. Spacecr. Sci. 4 (1) (2017) 1–19, https://doi.org/10.12989/aas.2016.4.1.001.
[29] Manual of MTS Load Frame, MTS systems Corporation, Box 24012, Minneapolis, Minnesota, USA 55424.
[30] 'MARC volume b: element library, Version K7', MARC analysis research corporation, USA, 1997.
[31] North Atlantic Treaty Organization, in: H.H. van der Linden (Ed.), Fatigue Rated Fastener Systems, North Atlantic Treaty Organization, 1985. AGARD-R-721.
[32] K.N. Smith, P. Watson, T.H. Topper, A stress-strain function for the fatigue of metals, J. Mater. 5 (1970) 767–778.
[33] C. Boller, T. Seeger, Materials Data for Cyclic Loading, Part D: Aluminium and Titanium Alloys, Elsevier Science Publishers, Amsterdam, The Netherlands, 1987.
[34] Y. Bombardier, M. Liao, G. Renaud, A new crack growth analysis tool for the assessment of multiple site fatigue damage, in: Proceedings of Aircraft Airworthiness & Sustainment Conference 2010 (AA&S2010) Austin, TX, USA, 2010.
[35] M. Liao, Y. Bombardier, G. Renaud, N.C. Bellinger, Advanced damage tolerance and risk assessment methodologies and tools for aircraft structures containing multiple-site and multiple-element fatigue damage, Proc. IMechE Part G: J. Aerosp. Eng. 226 (11) (2010) 1412–1423.

[36] C. Cheung, P.M. Gray, G.E. Mabson, K.Y. Lin, Analysis of fasteners as disbond arrest mechanism for laminated composite structures, in: Proceedings of 51st AIAA/ASME/ASCE/AHS/ASC Structures, Structural Dynamics, and Materials Conference, 12–15 April 2010, Orlando, Florida, Paper AIAA 2010-3024, 2010.

[37] G. Li, J.H. Chen, M. Yanishevsky, N.C. Bellinger, Static strength of a composite butt joint configuration with different attachments, Compos. Struct. 94 (5) (2012) 1736–1744.

[38] G. Li, Fatigue performance characterization of a composite butt joint configuration, Compos. Part A 51 (2013) 43–55.

Failure of joints in service

14

Richard Freeman
TWI Ltd., Cambridge, United Kingdom

14.1 Introduction

While media coverage of an air accident is very dramatic and is invariably associated with the loss of lives, the number of air accidents has been steadily decreasing over the last 30 years (Fig. 14.1: Aviation Safety Network 2019 statistics) and air travel is considered to be the safest form of travel. In 1998 the International Civil Aviation Organisation (ICAO) established a universal safety oversight audit programme, comprised of regular, mandatory, systematic and harmonized safety audits to be carried out by ICAO on all Contracting States. Since 1 January 1999, the Safety Oversight Audit (SOA) Section of the Air Navigation Bureau of ICAO has been conducting safety oversight audits of the civil aviation authorities of member countries in relation to personnel licensing, operation of aircraft, and airworthiness. The audits are designed to determine the status of States' implementation of the critical elements of a safety oversight system and the implementation of relevant ICAO Standards and Recommended Practices, associated procedures, guidance material and safety-related practices. In addition, in March 2006 the EU published a Community list of air carriers subject to an operating ban within the European Community. Bans and operational restrictions are only imposed based on evidence of violation of objective and transparent criteria. These criteria focus on the results of checks carried out in European airports; the use of poorly maintained, antiquated or obsolete aircraft; the inability of the airlines to rectify shortcomings identified during inspections; and the inability of the authority responsible for overseeing an airline to perform its task properly. Member States reported that five countries have an inadequate system for regulatory oversight. One important consequence of the black list will be to root out the practice of flags of convenience whereby some countries issue Air Operation Certificates to dubious airline companies (Aviation Safety Network safety assessment information).

While air travel is becoming even safer, it is worth reflecting on some high profile air accidents over the last 40 years involving the failure of joints or components to reflect on lessons learnt, and the importance of improved inspection techniques to ensure that safety continues to improve throughout the 21st Century. Five case histories are mentioned, with the information on the first four obtained from a review of aircraft structural integrity [1].

Welding and Joining of Aerospace Materials. https://doi.org/10.1016/B978-0-12-819140-8.00014-6
Copyright © 2021 Elsevier Ltd. All rights reserved.

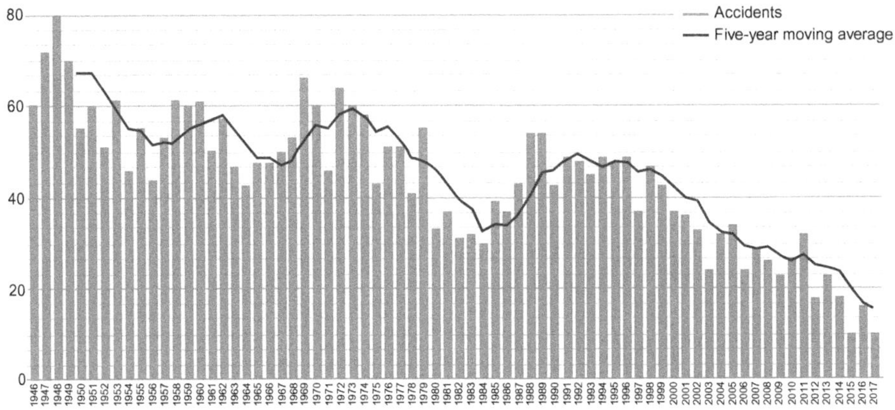

Fig. 14.1 Fatal aircraft accidents per year 1946–2019 (Aviation Safety Network).

14.2 DeHavilland Comet crashes

The DeHavilland Comet was the first commercial jet aircraft, and entered service is 1952. Its performance was much better than propeller driven aircraft, and quite apart from the increase in speed the aircraft was the first to operate at high altitude, with a cabin pressure differential almost double that of its contemporaries. Within two years of entering service however, two of the fleet disintegrated while climbing to altitude before the fleet was grounded. Subsequent testing and investigation of a test aircraft indicated that out of plane bending would have caused the principal stresses inside the shell to be significantly higher than forecast, and this could have contributed to early fatigue failure. It was also noted that the basic shell structure had no crack stopper straps to provide continuity of the frame outer flanges across the stringer cutouts. The cutouts created a very high concentration at the first fastener, which in the case of one of the lost aircraft was a countersunk bolt. The countersink created a knife edge in both the skin and outside doubler (Fig. 14.2). The early fatigue failure could be attributed to high local stresses, combined with the stress concentrations provided by the frame cutout and knife edge of the fastener hole. Once the fatigue crack initiated its growth went undetected until catastrophic failure of the pressure cabin.

The Comet accidents and subsequent investigations changed the fundamental design principles for commercial transport aircraft. The Comet aircraft was designed around SAFE-LIFE, which meant that the entire structure was designed to achieve a satisfactory fatigue life with no significant damage i.e. cracking. These accidents showed that cracks could sometimes occur much earlier than anticipated, owing to limitations in the fatigue analyses, and that safety could not be guaranteed on a SAFE-LIFE basis without imposing uneconomically short service lives on major components of the structure. These problems were addressed by the adoption of the FAIL-SAFE design principle in the late 1950's. While the structure is designed to achieve a satisfactory life with no significant damage, it is also designed to be inspectable in service and able to sustain significant and easily detectable damage

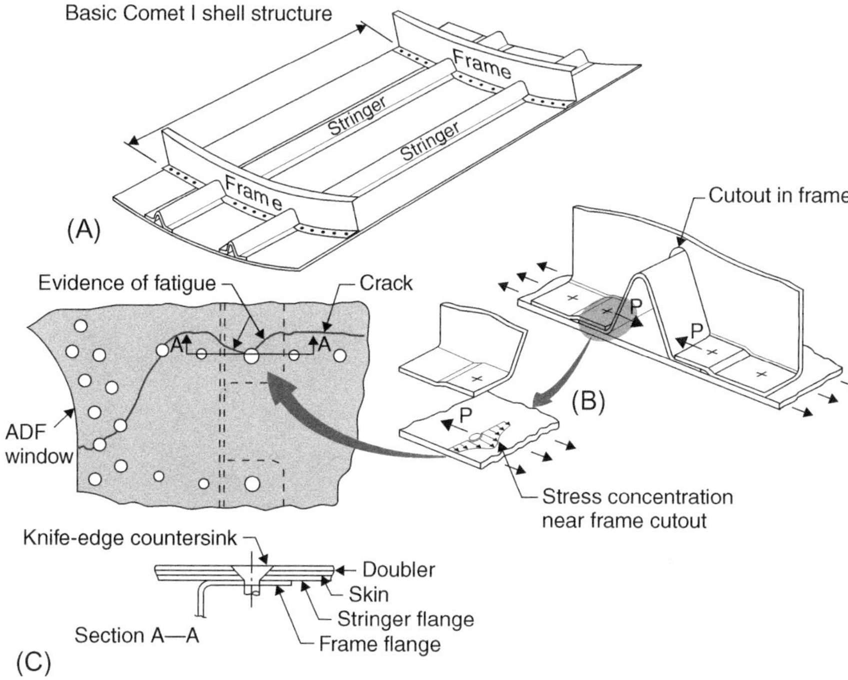

Fig. 14.2 Causes of DeHavilland Comet crashes. (A) Basic Comet I shell structure with no crack-stopper straps; (B) Cutouts showing high stress concentration at the first fastener; (C) Probable failure origin-countersunk bolt created a knife edge in both the skin and outside doubler.

before safety is compromised. These requirements were met mainly by employing structural design concepts having multiple load paths, with established residual strength requirements in the event of failure of one structural element or an obvious partial failure. This has led to mandatory full scale testing of aircraft.

14.3 General Dynamics F-111 crash

In 1964 the General Dynamics Corporation was awarded a contract for the development and production of the F-111 aircraft. In late 1969, just over a year from entering service, an aircraft lost the left wing structure during a low level training flight. It had accumulated only 107 airframe flight hours, and failure occurred while it was pulling about 3.5 g, less than half the design limit load factor. An immediate investigation revealed a flaw in the lower plate of the left hand wing pivot fitting, manufactured from high strength steel. The flaw had occurred during manufacture and remained undetected despite its considerable size: 23.4 mm × 5.9 mm (Fig. 14.3). A limited amount of fatigue crack growth had occurred in service before overload fracture of the plate, which resulted in immediate loss of the wing. Subsequent review of this

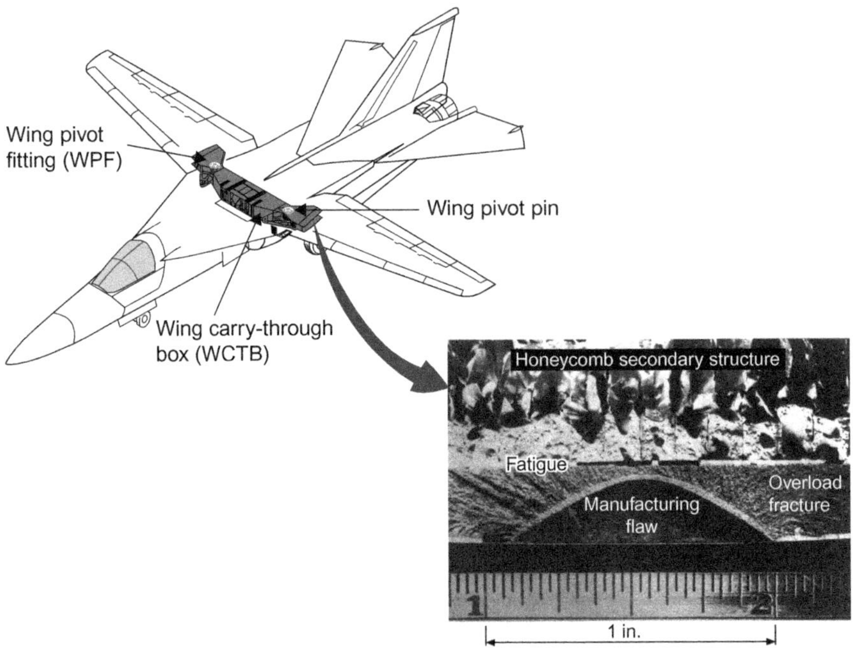

Fig. 14.3 Cause of General Dynamics F-111 crash, a manufacturing flaw in the high strength steel lower plate of the left hand wing pivot fitting.

incident led to proof testing of the entire wing carry though structure by periodically removing aircraft from service. This also led to mandatory guidelines from the US Air Force known as the Damage Tolerance philosophy, incorporated in Military Specification (MIL-A-83444).

14.4 Dan Air Boeing 707 crash

In May 1977 a Dan Air Boeing 707 freighter lost the entire right hand horizontal stabilizer just before it was due to land at Lusaka International Airport. The aircraft was manufactured in 1963 and had accumulated 47,621 airframe flight hours. The investigation traced the accident back to fatigue failure in the upper chord of the rear spar. Fatigue cracking began at a fastener hole owing to higher loads than anticipated in the design. The fatigue spread into the upper chord, with overall crack growth being accelerated by large intermittent tensile crack jumps. Fatigue crack growth finally gave way to overload fracture down through the entire rear spar, and this resulted in the stabilizer separating from the aircraft (Fig. 14.4). Although this configuration was intended to be a FAIL-SAFE design, the periodic inspection of the horizontal stabilizer had a recommended time of less than 30 min. This suggests visual inspection only, which would not have detected a partial failure of the upper chord of the rear spar. Once the

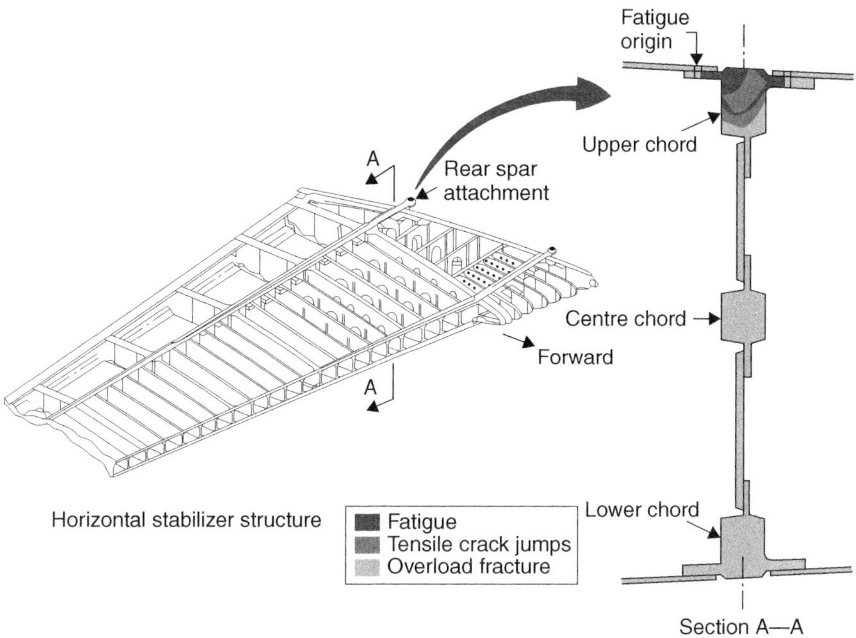

Fig. 14.4 Cause of Dan Air Boeing 707 crash, initial fatigue cracking at a fastener hole led to tensile overload failure of the entire rear spar.

upper chord had failed completely, enabling the damage to be detected visually, the structure could not withstand the service loads long enough for the failure to be detected.

The crash prompted airworthiness authorities to consider the fatigue problems of ageing aircraft, as it became very clear that existing inspection methods and schedules were inadequate and supplementary inspection programmes were needed to prevent older aircraft from becoming fatigue-critical. A further point that came out of the investigation was that the manufacturer modified the horizontal stabiliser design for the Boeing 707-300 series in order to increase the torsional stiffness. This was required due to an overall increase in aircraft weight. The material was changed from an aluminium alloy to a stainless steel for a large part of the top skin attached to the front and rear spars. This modification was not checked by a full scale fatigue test, which was not required by the contemporary regulations. After the crash a full scale test on a modified horizontal stabiliser reproduced the service failure.

14.5 Aloha Airlines Boeing 737 accident

In April 1988, Aloha Airlines 243, a Boeing 737-200, experienced an explosive decompression during climb at cruise altitude. About 5.5 m of the pressure cabin skin and supporting structure, aft of the cabin entrance door and above the passenger

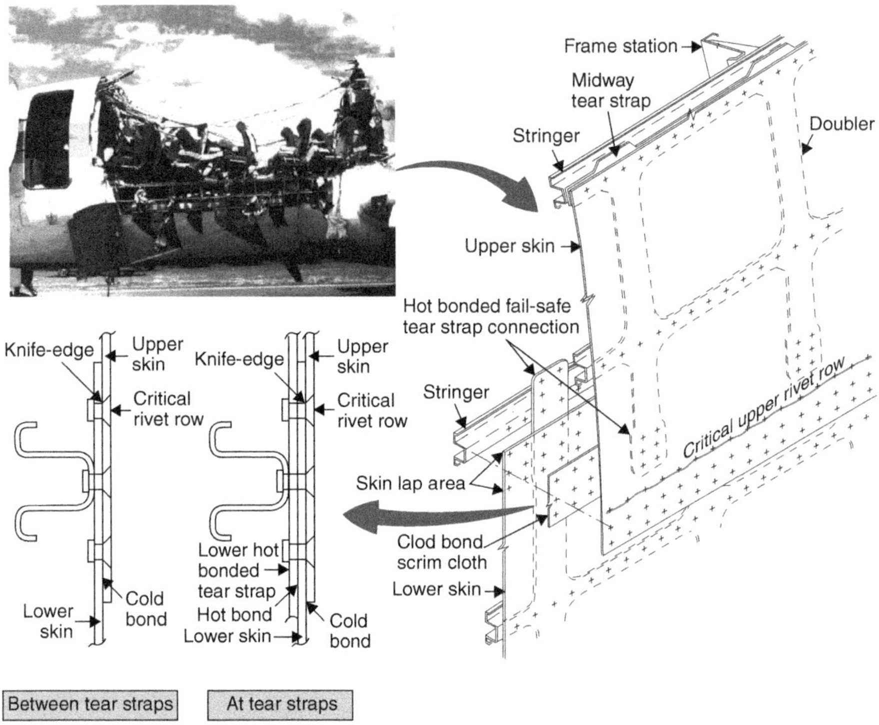

Fig. 14.5 Cause of Aloha Airlines Boeing 737 accident, multiple site fatigue damage occurred in the outer (upper) skin commencing from the knife edges of the rivet holes along the upper rivet row.

floorline separated from the aircraft (Fig. 14.5). Remarkably the damage did not result in the disintegration of the aircraft and a successful emergency landing was made, albeit with the loss of a flight attendant who was swept to her death. The aircraft had been manufactured in 1969 and had accumulated 35,496 airframe flight hours and 89,680 landings. Due to the short distance between destinations on some Aloha Airlines routes in the Hawaiian islands, the maximum pressure differential was not reached in every flight. Thus the number of equivalent full pressurization cycles was significantly less than 89,680. However the aircraft was nearly 19 years old, and was operating in a warm, humid, maritime environment.

The investigation showed the large loss of pressure cabin skin was caused by rapid link-up of many fatigue cracks in the same longitudinal skin splice. The fatigue cracks began at the knife edges of rivet holes along the upper rivet row of the splice, and this type of failure is called Multiple Site fatigue Damage (MSD). There were several factors relating to the incident, with the first one being the skin splice configuration. The pressure cabin longitudinal skin splice had been cold bonded using an epoxy impregnated woven scrim cloth, as well as riveting. This should have resulted in a safe and durable structure, whereby the pressure cabin loads would be transferred through the bonded splice as a whole, rather than the rivets only. However the early service history of production of Boeing 737's with cold bonded splices revealed difficulties with the bonding process. These problems resulted in random occurrence of bonds with low environmental durability (i.e. susceptibility to corrosion), and with some areas not bonded at all. Cold bonding was discontinued in 1972, after production of this aircraft, but well before the accident. Also, owing to the cold bonding problems Boeing issued service bulletins in 1972, 1974 and 1987 and the FAA issued an Airworthiness directive in 1987. These documents called for skin splice inspections at regular intervals, and repairs if necessary. However it is stated in the NTSB Air Accident Report (NTSB/AAR 89-03) that "proper eddy current inspection would have detected additional fatigue cracks in the holes of the upper rivet row of the lap joint".

The accident prompted manufacturers, operators and airworthiness authorities to collaborate and develop new regulations. The increased emphasis of Widespread Fatigue Damage (WFD) and the adoption of corrosion control programmes are two of the most important initiatives to come from the subsequent review.

14.6 United Airlines DC10 accident

In July 1989 United Airlines flight 232 departed Denver for a flight to Philadelphia via Chicago. The takeoff and the climb to the planned cruising altitude of 37,000 feet were uneventful. About 1 h and 7 min after takeoff, the flight crew heard a loud bang or an explosion, followed by vibration and a shuddering of the airframe. After checking the engine instruments, the flight crew determined that the No. 2 aft (tail-mounted) engine had failed. It was decided to conduct an emergency landing at Sioux City airport, Iowa but the crew had difficulty in controlling the aircraft and upon landing it skidded to the right of the runway and rolled to an inverted position. Fire fighting and rescue operations began immediately, but the aircraft was destroyed by impact and fire with 111 fatalities from the passenger and crew list of 296.

Investigation attributed the cause to an uncontained failure of the No. 2 engine stage 1 fan rotor disk assembly (NTSB/AAR 90-06). No. 2 engine fragments severed the No. 1 and No. 3 hydraulic system lines, and the forces of the engine failure fractured the No. 2 hydraulic system, rendering the aircraft's three hydraulic-powered flight control systems inoperative. Typical of all wide-body design transport aircraft, there are no alternative power sources for the flight control systems. Separation of the titanium alloy stage 1 fan rotor disc was the result of a fatigue crack that initiated from a type 1 hard alpha metallurgical defect on the surface of the disc bore. The hard alpha metallurgical defect was formed in the titanium alloy material during manufacture of the ingot from which the disc was forged. The hard alpha metallurgical defect was not detected by ultrasonic and macroetch inspections performed by General Electric Aircraft Engines during the manufacturing process of the disc. Indeed post crash inspection of the crack surfaces showed the presence of the fluorescent dye penetrant used during nondestructive inspection, indicating that crack was present and that it should have been detected during previous inspections. The metallurgical flaw that formed during initial manufacture of the titanium alloy would have been apparent if the part had been macroetch inspected in its final part shape. The cavity associated with the hard alpha metallurgical defect was created during the final machining and/or shot peening at the time of GE's manufacture of the disc, after GE's ultrasonic and macroetch manufacturing inspections. The hard alpha defect area cracked with the application of stress during the disk's initial exposures to full thrust engine power conditions and the crack grew until it entered material unaffected by the hard alpha defect.

The investigation and subsequent Airworthiness Directive revealed that, several other fan discs already in service from the same batch of ingots had started to exhibit initial cracking symptoms. The foundry also changed their manufacturing practice to use higher melting temperatures and a triple vacuum process to drive out gaseous elements during the production of ingots. The NTSB also recommended that research should be intensified in the non-destructive inspection field to identify emerging technologies that can serve to simplify, automate, or otherwise improve the reliability of the inspection process.

14.7 The importance of international standards

The previous section has shown how aircraft design principles, manufacturing practices and inspection techniques have developed over the last 40 years, unfortunately sometimes as a result of air accidents which have necessitated change. However with the development of more advanced welding and joining techniques, many of which were listed in Chapter 1, the importance of stringent welding and inspection specifications that meet the needs of the modern aircraft industry is even more apparent.

The AWS D17 committee was set up in 1993 to replace the US MIL-STD-2219 (Fusion welding for aerospace applications) and MIL-STD-1595A (Qualification of aircraft, missiles and aerospace fusion welders) specifications. D17.1 relates to fusion welding for the aerospace industry, D17.2 to resistance welding and 17.3 to friction stir welding. AWS D17.1 is the most mature of these specifications and is being widely used around the world by original equipment manufacturers and their supply chains. The D17 committee is also represented on the ISO/TC 44 SC14 committee, who have developed several international aerospace welding specifications for use around the world including:

- ISO 24934 Qualification test for welder and welding operators—Welding of metallic components
- ISO 17533 Welding for aerospace applications—Welding data in design documents
- ISO 17927 Welding for aerospace applications—Fusion welding of metallic components
- ISO 16338 Welding for aerospace applications—Resistance spot and seam welding
- ISO 19828 Welding and Brazing in Aerospace—Visual Inspection Requirements

ISO 25239 (Parts 1–5) for FSW of aluminium alloys has been in existence since 2011, and there is also an ISO specification for Friction Stir Spot Welding of Aluminium (ISO 18785 Parts 1–5) that was updated and released in 2018.

There are several standards bodies (including ISO, ASTM, SAE and AWS) working on the development of many specifications for the variety of additive manufacturing techniques and the associated raw materials (metal powder and wire). In addition the Nadcap Welding Task Group (administered by PRI) has developed a laser and EB powder bed additive manufacturing audit checklist and handbook, for auditing companies that want to get approved to use this process for the manufacture of aerospace components.

The author of this chapter has first hand knowledge of the integrity of these groups, having served for 20 years as the UK representative on the AWS D17 committee, for 10 years on the ISO/TC 44 SC14 group, and 9 years on the Nadcap committee. With the dedication of such engineers, combined with the natural risk averse nature of the aerospace industry, the safety of the air transport industry will always be paramount.

Reference

[1] R.J.H. Wanhill, Milestone Case Histories in Aircraft Structural Integrity, NLR report NLR-TP-2002-521October, 2002.

Further reading

NTSB/AAR-89/03—Air Accident Report. http://www.airdisaster.com/reports/ntsb/AAR89-03.pdf.

NTSB/AAR-90-06—Air Accident Report. http://libraryonline.erau.edu/online-full-text/ntsb/aircraft-accident-reports/AAR90-06.pdf.

Aviation Safety Network web site. http://aviation-safety.net/statistics/period.

Aviation Safety Network web site. http://aviation-safety.net/airlinesafety/enforcement/assessment.php.

Index

Note: Page numbers followed by *f* indicate figures and *t* indicate tables.

B

G

N

S

9780128191408